Bioelectromagnetism

Bioelectromagnetism
History, Foundations and Applications

Edited by
Shoogo Ueno & Tsukasa Shigemitsu

CRC Press
Taylor & Francis Group
Boca Raton London New York

CRC Press is an imprint of the
Taylor & Francis Group, an **informa** business

First edition published 2022
by CRC Press
6000 Broken Sound Parkway NW, Suite 300, Boca Raton, FL 33487-2742

and by CRC Press
4 Park Square, Milton Park, Abingdon, Oxon, OX14 4RN

CRC Press is an imprint of Taylor & Francis Group, LLC

© 2022 selection and editorial matter, Shoogo Ueno & Tsukasa Shigemitsu; individual chapters, the contributors

Library of Congress Cataloging-in-Publication Data
Names: Ueno, Shoogo, editor. | Shigemitsu, Tsukasa, editor.
Title: Bioelectromagnetism : history, foundations and applications / edited
by Shoogo Ueno & Tsukasa Shigemitsu.
Description: First edition. | Boca Raton : CRC Press, 2022. | Includes
bibliographical references and index.
Identifiers: LCCN 2021056504 | ISBN 9781032001210 (hardback) |
ISBN9781032019970 (paperback) | ISBN 9781003181354 (ebook)
Subjects: LCSH: Electrophysiology. | Biomagnetism.
Classification: LCC QP341 .B54 2022 |
DDC 572/.437—dc23/eng/20211123
LC record available at https://lccn.loc.gov/2021056504

ISBN: 978-1-032-00121-0 (hbk)
ISBN: 978-1-032-01997-0 (pbk)
ISBN: 978-1-003-18135-4 (ebk)

DOI: 10.1201/9781003181354

Typeset in Minion Pro
by codeMantra

Contents

Preface

This book is devoted to bioeletromagnetism. The title is *Bioelectromagnetism: History, Foundations and Applications*. The idea to write this book on bioelectromagnetism dates back to 2019. At that time, after a scientific meeting, we had talked with a few colleagues about bioelectromagnetism, its interesting and difficult aspects. Our discussion spanned a variety of topics: the long research history of bioelectromagnetism; science dates back to the publication of *De Magnete* by Gilbert; electro- and magneto-therapy; Galvani's discovery of bioelectricity; the relationship between electricity and magnetism; Faraday's induction law; Maxwell's theory of electromagnetic waves; biological and health effects; migratory birds use of geomagnetic fields; and others. At the end of our talk, it was decided to try to organize the wonderful aspects of bioelectromagnetism. Another major motivation was that one of the editors attended a workshop in Munich in November 2019, where he saw and heard the state-of-the-art findings of research on the effects of electromagnetic fields in the environment on flora and fauna. At this workshop, magnetoreception in animals and plants in the environment was also a topic of interest. This workshop was organized by the German Federal Office for Radiation Protection (BfS) as an international workshop, "International Workshop: Environmental effect of electric, magnetic and electromagnetic fields: Flora and Fauna" from November 5 to 7. This workshop covered the effects of anthropogenic fields of all frequencies on plants, invertebrates and vertebrates, possible mechanisms of action in living systems and gaps in knowledge.

The above led us to begin discussions with the intention of publishing a book. We approached a number of experts to make a book that would highlight various interests and the wonderful aspects of bioelectromagnetism. Naturally, we planned to ask Prof. Maffei of the University of Turin, who gave an interesting talk on magnetoreception in plants at a workshop in Munich.

Therefore, we decided to write this book. CRC Press/Taylor & Francis Group was kind enough to help and support our venture. It is fortunate that seven chapters arrived by the deadline of manuscript submission, although a few chapters did not arrive by the cutoff date. The manuscript that was not ready in time is expected to be available in an online version.

Although the research in bioelectromagnetism has been gradually developing and expanding into a variety of fields in engineering, biomedical engineering, life science, medicine and biology, this book with seven chapters is our attempt to reveal to readers the wonderful aspects and the specific regions related to bioelectromagnetism. In Chapter 1, it begins with general information which forms the framework for the understanding of this book. It is well known that the origin of bioelectromagnetism is the study of the relationship between electromagnetism and biological systems. It connects electricity, magnetism and electromagnetism with biology, medicine, chemistry, physics and engineering. Chapter 2 provides chronologically the fantastic and dramatic history of bioelectromagnetism from ancient times to the twenty-first century. Then the relationships between atmospheric electric phenomena, geomagnetism and biological systems are presented in Chapter 3 for understanding the electromagnetic environment around us. The central part of the book with three chapters is devoted to the wonderful aspects of bioelectromagnetism. It is designed to provide biological topics of current and growing interest on magnetic sensing in humans, animals and plants. The detection of magnetic fields by biological systems is an age-old question in bioelectromagnetism. The topics of recent advances in magnetobiology as a guide for bioelectromagnetism include: magnetic navigations; magnetotactic bacteria; and magnetic sense and magnetic responses of plants, birds, animals and humans. It mentioned also that in future, these researches are expected to progress in relation to quantum biology. Chapter 7 deal with the latest efforts to set safety guidelines from the viewpoint of human health in the electromagnetic environment that exists widely around us.

The chapter authors are experts in each area. They wrote their chapters with great pleasure to explain the background and recent advances in each field. We hope readers enjoy the essence of this expanding field. Through discussions in this book, past and today's research advances and a number of prospects in bioelectromagnetism are given for many students, researchers and specialists in biology, medicine, biophysics, electrical engineering and biomedical engineering.

The editors sincerely hope that this book will be used as a companion to our previous published books: *Biomagnetics: Principles and Applications of Biomagnetic Stimulation* and *Imaging* and *Bioimaging: Imaging by Light and Electromagnetics in Medicine and Biology*. Both books are published by CRC Press/Taylor & Francis Group in 2016 and 2020.

The following is an overview of the chapters covered in this book to give you a breadth of research in bioelectromagnetism.

Chapter 1 Introduction (Shoogo Ueno and Tsukasa Shigemitsu)

The term 'Bioelectromagnetism' became popular in scientific communities around 1980. This chapter begins with general information which forms frameworks for understanding this book. The origin of bioelectromagnetism is the study of the relationship between electromagnetism and biological systems. It connects electricity, magnetism and electromagnetism with biology, medicine, chemistry, physics and engineering. Today, bioelectromagnetism focuses on biological and medical studies of electromagnetic fields ranging from direct current (DC) to optical frequencies. These studies can be mainly divided into two issues: beneficial, such as various medical applications, and harmful, such as biological effects of electromagnetic fields with the setting of safety guidelines. Taking this into consideration, the discussion of this chapter will give us today's study of bioelectromagnetism and its future prospects, along with a discussion of the issues that bioelectromagnetism is facing. The electromagnetic environment generated by various devices that use non-ionizing radiation will be familiar to us all. Therefore, there is a need to clarify the impact of that electromagnetic environment on the global environment. In the future, quantum biology, which is a combination of quantum mechanics and biology, will occupy an important position in the research in bioelectromagnetism. We pointed out that the research in bioelectromagnetism will be more interdisciplinary than ever before and will require the coordination of many experts.

Chapter 2 The History of Bioelectromagnetism (Tsukasa Shigemitsu, Shoogo Ueno and Masamichi Kato)

This chapter presents how bioelectromagnetism has been gradually developing and expanding to a variety of fields in engineering, biomedical engineering, medicine and biology. This chapter provides the fantastic and dramatic history of bioelectromagnetism from ancient times to the twenty-first century. It deals with the chronological history of major subjects with scientists who have made physical, chemical, electrical and biological discoveries. In addition, since the late 1960s, research on the biological and health effects of man-made electromagnetic fields has been taken up. In this chapter, the history of this research is also presented. This chapter as far as possible includes the historical development of Japanese research. The emphasis is on how the long historical development of bioelectromagnetism reveals that today's scientific achievements in engineering, biomedical engineering, biology and medicine are shaped by solving the electromagnetic phenomena in biological systems.

Chapter 3 The Coupling of Atmospheric Electromagnetic Fields with Biological Systems (Tsukasa Shigemitsu and Shoogo Ueno)

There are various electromagnetic phenomena on the earth. Since the earliest time, biological systems have been exposed to electromagnetic fields produced by the earth itself, as well as electromagnetic fields associated with lightning discharges and solar activity. It is presumed that the evolution of organisms has been affected by these electromagnetic phenomena in various ways. This question of what relationship exists between electromagnetic phenomena on the earth and the evolution of life has been a matter of great interest. In this chapter, we briefly summarize how we are using natural electromagnetic phenomena to actively sustain life, with some examples of basic natural electromagnetic phenomena and their interactions with biological systems. In particular, the relationship between low-frequency electromagnetic fields caused by lightning discharges which are constantly occurring on a global scale and physiological changes in humans and animals is summarized. In addition to electromagnetic phenomena originating from atmospheric phenomena, man-made electromagnetic phenomena are expected to become ever more prevalent in our daily lives in the future. The need to harmonize global-scale electromagnetic phenomena with the environment is discussed. For example, it has been reported that the animals use actively the geomagnetic fields which include the magnetotactic bacteria, the behavior of animals, migratory birds and plants as magnetoreception and electroreception. In this chapter, the overview of the natural origins of electromagnetic phenomena on the earth, from the sun, atmosphere and cosmic phenomena is presented with discussions on their interaction with biological systems.

Chapter 4 Spin Control Related to Chemical Compass of Migratory Birds (Hideyuki Okano, Tsukasa Shigemitsu and Shoogo Ueno)

As one of the blue-light photoreceptor proteins, cryptochrome including flavin adenine dinucleotide (FAD) is widely distributed in nature, i.e., bacteria, plants and animals. Recently, the possibility that cryptochromes serve as highly sensitive magnetoreceptors has been suggested strongly, and then several research fields such as "Quantum Biology" are actively paying attention to this functional expression. The reaction for magnetic sensing is presumed to generate the radical pair between FAD and amino acid residue as the intermediate by photoirradiation, i.e., a radical pair mechanism for quantum-assisted magnetic sensing, which can detect even weak magnetic fields such as the geomagnetic field (approximately 50 µT). By using such magnetoreceptors in the retina, migratory birds are assumed to perceive visually in which direction to move. This chapter reviews historical background in theoretical physics and ethology, and outstanding results in several research fields to date.

Chapter 5 Magnetoreception in Plants (Massimo Elimio Maffei)

The geomagnetic field effects in plants have been observed. However, the impact of the geomagnetic field on plants is not well-understood and the mechanism of magnetoreception in plants is still not scientifically well-organized. The main models for magnetoreception are the magnetite model and radical pair model. The radical pair model is linked to the cryptochrome. Cryptochrome is a class of flavor-proteins which are sensitive to light and cryptochrome is found in a variety of plants and animals. Phytochrome and phototropin are also photoreceptor proteins in plants. The magnetic field effect in plants relates to the light-dependent plant process. In this chapter, the effects of the geomagnetic field on plants are reviewed and the possible mechanism of magnetoreception in plants is discussed with the involvement of photoreceptors.

Chapter 6 Geomagnetic Field Effects on Living Systems (Hideyuki Okano and Shoogo Ueno)

The evolution of life is affected by fluctuations and variations of geomagnetic field intensity, as well as changes in atmospheric oxygen level and ultraviolet (UV) radiation. When the geomagnetic field reverses, the geomagnetic field is estimated to weaken, and the changes in the last 200 years have been potentially diminishing declines. The magnetic poles are also moving, and it is estimated that the north pole has moved 1,100 km over 170 years. The simulation predicted that geomagnetic reversals could enhance the oxygen escape rate by 3–4 orders only if the magnetic field was extremely weak. Consequently, it is estimated that global hypoxia might gradually kill numerous species. The existence of the geomagnetic field is indispensible for the origin and evolution of life. This chapter presents fascinating topics cocnerning the relationship between the geomagnetic field declination and explosion and extinction of life, and the impact of geomagnetic field disturbance on human health.

Chapter 7 Safety Guidelines on Human Exposure to Electromagnetic Fields (Kenichi Yamazaki and Tsukasa Shigemitsu)

In order for people to live and work safely and securely in today's electromagnetic environment, it is necessary that electromagnetic field exposure have no effect. This chapter introduces a rationale of existing safety guidelines developed by international bodies such as ICNIRP (International Commission on Non-Ionizing Radiation Protection) and IEEE (Institute of Electrical and Electronics Engineers). In these safety guidelines, limit values are stipulated in terms of internal (in situ) electric field for frequencies lower than 10 MHz (ICNIRP) or 5 MHz (IEEE), and SAR (specific absorption rate) for frequencies higher than 100 kHz, based on biological effects of electric and magnetic fields, i.e., stimulation of nerves for low frequencies and temperature increase for high frequencies. Issues of computational dosimetry used to derive reference levels of electric and magnetic fields in these guidelines are also discussed.

Acknowledgments

We thank the chapter authors for their intellectual contributions. It is indeed great that the chapter authors concentrated on their chapters during the COVID-19 coronavirus pandemic all over the world. Further, we thank the staff of CRC Press/Taylor & Francis Group - in particular, Rebecca Davis, editor, and Dr. Kirsten Barr, editorial assistant in physics, for contacting us to discuss a book on bioelectromagnetism, and their consistent support and valuable advice during the editing processes. We also greatly appreciate the anonymous reviewers who gave valuable comments for the publication of this book.

Shoogo Ueno
Tsukasa Shigemitsu

Editors

Shoogo Ueno is Professor Emeritus of the University of Tokyo, Tokyo, Japan. He received his B.S., M.S. and Ph.D. (Dr. Eng.) degrees in electronic engineering from Kyushu University, Fukuoka, Japan, in 1966, 1968 and 1972, respectively. Dr. Ueno was Associate Professor with the Department of Electronics, Kyushu University, from 1976 to 1986. From 1979 to 1981, he spent his sabbatical with the Department of Biomedical Engineering, Linkoping University, Linkoping, Sweden, as a Guest Scientist. He served as Professor in the Department of Electronics, Graduate School of Engineering, Kyushu University, from 1986 to 1994 and subsequently served as Professor in the Department of Biomedical Engineering, Graduate School of Medicine, the University of Tokyo, Tokyo, Japan, from 1994 to 2006. During this time, he also served as Professor in the Department of Electronic Engineering, Graduate School of Engineering, University of Tokyo. In 2006, he retired from the University of Tokyo as Professor Emeritus. Since then, he has been a Professor with the Department of Applied Quantum Physics, Graduate School of Engineering, Kyushu University, and he served as the Dean of the Faculty of Medical Technology, Teikyo University, Fukuoka, Japan, from 2009 through 2012. His research interests include biomedical engineering and bioelectromagnetics, particularly in the biological effects of magnetic and electromagnetic fields, transcranial magnetic stimulation (TMS), and magnetic resonance imaging (MRI). He was the President of the Bioelectromagnetics Society, BEMS (2003–2004), and the Chairman of the Commission K on Electromagnetics in Biology and Medicine of the International Union of Radio Science, URSI (2000–2003). He received the *Doctor Honoris Causa* from Linkoping University (1998). He was the 150th anniversary Jubilee Visiting Professor at Chalmers University of Technology, Gothenburg, Sweden (2006), and Visiting Professor at Simon Frasier University, Burnaby, Canada (1994), and Swinburne University of Technology, Hawthorn, Australia (2008). He was named the IEEE Magnetics Society Distinguished Lecturer during 2010 and received the *d'Arsonval* Medal from the Bioelectromagnetics Society in 2010. Dr. Ueno is Life Fellow (2011) and Fellow of the Institute of Electrical and Electronics Engineers, IEEE (2001), Fellow of the American Institute for Medical and Biological Engineering, AIMBE (2001), Fellow of the International Academy for Medical and Biological Engineering, IAMBE (2006), and Fellow of the URSI (2017).

Tsukasa Shigemitsu received the B.S., M.Sc. and Dr. Eng. degrees in electronic engineering from Hokkaido University, Sapporo, Japan, in 1971, 1973 and 1979, respectively. He was an assistant researcher at the Max Planck Institute for Physiology, Dortmund, West Germany, from 1974 to 1976. In 1980, he joined the Central Research Institute of Electric Power Industry (CRIEPI), Tokyo. His main research interests covered issues concerning the investigations of the biological and health effects of extremely low frequency (ELF) and Intermediate frequency (IF) electromagnetic fields (EMF). He served three years in JAPAN EMF Information Center in Tokyo. He is a member of the Bioelectromagnetic Society and served as a member of the Board of Directors (2003–2006). He served as a chair and member of the International Union of Radio Science (URSI) Commission K Japanese Chapter on EMF (2006–now). He served as the secretary of the Evaluation Committee on Biological Effects of Electromagnetic Fields of the Institute of Electrical Engineers of Japan. He was appointed as a member of CIGRE EMF Export Group (2006–2013).

Contributors

Masamichi Kato
School of Medicine
Hokkaido University
Sapporo, Hokkaido, Japan

Massimo Emilio Maffei
Department of Life Sciences and Systems Biology
University of Turin
Turin, Italy

Hideyuki Okano
Advanced Institute of Innovative Technology
Saitama University
Saitama, Japan

Tsukasa Shigemitsu
Kashiwa, Chiba, Japan

Shoogo Ueno
Department of Biomedical Engineering,
    Graduate School of Medicine
The University of Tokyo
Tokyo, Japan

Kenichi Yamazaki
Central Research Institute of Electric Power
    Industry
Yokosuka, Kanagawa, Japan

1

Introduction

Shoogo Ueno

Tsukasa Shigemitsu

1.1 The Fusion of Physics and Biology

Foremost, the terminology which will be used and discussed later in this chapter will be introduced. Hans Christian Oersted showed that electricity and magnetism are two related mechanical phenomena. To explain this, he used for the first time the term "electromagnetism." Afterward, Michael Faraday gave an explanation and meaning to this phenomenon and used the term "electromagnetism" which soon became established as a common scientific term. After their findings and discussions were published, the scientific knowledge in electromagnetism improved. This led to a better understanding of the basic electromagnetic phenomena found in bioelectromagnetism. Electromagnetism is a part of science that includes electricity, magnetism and electromagnetic fields phenomena. Biolectromagnetism is the study of how electromagnetism interacts with biological processes produced by cells, tissues and organisms.

Bioelectromagnetism has been gradually developing and expanding to a variety of fields including engineering, biomedical engineering, life science, medicine and biology. This book attempts to present to the readers the various wonderful aspects and the specific fields related to bioelectromagnetism. In order to understand bioelectromagnetism, this introduction will provide the basic concepts and elements of electromagnetism, interaction between electromagnetic fields (natural and man-made) and biological systems and several applications such as biomedical imaging used in medicine and biology. Furthermore, the safety issues related to electromagnetism will be presented. Through the discussions in this book as well as the past and present research advances, a number of prospects for bioelectromagnetism will be suggested by many experts in biology, medicine, biophysics, electrical engineering and biomedical engineering.

Recently, an interesting letter written by Albert Einstein was discovered (Dyer et al., 2021). It was written on October 18, 1949, in response to a letter from Ghyn Davys who was residing in Hampshire, England. It is not known what was written in the letter to Einstein. However, in Einstein's reply, he noted Karl von Frisch's work and said that it would open up a new direction in physics by examining the behavior of bees, migratory birds and carrier pigeons to determine how they perceive external stimuli.

DOI: 10.1201/9781003181354-1

Einstein wrote in his letter, "It is thinkable that the investigation of the behavior of migratory birds and carrier pigeons may someday lead to the understanding of some physical process which is not yet known." It is known that Einstein had attended a lecture by Karl von Frisch at Princeton University in April 1949, a few months before he wrote his reply to Ghyn Davys. By attending this lecture, Einstein had become acquainted with Karl von Frisch who had published very important findings involving that honeybees could communicate the location of rewarding flowers with conspecifics via symbolic dance (von Frisch, 1949). For this discovery, he won the Nobel Prize in Physiology or Medicine in 1973, along with Nikolaas Tinbergen and Konrad Lorenz for their discoveries concerning organization and elicitation of individual and social behavior patterns.

The fusion of physics and biology predicted by Einstein has become a reality. A new research field was developed which combines physics (especially quantum mechanics) with biology, currently called quantum biology. One of the attempts to clarify animal behavior by quantum biology is that migrating birds may use geomagnetism to determine the direction of migration, and a radical pair model was proposed. In 2000, Ritz et al. published a hypothesis showing that birds can sense magnetic fields with the same strength as the geomagnetic field and weaker by considering a radical pair system model, and proposed that the blue-light photoreceptor protein, cryptochrome (CRY) found in the retina is the molecule most likely to act as a radical pair (Schulten et al., 1976: Ritz et al., 2000: Timmel and Henbest, 2004). This was constructed based on the concept of spin chemistry, in which the magnetic field controls the chemical reactions (Hayashi, 2004). CRY including flavin adenine dinucleotide (FAD) is widely distributed in nature, i.e., animals, plants and bacteria. The magnetic sensing ability is presumed to generate the radical pair between FAD and amino acid residue as the intermediate by radical pair mechanism. By having such magnetoreceptions in the retina, migratory birds are assumed to perceive visually in which direction to migrate. The possibility that cryptochromes serve as highly sensitive magnetoreceptions has been suggested. Currently, many other researchers are focusing on other mechanisms based on quantum mechanics (Al-Khalili and McRadden, 2014).

In 1944, Erwin Schrödinger, who is the founder of quantum mechanics and a Nobel Laureate, introduced the idea that living matter at the cellular level can be understood with the use of quantum mechanics in his book (Schrödinger, 1944). This book was based on public lectures delivered at Trinity College, Dublin, in February 1943. He wrote in his book:

> The living organism seems to be a macroscopic system which in part of its behavior approaches to that purely mechanical (as contrasted with thermodynamical) conduct to which all systems tend, as the temperature approaches the absolute zero and the molecular disorder is removed.
>
> *Schrödinger (1944, pp. 73–74)*

Schrödinger claimed that life is a quantum-level phenomenon capable of flying in the air, walking on two or four legs, swimming in the ocean, growing in the soil or, indeed, reading book (Al-Khalili and McFadden, 2014). Quantum biology has gained attention in recent years as a result of many experimental observations. Now, it is a growing area of interdisciplinary research fields investigating non-trivial quantum aspects of biological systems with the help of quantum physicists, chemists, biologists, biochemists and engineers among others.

Forthwith, bioelectromagnetism which is the subject of this book also has a long research history. Bioelectromagnetism can be regarded as a research field that has developed through the integration of classical physics and biology. Classical physics is a discipline based on the classical mechanics of Galileo Galilei and Isaac Newton with the electromagnetic theory of James Clerk Maxwell. It can explain the macroscopic phenomena found in our daily lives, and it is called classical physics. However, phenomena at the microscopic level that could not be explained by classical physics were gradually reported. Thus, a new physics to explain the phenomena at the microscopic level became necessary. Max Planck's discovery of the quantum of action in 1900 and Einstein's theory of the photoelectric effect in 1905 and theory of relativity led to the establishment of quantum mechanics as a new field in physics in the early twentieth century. Planck's and Einstein's theories were the originators of quantum mechanics. The formulation of quantum mechanics was made by Niels Bohr, Werner Heisenberg, Erwin Schrödinger and other physicists around the 1920s. Quantum mechanics is a mathematical theory that was introduced in order to understand the behavior of atoms and particles. For example, quantum mechanics is used to magnetically manipulate the spin of hydrogen atoms nuclei in the body for magnetic resonance imaging (MRI), which is used greatly today in medicine and biology. In bioelectromagnetism, for instance, quantum mechanics and classical physics now work together towards the clarifications of a wide variety of biological phenomena.

Electromagnetism or electromagnetic radiation of natural origins includes cosmic phenomena, the sun, the earth and the atmosphere. Lightning strikes occur in the atmosphere which produces electric field variations. Static electric fields are produced in the earth's atmosphere between the earth's surface and the ionosphere, known as the global atmospheric electric circuit. An atmospheric electric circuit is said to form between the negatively charged surface of earth and positively charged surface of the ionosphere. This structure has a resemblance to that of the spherical capacitor. The naturally originated electric field on earth is highest near the surface and can vary from about 100 to 150 V/m during fair weather conditions to several thousand V/m during severe weather conditions such as under thunderstorms.

The earth is a huge magnet. Its magnetic field is called geomagnetic field or geomagnetism, with strengths of about 60 μT at both poles and about 30 μT at the equator, and it can fluctuate with time scales of millisecond to hours. Geomagnetism is believed to be caused by convection movements in the liquid outer core of the earth. The huge current of hundreds of millions to billions of amperes of conductive fluid flows through it. This is called self-excited magnetohydrodynamic dynamo.

Life of all organisms on earth has been formed and has evolved under the influence of solar electromagnetic radiation. The sun emits electromagnetic radiation starting from the ultraviolet (UV) region, through the visible region until the infrared region, call optical radiation. About 45% of the radiation falls within the visible region wavelength (400–780 nm). The UV region is from below 100 to 400 nm, and this region is further subdivided into UVC (100–280 nm), UVB (280–315 nm) and UVA (315–400 nm). The infrared region is subdivided into IRA (780–1,400 nm), IRB (1,400–3,000 nm) and IRC (3,000–1 mm).

The possible interactions between electricity and magnetism with living organisms were known and used in ancient times to treat headache and gout by using the electric shocks generated by electric fish. The Swiss-born physician Paracelsus is believed to have used magnets to treat diarrhea, hernia, jaundice and bleeding. Afterward, the Italian physician Luigi Galvani discovered bioelectricity through his experiments with the lower limb muscles of frogs, which he claimed to have caused them to twitch when stimulated by electricity.

1.2 The Connection between Electromagnetism and Bioelectromagnetism

Since ancient times, electricity and magnetism were considered to be separate phenomena. In 1820, Hans Christian Oersted showed that electricity and magnetism are a combined mechanical phenomenon. To explain this phenomenon, he used for the first time the term "electromagnetism" in his paper (Oersted, 1820). Afterward, Michael Faraday immediately gave the explanation and meaning to this phenomenon and made active use of the term "electromagnetism," which became established as a scientific term within a year (Faraday, 1821, 1822). After Oersted's discovery, the scientific knowledge in electromagnetism was greatly improved. This improvement led to a better understanding of the basic electromagnetic phenomena in bioelectromagnetism.

Electromagnetic waves exist on earth. James Clerk Maxwell proposed theoretically the mathematical description of an electromagnetic wave in 1865. He considered that the components of the electromagnetic wave are electric and magnetic fields, independently. After the discovery of the electromagnetic

TABLE 1.1   Important Events in the History of Bioelectromagnetism

1493	Paracelsus – medical treatment with magnet as vital stones
1600	Gilbert – De Magnete
1786	Galvani – animal electricity from frog experiment
1793	Volta – Volta's pile
1830	Faraday – Induction law
1849	Du Bois-Reymond – medical application of induction coil
1864	Maxwell – the electromagnetic field theory
1891	Tesla and d'Arsoval – medical application of high-frequency current
1896	D'Arsonval – magnetophosphenes
1925	Schliephake – medical application of shortwave
1953	Yasuda – piezoelectricity of bone
1959	Collin – magnetic stimulation of nerves
1960	Human health effect of ELF electric field
1962	Bassett – electric stimulation of bone fracture
1979	Human health effects of ELF magnetic field
1985	Barker – magnetic motor coltex stimulation
1988	Ueno – magnetic stimulation of the figure-eight coil
2000	Human health effect of high-frequency field
2003	MRI medical application

theory by Maxwell and its formulation by Oliver Heaviside, the existence of the electromagnetic wave was discovered experimentally by Heinrich Rudolf Hertz in 1888. Maxwell equations are a set of four equations which describe an electromagnetic wave: Gauss's law, Gauss's law for magnetism, Faraday's law and Ampère's law. In 1897, using electromagnetic waves, Guglielmo Marconi was able to send radio transmission signals to a tugboat at a distance over 18 miles (29 km) from the Bristol Channel, England. In 1899, Marconi sent the first international wireless message from Dover, England, to Wimereux, France. Within 4 ycars, he sent a wireless message across the Atlantic Ocean. These historical events opened a new era for the use of electromagnetic waves.

Table 1.1 shows the discovery timeline and other important events in the history of bioelectromagnetism. They are divided into theory, instrumentation, stimulation and measurement in the field of the interactions between electromagnetic phenomena and the living body. Fundamental research studies on the medical application of magnetic phenomena were promoted, as seen by the magnetic stimulation of the nerve and the development of techniques used in clinical applications. As shown in Table 1.1, the history of bioelectromagnetism is nearly identical to that of electromagnetism. This means that the instrumentation based on bioelectromagnetic phenomena was quickly developed. These instruments were able to measure both electric and magnetic signals in biological systems.

After the first electromagnetic experiments on biological systems were conducted during the course of the nineteenth century, electromagnetism and bioelectromagnetism share the same long research history. Since then, what is the interaction between electromagnetism and living organisms, how electromagnetism affects living organisms and how living organisms use electromagnetic information been of interest to many scientists. These interests have led to numerous experiments and theoretical studies, and the concept of bioelectromagnetism was established through such research developments. The details of the history of bioelectromagnetism will be given in Chapter 2.

Although bioelectromagnetism has a long research history with help from electromagnetism, the term "bioelectromagnetism" gained popularity in scientific communities around 1980. The time around 1980 was when the Bioelectromagnetics Society (BEMS) was founded. The BEMS covers research related to electromagnetic phenomena ranging from static fields through the radio frequency field up to terahertz frequencies and acoustic energy with biological systems.

1.2.1 Electromagnetism

The properties of electromagnetic waves necessary for understanding the various effects of electromagnetic fields on biological systems will be briefly presented. They describe the fundamental aspects of electric and magnetic fields as well as the electromagnetic wave. To begin, consider the case where a direct current (DC) power supply is connected to a single wire stretched in the air. The wire is being charged by the electric charge from the power source, and electric lines of force are generated from the wire to the ground. Although the electric lines of force are invisible to the eye, their existence can be confirmed by the fact that when another charged particle is placed between the wire and the ground, it can receive either an attractive or repulsive force in the direction of the electric lines of force. The space where these electric lines of force exist, that is the field where the electric force act, is called electric field. Its strength is defined by the force that acts when a unit charge is placed there. Now, if an alternating current (AC) power source is connected to the same cable instead of a DC power supply, the polarity of the cable charge reverses direction with its frequency. Therefore, the direction of the electric lines of force is also reversing with its frequency, but the pattern of distribution remains the same. In short, an electric field is a region of space over which an electric charge exerts a force on charged objects in its vicinity. The unit of the electric field strength is Newton/Coulomb (N/C), and in practical use, the unit is expressed in Volt/Meter (V/m).

When a resistor is connected to one end of a wire and an electric current flows through the wire, the current creates magnetic lines of force around the wire. Magnetic lines of force are also invisible, but when a magnet is placed near the wire, its existence is confirmed by the experience of attractive or repulsive forces in the direction of the magnetic lines of force. The space where these magnetic lines of force exist or the field where they act is called magnetic fields. Its strength is proportional to the density of the magnetic lines of force. Even if an AC power supply is used instead of a DC power supply, the distribution pattern remains the same as in the case of a DC power supply, except that the direction of the magnetic lines of force reverses corresponding to the direction of the current. The units of magnetic field strength are Newton/Weber (N/Wb) and Ampere/Meter (A/m). Magnetic flux density, which is defined as the amount of magnetic field passing through a unit cross-section area, is used in place of magnetic field. The unit for magnetic flux density is Wb/m or Tesla (T) which is equal to 10^4 Gauss (G). If H and B are made to be the magnetic field and the magnetic flux density, respectively, it becomes

$$B = \mu H$$

where μ is the permeability in unit of Henry/Meter (H/m). In biological materials μ is equal to the free-space's value, i.e., $\mu_0 = 4\pi \times 10^{-7}$ H/m.

If an AC power supply is connected to two short wires, electric lines of force corresponding to the applied voltage are generated in the space between the two wires, and magnetic lines of force proportional to the current flowing through the wires (displacement current) are generated around the wires. It is clear that the electric and magnetic lines of force reverse their directions with the frequency of the power source. It is easy to imagine that if the tips of the two wires are spread out, the distribution pattern of electric and magnetic lines of force will spread outward. If the two wires are further spread out until they are lined up on the same line, there is no longer any space between the wires to confine the electric and magnetic lines of force, so further they will both spread outward. If the frequency of the power supply is chosen so that the length of the two wires is equal to a half wavelength, a resonant state occurs, and the magnitude of the voltage and the current flowing through the wires are maximized. This also maximizes the density of the electric and magnetic lines of force in the surrounding space. At higher frequencies, the electric and magnetic fields spread outward like waves. Using this property, high-frequency energy can be sent to distant places by a transmitting antenna, and the electrical vibrations radiated are called electromagnetic waves. In this sense, an electromagnetic wave consists of the oscillation of electric and magnetic fields in wave motion. In an electromagnetic wave, an electric field and a magnetic field coexist together and travel through a vacuum at the speed of light (3×10^8 m/s). They are characterized by their wavelength λ measured in meters (m) and frequency f measured in Hertz (Hz).

According to quantum mechanics, electromagnetic waves have properties of both a wave and a particle (photon). The photon energy of electromagnetic waves (U) is quantized with Planck's constant (h) times the frequency (f). This energy can be expressed in electron volts (1 eV = 1.6×10^{-19} J). The following relationships hold for the electromagnetic wave.

$$\lambda = c/f, \, U = hf, \, T = U/k = hf/k$$

where c is the speed of light (= 3×10^8 m/s), h is the Planck's constant (= 6.63×10^{-34} Js), k is the Boltzmann's constant (= 1.381×10^{-23} J/K), the photon energy U in Joule (J), and T in Kelvin (K). It can be observed that if the frequency increases or the wavelength decreases, the photon energy increases.

The electromagnetic spectrum is the distribution produced when the electromagnetic waves are decomposed into various frequency components. Different electromagnetic wave frequencies give different forms of electromagnetic radiation. Figure 1.1 shows the electromagnetic spectrum, which is usually divided into ionizing radiation or non-ionizing radiation along with the typical sources. Ionizing radiation is the radiation that occurs when its frequency is higher than 3,000 THz. It includes hard UV and higher frequency electromagnetic waves (X-rays, γ-rays). This frequency corresponds to a wave with a wavelength of 100 nm (1 nm = 10^{-9} m), a photon energy of 10 eV. Ionizing radiations cause severe damage to the gene of living materials. Non-ionizing radiation refers to the electromagnetic fields ranging from 0 Hz (static fields) to 3,000 THz. This wide frequency band, as seen in Figure 1.1, includes static electric and magnetic fields, extremely low-frequency electromagnetic fields (ELF-EMF), intermediate frequency electromagnetic fields, high-frequency field and visible light. The International Commission on Non-Ionizing Radiation Protection (ICNIRP) has published the definition of non-ionizing radiation (ICNIRP, 2020b). Non-ionizing radiation refers to electromagnetic radiation and fields with a photon energy lower than 10 eV, corresponding to frequencies lower than 3,000 THz and wavelengths longer than 100 nm. It is grouped into different frequency or wavelength bands, namely UV radiation (wavelengths 100–400 nm), visible light (wavelengths 400–780 nm), radiofrequency electromagnetic fields (frequency 100 kHz–300 GHz), low-frequency fields (frequencies 1 Hz–100 kHz) and static electric and magnetic fields (0 Hz). Non-ionizing radiations have no high energy to cause severe damage to chemical bonds. Non-ionizing radiations have different mechanisms of biological interactions depending on their frequencies or wavelength.

In addition to the naturally occurring electromagnetic radiations, man-made sources of electromagnetic fields produced after the electrification began to exist in the human environments. Now, electromagnetic fields on earth are consists of two components, electromagnetic fields generated from natural phenomena and man-made electromagnetic fields. These two differently originated sources of non-ionizing radiation exist together and around us. Electromagnetic radiations of broad frequency from the non-ionizing range are used in biomedical imaging in medicine and biology.

1.2.2 Bioelectromagnetism

Bioelectromagnetism is the area of science that studies the interaction between electromagnetic phenomena and biological systems. The knowledge in electromagnetism based on classical physics and quantum mechanics has led to understanding the basic phenomena in bioelectromagnetism. This is due since electromagnetism and bioelectromagnetism have advanced together until now and they will continue advancing together in the future. In the history of science so far, bioelectromagnetism has been basically developed by connecting electricity, magnetism and electromagnetism to biology, medicine, physics and engineering. MRI using static and time-varying magnetic fields and radiofrequency electromagnetic fields is a typical example. In this sense, bioelectromagnetism is significantly difficult

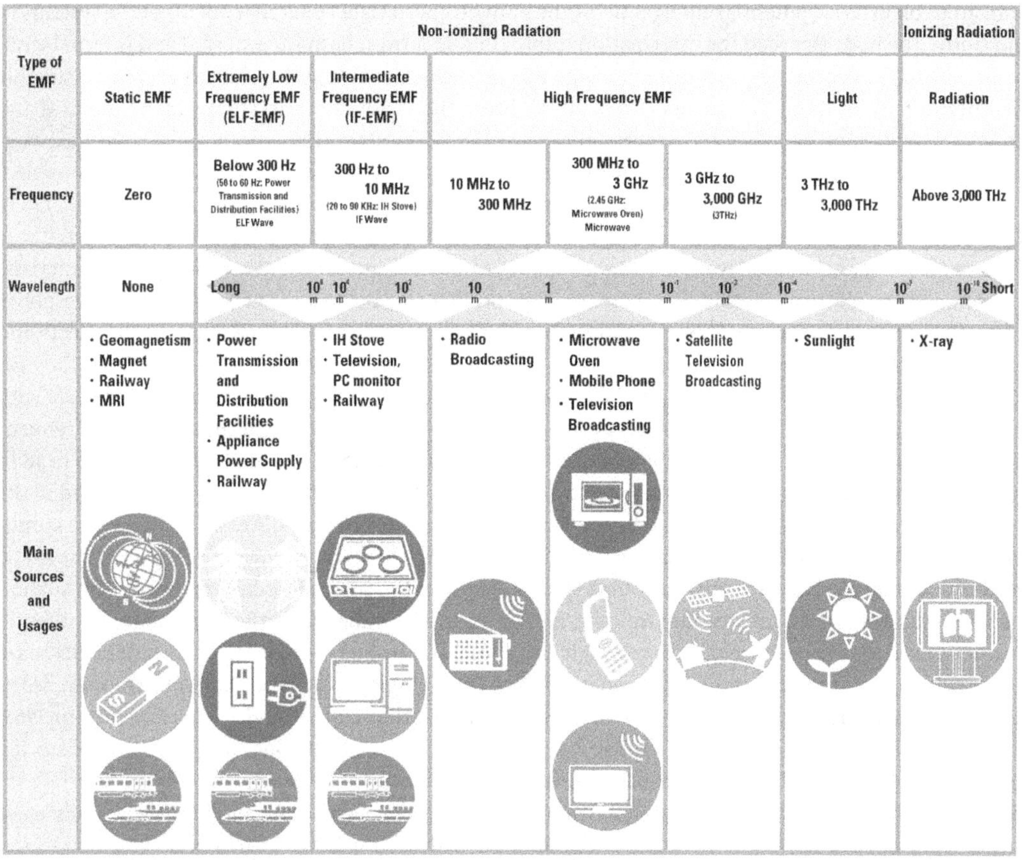

Type of EMF	Non-ionizing Radiation							Ionizing Radiation
	Static EMF	Extremely Low Frequency EMF (ELF-EMF)	Intermediate Frequency EMF (IF-EMF)	High Frequency EMF			Light	Radiation
Frequency	Zero	Below 300 Hz (50 to 60 Hz Power Transmission and Distribution Facilities) ELF Wave	300 Hz to 10 MHz (20 to 90 KHz: IH Stove) IF Wave	10 MHz to 300 MHz	300 MHz to 3 GHz (2.45 GHz: Microwave Oven) Microwave	3 GHz to 3,000 GHz (3THz)	3 THz to 3,000 THz	Above 3,000 THz
Wavelength	None	Long 10^6 m 10^4 m 10^2 m	10 m	1 m	10^{-1} m	10^{-3} m 10^{-4} m	10^{-5} m	10^{-8} m Short
Main Sources and Usages	· Geomagnetism · Magnet · Railway · MRI	· Power Transmission and Distribution Facilities · Appliance Power Supply · Railway	· IH Stove · Television, PC monitor · Railway	· Radio Broadcasting	· Microwave Oven · Mobile Phone · Television Broadcasting	· Satellite Television Broadcasting	· Sunlight	· X-ray

Note: The frequency unit "hertz (Hz)" represents the number of oscillations in a second, equal to the result obtained by dividing by the wavelength the speed, 300,000 kilometers per second (km/s) at which an electromagnetic wave propagates.
kilo- (k) = 10^3, mega- (M) = 10^6, giga- (G) = 10^9, tera- (T) = 10^{12}

FIGURE 1.1 The electromagnetic spectrum forms static field (0 Hz) through extremely low-frequency fields (ELF-EMF) high-frequency field (HF-EMF) to light. The typical sources are included. (Courtesy of JEIC, Tokyo.)

to understand because of the different types of electromagnetic phenomena that are involved in the concept of bioelectromagnetism. Thus, it is an interdisciplinary study that examines the electric, magnetic and electromagnetic phenomena produced by biological systems, and it also involves the measurement and stimulation associated with biological tissues. This means that the concept of bioelectromagnetism can be divided in different ways; here, it has been divided into two parts. The first one is the study of the interaction between electromagnetic fields up to 300 GHz and biological systems. The second one is the research field that measures electric and magnetic fields originated from bioelectric sources.

Looking back at the history of bioelectromagnetism, the basic research fields can be divided into the following four academic categories: bioelectricity, electrobiology, biomagnetism and magnetobiology. Bioelectricity is the study of the electrical phenomena generated in biological systems through the use of electrophysiological techniques. The electrical phenomena include properties inherent to the cells, such as membrane potential, action potential and the propagation of the potentials. Bioelectricity has also interest in the effects of biological systems by externally applied electricity. Here the word "effects" of external electricity means how the cells in biological systems respond to the applied or exposed fields. The difference in electrical potential existing between the inside part of a cell and the extracellular fluid surrounding it is called the membrane potential. It is a specialized function of the nervous system is to propagate membrane potential changes within a cell and to transmit them to other cells. The transmission of these potential changes helps the body to coordinate the activity of all of the systems in our body. The body can feed the information impinging on it from both the external and internal environments to the central nervous system, where it is processed, enabling the body to adapt in a suitable manner to both of its environments (Kato, 2006). From the study of bioelectricity, many medical and clinical applications such as electroencephalography (EEG), electrocardiography (ECG), electromyography and others have been developed. Moreover, electrobiology involves the study of how various electrical phenomena such as electric field, voltages and currents affect biological systems (Popp et al., 1989). In nature, it is known that some fish can detect the weak electrical changes generated by bait. These fish are known to have sensory organs that detect these weak electrical changes, and this characteristic is referred to as electroreception. It is now understood that this behavior is controlled by the response of mechanosensory organs to electric stimuli, based on electroreception experiments with insects and terrestrial animals.

Biomagnetism is the study of magnetic fields originating from biological systems. It also deals with magnetic phenomena of biological systems, which can be observed at different intensities and frequencies. For example, the so-called magnetophosphene is a visual sensation caused by exposing the head to a low-frequency (around 10–70 Hz) magnetic field of around 10–20 mT. This sensation is generated in the retina. The earliest magnetic stimulation was reported by Jacques-Arsene d'Arsonval. Magnetic stimulation is based on Faraday's law of induction. This law states how the change of an applied magnetic field induces an electric field with accompanying current in the tissue. The first magnetic stimulation of nerve was described by Alexander Kolin in 1959. Magnetic stimulation of the human brain and heart has been used for the purpose of both research and clinical treatment. The biological magnetic fields are extremely weak compared to the geomagnetic fields. The biological magnetic field of the human heart, called magnetocardiogram (MCG), was first detected by Gerhard Baule and Richard McFee in 1963. The biological magnetic field of the human brain called magnetoencephalogram (MEG) was detected by David Cohen in 1968. After these pioneering studies, using superconductive quantum interference devices techniques, the weak biological magnetic fields from the brain, heart and lung were easily measured from outside the body. Biomedical stimulation with ELF electric or magnetic fields was used first for clinical applications such as the healing of bone fracture. Fundamental research activities have been carried out for the healing promotion of various tissues. However, there is still no widely accepted mechanism by which ELF electric or magnetic fields can affect biological tissues. The well-known interaction between ELF electric or magnetic fields and biological tissues is the eddy currents induced in the tissues, which is a possible candidate mechanism. The other candidate is the interaction of the applied magnetic fields with an endogenous magnet such as magnetite. Because the brain is so important for human behavior, and because the functions of the brain inherently involve a great amount of electrical activity, since the beginning of bioelectromagnetism, it has been essential to examine the effects of magnetic fields, electric fields and currents, since they can induce electric fields and currents on the brain.

Magnetobiology involves the study of interaction between magnetic field and biological systems (Popp et al., 1989). Magnetobiology also studies the identification and sensitivities of biological organisms to weak magnetic fields. Magnetoreception, a category of magnetobiology, is known as the sensing of magnetic fields by biological organisms. It includes the magnetic navigation of migrating birds and other organisms, the magnetotactic behavior of some bacteria and the magnetoreception in humans. The study of magnetobiology and magnetoreception has made great advances in the last 10–20 years. In magnetobiology, the ability of birds, bees, turtles, reptiles, amphibians, plants and others to detect the geomagnetic field has been reported (Maffei, 2014, 2019; Wiltschko and Wiltschko, 1995). Certain crustaceans, lobsters and bony fish have also been found to use magnetism-related navigations (magnetic compass). In this sense, magnetic sensibility is common in the animal kingdom. Thus, there is no longer any doubt that animals can sense the geomagnetic field; however, how they sense it is still unknown. The geomagnetic field is weak and is unlikely to have a direct effect on the chemical reactions in the body. Currently, the main theory of the mechanism of action of the geomagnetic field is that magnetite and chemical compasses based on quantum biology are responsible for the magnetoreception. For this reason, the question has been raised as to whether humans can sense magnetism in the same way. This question has been addressed in outdoor and indoor experiments using human subjects. In the 1980s Baker and later in 2019 Kirschvink's group conducted experiments that showed that people can sense magnetism, but they have not yet been able to determine it clearly (Baker, 1980; Wang et al., 2019).

The study of magnetic-related phenomena also includes the change of magnetic properties of every material which occurs under high strength magnet. During the 1930s and 1940s, scientists have attempted to study the effects on living systems. However, they ignored the fundamental laws of electromagnetism. Significantly more intensive research in magnetobiology began in the 1960s with the research of space biology.

From these advances, considerable experimental and theoretical research studies involving the interaction between electromagnetic fields and living systems emerged, trying to elucidate what kind of relationship exists between electricity, magnetism and life, how electricity and magnetism give action to living systems, and how the living systems utilize electricity and magnetism. Furthermore, it is important to understand the interaction of electromagnetic fields with biological systems not only for medical applications but also for the protection from exposure to electromagnetic fields. In this way, bioelectromagnetism is an interdisciplinary science which combines biology, medicine, engineering, chemistry and physics.

1.3 Harmonization of Electromagnetic Fields with Biological Systems

Biological systems have been and are exposed to man-made electromagnetic fields. However, since ancient times, living organisms have been exposed to electromagnetic waves from cosmic rays, electromagnetic fields created by the earth, air ions and low-frequency electromagnetic fields that travel around the earth. There have been studies on how organisms interact with these natural electromagnetic phenomena and how they actively use the natural electromagnetic phenomena to sustain life. In particular, research has been conducted on the effects of the earth's electric field and the effects of air ions on living organisms and on how they are utilized by living organisms, and also on the relationship between low-frequency electromagnetic fields generated by lightning discharges with physiological changes in humans and animals. In addition, studies have shown that the geomagnetic field is used by migrating birds, fish and bees to determine their orientation.

Static magnetic fields can cause changes in the mass transfer due to the magnetic orientation of biological tissues and bio-macromolecules, changes in chemical reactions of radical pairs and effects on enzymes. Medical diagnostic equipment using static magnetic fields has been developed and now used clinically. Presently, it is believed that a static magnetic field below 2 T has no harmful effects on health. There are many animals and microorganisms that use weak static magnetic fields as an aid to their survival. Electric fish and cartilaginous fishes such as sharks or rays are known to be able to detect electricity.

Invertebrates such as the honeybee, fruit flies, some beetle species, ants and wild pollinators (e.g., Diptera, Hymenoptera, Lepidoptera, Coleoptera) have been used to study the effects of electromagnetic radiation. The studies on the responses of invertebrates to natural electromagnetic fields have demonstrated how they detect and orientate with electromagnetic fields, and what are the electromagnetic field effects on their behavior, physiological functions and reproduction. The laboratory studies are focused on physiological or developmental responses to short-term or acute exposure to electromagnetic fields. There are still no studies that examine effects on invertebrates of long-term exposure to electromagnetic fields. Today, there is a need for more ecological studies focusing on measuring the effects of electromagnetic radiation on wild communities. The ecological study of electromagnetic fields on vertebrates covers its wide spectrum range starting from hormone levels, other physiological parameters to behavior patterns. These studies might have ecological implications, but real ecological studies are extremely rare. The lack of ecological system studies is probably founded on the multiple environmental variables that affect ecological communities which make it difficult to identify the influence of particular electromagnetic fields. Furthermore, with the advance of our information societies such as the 5G networks, the need for research into the effects of electromagnetic fields on ecology systems has become of increased importance.

1.3.1 The Coupling of Atmospheric Electricity with Biological Systems

It is well known that there are various important processes occurring in the global electric circuit. Particularly, the charge separation in thunderstorms creates a potential difference between the highly conductive regions of the ionosphere and the earth's surface. An atmospheric electric circuit is said to form between the negatively charged earth's surface and the positively charged surface of the ionosphere. This structure has a strong resemblance to a spherical capacitor (Rycroft et al., 2012).

Negative air ions are believed to help purify the air. Negative air ions are present on the ground surface during clear weather and in large numbers near waterfalls, while the concentration of positive air ions increases when thunderclouds approach. Air ions can produce physiological effects; the negative air ions produced by atmospheric electrical phenomena tend to make you feel refreshed, while the positive air ions are said to cause discomfort. However, there is still not enough scientific evidence to clearly conclude these effects. Other physiological effects include those exerted by the electrostatic force of strong static electric fields on the hair and other parts of the body as well as the stimulation of nerve receptors on the skin surfaces; however, they have no harmful effects on health.

An intimate affinity of atmospheric electricity to biological systems exists. It shows how atmospheric electricity can govern human well-being (as well as for a wide array of organisms) and the processes they sustain; therefore, atmospheric electricity must be considered as a significant direct and indirect driver of biology (Cifra, 2021; Hunting, 2021a, b; Price et al., 2021).

There are various electromagnetic phenomena on earth. Since early times, living organisms have been exposed to the electric and magnetic fields produced by the earth, as well as the electromagnetic fields associated with lightning discharges and solar activity. Therefore, it is presumed that the evolution of organisms has been affected by these electromagnetic phenomena in various ways. In other words, the question of what relationship exists between electromagnetic phenomena on earth and the evolution of life has been a matter of great interest to both scientists and the general public. For example, it has been reported that animals actively use the geomagnetic field. These include magnetite-based magnetotactic bacteria, the behavior of honeybees and the migration of bird. It has also been reported that sharks and rays use the Lorenzini organ to sense electric fields that can control their behavior.

1.3.2 Safety Aspects of Electromagnetic Fields

According to the ICNIRP definition, non-ionizing radiation refers to electromagnetic radiation and fields with photon energy lower than 10 eV, which corresponds to frequencies lower than 3,000 THz. The groups of static electric and magnetic fields (0 Hz), low-frequency electromagnetic fields (1 Hz–100 kHz) and radiofrequency electromagnetic fields (100 kHz–300 GHz) are categorized into non-ionizing radiation (ICNIRP, 2020a).

The modern world is full of invisible electromagnetic waves. As mentioned previously, non-ionizing radiation is an integral part of our lives. There are a wide variety of electromagnetic environments ranging from low frequency, intermediate frequency to millimeter and terahertz waves in our surrounding (Figure 1.1). Since the electromagnetic environment of non-ionizing radiation is invisible to the eye, many people are concerned about their potential health effects. Non-ionizing radiation are now being used in information and communication technologies and are expected to be actively employed further. In the future, as technological innovation advances, different frequencies will be used for the new generation networks (e.g., 5G) and for wireless power transfer. As new frequency bands of millimeter and terahertz waves begin to flood our living environment, it will be necessary to conduct further research to assess the exposure to electromagnetic field generated from the new generation technologies and their impact on human health and global environment. Therefore, the safety study of non-ionizing radiation to humans, nature and the global environment is being called for. The effective use of non-ionizing radiation in the field of biotechnology is also expected to become increased. For this purpose and owing to the rapid development of new technologies, it is necessary to conduct basic research on various forms of non-ionizing radiation from the viewpoint of bioelectromagnetism and to promote and discuss the possible human risks for humans and also their beneficial use.

Here, a brief overview on the health issues of non-ionizing radiations will be presented. To start with, several reviews have been published on the biological and health effects due to the exposure to static electric and magnetic fields (IARC, 2002; ICNIRP, 2009b; Ueno and Okano, 2012; Ueno and Shigemitsu, 2007; WHO, 2006, 2007). Static electric and magnetic fields originate from both natural and man-made sources. Static electric fields are derived from the earth's atmosphere as a part of the global electric circuit. The naturally originated static electric field on earth is highest near the surface, ranging from about 100 to 150 V/m during fair weather to several thousand V/m beneath thunderclouds. Static electric field depends on temperature, relative humidity, altitude and other weather conditions. The natural static magnetic field of earth originates from the electric current flow in the liquid outer core of the earth. This field is called the geomagnetic field. The geomagnetic field is described by three components: total magnetic intensity, declination and inclination. The geomagnetic field fluctuates according to diurnal, lunar and seasonal variations. The total field intensity in Japan is around 50 μT. On the other hand, man-made sources of static electric and magnetic fields are found everywhere in our day life, industrial facilities, medical equipment and through power transmission systems (WHO, 1987, 2006). Static electric fields do not penetrate the human body but can induce a surface charge. This charge may be perceived through its interaction with body hair and by other phenomena such as discharge (microshock) at sufficiently high fields. Their perception in humans is dependent on various factors and can range from 10 to 45 kV/m. Static magnetic fields are not perturbed by the human body. There are three well-known mechanisms by which a static magnetic field interacts with biological systems: magnetic induction, magneto-mechanical effects and electron spin effects (ICNIRP, 2009a). Based on the evaluation of biological effect research, the WHO carried out a human health assessment of static electric and magnetic fields (WHO, 2006).

From the ICNIRP definition, low frequency (LF) is used to describe the fields with a frequency range from 1 Hz to 100 kHz (ICNIRP, 2003, 2010a, b). In this frequency range, the interaction of electric and magnetic fields with the human body induces electric fields and currents in the tissues. The demonstrated effect is the induction of magnetophosphenes, a perception of a faint flickering light in the periphery of the visual field. They are thought to result from the interaction of the induced electric field with electrically excitable cells in the retina. The threshold for induction of magnetophosphenes has been estimated to be low between 50 and 100 mV/m at 20 Hz. Epidemiological studies have suggested that long-term exposure to 50/60 Hz magnetic fields might be associated with an increased risk of childhood leukemia. Two pooled analyses indicate that an excess risk may exist for average exposures exceeding 0.3–0.4 μT. However, there is some degree of confounding and chance that could possibly explain these results. No biophysical mechanism has been identified, and results from animal and cellular studies do not support the observation that exposure to 50/60 Hz magnetic fields cause childhood leukemia. There is no substantial evidence for an association between LF magnetic field exposure and Parkinson's disease, multiple sclerosis and cardiovascular diseases. The evidence for an association between LF magnetic field exposure and Alzheimer's disease and amyotrophic sclerosis has not yet produced clear conclusion. Overall, research has not shown that long-term exposure to low-level LF magnetic fields has adverse effects on health. So, ICNIRP's analysis is that the currently existing scientific evidence that prolonged exposure to LF magnetic fields is directly related with an increased risk of childhood leukemia is too weak to form the basis for exposure guidelines. The perception of surface electric charge, the direct stimulation of nerve and muscle tissue and the induction of retinal phos-phene are the only well-established adverse effects thus serving as the basis for the exposure guidelines (ICNIRP, 2010b).

According to the ICNIRP's definition, radiofrequency electromagnetic field is used to describe fields in the frequency range of 100 kHz to 300 GHz. Due to the gradual use increase of radiofrequency-oper-ated devices such as mobile-phone, Wi-Fi, Bluetooth, radar, smart meters, medical equipment, etc., exposure levels of radiofrequency electromagnetic fields around us have also increased gradually. The understanding of the health, biological and environmental effects of electromagnetic fields in the fre-quency range of up to 300 GHz is advancing rapidly. In 2020, ICNIRP published a guideline for the fre-quency range of 100 kHz–300 GHz (IARC, 2013; ICNIRP, 2020b). This guideline aims to protect against adverse health effects relating to the exposure to radiofrequency electromagnetic fields, including from 5G technologies.

The development of mobile wireless technology has definitely evolved from 1 to 5G. In the fifth-gen-eration communication system (5G), three frequency bands are actually expected to be used: low fre-quency (700 MHz band), high frequency (3.4–3.8 GHz band) and extremely high-frequency millimeter (26 GHz and above) bands. Future telecommunications are planning to use radiofrequency electromag-netic fields working at frequencies above 6 GHz to the millimeter wave range (30–300 GHz) in order to achieve high-capacity, ultra-high-speed communication. The use of radiofrequency electromagnetic fields within these frequency bands urges that the human exposure and their health effects must be clarified quickly for a safe and effective use of 5G systems.

As a result of public concern regarding any possible adverse effects to human health exposure to radiofrequency electromagnetic fields above 6 GHz from 5G or other sources, the research group of Andrew Wood from Swinburne University of Technology, Victoria, Australia, reviewed 107 past experi-mental studies which investigated various bioeffects including genotoxicity, cell proliferation, gene expression, cell signaling, membrane function and other effects (Karipidis et al., 2021; Wood et al., 2021). This review included 31 epidemiological studies which investigated exposure to radar radiation of above 6 GHz. The review result showed no conclusive evidence that low-level radiofrequency electro-magnetic fields above 6 GHz at exposure levels below the ICNIRP occupational limits (ICNIRP, 2020b) were associated with biological effects relevant to human health (Karipidis et al., 2021). They pointed out that the studies reporting the effects came from the same research groups and that the results have not been independently reproduced (Karipidis et al., 2021). Furthermore, they recommended that the studies should improve their experimental design with attention to dosimetry and temperature control. In addition, as a companion paper, they pointed out from meta-analysis of in vitro and in vivo studies that the results do not confirm an association between low-level millimeter waves and biological effects (Wood et al., 2021). There are also several reviews focusing on high frequencies up to 100 GHz (Neufeld and Kuster, 2018; Simkó and Mattsson, 2019; Barnes and Greenebaum, 2020). The majority of studies concluded that the risk of adverse effects on humans is low even at high frequencies. Recently, in addi-tion to the current short-term exposure guidelines, Frank Barnes and Ben Greenebaum proposed an approach on how the long-term exposure to weak fields might be set (Barnes and Greenebaum, 2020). However, further research is still necessary to determine clearly whether if biological effects at higher frequencies indeed exist.

The Information Communication Technology is core technologies becoming part of the social infra-structure consisting of non-ionizing radiations. In the past, the safety of non-ionizing radiation in humans, animals, and on a global scale has been discussed. Still, the carcinogenic risk assessment of ion-ionizing radiation is being addressed by the IARC and the WHO, and basic scientific research is also underway.

At a time when environmental health researchers tackle serious global issues, there is an urgent need to address climate change and other environmental problems. The technological advances can alter the electromagnetic environments around us. Thus, it is necessary to consider and assess the possible biological effects, both beneficial and detrimental of the altered electromagnetic environment. It has been reported that the animals whose behavior is controlled by geomagnetic fields lose the ability to use their internal magnetic compass when radiofrequency electromagnetic fields are introduced. In today's society, as we enter the 5G era, various frequencies of electromagnetic waves will become even more prevalent in our surroundings, and with the rapid expansion of 5G networks, radiofrequency electromagnetic field will be truly ubiquitous in the environment. The effects of radiofrequency electromagnetic fields on honeybees are an increasing interest due to the rise of environmental exposure owing to electromagnetic fields (Favre, 2011, 2017). There is existing evidence of non-ionizing radiation effects on flora and fauna (Levitt et al., 2021a, b). For example, the global reduction in honeybees and other insects is plausibly linked to the increased radiofrequency electromagnetic fields in our environment. However, whether man-made electromagnetic field poses a significant threat to insect pollinators and the benefits they provide to ecosystems and humanity remains to be established (Vanbergen et al., 2019).

It has been established that the magnetic compass of migratory birds can be disrupted by the weak radiofrequency electromagnetic field background found in larger cities, but it is currently unclear which exact frequencies are likely responsible. Furthermore, some studies have suggested that fields emanating from power lines also affect the magnetic sense of vertebrates, but it is unclear whether this effect is specific to 50 Hz magnetic fields to harmonics or even to electric fields (Begall et al., 2008).

The spread of electric vehicles that do not emit oxygen dioxide is necessary against climate changes. It is expected that electric vehicles will be recharged without the need for a physical plug connection and that the possibility of wireless power charging or the ability to supply power while driving will be a great leap forward in convenience. The history of wireless power transmission goes back to 1890 when Nikola Tesla first attempted wireless power transmission. After that, microwave power transmission and RF-ID were studied. In 2007, a paper on wireless power transmission technology was published by a research group of the Massachusetts Institute of Technology, which triggered more research and development (Kurs et al., 2007). In order to transmit power in a wireless manner, electromagnetic waves must be radiated by the transmission antenna and received by a receiving antenna. Electromagnetic waves are generated between these antennas; therefore, it is necessary to understand the safety of these electromagnetic waves.

The world has been flooded with various devices and equipment that use electromagnetic waves, mainly of non-ionizing radiation. Thus, in order to protect the earth's natural environment and to maintain a sustainable society where nature and humans exist in harmony in the future, we need to know what kind of effect these man-made electromagnetic fields will have on the environment and whether the result of the electromagnetic flooding of the environment will interfere with the behavior of the wildlife.

In the future, electromagnetic waves will be more prevalent in our surrounding than ever before, since various information and communication devices using non-ionizing radiation (especially radiofrequency electromagnetic fields) will be in use. Therefore, there is a need to clarify the effects of electromagnetic radiation on human health as well as on vertebrates and invertebrates. Vertebrates and invertebrates are the species that make up the global environment, and the clarification of the effects of electromagnetic waves on them will help to solve the global environmental problems that we are facing today. Since the exposure to radiofrequency electromagnetic fields has been steadily increasing, many experts have rated cancer, heat-related effects, adverse birth outcomes, electromagnetic hypersensitivity, cognitive impairment, adverse pregnancy outcomes and oxidative stress as the most critical outcomes regarding radiofrequency electromagnetic field exposure (Verbeek, et al., 2021).

Since the first radio transmission signal broadcast made by Marconi in 1897, the exposure to man-made electromagnetic fields has steadily increased. We are now exposed to electromagnetic fields from a variety of sources. Due to their increased utilization, particular attention is given to the radiofrequency electromagnetic fields such as millimeter, terahertz radiation and optical radiation. The understanding of the interactions between electromagnetic fields (at all frequency ranges) with biological systems is necessary for the safety of humans and biological and ecological systems.

1.4 Applications of Electromagnetism in Medicine and Biology

The advancement in understanding the relationship between electromagnetism and physiology is creating non-invasive methods for medical treatment. Bioelectromagnetism and electromagnetic fields play important roles in biomedical engineering. Electromagnetic fields have a strong potential for medical and therapeutic applications. These applications include the use of pulsed magnetic fields, low-frequency electric and magnetic fields and radiofrequency electromagnetic fields. Shortwave and microwave diathermies have already been used.

As already mentioned, bioelectromagnetism can be divided mainly into two distinct categories. The first is focused on researching the beneficial effects of electromagnetic fields, which have strong potential in diagnostic and therapeutic applications in medicine. For example, MRI, a non-invasive medical imaging technique, uses a high-strength magnetic field, a rapid changing magnetic field and a radiofrequency electromagnetic field. The second focuses on the research of the interaction between electromagnetic fields and living systems, promoting the understanding of the biological and health hazards effects associated with the exposure to electromagnetic fields.

Ueno and Sekino edited a book on the recent advances of biomagnetics, and particularly, on the applications of biomagnetic stimulation and bioimaging (Ueno and Sekino, 2016; Ueno, 2020). The book reviews principles and applications of biomagnetic stimulation and imaging based mainly on the editor's original research which produced significant scientific and technical development in the field of the biomagnetics, such as transcranial magnetic stimulation (TMS), biomagnetic measurements and imaging of the human brain by magnetoencephalography (MEG) and by MRI, and the biomagnetic approach to treat cancers, pain and other neurological and psychiatric diseases such as Alzheimer's disease and depression. Since the time when Ueno and Sekino's book was published, the bioimaging and biosensing technologies have rapidly developed. Recently, Ueno edited a book which focuses on biomedical imaging and sensing technologies in medicine and technology (Ueno, 2020). This book covered scientific achievements and imaging technologies using electromagnetics and light. Ueno's book also reviewed the recent advances in electromagnetics in medicine and biology (Ueno, 2021). This book discussed the new horizons in bioelectromagnetics, particularly, the potential therapeutic treatment of brain diseases based on the effects of radiofrequency electromagnetic fields on iron ion release, and uptake into iron cage proteins, like ferritins.

TMS utilizes a magnetic field (1–10 kHz) to stimulate nerve cells for the treatment of mental and neurological diseases (Shigemitsu and Ueno, 2017; Ueno et al., 2019). Magnetic induction tomography (MIT) and MRI-based electrical properties tomography (MR-EPT) are non-invasive methods for characterizing electromagnetic properties of biological systems at operating frequencies in the range of 10 kHz–30 MHz and from a few MHz to a few hundred MHz. MIT is an imaging technique for mapping passive electrical properties such as conductivity, permittivity and permeability that could be quicker, more convenient and less harmful method for tomography in comparison to MRI and to computer tomography (CT) (Klein et al., 2020). This technique applies a magnetic field to induce eddy currents in the biological material through an excitation coil, which is mainly used in non-destructive inspection of in industrial products. MR-EPT is an imaging method that maps the electrical properties of biomaterials by the measurement of radiofrequency electromagnetic fields in MRI (Chi et al., 2020). This technique is of interest in clinical diagnosis, in particular, in ultra-high MRI due to the potential evaluation of patient-specific absorption rate (SAR).

In the long history of bioelectromagnetism, imaging techniques using non-ionizing radiation, such as MRI, have been developed and are now used in chemistry, physics, medicine and in engineering fields like non-destructive testing and security. The discovery of electromagnetic waves by Maxwell and Hertz contributed to these achievements. With the progress of electronic and optical technologies, there is now a lot of research on imaging using electromagnetic waves in the terahertz region, which is close to light. In the electromagnetic spectrum, a THz wave is typically defined as a radiation with a frequency between 0.1 and 10 THz (1 THz=10^{12} Hz) and a wavelength between 30 and 3,000 μm. THz waves are classified as non-ionizing radiation (Wilmink and Grundt, 2012). Over the past decades, THz sources, detectors, and transmission or reflection technologies have been developed and are widely used in different application fields such as chemistry, biomedicine, material science, security and communications. With the rapid development of THz technologies, the need to study and investigate the biological effects of THz radiations has become more important than in past.

Optical imaging with non-ionizing radiation makes it difficult to quantitatively assess deep tissue. This is due to its low penetrability into the living body. The penetration depth is related to the absorption and scattering of light. However, it has the advantage that there are no exposure afflictions like with ionizing radiation. Molecular imaging using near-infrared light is currently being carried out. Near-infrared radiation is absorbed in vivo mainly by the hemoglobin in the blood. Optical imaging can provide information on enzyme metabolism and circulation. Moreover, optical imaging using non-ionizing radiations in the optical region including the THz band has been studied in various fields. As mentioned previously, optical imaging has difficulty in imaging deep tissues because of the low penetrability of its electromagnetic waves into the living body. However, there are two non-invasive methods for visualizing inside the body: one is to irradiate light from outside the body and image the transmitted light (light transmission imaging). The other is to inject a fluorescent agent into the body and then detect and image the patterns emitted from outside the body (biofluorescence imaging). Near-infrared light with a wavelength of 700 nm or more has less absorption and scattering biological materials and water than UV and visible light. Because it has relatively high penetrability into tissues and because it can capture signals at a depth of several centimeters, near-infrared light can be applied from outside body, and the transmitted light can be imaged. For this reason, near-infrared light, which has a longer wavelength (700–900 nm) than visible light, is used in biofluorescence imaging.

Currently, CT and MRI are in practical use for non-invasive visualization of the inside of living body. It is hoped that the technology for imaging the inside of the body using non-ionizing radiation in the optical range will be developed, and in vivo imaging that can be used safely on living bodies will be utilized in the medical and biological research fields in the future (Pirovano et al., 2020).

1.5 Discussion

The term "bioelectromagnetism" became popular in scientific communities around 1980. This book starts with an overview of the historical developments of bioelectromagnetism starting from ancient times until the twenty-first century. Afterward, the book discusses past and current knowledge of the atmospheric electricity and geomagnetic fields with biological systems such as magnetic navigation, magnetoreception, magnetic sense, etc. and the safety issues regarding human health. Present-day bioelectromagnetism focuses on biological and medical studies of non-ionizing radiation ranging from static electric fields through low-frequency electromagnetic fields, radiofrequency electromagnetic fields up to optical frequencies. These studies can be divided mainly into two types, beneficial such as for the various medical applications and the harmful such as for the establishment of safety guidelines. By taking them into consideration, it will be possible to represent today's state on the study of bioelectromagnetism and its future prospects well to discuss the issues that bioelectromagnetism is facing.

The following text will provide an overview of the chapters covered in this book to illustrate the diversity of research in bioelectromagnetism. This book consists of seven chapters. It begins with general information which forms the objectives and frameworks for understanding of this book. It continues with the description of the origin of bioelectromagnetism and its definition as the study of the relationship between electromagnetism and living systems. It also explains the connections of electricity, magnetism and electromagnetism with biology, medicine, chemistry, physics and engineering.

The seven chapters of this book have been arranged into four parts. In Chapter 1 the historical discussion about the development of bioelectromagnetism which sets the fundamental background to then be able to introduce the topics regarding the interaction between nature and electromagnetism, and also for the biological and health effects and medical applications of electromagnetism, is shown. Chapter 2 presents how bioelectromagnetism has been gradually developing and expanding into a variety of fields such as engineering, biomedical engineering, medicine and biology. This chapter also explains the fantastic and dramatic history of bioelectromagnetism from ancient times until the twenty-first century. It chronologically details the history of the major subjects, the moments when scientists made relevant physical, chemical, electrical and biological discoveries, and a historical debate about possible human health effects due to exposure to electromagnetic fields. This chapter also includes, as far as possible, the historical development of Japanese research in bioelectromagnetism. The main emphasis of this chapter is to present how through the long historical development of bioelectromagnetism, today's scientific achievements in biomedical engineering, biology, medicine, etc. were shaped by elucidating the electromagnetic phenomena found in biological systems.

Chapters 3 and 6 will establish the introduction regarding the connection between atmospheric electricity, geomagnetic field and biological systems, and the relationship between the evolutions of life with the geomagnetic fields. As was already indicated, the electromagnetic phenomena of the atmospheres have affected the evolution of life. Furthermore, the evolution of life has also been affected by the fluctuations and variations of the geomagnetic field intensity as well as by changes in atmospheric oxygen levels and UV radiation. It has been estimated that following a geomagnetic field reversal, the geomagnetic field becomes weaker, and its reduction in the last 200 years has been substantial. Also, the earth's magnetic poles are moving, and it is estimated that the north pole has moved 1,100 km over the last 170 years. Furthermore, predictions based on simulations estimate that geomagnetic field reversals can enhance the oxygen escape rate by 3–4 orders if the magnetic field becomes sufficiently weak. Consequently, it is anticipated that the global hypoxia will gradually kill numerous species. Chapter 6 comprises fascinating topics regarding the "Cambrian explosion of life," the "extinction of Neanderthals," the "magnetic field deficiency syndrome," and the "magnetic storm and its related diseases," all of which are possible results of geomagnetic fluctuations.

Chapters 4 and 5 will present the recent advances in magnetobiology. Magnetobiology, a part of bioelectromagnetism, studies living systems' sensitivity to weak magnetic fields, such as magnetic navigation, magnetotactic bacteria, and the magnetic response of plants, birds, animals and humans. Both chapters focus on the magnetic navigation of birds and the magnetoreception in plants. Chapter 4 introduces the role of magnetic sensing in the migration of birds. Recently, the possibility that cryptochromes serve as highly sensitive magnetoreceptors has been strongly suggested, and several research fields such as quantum biology are actively paying attention to this functional characteristic. The research of this characteristic is also important because the blue-light photoreceptor protein, cryptochrome, which includes FAD, is widely found in nature, i.e., bacteria, plants and animals. Magnetic sensing is presumed to be the consequence of the reaction between FAD with the amino acid residue as the intermediate by photoirradiation, i.e., the radical pair mechanism for quantum-assisted magnetic sensing, which is able to detect weak magnetic fields such as the geomagnetic field (approximately 50 µT). By means of these magnetoreceptors found in their retina, the migratory birds are assumed to visually perceive in which direction to move. This chapter reviews historical background in theoretical physics and ethology, and outstanding results in several research fields to date. Chapter 5 introduces the magnetic sensing capability of plants. Effects of the geomagnetic field have been observed in plants. However, the impact of the geomagnetic field on plant is still not well-understood and the magnetoreception mechanism in plants is still not scientifically conclusive. The two main models for magnetoreception are the magnetite model and the radical pair model. The radical pair model is linked to cryptochrome, in a way similar to the magnetoreceptors found in migratory birds. It was previously mentioned that cryptochromes are a class of flavor-proteins sensitive to light that are found in a variety of plants and animals. Phytochrome and phototropin are also photoreceptor proteins found in plants. Magnetic field effects in plants are also related to light-dependent plant processes. In this chapter, the effects of the geomagnetic field on plants are reviewed and the possible mechanisms of magnetoreception in plants including the involvement of photoreceptors are discussed.

The fourth part of this book deals with the safety issues derived from biological and health risk assessment research developments. Chapter 7 introduces the concepts of existing safety guidelines developed by international bodies such as the ICNIRP and the Institute of Electrical and Electronics Engineers (IEEE/ICES). In these safety guidelines, limit values are stipulated in terms of internal (in situ) electric fields for frequencies lower than 10 MHz (ICNIRP) or 5 MHz (IEEE), and limit values for SAR (specific absorption rate) are stipulated for frequencies higher than 100 kHz, based on biological effects of electric and magnetic fields, i.e., stimulation of nerves at low frequencies and temperature increase at high frequencies. The discussions presented in this chapter can be used as a guide when evaluating daily life safety related to electric and magnetic phenomena.

1.6 Summary and Perspective

Bioelectromagnetism is an interdisciplinary research field that involves many areas of investigation. In this book, the introduction to the research history of bioelectromagnetism is presented first. Later, the relationship between atmospheric electric phenomena and geomagnetism with biological systems and the environment is shown. Afterward, the current status of research on magnetic sensing in cells, animals, plants and humans, which is expected to progress in the future in relation with quantum biology, is examined. The last chapter deals with the most recent efforts to protect human health and to improve the safety of the electromagnetic environment that exists around us.

The different areas of bioelectromagnetism are still developing and expanding from molecular and cellular levels to the human body interaction with electromagnetism. The authors of each chapter are experts in their respective area. All the chapters were created with great pleasure while at the same time attempting to clearly explain the background and recent advances in each field. In this sense, it is expected that the readers will be able to appreciate the essence of this growing field.

Although the important events in the history of bioelectromagnetism are chronologically structuralized in this book, the detection of electromagnetic phenomena by living systems is an age-old question. The research in bioelectromagnetism has attempted to clarify these relationships by relying in classical physics and life phenomena. For example, with the support of technology, methods for detecting electromagnetic phenomena from inside of the body such as ECG and EEG were developed, and they have made significant contributions to today's medicine.

Today, it is well known that quantum biology began with an idea of Erwin Schrödinger, who in February of 1943 gave a lecture about the fusion of quantum physics and biology at the Trinity College, Dublin, leading to the foundation of what is now quantum biology. During the 1900s, Max Plank and Albert Einstein advocated for the quantum theory, and in consequence quantum mechanics advanced greatly. Subsequently, quantum mechanics and biology would eventually become combined as Einstein had already predicted, and quantum biology was born. In particular, Schrödinger's book, *What Is life?*, published in 1944 based on his lecture of 1943, was a significant breakthrough that led to large developments in quantum biology, continuing until this day.

Life phenomena have a hierarchical structure that goes from a microscopic-level for electrons and atoms to the macroscopic level of molecules/cells, tissues, organs and living beings. In this context, quantum theory and quantum mechanics introduced quantum biology in order to clarify life phenomena at the microscopic level, explained by means of classical physics. For example, attempts have been made to try to explain that the magnetoreception of birds, based on geomagnetism for migration, is caused by a phenomenon of quantum entanglement in the retina. It is hypothesized that this phenomenon can be caused by the generation of radical pairs in the flavor-protein cryptochrome (FAD) upon sensing a weak magnetic field, a reaction that involves the transfer of electrons excited by light stimuli.

By using quantum theory and quantum mechanics, we will be able to elucidate phenomena such as differentiation, development and proliferation of cancer cells, DNA, mutation, photosynthesis, enzymatic reactions and magnetoreception, all of which are the targets of research at electronic and atomic levels, related to the biological effects of electromagnetic fields in bioelectromagnetism. Further, with the results from the research of these phenomena, it will be possible to elucidate biological phenomena at the macroscopic levels of tissues and living beings (Kim et al., 2021).

The 2018 Nobel Prize in Physics was awarded to Arthur Ashkin of Bell Laboratories, for the development of optical tweezers and their application in biological systems, using principles based on quantum mechanics. The optical tweezers technique can manipulate microscopic materials such as cells by means of light pressure, which is the force that light exerts on matter. In the future, it is expected that quantum sensors will be used in measurements on biological systems and that quantum optics will be applied for imaging biological systems. Quantum theory and quantum mechanics will continue to be utilized in the field of biology, thus promoting advances in bioelectromagnetism-related research.

References

Al-Khalili J and McFadden J (2014): *Life of Edge-The Coming of Age of Quantum Biology*. Bantam Press, London.

Baker RR (1980): Goal orientation by blindfolded humans after long-distance displacement: possible involvement of a magnetic sense. *Science* 210:555–557.

Barnes F and Greenebaum B (2020): Setting guidelines for electromagnetic exposures and research needs. *Bioelectromagnetics* 41:392–397.

Begall S, Červený J, Neef J, Vojtech O and Burda H (2008): Magnetic alignment in grazing and resting cattle and deer. *Proc Natl Acad Sci USA* 105:13451–13455.

Chi J, Guo L, Destruel A, Wang Y et al. (2020): Magnetic resonance-electrical properties tomography by directly solving Maxwell's curl equation. *Appl Sci* 10:3318. doi:10.3390/app10093318.

Cifra M, Apollonio F, Liberti M, et al. (2021): Possible molecular and cellular mechanisms at the basis of atmospheric electromagnetic field bioeffects. *Inter J Biometeorol* 65:59–67.

Dyer AG, Greentree AD, Garcia JE, Dyer EL, Howard SR and Barth FG (2021): Einstein, von Frisch and the honeybee: a historical letter comes to light. *J Comp Physiol A* 207:449–456.

Faraday M (1821): Historical sketch of electro-magnetism. *Annals Philos* 2:195–200, 274–290.

Faraday M (1822): Historical sketch of electro-magnetism. *Annals Philos* 3:107–119.

Favre D (2011): Mobile phone-induced honeybee worker piping. *Apidologie* 42:270–279

Favre D (2017): Disturbing Honeybees' behavior with electromagnetic waves: a methodology. *J Behav* 2:1010.

Hayashi H (2004): *Introduction to Dynamic Spin Chemistry: Magnetic Field Effects on Chemical and Biochemical Reactions*. World Scientific Publishing Co Ltd, Hackensack, NJ.

Hunting ER (2021): Atmospheric electricity: an underappreciated meteorological element governing biology and human well-being. *Inter J Biometeorol* 65:1–3.

Hunting ER, Matthews J, et al. (2021): Challenges in coupling atmospheric electricity with biological systems. *Inter J Biometeorol* 65:45–58.

IARC (2002): *Non-Ionizing Radiation, Part 1: Satic and Extremely Low-Frequency (ELF) Electric and Magnetic Fields. Vol 80. IARC Monographs on the Evaluation of Carcinogenic Risks to Humans*. International Agency for Research on Cancer, Lyon.

IARC (2013): *Non-Ionizing Radiation, Part 2: Radiofrequency Electromagnetic Fields. Vol 102. IARC Monographs on the Evaluation of Carcinogenic Risks to Humans*. International Agency for Research on Cancer, Lyon.

ICNIRP (2003): *Exposure to Static and Low Frequency Electromagnetic Fields, Biological Effects and Health Consequences (0–100 kHz)*. Matthes R, Vecchia P, McKinlay AF, Veyret B and Bernhardt JH, (Eds) International Agency for Research on Non-Ionizing Radiation Protection, Munich.

ICNIRP (2009a): *Exposure to High Frequency Electromagnetic Fields, Biological Effects and Health Consequences (100 kHz-300 GHz)*. Vecchia P, Matthes R, Ziegelberger G, Lin J, Saunders R and Swerdlow A, (Eds) International Agency for Research on Non-Ionizing Radiation Protection, Munich.

ICNIRP (2009b): Guidelines on limits of exposure to static magnetic fields. *Health Phys* 96:504–514.

ICNIRP (2010a): Guidelines for limiting exposure to time-varying electric and magnetic fields (1 Hz to 100 kHz). *Health Phys* 99:818–836.

ICNIRP (2010b): Fact Sheet LF on the guidelines for limiting exposure to time-varying electric and magnetic fields (1 Hz-100 kHz). *Health Phys* 99(6):818–836.

ICNIRP (2020a): ICNIRP statement: principles for non-ionizing radiation protection. *Health Phys* 118(5):477–482.

ICNIRP (2020b): Guidelines for limiting exposure to electromagnetic fields (100 kHz to 300 GHz). *Health Phys* 118:483–524.

Karipidis K, Mate R, Urban D, Tinker R and Wood AW (2021): 5G mobile networks and health-a state-of-the-science of the research into low-level RF fields above 6 GHz. *J Exposure Science & Environmental Epidemiology* 31:585–605.

Kato M (Ed.) (2006): *Electromagnetics in Biology*. Springer Verlag, Tokyo.

Kim Y, Bertagna F, D'Souza EM, et al (2021): Quantum biology: an update and perspective. *Quantum Rep* 3:1–48 Doi: 10.3390/quantum3010006.

Klein M, Erni D and Rueter D (2020): Three-dimensional magnetic induction tomography: improved performance for the center regions inside a low conductive and voluminous body. *Sensors* 20:1306. Doi: 10.3390/s20051306.

Kurs A, Karalis A, Moffatt R, Joannopoulus JD, Fisher P and Soljacic M (2007): Wireless power transfer via strongly coupled magnetic resonance. *Science* 317(5834):83–86.

Levitt BB, Lai HC and Manville AM (2021a): Effects of non-ionizing electromagnetic fields on flora and fauna, part 1. Rising ambient EMF levels in the environment. *Rev Environ Health*. Doi: 10.1515/reveh-2021-0026.

Levitt BB, Lai HC and Manville AM (2021b): Effects of non-ionizing electromagnetic fields on flora and fauna, part 2 impacts: how species interact with natural and man-made EMF. *Rev Environ Health*. Doi: 10.1515/reven-2021-0050.

Maffei ME (2014): Magnetic field effects on plant growth, development, and evolution. *Front Plant Sci* 5:445. Doi: 10.3389/fpls.2014.00445.

Maffei ME (2019): Plant responses to electromagnetic fields. In *Biological and Medical Aspects of Electromagnetic Fields* (Fourth Edition). Greenebaum B and Barnes F, (Eds), pp. 89–110. CRC Press, Taylor & Francis Group, Boca Raton, FL.

National Toxicology Program (2018a): *Technical Report on the Toxicology and Carcinogenesis Studies in Sprague Dawley (Hsd: Sprague Dawley ® SD®) Rats Exposed to Whole-Body Radio Frequency Radiation at a Frequency (900 MHz) and Modulations (GSM and CDMA) Used by Cell Phones*. NTP TR 595 U.S. Department of Health and Human Service ISSN: 2378-8925.

National Toxicology Program (2018b): *Technical Report on the Toxicology and Carcinogenesis Studies in B6C3F1/N Mice Exposed to Whole-Body Radio Frequency Radiation at a Frequency (1,900 MHz) and Modulations (GSM and CDMA) Used by Cell Phones*. NTP TR 596 U.S Department of Health and Human Services, ISSN: 2378-8925.

Neufeld E and Kuster N (2018): Systematic derivation of safety limits for time-varying 5G radiofrequency exposure based on analytical models and thermal dose. *Health Phys* 115:705–711.

Oersted HC (1820): Nouvelles experiences electro-magnétique. *Journale de Physique* 91:78–80.

Pirovano G, Roberts S, Kossatz S and Reiner S (2020): Optical imaging modalities: principles and applications in preclinical research and clinical settings. *J Nucl Med* 61:1419–1427.

Popp FA, Warnke U, Köning HL and Peschka W (Eds) (1989): *Electromagnetic Bio-Information*. Uran & Schwarzenberg, Munich, Wien, Baltimore, MD.

Price C, Williams E, Elhalel G and Sentman D (2021): Natural ELF fields in the atmosphere and in living organisms. *Inter J Biometeorol* 65: 85–92.

Ritz T, Adem S and Schulten K (2000): A model for photoreceptor-based magnetoreception in birds. *Biophys J* 78:707–718.

Rycroft MJ, Nicoll KA, Aplin KL, et al (2012): Recent advances in global electric circuit coupling between the space environment and the troposphere. *J Atmo Sol Terr Phys* 90–91:198–211.

Schrödinger E (1944): *What Is Life? The Physical Aspect of the Living Cell.* Cambridge University Press, Cambridge.

Schulten K, Swenberg CE and Weller A (1976): Magnetic field dependence of the germinate recombination of radical ion pairs in polar solvent. *Zeitschrift für Physikalische Chemie.* 101:371–390.

Shigemitsu T and Ueno S (2017): Biological and health effects of electromagnetic fields related to the operation of MRI/TMS. *Spin* 7(4):1740009.

Simkó M and Mattsson MO (2019): 5G wireless communication and health effects-a pragmatic review based on available studies regarding 6 to 100 GHz. *Int J Environ Res Public Health* 16:3406. Doi: 10.3390/ijerph16183406.

Timmel CR and Henbest KB (2004): A study of spin chemistry in weak magnetic fields. *Phil Trans R. Soc.Lond.* A 362:2573–2589.

Ueno S (Ed) (2020): *Bioimaging: Imaging by Light and Electromagnetics in Medicine and Biology.* CRC Press, Boca Raton, FL.

Ueno S (2021): New horizons in electromagnetics in medicine and biology. *Radio Sci* 56:1–7 Doi: 10.1029/2020RS007152.

Ueno S and Okano H (2012): Static, low-frequency, and pulsed magnetic fields in biological systems. In *Electromagnetic Fields in Biological Systems*, Lin JC, (Ed), pp. 115–196. CRC Press, Talyor & Francis Group, Boca Raton, FL, London, New York.

Ueno S and Sekino M (Eds) (2016): *Biomagnetics: Principles and Applications of Biomagnetic Stimulation and Imaging.* CRC Press, Boca Raton, FL.

Ueno S and Shigemitsu T (2007): Biological effects of static magnetic fields. In *Bioengineering and Biophysical Aspects of Electromagnetic Fields.* Barnes FS and Greenebaum B, (Eds), pp. 204–259. CRC Press, Boca Raton, FL.

Ueno S, Sekino M and Shigemitsu T (2019): Transcranial magnetic and electric stimulation. In *Biological and Medical Aspects of Electromagnetic Fields* (Fourth Edition). Greenebaum B and Barnes F, (Eds), pp. 345–405. CRC Press, Boca Raton, FL.

Vanbergen AJ, Potts SG, Vian A, Malkemper EP, Young J and Tscheulin T (2019): Risk to pollinators from anthropogenic electro-magnetic radiation (EMR): Evidence and knowledge gaps. *Sci. Total Environ* 695:133833. Doi: 10.1016/j.scitotenv.2019.133833.

Verbeek J, Oftedal G, Feychting M, van Rongen E, Sarfi MR, Mann S, Wong R and van Deventer E (2021): Prioritizing health outcomes when assessing the effects of exposure to radiofrequency electromagnetic fields: a survey among experts. *Environ Int* 146:106300. Doi: 10.1016/j.envint.2020.106300.

von Frisch K (1949): Die Polarisation des Himmelslichtes als orientierender Faktor bei den Tänzen der Bienen. *Experientia (Basel)* 5:142–148.

Wang CX, Hilburn IA, Wu DA, Mizuhara Y, et al (2019): Transduction of the geomagnetic fields as evidence from alpha-band activity in the human brain. *eNEURO* 6:e.0483-18.2019.

WHO (1987): *Magnetic Fields* (Environmental Health Criteria No.69), World Health Organization, Geneva.

WHO (2006): *Static Fields* (Environmental Health Criteria No.232), World Health Organization, Geneva.

WHO (2007): *Extremely Low Frequency Fields* (Environmental Health Criteria No.238), World Health Organization, Geneva.

Wilmink GJ and Grundt JE (2012): Terahertz radiation: sources, applications, and biological effects. In: *Electromagnetic Fields in Biological Systems.* Lin JC, (Ed), pp. 369–419. CRC Press, Boca Raton, FL, London, New York.

Wiltschko R and Wiltschko W (1995): *Magnetic Orientation in Animals.* Springer Verlag, Berlin, Heldelberg, New York.

Wood AW, Mate R and Karipidis K (2021): Meta-analysis of in vitro and in vivo studies of the biological effects of low-level millimeter waves. *J Expos Sci Environ Epidemiol* 31:505–613.

2

The History of Bioelectromagnetism

Tsukasa Shigemitsu

Shoogo Ueno

Masamichi Kato

2.1 Introduction

The bioelectromagnetism has been gradually developed, as seen from the document on the electric fish in ancient Egyptian mural, through the book *De Magnete* by Sir William Gilbert, the discovery of animal electricity by Luigi Galvani, the experiment of the connection between electricity and magnetism by Hans Cristian Oersted, the induction law by Michael Faraday, the electromagnetic theory by James Clerk Maxwell, the medical applications of high-frequency electric current proposed by Nikola Tesla and Jacques Arsène d'Arsonval, the study of biological and health hazard effects from exposures to electromagnetic fields, to the interaction between electromagnetic fields and living systems.

Before the seventeenth century, the science of electricity and magnetism was the natural magic. At the beginning of the seventeenth century, Gilbert published *De Magnete*. This book opened the era of modern science. The eighteenth century was an important period for basic and scientific developments of electricity and magnetism as a basis for bioelectromagnetism. The nineteenth century was greatly important in producing advances for the understanding of the biology behind bioelectromagnetism. In the twentieth century, the scientific achievements gained great progress in bioelectromagnetism.

The history of bioelectromagnetism is very much equivalent to the history of electromagnetism. The instruments for detection and measurement based on electromagnetic phenomena gradually developed. The developed equipment can measure electric or magnetic signals in living systems. It allows us to research the behavior of tissue at cellular and organic levels. Because the research that

DOI: 10.1201/9781003181354-2

measures bio-signals advanced with the developed equipment, bioelectromagnetism can offer opportunities for developing therapeutic and diagnostic applications and promote understanding of the biological interactions of electromagnetic fields with living systems. Now, bioelectromagnetism is mainly related to electricity, magnetism, bioelectricity (electrobiology), and biomagnetism (magnetobiology). Bioelectricity as a division of bioelectromagnetism is a fundamental process of all living systems and is the study of electrical phenomena generated in living systems. The electrical phenomena include inherent properties of the cells, such as membrane potential, action potential, and propagation of the potential. Biomagnetism deals with magnetic phenomena in all living systems, which can be observed at different intensities and frequencies. For example, the so-called magnetophosphene is a visual sensation elicited by exposing the head to a low-frequency magnetic field. The signal is generated in the retina. Magnetic stimulation of the human brain and heart has been used for the purpose of both research and clinical treatment. Using Superconductive QUantum Interference Devices (SQUID), the weak magnetic fields from the brain, heart, and lung can be measured from outside the body.

The long historical development of bioelectromagnetism reveals that today's scientific achievements in engineering, biomedical engineering, biology, and medicine are shaped by solving the electromagnetic phenomena in living systems. Many experimental and theoretical research studies in bioelectromagnetism emerged: (1) the relationship between electricity, magnetism, and life, (2) how electricity and magnetism give action to living systems, (3) how living systems utilize electricity and magnetism, and (4) how electromagnetic fields interact with living systems. In particular, it recently became important to understand the interaction of electromagnetic fields with living systems not only for medical applications but also for protection from exposure to electromagnetic fields. The discovery and important issues in the history of bioelectromagnetism can be mainly divided into theory, instrumentation, and measurement of interactions between the electromagnetic phenomena and living systems. Taking these categories into consideration, the description of the history of natural science of electricity and magnetism is needed to look back on the development of bioelectromagnetism.

In this chapter, through overviewing the history of bioelectromagnetism, the argument could mainly be divided into several topics, which got gradual success in bioelectromagnetism. This chapter as far as possible included the historical development of Japanese research in bioelectromagnetism. From Sections 2.1–2.5 the important events in the history of bioelectromagnetism are chronologically structuralized. We emphasize the description of history including the scientists who made physical, chemical, medical, and biological discoveries. Section 2.6 deals with the historical debate about possible human health effects due to exposure to electromagnetic fields: static, low-frequency, and radiofrequency (RF) field. This debate brings the fundamental approach for safety guidelines which will be presented in another chapter.

As major sources, we refer here three books: *Bibliographical History of Electricity and Magnetism* (Mottelay, 1922), *Electricity and Medicine: History of Their Interaction* (Rowbottom and Susskind, 1984), and *Galvani's Spark: The Story of the Nerve Impulse* (McComas, 2011).

2.2 Discovery and First Step in Bioelectromagnetism

The early history of bioelectromagnetism is not well documented. In 3000 BC, the Nile catfish (*Malapterurus electricus*) was written in an ancient Egyptian mural. This is the first recorded document on the electric fish. Thales of Miletus (625–547 BC), the Greek scientist, also an astronomer, one of the seven wise men of Greece, rubbed amber with cat fur, and recognized its power. He recorded first about the magnetic properties of ferric oxide, Fe_3O_4. Unfortunately, Thales left no writings. All were transmitted orally. The word "amber" means "electron" in Greek. In 341 BC, Aristotle (384–322 BC), the Greek philosopher, described about torpedo (electric fish), which gives electric shocks to humans. Theophrastus (372–287 BC), Aristotle's pupil and his successor at the Lykeum at Athens, described the attractive power of loadstone (magnet) and amber (Rowbottom and Susskind, 1984). Gaius Plinius Secundus (23–79 AD), known as Pliny the Elder referred to the electric torpedo in his "*Naturalis Historia (Natural History).*"

The electric discharge of certain fishes generates a strong average of 350 V with high of 650 V producing pain and shock in human. These fish include the Nile catfish, electric fish (*torpedo*), skate (*Mormyrus*), and the electric eel (*Electrophorus electrricus*). Later, the electric discharge from these fishes was introduced into medicine as a cure for headache and gout. As medical applications of the electric discharge of electric fish, Scribonius Largus (1–50 AD), the court physician to the Roman emperor Claudius (10 BC–54 AD), recommended first the use of torpedo over the scalp for curing headache and gout around 46 AD. This is the first written document on the use of electric discharge of the torpedo as electrotherapy. The first person known to have been cured by electricity was Anthero, who suffered from gutta (probably gout), a freed slave of the Emperor Tiberius Claudius Nero Caesar Augustus (42 BC–37 AD) (Cambridge, 1977). Claudius Galenus Galen (130–201), Roman physician, employed the electric shock from the torpedo to treat headache and epilepsy. In electrotherapy, Galen followed in the footsteps of Largus's medical treatment. In the late eleventh century, Ibn-Sidah, the Muslim physician of India, advocated placing live catfish on the brows of the patients (Delbourgo, 2006). The electric stimulation by electric fish spread to Africa. Jesuit missionaries in early modern Abyssinia (now, Ethiopia) reported that locals strapped patients to tabletops and shocked them with catfish as a method of expelling "Devils out of the human body"(Delbourgo, 2006). The use of electric fish to produce the electric discharge for electrotherapy was the most popular and was used until the seventeenth century.

The destructive power of high temperature on living tissues was well known in antiquity. In ancient ages, red-iron was used to treat cancer. The basic concept of hyperthermia is based on temperature rise. The source of the term "hyperthermia" is derived from two Greek words: "*hyper*" (over) and "*therme*" (heat). The hyperthermia refers to the induction of local heating in the human body. The temperature rise, hyperthermia, was referred to achieve therapeutic effects. Generally, the high-temperature rise is associated with the sun power for healing. An Egyptian chancellor, Imhotep (2655–2600 BC) was known first as the user of heat treatment. Two thousand years after his death, the status of Imhotep had risen as a god of medicine and healing. In ancient Greece and Rome, Parmenides (ca 500 BC, or 475 BC-unknown data of his death), the Greek philosopher, was convinced of the effectiveness of hyperthermia (Gas, 2011; Seegenschmiedt and Vernon, 1995). Hippocrates (460–370 BC), the famous Greek philosopher and physician, who is considered the father of medicine, used heat to treat breast tumors. Aulus Cornelius Celsus (25 BC–50 AD), a Roman encyclopedist, the author of the first medical work *De Medicina*, believed the curing effects of fever and described the use of hot baths in the treatment of various diseases. Hippocrates used the Greek words, or expressions "karkinos" (early tumor stage), or "karkinoma" (advanced tumor stage), meaning crab. These expressions were translated into the Latin language by Celsus as the term of "cancer." Anyway, the effectiveness of heat treatment had been increased.

Gaius Plinius Secundus was born in Como and was a Roman scholar in Italy. He wrote the encyclopedia, *Naturalis Historia* a 37-volume book which contained and condensed the collection of the knowledge of his time (Pliny the Elder, 77 AD, 1938). However, it contained a mixture of fact, fiction, folklore, and superstition. In his encyclopedia, Plinius regarded that a magnet has the power to cure human diseases. In chapter 25 of volume 36, he mentioned: "All magnets, incidentally, are useful for making up eye-salves if each is used in its correct quantity, and are particularly effective in stopping acute watering of the eyes. They also cure bums when ground and calcined." For amber, he mentioned in volume 37, chapter 11 that "For even today the pleasant women of Transpadane Gaul wear pieces of amber as necklaces, chiefly as an adornment, but also because of its medical properties. Amber, indeed, is supposed to be a prophylactic against tonsillitis and other affections." In addition, Plinius mentioned in the same volume that "amber is found to have some use in pharmacy, although it is not for this reason that women like it. It is of benefit to babies when it is attached to them as an amulet."

The above writing seems to imply that electric and magnetic phenomena have medical applications to cure human diseases. In his idea, sympathy and antipathy were the cause of magnetic phenomena. Plinius died in 79 AD during the rescue of his family from the eruption of Mount Vesuvius. This eruption destroyed the cities of Pompeii and Herculaneum.

From early times, the existence of the earth magnetic field was well known. In 2637 BC, Emperor Huang-ti of China used a compass in a battle. In 1110 BC, Taheon-Koung gave his crew a compass to sail from Cochin to Tonquin. The first Western written document is by Alexander Neckam (1157–1217), a monk in England. In 1186, he described first how the working of a compass "showed mariners their courses when the Polar Star is hidden" (Mottelay, 1922). In 1254, Roger Bacon (1220–1292), a philosopher and a Franciscan monk in England, dealt with a magnet and described its properties in his *Opus Minus*. In 1497, Vasco da Gama (1469–1524), a Portuguese navigator, used the compass for his trip to the Indies. In 1581, Robert Norman (fl. 1560), a maker of compass needles, mariner, at Wapping of England, rediscovered the dip or inclination to the earth magnetic needle and was the first to measure them (Sarkar et al., 2006). Magnetic inclination and variations were known before Norman's work. As we mentioned above, magnetic phenomena were well known before the seventeenth century even in antiquity and the Middle Ages. Many scientists including natural philosophers, physicians, physicists, etc. shared their interest with magnets, iron, and compasses, which led to the study of electricity, magnetism, electromagnetism, and further to bioelectromagnetism.

The Swiss doctor Philippus Aureolus Theophrastus Bombastus von Hohenheim, known as Paracelsus (1493–1541), physician, alchemist, a lay theologian, used magnets to treat various diseases such as diarrhea, hernia, jaundice, and bleeding. He considered magnets to be vital stones. His explanation was that the invisible fluid from the earth and the stars acts on the human body, and like a magnet, the human body attracts effluvia of both, health or disease. Under the condition in which a magnet is placed onto the pain point of the patients, the baneful qualities of astral and terrestrial fluids are drawn out from the human body, and are returned to their original point with swirling through the atmosphere (Tatar, 1978). Paracelsus traveled widely through Europe including England, Ireland, and Russia, and wandered Egypt and Arabia. Through traveling worldwide, he got encyclopedic knowledge. Once he was appointed as a licensed physician, to lecture medicine at the University of Basel, he contacted with Desiderius Erasmus (1466–1536), Dutch philosopher and Christian scholar, Rotterdam. Paracelsus means "greater than Celsus." Aulus Cornelius Celsus (25 BC–50 AD) was a famous Roman encyclopedist and doctor. It is frequently said that Paracelsus was a pioneer of the medical revolution of the Renaissance.

Giambattista della Porta (1535–1615) was born at Vico Equense, Kingdom of Naples (now Italy), and became an Italian natural philosopher. In 1558, he published works with his first edition entitled *Magiae Naturalis (Natural Magic)*. He became historically famous because of this work. His work discussed many subjects including demonology and magnetism. In 1589, he published the second edition with 20 books in Naples (Della Porta, 1589). In particular, chapter 56 in the seventh book of this second edition was entitled *The Wonders of the Loadstone*, which dealt with magnetism. The content of this seventh book included the learnings of Plinius, with many observations made by him. For Della Porta, magnetism was still a mysterious and magical learning. Interestingly, the content of this seventh book gave the influence to Gilbert's book, *De Magnete*. Gilbert made references to this seventh book when he wrote his famous book. The first and second chapters were referred to works of Aristotle, Galen, Plinius, Epicurus, Lucretius et al. In chapter 56, Della Porta talked about loadstone as follows:

Our ancestors attribute to iron and loadstone "an understanding of venerious actions, and that they are one in love with the other, and when they turn their backs, they hate one the other, and drive one the other off, and that they contain in them also the principles of hatred" (Della Porta, 1589).

It is well known that the science of electricity, magnetism, and electromagnetism started from the time when Sir William Gilbert (1544–1603), a British physician to Queen Elizabeth I of England, published the famous book *De Magnete, Magneticisque Corporibus, et de Magno Magnete Tellure; Physiologia, Noca, Pluribus et Argumentes et Experimentis Demonstrata* in Latin in 1600. This title is now shortened into *De Magnete* (Mottelay, 1893). In Figure 2.1, the workman is directed to the north and there he hammers the hot iron so that it will expand or elongate in a north direction (Benjamin 1898; Mottelay, 1893). Gilbert was the eldest of the five sons of Jerome Gilbert of the prestigious post of Recorder, in Colchester, northeastern of London with 75,000 residents. Today, Sir William Gilbert is regarded as the founder of modern science. Gilbert discovered the Earth is a giant magnet. He collected well-known knowledge about magnetism in his time. *De Magnete* is subdivided into six books. In the first book, it starts with a historical survey of loadstone and ended with the announcement that loadstone and iron ore are the

FIGURE 2.1 Magnetizing hot iron by hammering (Gilbert's Blacksmith, From Mottelay 1893, English version translated from original, Dover edition, 1958.)

same, and the theory of Earth's magnetism. Through the second book, the systematic study of amber, a distinction between the attractive properties of magnet and amber was discussed. Many magnetic and electrostatic experiments were reported. Book three was concerned with the directive properties of magnets, with the magnetization of needles and the distribution of magnetism in a terrella. The terrella is little Earth and formed as a spherical magnet, which serves as a model. From this terrella model, Gilbert drew the analogy between the Earth's magnetic field and terrella's magnetic field. He introduced the Earth's magnetism with more detail in book four, and showed declination and magnetic dip with the description of an instrument such as the action of a loadstone, and the degree of dip below the horizon is developed in book five. The final book six concerned stellular and terrestrial motions associated with magnetism. He coined the term "electric" to denote the attractive property of substances like amber-electrostatic phenomena. The term "electric" comes from elektron, the ancient Greece. In his book, Gilbert gave many examples of the therapeutic use of magnets (Rowbottom and Susskind, 1984). Gilbert's assertions were based on rigorous experimentation. According to Humboldt, Gilbert used for the first time the words electric force, electric attraction, and electric emanations. Although the original book and scientific work by Gilbert belonged to the Royal College of Physicians as his bequest, they were destroyed in the Great Fire of London in 1666.

Otto von Guericke (1602–1686), Major of Magdeburg, physicist and inventor, Germany, had studied Gilbert's experiments further, and discovered experimentally the phenomena of static electricity. Using a rotating sulfur sphere with an iron axle, and mounted on a revolving axis, a frictional electric machine was first constructed by him in 1663. When the sulfur sphere rotated, and rubbed, static electricity was generated with the emission of both sound and light. When rubbed by a cloth pressed against it by the hand, Guericke heard for the first time sound and saw for the first time light in artificially exited electricity (Mottelay, 1922). In addition, he was historically a well-famed scientist for his research on the nature of vacuum (Magdeburg Hemispheres). The Middlesex Hospital (England) was probably the first hospital to purchase an electric machine in 1767 (Cambridge, 1977).

Renē Descartes (1595–1650), the French physicist, mathematician, and philosopher, theorized that the magnetic poles were on the central axis of a spinning vortex of fluids surrounding each magnet. The fluids entered by one pole and left through the other. Sir Isaac Newton (1642–1727) improved the electric machine by substituting with a glass globe the globe of sulfur used for the electric machines of both von Guericke and Robert Boyle (1627–1691), an Irish natural philosopher and chemist. In his work book "*Optiksi*," Newton had posited the existence of a single universal ether which would explain everything from light, heat, and gravity through electricity and magnetism to nervous impulses and vision (Tatar, 1978). Ether is, according to Newton, a thin subtle matter much finer and rarer than air (Mottelay, 1922). With the development of the electric machine, the experimental electrophysiology began and developed in the middle of the seventeenth century. Jan Swammerdam (1637–1682), a Dutch anatomist, natural philosopher, developed a neuromuscular preparation in 1664. He studied the volume change of frog muscles during contraction. The muscle was immersed in water and caused to contract by pressure applied to the nerve. The level of the water fell slightly as the muscle contracted, thus showing that the volume of the muscle had decreased rather than increased (Rowbottom and Susskind, 1984). His experiments remained unknown until Galvani's discovery of animal electricity. Through Swammerdam, the Germans claim to be the origin of what has been called galvanism (Mottelay, 1922). Experimental explanation for the electric nature of action potential was provided many years later.

2.3 Development of Bioelectromagnetism

During the period from the beginning of the eighteenth century to the first half of the nineteenth century, the study of electricity and magnetism had rapidly developed which made progress in electromagnetism and bioelectromagnetism. These developments led also to the application of electromagnetic phenomena in industry and medicine (Fleming, 1921). Then, the total electrification concept of the world gradually proceeded and humans began to live in an anthropogenic electromagnetic environment in addition to the naturally originated electromagnetic environment. The essential history of electricity and bioelectricity is briefly introduced to understand the source of bioelectromagnetism. The research progress of electricity during the Edo period in Japan is reviewed shortly as an interesting topic of electricity.

2.3.1 Development of Electricity

Ewald Georg von Kleist (1700–1748), a German inventor, dean of the cathedral chapter of Kamin in Pomerania, Germany, invented a primitive form of a condenser in 1745. His condenser collected electric charge from the atmosphere. The condenser contained mercury or alcohol. At the same time, Pieter van Musschenbroek (1692–1761), professor of physics at the University of Leyden, the Netherlands, refined it with caution. His great caution was that a glass jar containing a metal rod was filled with water, which can store electric charge. This invention became to be the modern condenser, called Leyden jar after its birth place, Leyden. This Leyden jar provided to be a useful tool for the research of electricity until the discovery of Voltaic piles. Van Musschenbroek chose water; the most readily procured non-electric, and placed some in a glass bottle. As shown in Figure 2.2, no important results were obtained until Cunaeus, a pupil of Musschenbroek, an eminent philosopher of Leyden, burgess of Leyden, who while holding the bottle, attempted to withdraw the wire which connected with the conductor of a powerful electric machine. He received a severe shock in his arms and chest just after touching the wire dipped in the water (From Deschanel, 1876; Mottelay, 1922).

FIGURE 2.2 Discovery of the Leyden jar in Musschenbroek's laboratory. The static electricity produced by the rotating glass sphere electrostatic generator was conducted by the chain through the suspended bar to the water in the glass (From Deschanel, 1876.)

From 1773, John Walsh (1725–1795), fellow of the Royal Society, England, sent a letter to Benjamin Franklin (1706–1790), the American editor, philosopher, and statesman. He reported in his letter that a visible spark and shock from the torpedo is the same sensation from a shock of a Leyden jar. In 1774, Walsh published the results of the torpedo's electricity together with the anatomical observations on its electric organs sketched by John Hunter (1728–1793), a Scottish anatomist and surgeon (Figure 2.3) (Walsh, 1773–1774; Rowbottom and Susskind, 1984). His observation is that the electricity of the animal is generated by organs located on each side of the cranium and gills, somewhat resembling a galvanic pile, and consisting wholly of perpendicular columns reaching from the upper to the under surface of the body (Mottelay, 1922). John Hunter was the teacher of Edward Jenner (1749–1823), a British physician, and contributor of the smallpox vaccine.

Toward the end of the eighteenth century, unreasonable forms of medical treatment had popularity in Europe and North America. The doctrine of "animal magnetism" by Mesmer and "metallic tractor" by Perkins were introduced in relation to electro- and magneto-therapy in their day. Both were very popular, but general practitioner frowned on these medical treatments (Rowbottom and Susskind, 1984). Scientists in the eighteenth century regarded invisible fluids as the source of magnetism. Nearly 200 years after the death of Paracelsus, Franz Antoine Mesmer (1734–1815) was born in the village of Iznang near Lake Constance. He accepted Paracelsus's idea into his concept, and developed his theory, later called "Mesmerism." He studied medicine in Vienna. In 1765, his medical dissertation entitled *On the influence of the planets upon the human body* discussed that the influence of the stars and planets on the human body, health, and disease was by means of an invisible fluid (Parent, 2004). His theory derived from the concept that the invisible fluid existed in the entire universe and infused both matter and spirit with its vital force. After the publication of his dissertation, Mesmer met Maximilian Höll (1720–1792), a Jesuit priest and astronomer of the University of Vienna, Vienna, Austria. Using magnets, Höll experimented

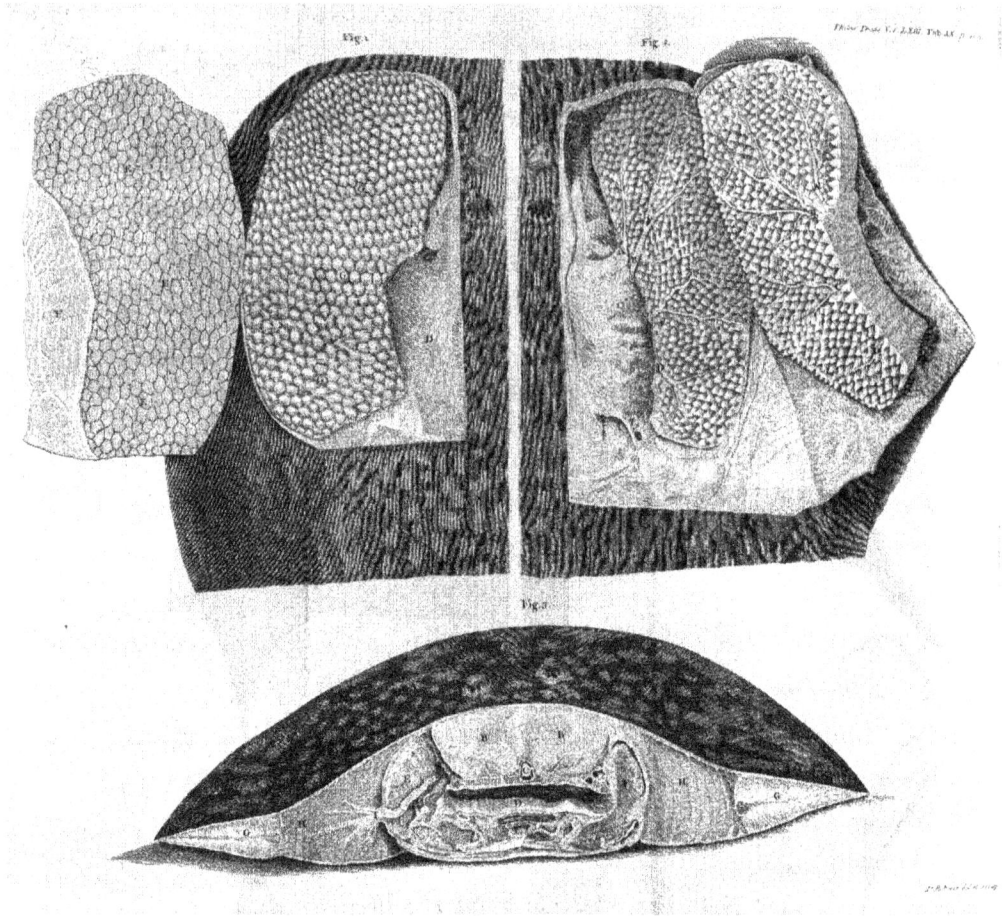

FIGURE 2.3 Dissections of the electric organ of French torpedo by John Hunter. Above: the right electric organ divided horizontally into two nearly equal parts at the place where the nerves enter. Bottom: A perpendicular section of just below inspiratory openings (From Walsh, 1773–1774.)

to cure people. Mesmer got some magnets from Höll. Soon, Mesmer assumed that his invisible fluid had connections with electricity and magnetism. First, Mesmer found that he could cure people by channeling the magnetic influence. Later he discovered that he could achieve the same results by the power of touch alone and he abandoned the use of actual magnets (Rowbottom and Susskind, 1984). This invisible fluid could be used for medical treatment with the help of magnets. The treatment developed into the idea of magnetotherapy. He gained first fame in Vienna and later went to Paris. In 1784, the doctrine of "animal magnetism" introduced by Mesmer was investigated by a Royal Commission of the French King of Louis XVI. The members of this commission consisted of four from medical circles and five from the Academy of Science including Benjamin Franklin, Antoine-Laurent de Lavoisier (1743–1794), and Joseph-Ignace Guillotin (1738–1814). Lavoisier was the center of the eighteenth century chemistry and was later convicted and guillotined in 1794. Guillotin introduced and designed the execution apparatus "guillotine" named after him. The official report of the investigation by a Royal Commission said that *the crises provoked by the doctor principally to the "imagination" or, more accurately, to the suggestibility of the patients,* and *patients without much imagination simply imitated the behavior of others, the imagination works wonders; magnetism yields no results.* So, the commissioners alerted their Majesty to the *highly pernicious* influence of animal magnetism (Franklin, 1784; Tatar, 1978). Mesmer escaped from

Paris and wandered through France, Switzerland, Austria, and Germany to search for an environment more sympathetic to his views and finally died in Meersburg, Germany, in 1815.

Historically, the most interesting and curious topic is the case of the metallic tractor. Elisa Perkins (1741–1799), practitioner of Connecticut, USA, who developed a unique form of electrotherapy in 1795. He named it metallic tractor. It consisted of pairs of two three-inch metallic rods made of metallic alloys (zinc, copper, gold, iron, platinum, and silver). After obtaining North American and England patents for his metallic tractor in 1796 and in 1798, he and his son, Benjamin Douglas Perkins (1774–1810), sold it in North America and in England. At the end of the eighteenth century, it was immensely popular for electrotherapy. Medical treatment with metallic tractor was caricaturized by James Gillray (1757–1815), a British caricaturist and printmaker (Colwell, 1922, p. 49). Elisa Perkins died of yellow fever during New York's epidemic in 1799. John Haygarth (1740–1827), Bath physician, England, investigated the efficacy of medical treatments of the metallic tractor. He made a wooden tractor. The shape of it was strongly similar to the shape of the metallic tractor. He treated this wooden tractor as a dummy for medical treatment. He reported its results by comparing the result of the medical treatment with the metallic tractor. The comparison of the results between the metallic and wooden tractors gave no significant differences, which led to the conclusion that medical treatment with the metallic tractor exerts no medical curing effect. Haygarth and his friend, Dr. William Falconer (1744–1824), treated five patients with wooden tractors. Four patients gained relief. They used the metallic tractor on the same five patients. Four patients reported relief. This treatment was the first scientific documentation of the placebo-controlled trial in 1779 (Haygarth, 1800). Haygarth's descendants reside in Perth, Australia.

From a more scientific point of view, in the eighteenth century, Stephen Gray (1666–1736), pensioner of the Charter House in London, England, theorized that electricity can flow like a fluid. His investigation led to the discovery of the principle of electric conduction and insulation. Then, he proved that electricity can be excited by the friction of feathers, hair, linen, paper, silk, etc. Charles François de Cisternay du Fay (1698–1739), a physicist, superintendent of gardens of the King in Paris, France, discovered that there are two kinds of electricity. He termed them *resinous* (−) electricity and *vitreous* (+) electricity in 1733. The first was produced on amber, copal, gum-lac, silk, thread, paper, etc. The second was produced on glass, rock crystal, precious stone, hair of animals, wool, etc. For this difference, Lichtenberg (1744–1799), professor of Experimental Philosophy at the University of Göttingen, Germany, proposed the terms *positive* electricity and *negative* electricity. He showed the condition of electrified surfaces by dusting them with power which was later called "Lichtenberg Figure" (Mottelay, 1922). Friedrich Wilhelm Heinrich Alexander von Humboldt (1769–1859), natural historian and explorer in Germany, hired Lichtenberg as a private tutor who taught him about electricity.

Abbè Jean-Antoine Nollet (1700–1770), member of the court of Louis XV and a professor at the French Royal Children, co-worker of du Fay, made many discoveries and performed many experiments. He demonstrated electroshocks by using electric machines and Leyden jars. During the month of April in 1746, in Paris, he shocked 180 of the King's Guards in an instant in the presence of the French King, Louis XV (Figure 2.4) (Figuier, 1865, p. 262). At the Carthusian Convent, he formed many person-trains by a line of monks stretched for more than 1.6 km, and administered a shock to this human train, and they instantaneously convulsed with the shock.

After thinking that the electric discharge of a Leyden jar is analogous to the phenomena produced by a lightning, Benjamin Franklin opened the theory of one-fluid electricity, and discovered the identity of electricity and lightning. He introduced the concept of positive and negative electricity. In the past, du Fay considered two distinct species of electricity, *vitreous* and *resinous*. Franklin conceived them to be two different states of the same electricity, called *positive* and *negative*. Franklin's idea is the foundation of the present theory of electricity. He began to investigate the relationship between two electrical phenomena. During the month of June in 1752, during the approach of a summer thunderstorm, he and his son flew a kite with metal tip, and charged a Leyden jar through this metal tip from the clouds, which demonstrated that the lightning is an electric discharge. In a letter to Peter Collinson (1694–1768),

FIGURE 2.4   The Abbè Nollet and the French King's Guards (From istock, and from Figuier, 1865.)

a gardener, Fellow of the Royal Society of London, dated October 19, 1752, Franklin described this kite experiment as follows:

> Make small cross of two light strips of cedar, the arms so long as to reach to the four corners of a large thin silk handkerchief when extended. Tie the corners of the handkerchief to the extremities of the cross, so you have the body of a kite which, being properly accommodated with a tail, loop and string, will rise in the air like those made of paper; but, this being made of silk, is fitter to bear the wet and wind of a thunder-gust without tearing. To the top of the upright stick of the cross is to be fixed a very sharp-pointed wire, rising a foot or more above the wood. In the end of the twine, next the hand, is to be tied a silk ribband, and where the twine and silk join a key may be fastened. The kite is to be raised when a thunder-gust appears to be coming on, and the person who holds the string must stand within a door or window, or under some cover, so that the silk ribband may not be wet, and care must be taken that the twine does not touch the frame of the door or window. As soon as any of the thunder clouds come over the kite, the pointed wire will draw the electric fire from them, and the kite with all the twine will be electrified, and the lose filaments of the twine will stand out every way and be attracted by an approaching finger. When the rain has wet the kite and twine so that it can conduct the electric fire freely, you will find it stream out plentifully from the key on the approach of your knuckle. At this key, the phial (Leyden jar) may be charged, and from electric fire thus obtained spirits may be kindled, and all the other electric experiments

be performed which are usually done by the help of a rubber glass globe or tube, and thereby the sameness of the electric matter with that of lightning completely demonstrated.

Franklin (1752)

From this experiment, Franklin discovered the atmospheric electricity. Then, he introduced the concept, using erected pointed iron rods, that buildings could be protected from lightning (Mottelay, 1922). George Wilhelm Richmann (1711–1753), a German physicist, member of St. Petersburg Academy of Sciences, constructed an apparatus for obtaining atmospheric electricity according to Franklin's idea. Unfortunately, he was killed by lightning during a kite experiment with a sharp shaped rod in a thunderstorm. Historically, he is believed to be the first victim of the lightning experiment with electricity as shown in Figure 2.5.

FIGURE 2.5 George Wilhelm Richmann and his engraver during the electrocution in St. Petersburg (From istock.)

Edward Bancroft (1744–1820), an American scientist and politician, had lived in Guiana for some years before coming to England, and in 1769 showed that the shock from a torpedo, *gymnotus electricus*, is electrical in nature and the characteristics of the electric discharge of the torpedo are similar to those from a battery of a Leyden jar. The similarity between the electric discharge of the electric fish and the shock by the Leyden jar was found. This similarity led to the use of the Leyden jar in electrotherapy. The Leyden jar made it possible to demonstrate the electric stimulation of nerves and muscles, which is the reason for its therapeutic application instead of electric fish.

During the second half of the eighteenth century, the law of electrostatic force was formulated. Joseph Priestley (1733–1804), a chemist and English Presbyterian, formulated the inverse-square law of electrical attraction and repulsion in 1767, then Henry Cavendish (1731–1810), a British chemist and physicist, experimentally demonstrated it in 1772, and finally Charles Augustin de Coulomb (1736–1806), a French physicist, completed the electrostatic force law in 1785.

2.3.2 Development of Bioelectricity

2.3.2.1 Early Stage

Through the 1780s and 1790s, Luigi Aloisio Galvani (1737–1798), an Italian anatomist and physicist, professor of Anatomy at the University of Bologna, Italy, discovered bioelectricity-animal electricity. With assistance of his wife, Lucia Galeazzi (1743–1788), and his two nephews, Camillo Galvani (1753–1828) and Giovanni Aldini (1762–1834) who later became professor of Experimental Physics at Bologna, Galvani performed famous experiments in neuromuscular stimulation. He published its results in Latin in 1791, which led him to be the founder of electrophysiology. The publication was divided into four parts. This work has been translated into many languages. In case of English, the translated title was *Commentary on the effect of electricity on muscular motion* (Green, 1953). According to this publication, in the first part, he investigated the effects of artificial electricity on the nerves and muscles of frogs. On January 26, 1781, a dissected and prepared frog by Galvani was placed on a table on which was an electric machine, but at some distance from it and not connected with it in any way. Violent muscular contraction occurred when an assistant happened to touch the femoral nerve with a scalpel. Galvani began immediately to investigate this unexpected phenomenon (Rowbottom and Susskind, 1984).

The most-cited second part of his publication is the results of the investigation under the title of *The effects of atmospheric electricity on muscular motion*. This part is only short. Galvani attached the nerve of the frog leg to the side of the house, and grounded nerves in an adjacent well. Contractions were obtained when every flash of lightning occurred. By August and September of 1786, he was trying to obtain contractions from changes in atmospheric electricity during ordinary calm weather in open space. He suspended frog preparations from an iron railing in his garden by brass hooks inserted through the spinal cord. He happened to press the hook against the railing when the leg was also in contact with the iron railing. He found that there were contractions of the muscles. His thought was that the contractions were due to the slow accumulation of atmospheric electricity within animal, which was suddenly discharged when the hook was pressed against the iron railing. Observing frequent contractions, he repeated the experiments in a closed room (Rowbottom and Susskind, 1984).

The third part of his publication came to new discovery under the title of *The effects of animal electricity on muscular motion*. The earliest results are related below:

> But when I had transported the animal into a closed chamber and placed him on an iron surface, and had begun to press against it the hook fixed in his spinal cord, behold the same contractions and the same motions! Likewise continuously, I tried using other metals, in other places, other hours and days; and the same result; except that the contractions were different in accordance with the diversity of metals, namely more violent in some, and more sluggish in others. Then it continually occurred to me to employ for the same experiment other bodies, but those which transmit little or no electricity, glass for example, gum, resin, stone, wood, and those which are dry; nothing

similar occurred, it was not possible to observe any muscular motions or contractions. Results of this sort both brought us no slight amazement and began to arouse some suspicion about inherent animal electricity itself. Moreover both were increased by the circuit of very thin nervous fluid which by chance we observed to be produced from the nerves to the muscles, when the phenomenon occurred, and which resembled the electric circuit which is discharged in the Leyden jar.

Green (1953)

This third part is most important, from which Galvani showed the existence of animal electricity with the demonstration of muscular contractions in frogs without the use of metals. This led to the controversy with Volta. Volta judged that Galvani's conclusion was a wrong idea. At the end of this third part, Galvani insisted on the existence of animal electricity, perfectly based on the experimental results. The fourth part of his publication begins as follows: *From what is known and explored thus far, I think it is sufficiently established that there is electricity in animals, which, with Bartholinus and others, we may be permitted to call by the general name of animal electricity* (Green, 1953). Galvani's theory was finally summarized as follows:

it would perhaps be a not inept hypothesis and conjecture, nor altogether deviating from the truth, which should compare a muscle fiber to a small Leyden jar, or other similar electric body, charged with two opposite kinds of electricity; but should liken the nerve to the conductor, and therefore compare the whole muscle with the assemblage of Leyden jars.

Green (1953)

The conjectures in fourth part are summarized: one of his suggestions is the use of electricity for the cure of certain nervous diseases such as various forms of paralysis (Potamian and Walsh, 1909). These experiments by Galvani became the center of electrophysiology and bioelectricity. Although Galvani showed animal electricity by his own experiments using animal tissue, Volta repeated Galvani's experiments and confirmed the conclusion of Galvani's experiments, existence of animal electricity. However, Volta did not believe Galvani's explanation. The observed muscle contraction was the result of electricity generated by two dissimilar metals. Volta attempted to show that the combination of two different metals generates the electric current. Volta proposed in 1794, the theory of "metallic electricity," in which electricity was generated by the contact between two kinds of metal in electrolyte. Galvani's work published in 1791 caused a great sensation. As mentioned above, Alessandro Guiseppe Antonio Anastasio Volta (1745–1825) was inspired by Galvani's work. He was born in Como, Lombardy, Duchy of Milan in 1745. At the time when Galvani published his work, Volta was already a well-known professor in his work for the invention of the electrophorus and straw electrometer, etc. He was a professor of Natural Physics at the University of Pavia, in 1779. He traveled much through Europe, first from Switzerland to Paris where he met Antoine Lavoisier and Pierre Simon de Laplace (1749–1827), a French astronomer and mathematician, and to Germany, and to England where he met Joseph Priestley. Volta invented and built the first battery called the Volta battery (Voltaic pile). A Voltaic pile produces a continuous electric current. This pile consists of an equal number of copper and zinc discs separated by circular plates of cloth, paper, or pasteboard soaked in salt water or dilute acid (Mottelay, 1922). Volta demonstrated his battery before Napoleon I (1769–1822) in Paris. Napoleon made Volta a count and senator of the kingdom of Lombardy. Through the observation of the contraction of isolated frog nerve-muscle preparations, Galvani found that the electricity was generated within the body of the animal, and called this electricity, animal electricity. Through the discussion between Galvani and Volta, Volta invented the voltaic pile.

Giovanni Aldini, his mother was the sister of Luigi Galvani, was accepted as a professor of experimental physics with no medical background, at the University of Bologna in 1798. He was also a researcher in the engineering field such as the construction of lighthouses and their illumination, and protection of human life and materials from destruction by fire. He applied the first electric stimulation from a

Voltaic pile to mammalian brains. This gave the observation that stimulation of the corpus callosum and cerebellum triggered motor responses. He used this method in the resuscitation of almost dead people.

Aldini traveled through Europe to demonstrate the existence of animal electricity and the usefulness of galvanism in the field of medicine (Parent, 2004). In London at Mr. Wilson's Anatomical Theatre on January 17, 1803, Aldini tried to reanimate human corpses using electricity with a Voltaic pile of 100 and 20 copper and zinc couples, respectively. His experiment produced powerful muscular contractions upon the head of decapitated oxen. The contractions were produced by introducing into one of the ears a wire connecting to one of the battery's pole and into the nostrils or tongue another wire communicating with the other battery pole. The eyes repeatedly opened and rolled in their orbits while the ears would shake, the tongue move, and the nostrils dilate very perceptibly (Mottelay, 1922). These scandals inspired Mary Shelly (1797–1851) to write in 1818, the famous book *Frankenstein, or the Modern Prometheus*, which was a novel about the revitalization of dead people. Her name as the author was not mentioned in original edition.

Humboldt learned about Galvani's experiment from Volta during the visit to him in 1795. He repeated Galvani's and Volta's experiments. During a stay in London in 1776, young Humboldt studied with Cavendish the electricity of torpedo. Interestingly, Johann Wolfgang von Goethe (1749–1832), the greatest German poet, joined Humboldt's galvanic experiments. Over a 5-year period, Humboldt conducted over 4,000 galvanic experiments and evaluated the roles of Galvani's experiments. The existence of animal electricity with his own theory of the formation in the nerves of a galvanic fluid distinct from electricity appeared in his work in 1797. Humboldt explained that the excitability of the muscle nerve is not due to galvanic current and referred galvanic current to the intrinsic animal force (Finger et al., 2013a, b). He along with Aimé Jacques Alexandre Bonpland (1773–1858), a French explorer and botanist trained in medicine, described the sensation of pain produced by eels. As young explorers, they explored the jungles of South America in 1800. During their travels in Guiana, they captured electric fish as experimental specimens, and called them "torporific eel," Gymnoti (*trembladores*), and compared their electric shock to that of electricity. He described the difficulty to capture torporific eels in his book (Humboldt, 1826). From this book, in order to capture them, the natives procured about 30 horses and guided them into the pond where torporific eels live. The horses received the electric shock from the eels, and the eels were captured in a state of exhaustion. After that, Humboldt and Joseph Louis Gay-Lussac (1778–1850), a French physician and chemist, together conducted torpedo and eel experiments. This experiment brought the foundation of bioelectricity. Humboldt was a patron to physiologists, Johannes Peter Müller, Emil Heinrich du Bois-Raymond, and Hermann Ludwig Ferdinand von Helmholtz. He supported these three physiologists opening a new era in the field of electrophysiology.

Henry Cavendish examined with Joseph Louis Gay-Lussac the electrical properties of torpedo, and had observed that the animals must be irritated previous to the shock, preceding later a noticeable convulsive movement of the pectoral fins that electrical action produced minimal injury to the brain of the fish; also that a person accustomed to electric discharges could with difficulty support the shock of a vigorous torpedo about 14 inches long; that the discharge can be felt with a single finger placed upon the electrical organs; and that an insulated person will not receive the shock if the fish is touched with a key or other conducting body (Mottelay, 1922). Interestingly, Cavendish experimented torpedo electricity with artificially created electric fish and gained insight into electric fishes (1776). He was convinced that Walsh's research on the torpedo and its power was just electric. Cavendish modeled the torpedo after shoe leather. The backbone of his model fish was made by cutting and layering thick leather, similar to the size and shape of a torpedo. The electric organs were represented by barbell-shaped pieces of pewter attached to each side of the leather backbone (Schiffer, 2003). He connected them to wire passing through long glass tubes along both sides of the torpedo's tail. For electrifying this artificial fish, he employed a large battery. Using this artificial fish, he carried out many experiments, for example, to compare the electric shock of this fish to that of a real torpedo. With the help of Walsh, Cavendish showed that the artificial fish produced the same sensation as that produced by a real torpedo.

Michael Faraday (1791–1867), a great distinguished British chemist and natural philosopher, was interested in bioelectromagnetism, in particular, in bioelectricity during his Grand Tour to Europe. In 1813, Faraday, Sir Humphry Davy (1778–1829) and his wife traveled from Turin to Genoa. Faraday was much interested in several water-spouts which he saw in the bay of Lerici, and then in Florence. In Genoa, they visited Domenico Viviani (1772–1840), an Italian botanist and naturalist, professor at University of Genoa. At that time, Prof Viviani caught the electric fish torpedo in the bay of Lerici, and tried to investigate whether the electric discharge of the torpedo is strong enough to decompose water (Hamilton, 2002). Faraday studied with him the electric discharge from the torpedo. The results of his torpedo experiments proved that the identity of the electric power of the *gymnotus* or the torpedo was common electricity (Faraday, 1839). He concluded that a single medium electrical discharge of the fish is at least equal to the electricity of a Leyden battery of 15 jars, containing 3,500 square inches of glass coated on both sides, discharged to its highest degree (Motteday, 1922). During their Grand Tour, they visited from England, to France, Belgium, Italy, Switzerland, Germany, and met André Marie Ampère (1775–1857), a French physicist, professor at the École polytechnique, Dominique François Arago (1786–1853), a French physicist and astronomer, Joseph Louis Gay-Lussac, Alessandro Guiseppe Antonio Anastasio Volta, Jean-Baptiste Biot (1774–1862), a French physicist and mathematician, Sir Benjamin Thompson (Count von Rumford) (1756–1819), the founder of Royal Institution of Great Britain, Wilhelm Heinrich Alexander von Humboldt, etc. In their Grand Tour, various experiments were made by Sir Davy at each place, on iodine, and on the electricity of the torpedo fish. While at each place Faraday found some opportunities for helping satisfy his craving for improvement (Jerrold, 1891). In their Grand Tour course, in Milan on June 17, 1814, he visited Volta who was nearly 70 years old. Faraday, Sir Davy and his wife left London on October 13, 1813, and returned to England on April 23, 1815. Soon, Faraday engaged as Davy's assistant at the Royal Institution. The time when they took their Grand Tour corresponded to the time of the Napoleonic Wars.

2.3.2.2 The Developing Stage

Until the nineteenth century, electricity and magnetism were considered as separate physical properties. In 1820, Hans Christian Oersted (1777–1851), professor of Natural Philosophy at the University of Copenhagen, Denmark, observed that an electric current flow through a wire would move a compass placed beside it. This showed that an electric current produced a magnetic field which means that there is a connection between electricity and magnetism. His great discovery made a new era which opened the scientific development in electromagnetism and bioelectromagnetism. Oersted sent a copy of his experiments to Berlin, Paris, London, and other places. Soon, Oersted was offered membership in the Royal Society and received the Copley medal from its Society in late 1820. Within the next 45 years, this great discovery was repeated by many scientists in the world.

Oersted's work was published in a Latin pamphlet dated July 21, 1820. Only 3 months after publication, on October 20, 1820, Jean Baptiste Biot (1774–1862) and Félix Savart (1791–1841), a French physicist, gave together the theoretical concept of Oersted's work, called later, the law of Biot-Savart. This law states that the force between a wire with electric current flowing and a magnet pole is inversely proportional to the distance between them. Later, Ampère generalized it and developed the mathematical theory of electrodynamics (Ponchon et al., 2020). Soon after, the translation of Oersted's experiments into English appeared in the Annals of Philosophy, and Faraday obtained a copy of Oersted's paper from Sir Davy in 1821 and began to repeat the experiments.

In 1824, Dominique Arago observed that when a magnetic needle was oscillating above or close of a non-magnetic body (such as water or a metal), it gradually oscillated in arcs of less and less amplitude, and he made a circular copper plate to revolve immediately beneath a magnetic needle or magnet, freely suspended so that the latter would rotate in a plane parallel to that of the copper plate, and he found that the needle tends to follow the circumvolution of the plate; that it will deviate from its true direction; and that by increasing the velocity of the plate the deviation will increase until the needle passes to the opposite point, when it will continue to revolve, and last rapidly so that the eye will be unable to distinguish it (Mottelay, 1922). However, this phenomenon remained inexplicable.

In September 1831, Faraday tried to solve the problem of producing electricity from magnetism. Through a series of experiments, he discovered electromagnetic induction, the results which were presented to the Royal Society on November 24, 1831. His successful results of electromagnetic induction spawned through Europe. Faraday's law of electromagnetic induction was the basis of the transformer, which is the converter of mechanical energy into electric energy. Joseph Henry (1797–1878), an American physicist, the first secretary of the Smithsonian Institution, Washington, DC, also discovered that a change in magnetism can produce electric currents. Electromagnetic induction was discovered independently by Faraday and Henry. Later, this invention became a very important contribution to bioelectric research.

Apart from Faraday's contribution, the discussion turned on to Oersted's discovery and the progress of instrumentation. After Oersted's discovery, it gradually began the development of the instrumentation which can support the study of electrophysiology in bioelectromagnetism. Oersted's finding was that the deflection of the needle could be used to indicate electric current strength, the appearance of the magnetic effect on an electric current, which led to the appearance of the first galvanometer. Two months after Oersted's discovery, in September 1820, Johann Salemo Christoph Schweigger (1779–1857), professor of Chemistry at University of Hall, Germany, using of wire helix and a compass needle, developed an instrument for detecting and measuring electric current, called electromagnetic multiplier (multiplicator) or galvanometer. The needle was placed at the center of a rectangle consisting of many turns of insulated wire. Many electrical instruments depended on the operation of this simple principle and were employed for the measurement of electricity. Schweigger's instrument was refined by Leopoldo Nobili (1784–1835), a preeminent Italian physicist, professor of Physics at University of Florence, Italy, in 1825 (Nobili, 1825). The refined instrument was called astatic galvanometer. This astatic galvanometer employs a double coil of 72 turns wound in a form of figure-eight. The effect of the geomagnetic field was compensated by placing two identical magnetic needles connected on the same suspension with opposite polarity. This became the most important instrument for measuring the electric current strength in the field of electrophysiology (Possenti and Selleri, 2017). Using the astatic galvanometer, Nobili measured and recorded animal electricity such as the electric current from the neuromuscular preparation of frogs. Nobili's astatic galvanometer opened the instrumental era of electrophysiology.

As can be seen above, the history of bioelectromagnetism ran closely in parallel to the developmental history of the measuring instruments of bioelectric signals. With the development of the instruments, electrophysiology became its own field in the first half of the nineteenth century.

Faraday predicted experimentally the existence of electromagnetic waves after the discovery of electromagnetic induction. James Clerk Maxwell (1831–1879), physicist and mathematician at Scotland, published the theory of electromagnetic waves in 1864. This theory predicted the existence of a whole spectrum of electromagnetic waves with the speed of light (3×10^8 m/s), and Heinrich Rudolf Hertz (1857–1894), German physicist, professor at the University of Karlsruhe, demonstrated experimentally the existence of it, which contributed to the age of electromagnetic waves technology. In 1888, Hertz won the Matteucci Medal which was named after Carlo Matteucci. In the 1830s, Carlo Matteucci (1811–1868), an Italian physiologist, a graduate of University of Bologna and professor of Physics of University of Pisa, had shown that the discharge of the electric organs of the ray was controlled by a special structure, the electric lobe. At the time when he published the work on the electric fish, he also repeated Galvani's and Nobili's animal electricity experiments. Using Nobili's astatic galvanometer, he recorded electrical activity from the heart of a frog in 1842. He developed the idea of animal electricity and found that during the leg's tetanic contractions, the measured current was driven by a difference in electrical potential between an injured and uninjured nerve. He was a pioneer in the study of bioelectricity. Through the experiments by Galvani and Matteucci, the action potential was discovered in cardiac muscle and nerves. In particular, Matteucci's work came to the attention of Johannes Peter Müller (1801–1858), a German physiologist, professor at the University of Berlin. Müller found the capacitive properties of tissues and the anisotropy of muscle conductance.

In Berlin, Müller proposed the theory of specific nerve energies. His proposed doctrine was that a particular sensation depends on the nature of the sensory receptors that had been stimulated, and that it was the function of a receptor to convert the energy of the stimulus into impulses (action currents) in the nerve fiber (McComas, 2011). He also developed the concept of electric signal propagation through the nerve. Müller's pupil and assistant, Emil du Bois-Reymond (1818–1896), 8 years younger than Matteucci, a Swiss-German physiologist, professor at the University of Berlin, started the investigation of animal electricity using electric fish, and moved to the study of current impulse arising from nerves and muscles of frog legs using of galvanometer in 1842. In order to do, he invented the galvanometer with 4,650 windings of a wire (1 km long and 0.17 mm in diameter) and developed a more sensitive galvanometer with over 24,000 windings and 5 km long wire (Rowbottom and Susskind, 1894). du Bois-Reymond was the first to measure the potential difference accompanying nerve and muscle excitation. He repeated these observations and discovered the action potential. This galvanometer was used for several decades in the research field of neurophysiology. Hermann Ludwig Ferdinand von Helmholtz (1821–1894), a German physicist and physiologist, professor at the University of Berlin, was the first to measure during his work in Heidelberg, the conduction velocity of a nerve cell axon in 1850s. He showed the propagation of action potential, by measuring the delay time between an electric stimulation of the nerve and muscle contraction of frogs. The values of action potential propagation were experimentally in the range of 25–40 m/s.

The phenomenon of electric current flowing through a nerve cell with a certain resistance and capacity is called electrotonus, which was first studied around the end of the nineteenth century. In electrophysiology, tonus is the slight contraction of a muscle. Electrotonus refers to the altered electrical state inside a neuron and between cardiac muscle cells or smooth muscle cells from passive electric current. The research of electrotonus began mainly in Germany in the middle of the nineteenth century. At this time, the long distance telegraphic cable laying work began between the continents. In 1855, William Thomson (later, Lord Kelvin, 1824–1907), a British physicist, professor in the University of Glasgow, proposed a theoretical model called the cable theory (Thomson, 1854–1855). This theory provided the understanding of electrical propagation of the Atlantic submarine (undersea) telegraphic cable. The proposed theory takes only the capacitance and resistance of a cable into the calculation and it was later corrected by Heaviside. Oliver Heaviside (1850–1925), a British physicist and electrical engineer, applied the cable theory to analyze the submarine (undersea) telegraphic cable in 1876. From his cable theory, the electrical characteristics of the excitation of nerves and muscle cells could be analyzed. Oliver Heaviside used for the first time the terms, impedance (1886), conductance (1885), permeability (1885), admittance (1887), and permittance (susceptance). The terms are used not only in electrical engineering, but also in neurophysiology. The properties of nerve fibers can be derived from the solution of the cable theory because the electric properties of the nerve fibers are similar to that of the submarine (undersea) telegraphic cable. If we consider a simple nerve fiber, we can use the submarine undersea cable theory as a model to represent the nerve fiber in form of an electric circuit. The resistance of the protoplasm, the resistance of the cell membrane, and the capacity of the cell membrane can be used to describe nerve fibers by the cable theory. The cable theory can be applicable to the characterization of electric conduction along a nerve fiber. Excitation of nerve and muscle cells can also be analyzed using the cable theory.

Until the early twentieth century, the advantage of applying the cable theory to nerve fiber conduction was not recognized. In 1945, Kenneth Stewart Cole (1900–1984), an American biophysicist, Alan Lloyd Hodgkin (1914–1998), a British biophysicist, and William Albert Hugh Rushton (1901–1980) developed the mathematical theory of nerve fiber conduction based on the cable theory. The measurement technique for the electric activity of a neuron was improved in the 1950s. In order to understand the electric properties of neuron, the cable theory may also be suitable.

Julius Bernstein (1839–1917) studied medicine at the University of Berlin as a pupil of Emil du Bois-Reymond and trained for a time under Helmholtz as an assistant to record the action potential. The available instruments for the measurement and recording of the action potential were galvanometers. The galvanometer was still slow to respond, and its response time was in the order of seconds. At the time when Julius Bernstein was at the University of Halle, Germany, he developed more sensitive equipment with the combination of galvanometers and differential rheotomes and recorded the rest and action potential time course in the range of several tens of microseconds and conduction velocity of nerve impulse from isolated nerve fibers of frogs. A differential rheotome is an instrument that enables electrical connections to be made for extremely short, but nevertheless precise time intervals (McComas, 2011). He showed that the action potential has a rapid rise and a rather slower decay, the total duration being less than a millisecond (McComas, 2011). This differential rheotome consists of three parts: a turntable with a diameter of 20 cm, a Ruhmkorff induction coil for a stimulator, and recording is done with a galvanometer. Bernstein's investigation was facilitated by the use of Lippmann's capillary electrometer. This meter is a more sensitive instrument than the astatic galvanometer. This capillary electrometer was used in the measurement of the action potential of nerves. Gabriel Jonas Lippmann (1845–1921), a French physicist, invented the capillary electrometer in 1873, which is a tube of ordinary glass, 1 m long and 7 mm in diameter filled with mercury, and the capillary's tip point is immersed in dilute sulfuric acid. Lippmann was the winner of the Nobel Prize in Physics in 1908 for his method of reproducing colors photographically based on the phenomenon of interference.

In 1902, after many important contributions to electrophysiology, Bernstein assumed that the membrane is permeable to a single ion and proposed the primitive semipermeable form of the membrane theory of action potential across the cell membrane. Cell membrane is able to selectively pass a certain kind of ions. According to this theory, the semipermeable membrane surrounding the cell upon excitation becomes permeable to potassium ions which enter the cell from the surrounding tissue fluids and give rise to a state of polarization on the surface of the cell (Rowbottom and Susskind, 1984). His experiments were facilitated by the use of Lippmann's capillary electrometer. However, the formation of the membrane theory of Bernstein was still incomplete. His hypothesis was that the potential difference across the cell membrane was maintained by the difference of ion concentrations, permeable to potassium ions, and impermeable to intracellular anions and sodium ions, which is the concept of a semimembrane. His membrane theory was the basis for the evaluation that the transmembrane voltage was as proportional to the logarithm of the concentration ratio of the potassium ions, which was expressed by the Nernst equation. He was recognized as the founder of the membrane theory.

The German school made great contributions in physiology, in particular in electrophysiology from the second half of the nineteenth century. Historically, Bernstein's membrane theory was a shift in electrophysiological research into bioelectric phenomena and laid the background for bioelectromagnetism. This background led to international research which began in the 1930s by Alan Lloyd Hodgkin, Andrew Fielding Huxley (1917–2012) (England), Howard James Curtis (1906–1972) (USA), Herman Paul Schwan (1915–2005), and Kenneth Stewart Cole and John Carew Eccles (1903–1997) (Australia). The development of electrophysiology gave significance to the applications of electricity in medicine, to electrotherapy and magnetotherapy.

As a basis for bioelectromagnetism, two works by Japanese scientists are introduced. The two discoveries contributed greatly to the development of electrophysiology. Gen-ichi Kato (1890–1979) was born in Okayama Prefecture, Japan, physiologist and professor at Keio University (Kato, 1970). In 1923, he proposed the decrementless conduction of the nerve impulse in the presence of a narcotic agent. Three years later, the XII International Physiological Congress was held in Stockholm. Kato and his assistant attended this Congress to perform an experimental demonstration of decrementless nerve conduction with Japanese toad (*Bufo vulgaris Japonicus*). This Japanese toad has long nerves due to its large size. During this travel from Japan through Siberia to Stockholm, all of the Japanese toads turned out to be dead. So, they used Dutch toads instead of Japanese toads. Either way, they were successful in the demonstration. In addition, Kato and his group made the successful isolation of a single nerve fiber in conducible state in 1930. In 1935, Ivan P. Pavlov (1849–1936), a Russian physiologist and 1904's Nobel Prize Laureate, nominated Kato as a nominee for the Nobel Prize in Physiology or Medicine for his work on isolation of single nerve fibers and muscle fibers and for the demonstration of the existence of reflex excitatory fibers and reflex inhibitory fibers.

Ichiji Tasaki (1910–2009) was the pupil of Gen-ichi Kato, Japanese-born American biophysicist. He was first working in Japan and later worked as a visiting scientist at research institutions in Switzerland and England. Then he went to the United States and joined the National Institute of Health (NIH) in 1953. During his scientific carrier in 1939, he studied anesthetics on isolated nerve fibers, discovered the insulating function of myelin sheaths (the nodes of Ranvier), and showed that action potential can jump actually over the anesthetized region between nodes of Ranvier; later he called this process saltatory conduction (Tasaki, 1939 a, b).

Historically, the idea of saltatory conduction of the action potential in myelinated axons seems to have originated in 1925 with Ralph Stayner Lillie (1875–1952), professor of General Physiology at the University of Chicago, USA (Lillie, 1925). He modeled the nodes by enclosing iron wire in a glass insulant tube containing dilute (70%) nitric acid with periodic breaks. He showed that electric conduction occurred faster in saltatory fashion.

Gen-ichi Kato and his pupil, Tasaki, introduced the technique with myelinated axons of frog nerves. Huxley and Robert Stämpfli (1914–2002), professors at the University of Berne, Switzerland, developed the technique to measure the resistance and capacitance of myelin. By drawing single myelinated nerve fibers isolated from the sciatic nerve of frogs through a glass capillary and measuring the current flowing around the outside of the fiber during the passage of an action potential, Huxley and Stämpfli proved exactly the existence of saltatory conduction of the action potential from node to node in isolated frog nerves (Huxley and Stämpfli, 1949). Their work confirmed Tasaki's findings. The concept of the saltatory conduction is featured in the textbooks of physiology.

Conduction velocity of action potential in myelinated fiber depends on distance of neighboring Ranvier nodes, as duration of action potential is about 1 m/s in any nerve fibers. Distance of neighboring Ranvier nodes depends on diameter of nerve fibers; distance is longer in thicker fibers. There are myelinated and unmyelinated fibers with different diameters in mammalian nerves. In unmyelinated fiber impulse conduction is continuous, and therefore conduction velocity is much slower than myelinated fibers. Masamichi Kato, professor at Hokkaido University, measured conduction velocity of human ulnar nerve of the arm with results of about 67 m/s (Kato, 1960).

The properties of ion channels and the propagation of nerve signals were obtained from the study of the current clamp and the voltage clamp in electrophysiology. In 1939, using extracellular electrodes, Kenneth Steward Cole and Howard James Curtis, both at Columbia University, performed measurement by alternating current impedance over a wide frequency range on the giant axon of a stellar nerves of Atlantic squid, *Loligo pealei* which is about 0.5 mm in diameter and its segment is 3–8 cm long (Cole and Curtis, 1939). They demonstrated the rapid fall in membrane resistance during the development of the action potential. Soon after, the intracellular electrode was developed by Cole, Curtis, Hodgkin, and Huxley, by inserting the electrodes directly into the squid axon, direct recordings of action potential were performed.

In 1949, the voltage-clamp technique was designed by Cole and employed by Hodgkin and Huxley to produce the ionic theory of membrane excitation. They demonstrated that membrane excitability is determined by passive ion flux according to their electrochemical gradient (Verkhratsky et al., 2006). During the 1950s, Hodgkin and Huxley began to understand the electrical nature, action potential in the axon of the giant squid by using their developed intracellular electrode. This electrode could be inserted into the squid axon. They recorded the first time directly action potential. From such experiments, they discovered the ionic mechanism of excitation and inhibition in nerve cell membranes and developed the mathematical model of the activation process. In 1952, Hodgkin and Huxley published a series of five papers in the *Journal of Physiology* (Hodgkin and Huxley, 1952a–d; Hodgkin et al., 1952). The characterizations of the changes during action potential were described in the first four papers and in the last paper the mathematical model was presented. The model explains the ionic mechanism underlying the initiation and propagation of action potential in the squid's giant axon. Sir John Carew Eccles, Australian physiologist, investigated the synaptic transmission of ions, the behavior of the cell membrane. These three specialists won the Nobel Prize in Physiology or Medicine in 1963 for their discoveries concerning the ionic mechanisms involved in excitation and inhibition in the peripheral and central portions of the nerve cell membrane. These scientific contributions opened further the era of neurophysiology and electrophysiology. It brought the new patch clamp to the study of excitable cells as a specialized form of the voltage clamp. During the late 1970s and early 1980s, Bert Sakmann, German cell physiologist at the Max Planck Institute of Neurobiology, and Erwin Neher, German biophysicist, University of Göttingen and Max Planck Institute for Biophysical Chemistry, developed a new patch clamp technique. They constructed an extracellular small tip (~1–2 μm in diameter) electrode with a polished surface and pressed it to the surface of an isolated skeletal muscle fiber of a frog and electrically isolated a small patch of membrane (<10 μm²) which laid beneath the opening of the electrode. With this experimental approach, they were able to monitor single channel ion currents generated by the opening of nicotinic acetylcholine receptors (Verkhratsky and Parpura, 2014). Sakmann and Neher received together the Nobel Prize in Physiology or Medicine in 1991 for their discoveries concerning the function of single ion channels in cells.

2.3.3 Pioneers of Electricity in Japan

During the Edo period (1600–1868) in Japan, the contact with the outer world had been prohibited by the national seclusion policy promulgated in 1639. Dutch, Chinese, and Korean were the only foreigners with permitted access to Japan. Dejima, a man-made small island in Nagasaki harbor, was reclaimed in 1641, and there was the only place opened to the outside. Korean accessed only through Tsushima Island to Japan. Through Dejima Western culture was introduced to Japan. From the time when Dejima was reclaimed, the Dutch remained trading with Japan until 1854. In 1720, the 8th Shogun, Yoshimune Tokugawa (1684–1751) lifted the ban on the import of Western books, except for the ban on Christianity. As a result, Western books written in Dutch were gradually imported from the first half of the eighteenth century. The time around the year 1720 corresponded to the time when the first edition of *Robinson Crusoe* (1719), *A Journal of the Plague War* (1722) both by Daniel Defoe (1660–1731), *Lettres Persanes* (1721) by Charles Montesquieu (1689–1755), and the first edition of *Gulliver's Travels* (1726) by Jonathan Swift (1667–1745) were published. Swift presented satirically in his novel that the flying island city, Laputa levitated through the use of magnet. During this time, Sir Isaac Newton (1642–1727) passed away, and there were many historical well-known active scientists and artists, Anton von Leeuwenhoek (1632–1723), Edmond Halley (1656–1742), Gottfried Leibniz (1646–1716), Johann Sebastian Bach (1685–1750), etc. These were the founders of modern natural science and music.

   Due to the prohibition of direct contact between Japanese and the outer world, Western science was imported only through Dutch books, called "Ran-gaku." Through the narrow window of the island of Dejima, Ran-gaku (Dutch Studies) was popular study of Western science during the Edo period. The import was more active from the second half of the eighteenth century. During this and the following times, a wide range of sciences such as physics, chemistry, mathematics, technology, biology, and medicine gained importance for the development of Japanese society.

   To contribute to the development of Japanese society, historically recognized are three persons who will be introduced (Reischauer and Tsuru, 1983). First is William Adams (1564–1620). He was born in Gillingham, Kent, and a British navigator. He came to Japan in 1600. The time when he arrived was the time before the national seclusion policy. After a series of mishaps, his disabled ship, the Liefde (belonging to a Rotterdam trading firm in the Netherlands), reached the Usuki province (Now, Oita prefecture), Kyusyu. The Liefde is no longer in existence, but the wooden carving figure of Erasmus that hung on the stern of the Liefde is still in existence as an important Japanese cultural property. Adams became soon a key advisor to the 1st Shogun, Ieyasu Tokugawa (1543–1616). He employed Adams as commercial agent, interpreter, and shipbuilder. Ieyasu named Adams "Miura Anjin," after the location of Miura Peninsula near Edo (now Tokyo) of the estate given by Ieyasu. In the first several years of the seventeenth century, he contributed greatly to the Japanese society. In 2020, it was reported that the human remains believed to be of Adams were found in Hirado, Nagasaki (Mizuno et al., 2020). The second person is Engelbert

Kämper (1651–1716), physician and historian, who was born at Lemgo in the duchy of Lippe (now North Rhine-Westphalia), Germany. Before coming to Dejima, he received excellent medical and humanistic education. In 1690, he arrived at Dejima and got the position of physician at the Dutch East India Company. There, he taught Western medicine. In 1691 and 1692, he accompanied the annual tribute mission by the Dutch East India Company chief to the Shogun court at Edo. Through these mission journeys, he obtained and described precise information about Japan. Through the accurate and valuable description of Japan, he introduced greatly Japan to the Europeans. In 1692, he returned to Europe from Nagasaki via Batavia (now Jakarta). He then published the two-volume book *History of Japan* which was the first systematic introductory work about Japan.

The third person is Philipp Franz Balthasar von Siebold (1796–1866). He was born in Würzburg, Germany. After studying medicine at the university there, he entered the Dutch government service as an army doctor. In 1823, he was appointed physician at Dejima. There, he taught Western natural and medical sciences and physics during his 5 year stay in Japan. In particular, he established a boarding school, Narutakijuku, at Narutaki (a suburb of Nagasaki). There, he also treated Japanese patients. In exchange for Dutch books, he got a map of Japan from Kageyasu Takahashi (1785–1829), the shogunate astronomer. The map of Japan was made by Tadataka Ino (1745–1818), the famous cartographer and land surveyor. In 1830, Siebold returned to the Netherlands and settled at Leyden. After 30 years, in 1859, he returned to Japan with his son Alexander George Gustav von Siebold (1846–1911), a German interpreter, and worked for 2 years on behalf of the Dutch East India Company (Reischauer and Tsuru, 1983). Siebold contributed importantly to the development of Japanese science through teaching Western science in the Edo period. His collection of Japanese materials became the foundation of the National Museum of Ethnology at Leyden.

Although the above three persons had encyclopedic knowledge about Japan, there were many other European who taught Western science to the Japanese. Below, three famous Japanese persons in the Edo period are introduced. Gennai Hiraga (1728–1780) was born as a lower-rank samurai in Sanuki province (now Kagawa Prefecture), Shikoku. He was known as being a scholar, scientist, geologist, artist, etc., and has been described as being the "Leonardo Da Vinci of Japan." He went first to Nagasaki in 1752, in order to study Ran-gaku for 1 year. Around 1770, he obtained a broken rubbing electrostatic generator at Dejima. After some years of effort, he restored it to its original state, a kind of Leyden jar, which he later called "Elekiteru" in Japanese (Figure 2.6). Using this restored Elekiteru, he gave shocks and electrical discharges to humans as an advertisement. He killed his disciple with a sword and died in prison in 1780. During Hiraga's times, Japanese industries were not developed enough to use an electrostatic generator.

After about 60 years since Nollet's many-person train discharge chain shock experiment in 1746 and from Franklin's kite experiment in 1752, Sokichi Hashimoto (1763–1836) conducted the same experiments. He was born in Osaka as the son of a small merchant. In his childhood, he was educated in Ran-gaku. After learning Ran-gaku, he started his carrier as a physician. During translating the Dutch books into Japanese, he obtained knowledge of electricity and read about the kite experiment carried out by Franklin including many other experiments on static electricity. Following these Dutch books, he confirmed by himself static electricity (Hashimoto, 1984). One of which was the replication of Franklin's kite experiment. Under his leadership, Kikuta Naka, pupil of Hashimoto, used a pine tree with 40 m height (Figure 2.7), and then they proved that a thunder is electricity. As an advertisement, Hashimoto demonstrated shock experiments as shown in Figure 2.8. This was the Japanese version of Nollet's fantastic demonstration of the many-person train discharge chain shock by a Leyden jar.

The third person is Shozan Sakuma (1811–1864), who was born as the son of the middle-class samurai in Shinano province (now Nagano prefecture) and was known as a scholar and politician. He started to study Ran-gaku when he was 33 years old. Through the study of Ran-gaku, he learned about electricity and developed Japan's first telegraph and seismometer, electrical machines based on Elekiteru, and the Daniel cell and medical instruments with electricity. He was a famous politician rather than scientist. These historical and well-known Japanese persons gained knowledge about electricity through learning Ran-gaku in the Edo period and made many experiments concerning electricity.

FIGURE 2.6   Elekiteru by Gennai Hiraga, replica (Courtesy of Postal Museum Japan, Tokyo.)

2.4 Development of Electric and Electromagnetic Stimulations

As a subdivision of bioelectromagnetism, this section introduces electric and electromagnetic stimulations. The historical overview focusing on the effect of electricity on plant growth is shortly described. Next, very important historical developments of electric and electromagnetic stimulations in basic research and early medical applications are presented as treatments of electrotherapy and magnetotherapy.

2.4.1 Effect of Electricity on Plant Growth

In the middle of the eighteenth century, after the development of the Leyden jar, the effect of electricity on the growth of plants was examined in open spaces. The earliest experiments on plant growth were made by Maimbray, in Edinburgh, England, which were reported in 1746. After the electrification of two myrtles with an electrostatic generator during the whole month of October, he found that they put forth small branches of some inches in length that blossomed sooner than other shrubs of the same kind without electrification. He attributed this effect to the influence of electricity (Solly, 1845). Pierre

FIGURE 2.7 Japanese version of Franklin's kite experiment (Courtesy of Electric Power Historical Museum, Tokyo.)

FIGURE 2.8 Japanese version of many-person train discharge chain shock experiment by Elekiteru (Courtesy of Electric Power Historical Museum, Tokyo.)

Bertholon (1741–1800), professor of Natural Philosophy at Montpellier, France, made the first of many experiments of electric stimulation on plant growth. His developed "electro-vegetometer" was used for the collection of atmospheric electricity by means of an antenna arrangement (Schiffer, 2003, p. 247). Using this arrangement, he observed the increased growth and fertility of treated plants under the influence of electricity compared to untreated plants. In 1783, Bertholon attempted for the first to apply the electric field generated from the open air electrified network on plants and observed that the fresh weight of the plants grew more under the network than the control samples. Many other investigators also tried to apply electricity on plants.

Maimbray's result was confirmed by Abbè Nollet. He electrified two pots containing growing seeds for 15 consecutive days. The electrified pot put forth sprouts earlier. He carried out the same experiments on several plant species. Nollet showed that seeds in electrified metal containers germinated rapidly (Schiffer, 2003, p. 109). In a series of many experiments using seeds, birds, and cats, he established the growth enhancement by electrification which became lately the "electro-culture." Abbè Menon, principal of the College of Bueil at Angers, France, made a gentle electrification on the animal system for the determination of physiological effects. He chose cats, pigeons, and chaffinches. After electrification for 5 or 6 hours at a time, he found that one cat was 65 or 70 grains lighter than the other, the pigeon from 35 to 38 grains, and chaffinch has lost 6–7 grains (one grain is 0.0648 g). He concluded that electricity augments the slow, continuous perspiration of animals. The same was found in human. He found a loss in weight of several ounces (1 ounce is 28.35 g) when persons of 20–30 years old were electrified for 5 hours (Mottelay, 1922).

In 1747, Jean Louis Jallabert (1712–1768), professor of Philosophy and Mathematics at Geneva, Italy, improved Nollet's experiments. He demonstrated that the electrified plants grow faster and have more firm stems than the non-electrified ones. He electrified various plants for 2 hours every day, and exposed them to open air after electrification, and found that all of them grew rapidly and flourished remarkably (Solly, 1845). Edward Solly (1819–1886) was a British chemist, and in 1845 he became professor of Chemistry at Addiscombe College. He performed in 1845 experiments where grains, vegetables, and flowers were grown between electrodes, which were buried into the ground and about 200 feet apart. In this treatment, he got a beneficial effect in 19 cases, 16 harmed cases, and 35 cases with no effect. He concluded that electricity has no marked effect on plant growth. Many experiments have been used for passing to collect electricity through plants. The results of these experiments were doubtful. On the other hand, numerous experiments have been done, in which a potential gradient around the plants with a network of wires was created. A wire network was placed at various distances above the treated plants. This arrangement produced an electric field between the wire network and the ground. This means that the plants, growing in the ground under the wire network, were exposed to an electric field. In 1878, Louis Grandeau (1834–1911), professor at the Collège de France, investigated the effect of atmospheric electricity on the growth of tobacco, corn, and wheat. In order to investigate the effect, there were two parts which divided into the uncovered area (natural condition) and a covered area with wire net (shielded from atmospheric electricity). The uncovered plants grew 50%–60% higher than the shielded plants.

The Finnish scientist Karl Selim Lemström (1838–1904), professor of physics at Helsinki, noticed that the vegetation looked green and healthy during his visits to the Polar region such as to Spitzbergen, the north of Norway and the Finnish Lapland, and wondered about it. His suspicion was that the looking green and healthy vegetation might be due to the electric currents from the aurora borealis. To confirm this, he performed an agricultural experiment on the effects of electrostatic fields on plants in different places in Europe such as Finland, Germany, Sweden, and Burgundy (Lemström, 1904). His study showed that the electrical discharge from an aerial system of wire network points stimulated the growth of plants such as potatoes, carrots, and celery and that there is an average yield increase of 45% compared to the control samples within 8 weeks. He proposed that the best times for the application of the static electric field were for 4 hours in the morning and another 4 hours during the late afternoon. The application of electricity was for the whole day during cloudy weather, and during nights of moist weather. This result led to the application of electricity in horticulture and agriculture, later called "electro-culture."

In 1908, Oliver Joseph Lodge (1851–1940), a British physicist and inventor, Principal of the University of Birmingham, and his son, Lionel Lodge (1883–1948), the founder of the Lodge Fume Deposit Company (now, Lodge Copttrell Ltd.), with help from J. E. Newman and R. Romford, repeated the study of the discharge of the electricity on plants made by Lemström (Lodge, 1908). Thin wires were connected to a high voltage generator of a hundred thousand volts. The wires were supported on posts and under, the experimental area was of about 40 acres, 20 of which were electrified with a wire network 15 feet above ground. As preliminary results, they summarized the results of the electrified treatment compared with the non-electrified treatment. The product (ear and straw etc.) of electrified plants such as wheat was better than that of the non-electrified. Oliver Lodge was known as a person who developed the first radio wave detector called coherer. The coherer was the first radio reliever, a primitive form of radio signal detector. During the 1900s–1930s, in England an organized committee investigated the effects of the static electric field on plant, later called "electro-culture" (Blackman, 1924a, b; Briggs et al., 1926). Vernon Herbert Blackman (1872–1967) and his co-workers at the Imperial College of London reported using an aerial system similar to that of Lemström, with voltage gradients of 20–40 kV/m, the increase of about 50% for several plants. A current of $10^{-11} \sim 10^{-8}$ A per plant stimulated growth. Using wheat, barley, and oats, they performed 18 field trials, but only 14 gave a significant increase in dry weight. Blackman also noticed no significant changes in soil nitrogen after the application of electric fields to oats. However, it was impossible to reach definite conclusion. From the USA, Lyman James Briggs (1874–1963), physicist, Office of Biophysical Investigations, Bureau of Plant Industry, and his co-workers performed a series of experiments over a period of 19 years (Briggs et al., 1926). Their data showed no increased growth caused by electricity (Lund, 1947). Briggs pointed out that there are several uncontrolled factors, such as the variations of soil, moisture, temperature, and light. These factors vary from experiment to experiment, even within the same experiments.

As for Japanese research, Motoharu Shibusawa (1876–1975), professor at the Tokyo Imperial University, and his co-workers reviewed the results of Lemström's research, and studied the effects of electric discharges on the rate of growth in plants in 1927 (Shibusawa and Shibata, 1927; Shibusawa, 1963). They conducted their experiments in a greenhouse and applied three kinds of currents: 21 kV (50 Hz), 10–15 kV (DC), and 13 kV (130 kHz) to various plants (Figure 2.9). As a result, they found that (1) two experiments with buckwheat showed an average yield increase amounting to 9.8% and 8.0%, respectively, for the electrified samples. (2) The first experiments did not show differences between the electrified and control plants. In the case of tobacco plants, the dry weight increased 21.7% compared to the controls. They concluded that further experiments are needed.

Through electrical treatment, "electro-culture" was used to accelerate growth rates, increase yields of plants, and improve the quality of crops. The electrical treatment includes the setup of the electrified wire network with DC and AC voltages over the plants and the application of an electric current through the soil. Although there had been many experiments, the beneficial effects of the electrical treatment in the electro-culture were still controversial. Due to this controversy and the lack of economical merits, the electro-culture was discontinued in the 1930s.

Apart from large-scale beneficial studies of the electricity in horticulture and agriculture, Lawrence E. Murr, the Pennsylvania State University, investigated the lethal effects of electric fields on plants from the point of view of plant physiology (Murr, 1963, 1964a, b, 1965, 1966). He conducted studies on the effects of electrostatic fields on parameters such as plant's germination, growth, and yield; plant and leaf damages; and metabolism. The field strength varied between 0.25 and 500 kV/m. He observed changes in plant morphology such as leaf tip burning and discoloration at high electrostatic fields. Murr studied the effects of high electric fields on leaves of orchard grass and reported burning of their leaf tips. Murr included the effect of corona current and suggested that an over-respiration process was causing the burning. From the point of view of plant physiology, Murr started a series of experiments on the effects of electrostatic fields on plants. From the results of these experiments, he suggested the concept of "lethal electrotropsim." In his experiment, the electric field was generated by two aluminum lattice electrodes, and a pot with orchard grass seedlings was fixed to the lower electrode, with the upper

FIGURE 2.9   Plant growth experiments (From Shibusawa, 1963.)

electrode no more than 10 cm from the tip of the plant, allowing the variation of the field strength. When the seedlings were continuously exposed to the electric field, the tips of the seedlings turned brown, as if they were burning, which Murr observed to be the same as a situation of mineral deficiency. The relationship between the plant's response to the electric field and corona current was then examined. Murr classified the effects on the plants according to the range of currents flowing in them. The results showed that currents flowing through the plants above 10^{-5} A resulted in damage to the plant and leaves; $10^{-8} \sim 10^{-6}$ A resulted in leaf damage and negative effects of reduced dry weight; $10^{-15} \sim 10^{-9}$ A resulted in positive effects such as growth promotion and increased dry weight; and below 10^{-16} A had no effect (Murr, 1966). Francis Xavier Hart, a professor at the University of the South, Sewanee, Tennessee, USA, believed that the effects were due to the water loss from the leaf tips when corona current flowed caused by the high electric fields (Hart and Schottenfeld, 1979).

In the next steps, more interest in other than static electric field, started as if had been shown about the effects of air ions on plant growth (Bachman and Reichmanis, 1973; Krueger et al., 1962). Krueger et al. showed that air ions rather than electric fields were important for plant growth. Albert Paul Krueger (1902–1982) was a well-known bacteriologist of air ions and professor at the University of California, apart from the application to electro-culture, a few studies on its beneficial effects appeared in the 1960s and 1980s (Kotaka and Kruger, 1968; Pohl and Todd, 1981; Yamaguchi, 1983). They were conducted to investigate the effects of air ions on plant growth and production with the presence of the static electric fields. In order to put Kruger's results into practice, Herbert Ackland Pohl (1916–1986), a professor at the Oklahoma State University, observed growth effects in Persian violets as well as in seedling geraniums in green houses with high concentrations of negative ions (4 pA/cm^2) and showed that exposure to these ions resulted in longer grasses, earlier flowering time and increased flowering numbers. The experiment was conducted using an air ion generator with needle electrodes (Pohl and Todd, 1981). Frank M. Yamaguchi from the Research & Development, General Agriponics, Inc., USA, conducted hydroponic cultivation experiments of tomato plants (*Lycopersion esculentum P. Miller*) under high concentration of negative ions in a green house, using a negative ion generator with corona discharge. He reported that the time from sowing to harvest was shortened and that the yield was increased. For plant growth, ozone, nitrogen oxides, and charged dust generated by corona discharge also affect growth. These effects need to be considered before the growth-promoting effect of air ions can be put into practice. The rapid need for electricity is expanding. Further, studies on the biological and health effects of power generation and transmission are needed. This need drives researcher to investigate the biological and health effects of the combination of air ions and DC electric fields.

2.4.2 Electric and Electromagnetic Stimulations

2.4.2.1 Electric Stimulation

Judging from the historical work, there have been three eras in the medical application of electricity (Rowbottom and Susskind, 1984). As we mentioned previously, the Romans were already using the electric discharge of the electric fish, torpedo, to treat headache and gout as electrotherapy. The mysterious power of magnets was also used in the treatment of diseases. These are referred to as the first era. As for the next era in electrotherapy, the medical application of electricity was developed with the discovery of the Leyden jar. It began with the discovery of animal electricity by Galvani and the invention of the Voltaic pile by Volta. Luigi Galvani and his nephew, Giovanni Aldini used electric stimulation to cure melancholy in patients. Aldini treated Luigi Lanzarini, a 27-year-old farmer suffering from melancholy madness (Fitzgerald, 2014). Lanzarini had been committed to Santo Orsola Hospital, in Bologna, on May 17, 1801. After the first application of direct current on his head, the mood of the patient improved progressively. Lanzarini was considered completely cured several weeks after the beginning of the treatment (Cambiaghi and Parent, 2018). In January and February 1802, Aldini applied Galvani stimulation on the bodies of three criminals executed by decapitation close to Bologna's of Justice (Cambiaghi and Parent, 2018). Marked muscular contractions of various types resulted from the application of electric arc on different parts of these corpses were observed, and Aldini noted that such effects were still elicitable up to 3 hours after death (Cambiaghi and Parent, 2018). From Aldini's study, the use of electric stimulation in the form of electroconvulsive therapy (ECT) was derived.

The most recent era started after the discovery of Faraday's induction law. In 1831, Faraday discovered the induction coil. The induction coil became the tool for the stimulation of excitable tissue. It delivered shocks in humans. After Aldini's experiments, in 1835, Guillaume Benjamin Armand Duchenne (1806–1875), a French scientist, suddenly decided to try the effects of electropuncture while attending patients suffering from neuralgia. Then, he observed, using electricity, not just for the treatment but also for diagnosing the causes of muscle weakness that affected many of his patients on the wards of the Salpêtrière Hospital in Paris (McComas, 2011).

In 1870, Gustav Theodor Fritsch (1838–1927), a German physiologist, and Julius Eduard Hitzig (1838–1907), a German neurophysiologist, applied constant galvanic electric stimulation to locate the primary cerebral cortex of dog and reported five punctate "centers" that could be distinguished in the anterior cortex of the dog, stimulation which led to contractions of the contralateral muscles in the neck, legs, and face (Millett, 1998). Investigators in Britain, France, and America realized significant cortical excitability and they sought the confirmation of the observations of Fritsch and Hitzig. Three years later, Sir David Ferrier (1843–1928), a British neurologist, professor at the King's College Hospital, conducted a series of experiments, using Faradic electric (or induction) current, to stimulate movements from various cerebral "centers" in cats, fogs, rabbits, guinea pigs, and birds. Then, he moved on to conduct at least 13 experiments on the macaque monkey brain with Faradic electric current.

In 1874, Roberts Bartholow (1831–1904), an American physician, was the first to report the effects of the Faradic electric current stimulation to the cortex in conscious humans. He elucidated functional and anatomical localization of human motor areas (Cambiagihi and Sandrone, 2014). In nonhuman primates, Stamm and Rosen tested the effect of electric stimulation on delayed response tasks. They provided evidence that prefrontal and inferotemporal cortices are crucial for short-term memory (Stamm and Rosen, 1969). The abovementioned original studies gave further research on electric stimulation.

Svante August Arrhenius (1859–1927), a Swedish chemist, professor at Stockholm, performed stimulating effect experiments with electricity on the growth in children. This appeared as an interesting article in *The New York Tribune*, April 1912:

Paris, April 5- The Swedish doctor, Svante Arrhenius, has concluded some interesting experiments in Stockholm to test the effect of electricity on the growth of human organism. According to the "Matin" two groups of fifty children, roughly corresponding in age, health, weight, height and intelligence, were chosen from among the pupils of the Swedish communal schools. One of these groups of fifty children was set to work in a room which was filled with an electric installation which passed high currents into the atmosphere from the wires in the walls, floor and ceiling. The other group was set to work in an ordinary classroom. None of the children or the teachers knew that the experiment was being made. At the end of six months the children who had lived in the electrified atmosphere were found to have grown on an average three-quarters of an inch most marked manner, and they completely outclassed the non-electrified children in a competitive examination. The electrified teachers declared that their own powers of resistance to fatigue had been increased by the treatment.

New York Tribune (1912)

Arrhenius was one of the founders of physical chemistry, and won the Nobel Prize in Chemistry in 1903, in recognition of the extraordinary services he has rendered to the advancement of chemistry by his electrolytic theory of dissociation.

In the middle of the twentieth century, Iwao Yasuda (1909–1983), professor at the Kyoto Prefectural University of Medicine, and his co-worker applied pressure to bones, and measured the piezoelectricity of bones (Yasuda, 1954; Fukada and Yasuda, 1957), and identified the relationship between mechanical loading (pressure) and the piezoelectricity in bones. Piezoelectricity is the production of electricity and/or electrical polarization of bone in response to applied pressure. Their result was soon confirmed by C. Andrew L. Bassett (1924–1994), professor at the University of Columbia, and Robert Otto Becker (1923–2008), professor at Upstate Medical Center in State University of New York, Syracuse, and Director of Orthopedic Surgery at the Veterans Administration Hospital (Bassett and Becker, 1962). Clinical attempts have been made to apply direct and/or low-frequency electromagnetic fields to promote the healing of fractured bones. Since then, many therapeutic studies have been published (Bassett et al., 1974, 1981). Yasuda is acknowledged as the pioneer in the field of research on electrical currents in bones.

However, until the middle of the 1980s, most clinical studies did not use double-blind, randomized, placebo-controlled studies. Barker et al. first published a series of double-blind, randomized, placebo-controlled studies on bone healing by pulsed magnetic field (1984). The results were generally positive. Binder et al. published an article to test the usefulness of pulsed electromagnetic fields for the treatment of persistent rotator cuff tendinitis in a double-blind controlled study (1984). This therapeutic study showed that pulsed electromagnetic fields may be useful in the clinical treatment. Since then, many clinical and laboratory animal experiments with electromagnetic fields have been published, and it appears that the efficacy of electric and magnetic field therapy has been established.

Potential techniques of electric stimulation such as direct electric current, capacitive coupling, and inductive coupling have been studied. In the case of direct electrical current, two electrodes must be in direct contact with the surface of the skin surrounding the tissue. Capacitive coupling has no direct contact with the body. In the case of inductive coupling, a time-varying magnetic field induces an electrical field that produces a current in the conductive tissue. Electric stimulation techniques due to the application of electromagnetic fields from outside the tissue have been applied to recover delayed union, nonunion, acute fracture, and joint arthrodesis. Electric stimulation can also be defined as the significant and statistical enhancement of proliferation, differentiation, regeneration, and remodeling of biological cultured tissues. Modern approaches in this field began with the use of electromagnetic devices by orthopedic surgeons who attempted to heal bone fractures. This field developed progressively with commercial pulsed devices approved by the Food and Drug Administration (FDA). In 1979, FDA-approved devices with pulsed magnetic fields were designed to stimulate bone growth.

2.4.2.2 Electromagnetic Stimulation

In the past, electricity could be easily generated and controlled, which led researchers to investigate the effect of electricity on biological systems. After the demonstration of the existence of the electromagnetic waves (called Hertzian wave) made by Heinrich Rudolf Hertz, the researchers were engaged in developing methods for the production of Hertzian waves, and they suggested various applications of the newly developed method. Naturally, the researchers who had interest in the biological effects of electricity soon shifted their interest to the biological effects of electromagnetic waves. So, the existence of electromagnetic waves led to the development of the research of electromagnetic stimulation.

In 1902, Berthold Beer (1859–1922), Vienna, Austria, reported that phosphenes could be produced by applying a magnetic field to the head (Beer, 1902). He and Adrian Pollacsek (1850–1921), both psychiatrists, got the first patent application for use of magnetic coils to treat psychological disorders, depression, and neuroses, known as magnetic field therapy.

Since the late nineteenth century, the inductive brain stimulation had been known. In 1896, d'Arsonval attempted for the first time to stimulate the retina with a non-invasive applied magnetic field. When a human head was placed within a strong time-varying magnetic field generated by a large coil carrying 30 A at 42 Hz, 110 V, human perceived a flickering visual sensation. d'Arsonval called this phenomenon "magnetophosphene." The retina is known to be very sensitive to stimulation by induced current. This phenomenon is the visual sensation induced in the human brain. This was the first magnetic inductive stimulation effect to the retina of the nervous system.

Unaware of d'Arsonval's experiments, Silvanus Phillips Thompson (1851–1916), professor of Physics at the City and Guilds Technical College in England, induced magnetophosphenes (Thompson, 1910). He constructed a coil consisting of 32 turns of stranded copper wire (cross-sectional area of 0.2 square inch), which was wound on the cylinder with an internal diameter of 9 and 8 inches in length. He applied up to 180 A with 50 Hz, which produced an rms field strength of 1,000 Gauss at the center of the coil. The magnetic field strength at the mouth of the coil was about two-thirds of this value. During the experiments, the subjects closed their eyes in a darkened room, and inserting the head into the coil produced "a faint flickering illumination, colorless or of a slightly bluish tint." Even in daylight with the eyes open, the visual sensation of flickering superposed the ordinary vision.

Carl Edward Magnusson (1872–1941), professor of Electrical Engineering, and H.C. Stevens, University of Washington, carried out the phosphene studies with a large coil which was lowered over the subject's head (Magnusson and Stevens, 1911, 1914). When direct current flow in the coil was given, no phosphene was perceived. However, when direct current flow was initiated, and interrupted, phosphene was seen. Further, they carried out experiments to check the importance of frequency and field strength in producing phosphene. The phosphene appeared when the magnetic field frequency was below 25 Hz, and appeared brighter with a 20–30 Hz current at a given field strength.

More than 30 years after, Horace Basil Barlow (1921–2020), a British visual neuroscientist, University of Cambridge, and his co-workers described magnetophosphenes. Barlow was the great-grandson of Charles Darwin. Using alternating magnetic fields of variable frequency, they compared the properties of the sensations thereby produced, with those produced by passing sinusoidal electric currents through the head (Barlow et al., 1947). The magnet had a coil of 397 turns of 16-gauge copper wire. The coil had 10.5 cm of inner diameter and 20.7 cm of outer diameter and was 7.3 cm long. A laminated core of 5.3×2.9 cm in cross section and 37 cm long was placed inside the winding. The subject was seated with his temple close to the core of the magnet. In this situation, they could obtain up to 900 Gauss with 20 A (60 Hz). Phosphenes were perceived at 20 A of a 60 Hz sinusoidal current. To illustrate the phosphenes with electric stimulation of the retina by passing electric current through the head, Barlow placed an active electrode on the side of the forehead and a reference electrode on the back of the forearm. These electrodes were copper discs 3 cm in diameter covered with cloth soaked in saturated NaCl solution.

The currents used never exceeded 1 mA. Barlow stated that the phosphene was usually colorless, but occasionally, they appeared bluish and sometimes slightly yellowish. No phosphene could be perceived when the core of the coil was placed over the occipital area, the site of the visual cortex (Geddes, 2008).

The electromagnetic stimulation began from the works of Nicola Tesla (1856–1943), a Serbian-born American inventor and electrical engineer, and Jacques Arsène d'Arsonval (1851–1940), a French physician and physicist, professor of Medicine of the Collège de France. d'Arsonval was a pupil of Claude Bernard (1813–1878), a French physiologist, professor at University of Paris. Now, Tesla and d'Arsonval are considered as the father of the electromagnetic therapy. Nikola Tesla paid attention to the application of high-frequency electric current and invented the use of it for therapeutic applications. He observed that high-frequency electric current would rise the temperature of living tissue. Tesla pointed out in his article entitled *Message with currents of high frequency* published in *The Electrical Engineers* in December 1891 as following:

> The human body is, in such a case, a fine conductor, and if a person insulated in a room, or no matter where, is brought into contact with such a source of rapidly alternating high potential, the skin is heated by bombardment. It is a more question of the dimensions and character of the apparatus to produce any degree of heating desired....Without vouching for all the results, which must of course be determined by experience and observation, I can at least warrant the fact that heating would occur by the use of this method of subjecting the human body to bombardment of alternating currents of high potential and high frequency such as I have long worked with.
>
> *Tesla (1891)*

Tesla noted in this article that a person contacted with a source of high-frequency electric current would experience heating, without saying whether it would be beneficial or not.

Next, d'Arsonval carried out human experiments. In 1891, d'Arsonval suggested the use of high-frequency electric current in medicine. He passed high-frequency electric current (3 A) through himself and found that the electric current above 10 kHz produced a sensation of warmth without any sensation. As a result of these experiments, he introduced the "auto-conduction" in which a human stands under the influence of a high-frequency electromagnetic field.

The d'Arsonval's solenoid for auto-conduction is shown in Figure 15 in the book (Rowbottom and Susskind, 1984, p. 133). The figure shows a patient isolating from all contact with the current-carrying wire. During 1894 and 1895, the d'Arsonval research group made a clinical assessment of 75 patients suffering from various ailments. Each of them was placed in the solenoid for 15–20 min/day (Rowbottom and Susskind, 1984). Through these clinical treatments, d'Arsonval proposed the diathermy.

The basic concept of hyperthermia is based on the temperature rise. Hyperthermia generally can be defined as a therapy to rise the temperature in tumor tissues between 41°C and 45°C. The modern era of hyperthermia began in 1866. The first paper on hyperthermia cancer therapy is reported in 1866 by Carl D.W. Busch (1826–1881), a German surgeon (Busch, 1866). His description was the case of a 43-year-old woman patient and a spontaneous regression of advanced multiple sarcoma on her face following a streptococcal skin infection with fever. This was the first report case stating that high temperature can selectively kill cancerous cells without affecting the health. This report led to a more increased interest in hyperthermia (Gas, 2011).

The therapeutic application of high-frequency and short waves led to diathermy. In 1906, Karl Franz Nagelschmidt (1875–1952), a German physician, demonstrated the possibility of deep heating the arms and chest of the human body using high-frequency energy for therapeutic applications and introduced the concept of diathermy. He stated that heating was also produced in deeper-lying tissues by the application of high-frequency electric current (Westermark, 1927). Walther Hermann Nernst (1864–1941), a German physician and chemist, showed that the heat produced in the tissues by high-frequency electric current was purely an effect of the resistance and that this heat production follows the law of electric heating (Westermark, 1927). Nernst received in 1920 the Nobel Prize in Chemistry in recognition for his work on thermochemistry. Then, Nagelschmidt produced his diathermy apparatus with a 50 Hz, 120 V supply. The voltage was transformed to 2,000 V. As long wave diathermy, spark-discharges from Tesla and Oudin coil machines were used. Its frequencies were limited to 0.5–3.0 MHz. Short wave diathermy with frequencies up to 100 MHz was introduced in 1920, and with frequencies of 100–3,000 MHz in 1930. Paul Marie Oudin (1851–1923), a French physical therapist, applied current directly to the body surfaces with electrodes.

In 1906, de Forest (1873–1961), an American inventor, invented the triode tube. From 1920, the triode tube was used as a component of the diathermy apparatus. The invention of the triode tube allowed frequencies to be increased to 10–300 MHz. As a result, therapy called short wave therapy. This led to electromagnetic energy with inductive coils to be applied directly to the body. From the 1940s, microwave was being used for therapy. In 1921, George L. Rohdenburg, Columbia University, and Frederick Prime were the first to analyze the combined effects of heat and radiation in mouse Crocker sarcoma and some spontaneous breast tumors using temperatures between 42°C and 46°C. They revealed a definite synergistic effect above 42°C for combined treatment (Rohdenburg and Prime, 1921: Seegenschmiedt and Vernon, 1995). In 1924, Antonin Gosset (1872–1944), a French surgeon, tried for the first time the effects of short wave (with wavelength of 2 m) on tumors, on living plant cells and observed that the plant tumors were destroyed without harmful effects. In 1921, Albert Wallace Hull (1880–1966), an American physicist and electrical engineer, invented the Magnetron oscillator, which produces ultrashort waves. It produces local heating in tissue and was soon used as a piece of medical apparatuses in ultra-short wave therapy. After the 1920s, with the introduction and expansion of the medical use of diathermy, the debate over thermal and non-thermal effects began.

In the 1920s, Erwin Friedrich Karl Victor George Henrich Schliephake (1894–1995), a German physician, professor at the University of Jena, became the first patient to be treated with a short wave of 3 m in wavelength. In 1925, he recovered by himself from the suffering of a painful nasal furuncle after turning the field on. This short wave field killed small animals such as flies, rats, and mice. In the 1930s, Schliephake investigated the biological effects of short wave treatment on various tissues and introduced it in clinical applications. In 1932, Schliephake published the monograph *Kurzwellen-Therapie: Die medizinische Anwendung kurzer elektrischen Wellen* which was the first commercially available work of non-thermal technology. This German book was translated into English. The short wave technology spread rapidly over the world. Diathermy, a treatment using electromagnetic waves, became a part of hyperthermia as cancer therapy after World War II.

Nils Westermark (1892–1980), worked at Radiumhemmet in Solna, Sweden, showed that two malignant rat tumors (Flexner-Jobling carcinoma and Jensen sarcoma) regressed completely after high-frequency electric current heating (Westermark, 1927). Both tumors were caused to disappear if exposed to a temperature of 48°C. Total tumor regression occurred to an equal extent after 180 minutes at 44°C and after 90 minutes at 45°C. On the other hand, skin and normal tissues were unaffected under the thermal conditions which were lethal to the malignant tumor tissue. However, the author pointed out that the differential heat sensitivity between tumor and normal cells diminished at a thermal level above 42°C (Seegenschmiedt and Vernon, 1995). The relationship between the times required at different temperatures for these tumors to cure was well in accord with the well-known Arrhenius relationship. He concluded that these tumors can be healed by heat treatment without destruction being caused to surrounding tissues and that the tumors may be more sensitive to heat than healthy tissues. This result was more pronounced at low temperatures (44°C–45°C) than at high ones (47°C–48°C).

In 1928, Joseph Williams Schereschewsky (1873–1940), an American physician, Harvard Medical School, observed the effects of short waves on tissue cells and on sarcoma in mice (Schereschewsky, 1928, Schereschewsky and Andervont, 1928). He and his co-workers investigated the effects of electrical current of high frequency (8.3–135 MHz) in transplanted mouse sarcoma tumors. As a result, tumor growth delay or inhibition and then complete disappearance of some tumors were observed. Its greatest effects were at a frequency range of 66–68 MHz. From these observations, the therapeutic application of short wave diathermy began. Historically, in the 1920s and 1930s, therapeutic application of short wave became more popular in the USA and Germany. It returned to the times when d'Arsonval used the induction method by high-frequency electric current. The medical and military uses of radiowaves and microwaves began rapidly in the 1920s after these technical developments.

From 1930 to 1932, Takashi Minoshima (1895–1990), professor at Hokkaido Imperial University, Sapporo, studied photochemistry at the University of Leipzig, Germany. During this time, Minoshima visited the laboratory of professor Sieminsky, an electrophysiologist, at the University of Vienna, Austria. Minoshima saw experiments in which the frog's belly and back were placed on electrodes, separated by air, and the samples were irradiated with short waves, which caused the frog's spinal reflexes to disappear. When the radiation stopped, the reflexes returned to normal. Minoshima was strongly impressed with the action of short wave stimulations. After returning to Japan, he formed a research group. He and his co-workers conducted basic research on cancer treatment and disinfection, and on the effects of ultrashort wave and on the physiological effects in animals. In 1940, he organized the Ultrashort-Wave Laboratory, and in 1943 the Research Institute of Ultrashort-Wave was established (HUSCAP, 1980). There, electrophysiological studies were conducted, including studies on blood and biological tissue conductivity, cervical sinus nerve activity, and oil-water nerve models. This institute has been reorganized two times, into the now Research Institute for Electronic Science passing through the Research Institute of Applied Electricity (RIAE). Since the 1970s, research on the biological effects of electromagnetic fields has been carried out under the leadership of professors Goro Matsumoto (1923–2009) of the RIAE and Masamichi Kato of School of Medicine, Hokkaido University.

2.5 Progress of Bioelectromagnetism in the Twentieth Century

It has passed over 200 years since the discovery of animal electricity by Galvani and the discussion between Galvani and Volta. During these times, many researchers had engaged in the research of bioelectromagnetism. In addition, the invention of instruments brought about great innovation in bioelectromagnetism. After entering the twentieth century, there have been many instrumental developments. It advanced from measuring myocardial action potential with a string electrometer by Einthoven in 1903, the invention of the vacuum tube triode amplifier by Lee de Forest in 1907, the cathode-ray tube (later called Braun tube) by Karl Ferdinand Braun (1850–1918), a German inventor, professor at University of Straßburg, in 1897, and recordings of nerve action potential by Erlanger (1874–1965), an American physiologist, professor at Washington University, in 1929. The use of bioelectric and biomagnetic phenomena has been the majority in bioelectromagnetism. Their successful use is parallel to the invention and development of the instruments.

2.5.1 Bioelectricity

Bioelectricity is a fundamental process of all living systems and is the study of electrical phenomena generated in living systems. The electrical phenomena include inherent properties of the cells, such as membrane potential, action potential, and propagation of the potential.

2.5.1.1 Bioelectric Source

Historically, since the great Italian scientists such as Galvani, Volta, and Matteucci carried out well-known experiments on nerve and muscle excitation, there had also been many physiologists. In particular, scientists of the school of the University of Berlin, such as Müller and his pupils du Bois Reymond, Helmholtz, etc. had produced great contributions. Two of the greatest pupils followed Müller's study, du Bois-Reymond detected the action current of the nerve, and his colleague, Helmholtz determined the velocity of the action current. Before the end of the nineteenth century, the electrical nature of the action potential was established from the studying of the German school.

About 45 years after Matteucci's experiment, Augustus Desirè Waller (1856–1922), British physiologist in 1887, showed the possibility to record the waveforms of the human heart potential from the body surface (Rowbottom and Susskind, 1984). The electrical activity from the human heart was recorded as an electrocardiogram (ECG) with a highly distorted, using Lippmann's capillary electrometer and placing electrodes on the body surface, chest, and back. This was the first recorded ECG in a clinical and physiological setting. Inspired by Waller's work, Willem Einthoven (1860–1927), a Dutch physiologist, University of Leyden, refined the recording technique. Waller used five electrodes, one on each of the four extremities and the mouth, to record ECG. Einthoven reduced the number of electrodes to three from five, which is today's way to construct the Einthoven's triangle. Using a quick quartz string galvanometer, he presented the first direct record of high-quality clinical ECG in 1908. This great work brought him to be a pioneer in electrocardiography. As shown in Figure 2.10, the first electrodes were cylinders filled with electrolytes, which means that subject's extremities need to bathe in saline solution filled the tubs (Barron, 1950). The arc lamp, electrode jars, time motor, and falling plate camera are shown in Figure 2.10. Further, Einthoven's triangle as the basis for recording of ECG expanded rapidly in the world. Einthoven talked about the production and sale of ECG recorder of quartz string galvanometer with Horace Darwin of the Cambridge Scientific Instrument Co. Ltd. of London. Horace Darwin (1851–1928) was the founder of this company and the youngest child of Charles Darwin (1809–1882), a British naturalist, geologist, and biologist. After talking, Horace Darwin sold the first commercial ECG machine. Einthoven received the Noble Prize in Physiology or Medicine in 1924 for his discovery of the mechanism of the electrocardiogram.

PHOTOGRAPH OF A COMPLETE ELECTROCARDIOGRAPH, SHOWING THE MANNER IN WHICH THE ELECTRODES ARE ATTACHED TO THE PATIENT, IN THIS CASE THE HANDS AND ONE FOOT BEING IMMERSED IN JARS OF SALT SOLUTION

FIGURE 2.10 The world first model of ECG machine, 1911–1912. This first model was installed in a basement at University of Lewis (From Almay photo.)

After the demonstration of Du Bois Reymond on the electrical nature of the action potential, Richard Caton (1842–1926), British physiologist at the Royal Infirmary School of Medicine in Liverpool, investigated the spontaneous electrical activity of animal brain in rabbits, dogs, and monkeys in 1875. He placed electrodes on the surface of the animal's brain and measured the potential changes. In particular, he found that shining a light toward the animal's eyes caused a change in the electrical potential of a certain part of the animal brains. This is believed to the earliest record of an electroencephalogram (EEG). In 1890, Adolf Beck (1863–1942), a Polish physiologist at Cracow, described the discovery of spontaneous fluctuations in the electrical activity of animal brains in rabbits and dogs. He was nominated three times for the Nobel Prize in physiology or Medicine, but he never received it. Beck's finding was done independently of Caton's observation (Beck, 1890).

In 1924, Hans Berger (1873–1941), professor, psychiatrist at the University of Jena, Germany, using a double-coil galvanometer made by Siemens, succeeded in finding the decrease a human brain's electrical activity following sensory stimulation. This was the first record of the electrical activity of the human brain, called electroencephalogram (EEG) from the surface of the skull. In 1931, he identified major α (8–12 Hz) and β (13–60 Hz) rhythms. He applied this new technique to the study of epilepsy and to the effects of activities such as mental effort. In 1941, unable to cope with the growing realization that he was mentally ill, he hanged himself (Rowbottom and Susskind, 1984). After visiting Berger's laboratory from England, William Grey Walter (1910–1977), an American-born British neurophysiologist, improved the original Berger's EEG machine with an automatic frequency analyzer. As mentioned above, physiological understanding of the heart and brain including heart rhythm (ECG) and brain rhythm (EEG) appeared in the first half of the twentieth century. This understanding with the advance of electronic technology brought the introduction of transistorized pacemakers and defibrillators and use in clinic in the second half of the twentieth century.

2.5.1.2 Electric Properties of Biological Systems

In order to understand the interaction between electromagnetic fields and biological systems in bioelectromagnetism, the knowledge of the electric (dielectric) properties of biological systems is the key for providing experimental and theoretical insights into the bioelectromagnetic phenomena. Elucidation of the electrical properties of living organisms is important for measuring the ECG, EEG, etc. generated by living organisms, for the response of living organisms to external stimuli, and for clarifying the safety of medical devices. As mentioned in previous sections, the use of electricity and magnetism has a long history in electromagnetism. From the nineteenth century, the electric properties of tissues and cell suspensions were investigated. The theory to explain bioelectric phenomena was gradually developed.

In the 1870s, Ludimar Hermann (1838–1914), a Berlin-born physiologist and an assistant to Emil du Bois Reymond, was concerned with the electric and chemical actions which take place during muscle and nerve activity (Rowbottom and Susskind, 1984) and contributed greatly to the model from which the electric properties of the nerve fiber are described in terms of resistances and capacitors. This means that the membrane of the nerve fiber could be represented by a number of resistors and capacitors in parallel (McComas, 2011). This model becomes very useful for neurophysiologists. Herman determined the resistance anisotropy-ratio of muscle and nervous tissues with DC (Hermann, 1872). Further, based on the cable theory, Hodgkin and Rushton developed the mathematical theory of nerve fiber conduction. Hermann coined the term of format in acoustic phonetics.

Early electrophysiological research studies included the study of the electrical conductance of blood and tissue. The study of excitability and contractility of membranes was also included. Contractility was recognized to be caused by the nonlinear properties of the membranes above certain threshold values. Stewart, Philippson, and Höeber carried out experiments on red blood cells. George Neil Stewart (1860–1930) was Scottish-Canadian physiologist, professor at Western Reserve University. He studied 1 year as a post-graduate with Emil du Bois-Reymond in Berlin in 1886–1887. He measured the resistance of blood at low frequency range (Stewart, 1899). After the introduction of the Wheatstone bridge and Kohlrausch bridge, the determination of the electrical properties of blood was introduced first by Höber. In 1912,

Rudolf Höber (1873–1953), a German physician, investigator, examined the frequency dependence of the conductivity of blood at low (100–200 Hz) and high (~MHz) frequencies. He presented that the interiors of red blood cells and muscle cells contain conducting electrolytes, and that each conducting core is contained within an insulating membrane (Pethig and Schmueser, 2012). He showed that the impedance decreased with increasing frequency and estimated the resistivity of the interior of the erythrocyte (Höber, 1910, 1912, 1913). He also developed the concept of β-dispersion in suspended red blood cells which later generalized to muscle tissue. Höber with his Jewish wife had left Germany before World War II and got a position in the physiology department at the University of Pennsylvania.

After Höber's investigation, Maurice Philippson (1877–1938), professor of zoology and physiology at Brussels University, measured animal tissue impedance as a function of frequency (500 Hz–3 MHz) and found that the capacitance of animal tissues varied roughly as the inverse square root of the frequency (Philippson, 1920, 1921). The capacity of vegetable tissues varied as the inverse fourth root of the frequency. The magnitude of the impedances decreased with frequency. This polarization capacitance was similar to that found for the metal/electrolyte interphase (Grimnes and Martinse, 2000). He reported the β-dispersion by red blood cells (Pethig and Schmueser, 2012). Historically, equivalent circuits have been proposed by many scientists. Maurice Philippson presented an equivalent circuit of red blood cells consisting of the protoplast resistance in series with a parallel combination of the membrane resistance and capacitance.

The next important step was made by Hugo Fricke (1892–1972), Cleveland Clinic Foundation, Cleveland, and Sterne Morse, who proposed an equivalent circuit of red blood cell suspensions. Assuming that the cell membrane is electrically as a thin dielectric layer, and analyzing the passive electrical properties of canine red blood cells, they hypothesized that the reasonable thickness of the membrane is around 3.3 nm and that the electrical capacity of the membrane is of 1 μF per square centimeter. Their equivalent circuit consisted of the resistance around the cell, resistance of the cytoplasm, and the capacitances of the membranes (Fricke, 1925; Fricke and Morse, 1924, 1925a, b). In this circuit, the membranes have a high reactance at low frequencies and a low reactance at high frequencies. Kenneth Stewart Cole and Richard F. Baker, Columbia University, developed an equivalent circuit with the resistance of the cytoplasm, the inductance of the membrane and in series with a parallel combination of capacitance of the membrane after discovering an inductive reactance within the membrane structure (Cole and Baker, 1941: Pethig and Kell, 1987; Pethig and Schmueser, 2012). Tissue can be modeled as an electric circuit with resistive and capacitive properties.

Kenneth Stewart Cole, American biophysicist, Columbia University, was trained as biophysicist and spent time with Peter Debye (1884–1966) in Leipzig. In 1941, Cole and Robert H. Cole (1914–1990), K. S. Cole brothers, chemist, the professor at Brown University, published a paper in which they introduced the famous Cole-Cole equation (Cole, 1928a, b; Cole and Cole, 1941). They opened the way to treat analytically and mathematically the tissue conductivity and permittivity. K. S. Cole first derived the electric impedance of a suspension of spheres each having a homogeneous non-reactive interior and a thin surface layer with both resistance and reactance (Cole, 1928a). His discussion was that the equivalent circuit has the form of a capacitor in parallel with resistors. Impedance is the ratio between voltage and current. In a companion paper, he presented the results for the measurements of the electrical impedance of suspensions of small *arbacia* eggs in sea water (Cole, 1928b). The results were in accordance with the theory presented in the previous paper. Their introduced Cole-Cole plot is a useful method which shows the behavior of tissue impedance as a function of frequency. This is a complex plane locus of real components versus imaginary components with the frequency parameter (Cole and Cole, 1941). Now, it is a simple presentation. It has been successfully applied to a wide variety of tissues. The Cole-Cole plot is an empirical model of measured data. However, it does not give any information about the underlying electrical phenomena being measured.

During the 1920s and 1940s, great interest in the electric properties of cells and tissues increased rapidly among researchers. Cole, Curtis, and others applied the potential theory based on Maxwell's work to cell suspensions (Cole and Curtis, 1939). They also collected data of the electric properties of cells and tissues from kHz to about 100 kHz. After World War II, the data was extended from below the kHz range to the microwave frequency range. From these investigations, the concept of dispersion was developed.

The dielectric (electric) properties of tissues are usually described in terms of electrical conductivity and the relative dielectric constant called permittivity. These two parameters depend strongly on the frequency and type of tissues. The dielectric constant is represented as ε (F/m), and since the dielectric constant of free space is ε_0 (= 8.854×10^{-12} F/m), the relative permittivity is defined as $\varepsilon_\gamma = \varepsilon/\varepsilon_0$. The electrical conductivity is represented as σ (S/m). There are various electrical properties related to biological cells and tissues, which can help in understanding the characteristics of them. The relative permittivity ε_γ and the electrical conductivity σ of tissues change with frequency. An overview of the dielectric properties of biological tissues has been presented in terms of their relaxation mechanisms. The dielectric properties of biological tissues have been categorized according to relaxation regions that are related to the sizes of the composing cells and ions. Through the measurements of the conductive and capacitive properties of biological tissues, Schwan introduced first the concept of dispersion. There are relaxation characteristic regions allowing the frequency of dispersion.

Figure 2.11 shows data on the dielectric properties of muscular tissue as a function of frequency which indicates three unusual features. For relative dielectric constants of biological tissues, there are three break points; and they occur: (1) below a few kHz, (2) in the frequency range from tens of kHz to tens of MHz, and (3) in the microwave range, about 20 GHz. At each of these frequencies, a dispersion phenomenon of a rapid decline of relative dielectric constant and rapid increase in electrical conductivity occur. These are referred to, respectively, as α-, β-, and γ-dispersions. At frequencies below a few kHz, an ionic diffusion process in cellular membranes occur which allows the dielectric relaxation in the α-dispersion. Bioelectric studies of biological tissues below a few kHz were difficult because electrode polarization at these frequency regions is significant. For this reason, the mechanism of the α-dispersion of biological tissues was not well understood. It is believed to be associated with a counterion layer (electrical double layer) polarization in the tissues (Foster and Schwan, 1989). The high dielectric constant observed at low frequencies is a result of the complex and non-uniform structure of biological organisms. The origin of the α-dispersion is a relaxation phenomenon of cell membranes; it is related to the permeability of membranes and to the diffusion process of ions in these complex structures. In the frequency range from tens of kHz to tens of MHz, the β-dispersion region becomes more evident in response to the relaxation from the polarization of cellular membranes and organic macromolecules. Although the origin of the β-dispersion is not well understood, it is due to the inability of the polarization of cellular structural components, including cell membrane, which act as barriers of ion flow. The β-dispersion also comes from the polarization of organic polymers and proteins. At the GHz regions, about 20 GHz, the γ-dispersion is caused by the polarization of water molecules, both free and bound ones, which are common in biological systems.

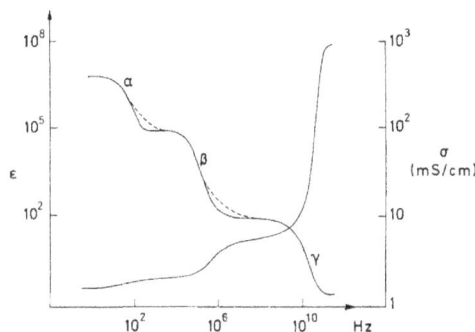

FIGURE 2.11 The dielectric constant ε and the conductivity σ of muscle are shown as a function of frequency of electromagnetic radiation (From Schwan, 1988.)

Herman Paul Schwan (1915–2005), born in Aachen, Germany, later German-born-US biophysicist, professor at the University of Pennsylvania. Schwan studied mathematics and physics at Göttingen and Frankfurt. Before World War II, in 1937, Schwan started his carrier at the time when the electrical properties of tissues and cells had attracted the scientific attention at the Kaiser Wilhelm Institute for Biophysics (Max Planck Institute of Biophysics after 1948) where his main research areas were the biological effects of electromagnetic radiation (both non-ionizing and ionizing) including hazards and safety standards for microwaves, blood and blood serum conductivity in low-frequency, heating and body tissue properties in ultra-high-frequency, tissue relaxation, and electrode polarization (Grimmes and Martinsen, 2000). He was employed as Boris Rajewsky's technician. Boris Rajiewsky (1893–1974) was a Russian biophysicist at the University of Frankfurt am Main. In early 1946, Schwan was awarded his Dr. Habilitation by his work entitled "The determination of the dielectric properties of semiconductors, especially biological substances in the decimeter wave range." In 1947, he came to the United States, and joined the University of Pennsylvania in Philadelphia in 1952. Schwan is the founder of biophysical studies related to the dielectrical properties of cells and tissues which gave the understanding of the effects of the electromagnetic fields on biological systems and the health effects of non-ionizing electromagnetic fields. In 1985, he was recognized as the first recipient of the d'Arsonval Medal of the Bioelectromagnetics Society for his lifelong contributions to the understanding of microwave radiation. One of Schwan's earlier theories predicted that electromagnetic radiation of 0.9 GHz or below would be better for therapeutic diathermy than 2.45 GHz. The understanding of the interactions between electromagnetic fields and biological systems must be based on the knowledge of electric properties of the tissues. This understanding has led to many applications in biomedical engineering, agriculture, etc.

2.5.1.3 Electric Field-Force Effect

The electric field is very important for biological studies. Several electric field-force effects can be induced in cells. These effects are illustrated in Figure 2.12 (Schwan, 1988). The term "force effect" of an electric field means how the cells and bioparticles respond to the applied electric fields. Electric-force effects are the basis of bioelectric phenomena such as electroporation, electrofusion, electrorotation, pearl chain formation, and traveling wave dielectrophoresis.

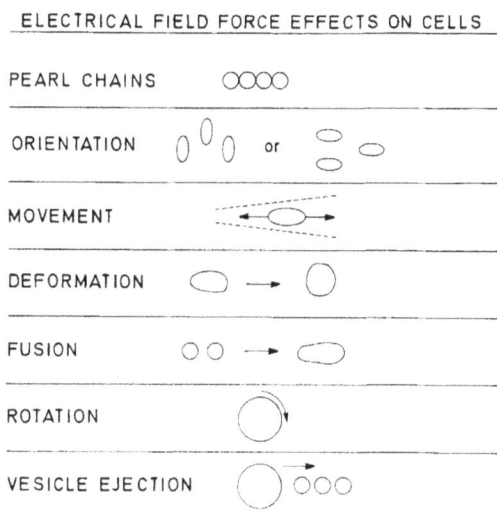

FIGURE 2.12 Electric field-force effects can be induced in cells and biologically simulating particles by time-varying electric fields (From Schwan, 1988.)

The earliest observations of phenomena similar to irreversible electroporation could be back to the 1700s. Irreversible electroporation may have been observed in 1754 when Abbè Nollet studied the discharge of static electricity. Using this discharge, he applied an electric spark to human and animal skin. He reported the formation of red spot on the spark-applied areas. These can be explained as damage to the capillaries by irreversible electroporation (Vanbever and Prèat 1999). In the 1800s, high voltage discharge had been used in order to purify river water (Deipolyi et al., 2014). In 1898, Alphonso David Rockwell (1840–1933), an American electrotherapist, found that *Under the discharges of the Leyden jar the red corpuscles change their shape and lose their color.* This may be homolysis induced by irreversible electroporation (Rockwell, 1903). In 1954, Stämpfli published the first report about the breakdown of the plasma membrane by the direct application of a high intensity electric field (Stämpfli, 1954). Stämpfli reported of electroporation-related phenomena, the irreversible and reversible of breakdown of the excitable membrane of Ranvier node of frog (Stämpfli, 1957; Stämpfli and Willi, 1957). He reported that membrane breakdown is irreversible under certain conditions, whereas in other cases, it is reversible, and he compared this breakdown phenomenon to that of a dielectric/capacitor. In 1956, Bernhard Frankenhaeuser (1915–1994) and Lennart Widén, both at the Karolinska Institute, Stockholm, attempted to explain the change in normal nerve conductivity behavior. This change was anode break excitation when electric pulses are applied on nerve nodes. The authors stated that

> in experiments with bipolar stimulation of the desheathed nerve, with both stimulating electrodes well away from the cut ends of the fibres, it was found that stimuli from threshold strength to about ten times threshold strength with durations from less than 1 ms to more than 100 ms elicited a discharge in the nerve at make but not at break.

Frankenhaeuser and Widén (1956)

J. H. Sale and W. Allan Hamilton, from the Unilever Research Laboratory and University of Aberdeen, Scotland, published three seminal papers which became the basis of irreversible electroporation (Hamilton and Sale, 1967; Sale and Hamilton, 1967a, b). They reported in the first paper the non-thermal effects of high DC pulsed electric field on killing the bacteria (inactivation of microorganisms). They exposed vegetative bacterial and yeast preparations (*Escherichia coli, Staphylococcus aureus*, etc.) to DC pulsed electric fields of 25–30 kV/cm. The evaluation was that the non-thermal bactericidal effect was to minimize the temperature rise. In this experiment, very short pulses between 2 and 20 µs at several second intervals were used. In the second paper (1967a), they tried to elucidate what is the mechanism by which the high DC pulsed electric field kills the cells. Their conclusion was that the irreversible loss of the membrane's function as a semipermeable barrier was the cause of cell death. In the last and third paper (1967b), they suggested that the transmembrane potential induced by the external field may cause "conformational changes in the membrane structure resulting in the observed loss of its semipermeable properties." They concluded that the cell membrane is damaged when transmembrane voltages of around 1 V are reached.

In the 1970s and 1980s, key research studies brought the reversible electroporation into the field of biotechnology and medicine, and they focused on the fundamental understanding of the mechanisms involved. Eberhard Neumann, Max Planck Institute for Biophysical Chemistry, Göttingen, Germany, and Kurt Rosenheck, professor at The Weizmann Institute of Science, Rehovot, Israel, investigated the reversible electroporation of cell membranes in 1972 (Neumann and Rosenheck, 1972). They used pulsed electric fields of about 18–24 kV/cm with about 150 µs duration and showed the production of the reversible permeabilization of the cell membrane of bovine adrenal medullary chromaffin granule cells as a vesicular model system. The main soluble constituents of these vesicles are catecholamines (CA; epinephrine (E), norepinephrine (NE)), ATP, and proteins. Release of part of the CA and ATP content from the isolated granules of the bovine medulla occurred readily at higher temperatures. Temperature increase accompanying the electric impulse was smaller than 6°C. The impulse-induced release of CA and ATP was non-thermal (Neumann and Rosenheck, 1972). This paper triggered many research groups to begin the study of electroporation.

In 1974, Ulrich Zimmermann, professor at the University of Würzburg, Germany, concluded that membrane permeabilization occurs when the transmembrane potential reaches a threshold of 1V (Zimmermann et al., 1974). The emphasis was that the result is relevant to irreversible electroporation parameters. The effects are not thermal. Zimmermann used reversible electroporation to produce fusion between cells after exposure to high electric field pulses (Zimmermann, 1982). In 1982, Neumann and his co-workers coined the term electroporation to describe the membrane breakdown induced by electric fields. They introduced first the use of reversible electroporation to insert genes into cells by pulsed electric fields (Neumann et al., 1982).

Mitsugi Senda (1929–2016), professor at Kyoto University, Japan, and his co-workers reported the first study of the electrofusion with two kinds of plant protoplasts (Senda et al., 1979). They brought mechanically two protoplasts isolated from cultured cells of *Rauwolfia serpentina* through two glass capillary microelectrodes in contact with the adhering protoplasts. To them, they applied an electric impulse with 5 and 12 µA for a few milliseconds generated by an electric stimulator (Nippon-Koden Co. Ltd., MSE-3). Fusion was immediately induced. This phenomenon seemed to be related to transient changes in the membrane state. This was a very important milestone in the development of the electrofusion technique. Zimmermann and Scheurich also had a pioneer approach in electrofusion (Zimmermann and Scheurich, 1981). Electrofusion is the use of electric pulses to combine cells, allowing the mixing of their intracellular contents. The electrofusion has two steps: first, the connection of two separated closed-physical contact cell membranes is achieved by dielectrophoresis of the cells in non-uniform AC electric fields. Next, electrofusion is triggered between adjacent cells by the application of high pulsed electric fields with very short duration.

There have been papers published on cell movement in nonhomogeneous alternating and in rotating electric fields. These phenomena may be used in cell fusion and cell separation (Zimmermann, 1982). Fundamentally, pulsed electric fields with duration from the order of milliseconds to micro-seconds and with amplitudes of the order of hundreds to thousands of volts per centimeter was applied across cells. In the 1990s, the electroporation had extended to include nano-second pulsed electric fields. Nano-second pulsed electric fields were designed to induce predominately intracellular effects without heating. In 1997, Karl H. Schoenbach, professor at the Old Dominion University, and his co-workers reported on in vitro study, using very high voltage pulses of submicrosecond duration (Schoenbach et al., 1997, 2001). After the publication of this paper, studies on the use of pulsed electric fields with the duration of some nano-seconds or tens of nano-seconds were published. The concept of these researches came from the ideal that ultra-short pulses could be able to induce electroporation of intracellular membranes structures without disturbing the cell membrane. Electroporation is also used for food sterilization and for transfection of yeast, bacteria and plant protoplast in biotechnology. The reversible electroporation is a method for drug delivery into tumor cells. In biotechnology, it is useful for microbial deactivation.

If a non-uniform AC electric field is applied to a cell, it experiences a force which can cause it to move. This effect is known as dielectrophoresis. The term dielectrophoresis (DEP) refers to the lateral movement of non-charged dielectric particles in a non-uniform AC electric field. Dielectrophoresis is observed when cells are suspended in low conductivity medium and electric fields are applied to the cells using microelectrodes (20–200 µm in diameter), and AC voltages of 2–20 V_{p-p} with a frequency in the range of 10 kHz–100 MHz. Herbert Ackland Pohl, professor at Oklahoma State University, is the pioneer of the study of dielectrophoresis and its application to the study of cells. In 1966, Pohl and his co-workers reported the first demonstration of dielectrophoresis (DEP) of living and death yeast cells (Pohl and Hawk, 1966). The theory of DEP of biological particles was presented by Pohl. However, it is said that his approach cannot be considered to be satisfactory due to the failure of giving reasonable results in the low-frequency region. The correct theory for the possible model of cell membrane was given by Schwan.

Dielectrophoresis is simply the movement of cells in inhomogeneous electric field. During dielectrophoresis, charged dipoles are induced in the cell when a weak AC current passes through the membrane suspension causing the cells between the electrodes to become aligned in chain-like aggregates, referred to as "pearl chain formation" (Van Wert and Saunders, 1992). The pearl chain formation was probably first observed by John Kerr (1824–1907) about 150 years ago. Kerr was a Scottish physicist, Mathematical Lecturer of the Free-Church Training College, Glasgow, and the term of "Kerr effect" is named after him. Kerr reported electric birefringence observations in various liquids (Kerr, 1875). In this paper, he presented about electric field-effects on dust particles (dielectrified body). When the particles in benzene and carbon disulfide were numerous enough, they formed a chain between the electrodes. The chain breaks up violently at the instant discharges. When the particles are few and of the same forms, they do not produce a chain, and they are presented as a set of sparkling points, which dart hither and thither through the ventral parts of the electric field (Kerr, 1875; O'Konski, 1981). In 1927, Ernst Muth, University of Halle, observed the phenomenon of pearl chain formation by exposing fat emulsions to high-frequency AC electric field (Muth, 1927). Ten years later, Paul Liebesny (1881–1962), physician, New York, was the first to demonstrate the pearl chain formation of erythrocytes in a high-frequency electric field (Liebesny, 1938).

The pearl chain formation is the phenomenon where two dielectric particles in a homogeneous electric field will be attracted to each other forming a dipole and will be oriented in the direction of the electric field. Cells will tend to form pearl chain in the direction of the electric field. In a gradient electric field, the pearl chains will protrude from the surface of electrode (Grimnes and Martinsen, 2000). William Krasny-Ergen presented a theory of the pearl chain formation (Krasny-Ergen, 1936, 1937). He proposed a theory that explains the pearl chain formation of dispersed particles in terms of potential energy. However, he made no comparison between theoretical results and experimental data. Later, Schwan and his colleagues developed a general theory to account for the pearl chain formation of spherical and nonspherical particles in AC fields (Saito and Schwan, 1961; Schwarz et al., 1965; Saito et al., 1966; Schwan and Sher, 1969). They studied for the first time the pearl chain formation experimentally and theoretically. Masao Saito (1934–2016), a visiting scientist, later professor at the University of Tokyo, pointed out that the time constants of pearl chain formations were of the order of a second for a particle with radius 1 μm, and they were proportional to the cube of the radius. At low fields, the time constants were not strongly dependent on the field intensity, but at higher fields, they are inversely proportional to the square of the field strength. Large error of the threshold field strength in experimental works may occur if the time constants for the pearl chain formations are in the order of hundred of a second, or minutes when the particles measure a few microns or more in size. Schwan pointed out that a minimal field strength is needed to cause field effects and termed this minimal field value as the threshold field strength E_{th}. Saito and Schwan presented an equation for E_{th} for pearl chain formations based on an expression derived for the potential electrical energy of a particle suspended in a medium of different dielectric properties (Saito and Schwan, 1961; Schwan, 1989). Schwan noted that it could not be applied with experimental data casting, doubt on the validity of the Saito-Schwan expression for E_{th} (Schwan, 1989). Later, Sauer pointed out that the equations of the theoretical considerations of Krasny-Ergen, Saito and Schwan gave only qualitative and semi-quantitative analyses for the pearl chain formation in special case when the medium and the particles have no dielectric losses (Sauer, 1983). Sauer calculated the forces on two particles in an electric field in the case when the medium and the particles have dielectric losses. His calculation successfully predicted the trajectories of the particles during the process of the pearl chain formation.

Electrorotation refers to the rotation of particles in electric field. A. A. Teixeira-Pinto, a postdoctoral Fellow from Portugal in the New England Institute for Medical Research, Connecticut, and his co-workers noted not only particle pearl chain formation but also particle orientation and observed first that an *Euglena* cell and an *amoeba* began to rotate when they approached each other in an electric field in the radio frequency range (Teixeira-Pinto et al., 1960). Later, A. A. Füredi and I. Ohad, The Hebrew University of Jerusalem, Israel, reported the behavior of human erythrocytes in a high-frequency electric field (Füredi and Ohad, 1964). Because longer exposure (over 10 seconds) causes damage to cells by heating for erythrocytes, it was applied for 1–5 seconds. Erythrocytes showed a reversible elongation which was accompanied by a rotatory motion. Old erythrocytes do not elongate or rotate but can form chains oriented in the direction of the field. Electrorotation is used to differentiate between viable and non-viable biofilms of bacteria (Grimnes and Martinsen, 2000). Particles suspended in a liquid will experience torque in a rotating E-field (Arnold and Zimmerman, 1982).

Sen-ichi Masuda (1926–1995), professor at the University of Tokyo, and his co-workers were the first to report on the manipulation of biological particles using rotating traveling electric fields (electric curtain) with three-phase voltage frequencies between 0.1 and 100 Hz (Masuda et al., 1987, 1988). Using relatively low frequencies in his original papers, the technique of electrophoresis was predominant. The electrical traveling waves in non-uniform AC electric fields can also be used for the manipulation of particles in 1988. An important point is that the selected particles move in a stationary supporting fluid in the separation process. His technique was improved to use traveling-wave dielectrophoresis with higher frequencies to separate viable yeast cells from non-viable yeast cells (Talary et al., 1996). Employing this traveling field dielectrophoresis, Pethig's group of the University of Wales, UK, separated red and white blood cells and selectively viable yeast cells (Burt et al., 1998).

Schoenbach and his co-workers at Old Dominion University, USA, discovered the induction of apoptosis by nano-second pulsed electric fields in cancer cells, in 1999. Since these observations, there has been increasing numbers of studies of the electrical behavior of cell membranes by nano-second pulsed electric fields. For example, the electroporation is used as a universal tool for drug deliveries and genes into living cells for electrochemotherapy and gene therapy in medical application (Schoenbach et al., 2001, 2004).

2.5.2 Biomagnetism

Biomagnetism is simply the study of the interactions between magnetic fields and biological systems. Various biomagnetic phenomena for different magnetic field strengths and their frequencies are shown in Figure 2.13. It is important to know the strengths and frequencies of the magnetic fields involved in biomagnetic phenomena. For example, the so-called magnetophosphene is a visual sensation elicited by exposing the head to a low-frequency (around 10–70 Hz) magnetic field of around 10–20 mT. The signal is generated in the retina. Magnetic stimulation of the human brain and heart has been used for the purpose of both research and clinical treatment. Using SQUID magnetometers, the very weak magnetic fields from the brain, heart, and lung can be measured from outside the body. Since the beginning of the history of biomagnetism, there are now Magnetic Resonance Imaging (MRI), functional MRI (fMRI), Transcranial Magnetic Stimulation (TMS), etc. related to medical applications.

The understanding of biomagnetism is based on the concept of magnetism. Magnetism is mainly characterized by three variants, ferromagnetism, paramagnetism, and diamagnetism. Ferromagnetism has its own magnetic poles and exerts force on each other as a magnet. The magnetic susceptibility of ferromagnetism is positive and large. Paramagnetism produces weak magnetization parallel to the direction of the magnetic field. Diamagnetism is a substance that exhibits weak magnetization in the opposite direction to the magnetic field and a force proportional to the product of the magnetic field strength and its gradient is produced in the direction of repulsion to the magnet. The magnetic susceptibility of diamagnetism is negative in the order of 10^{-6} to 10^{-5}(SI). The magnetic susceptibility of paramagnetism is positive in the order of 10^{-5} to 10^{-3}(SI).

2.5.2.1 Magnetic-Related Phenomena

In the past, the magnetic field strength of a permanent magnet was too weak to demonstrate the magnetic-related phenomena of diamagnetism and paramagnetism. Around the 1980s, the magnetic field strength of a permanent magnet increased drastically due to the development of neodymium magnets. This opened a new era for the study of the magnetic properties of every material in high magnetic fields. In the presence of a neodymium magnet, the magnetic properties of diamagnetic materials can be observed with the naked eye. With neodymium magnets, apples, eggs, carbon, bismuth, etc. can be levitated. A neodymium magnet is a rare earth magnet ($Nd_2Fe_{14}B$) consisting mainly of neodymium, iron, and boron. It was invented by Masato Sagawa and others at Sumitomo Special Metals (later, Hitachi

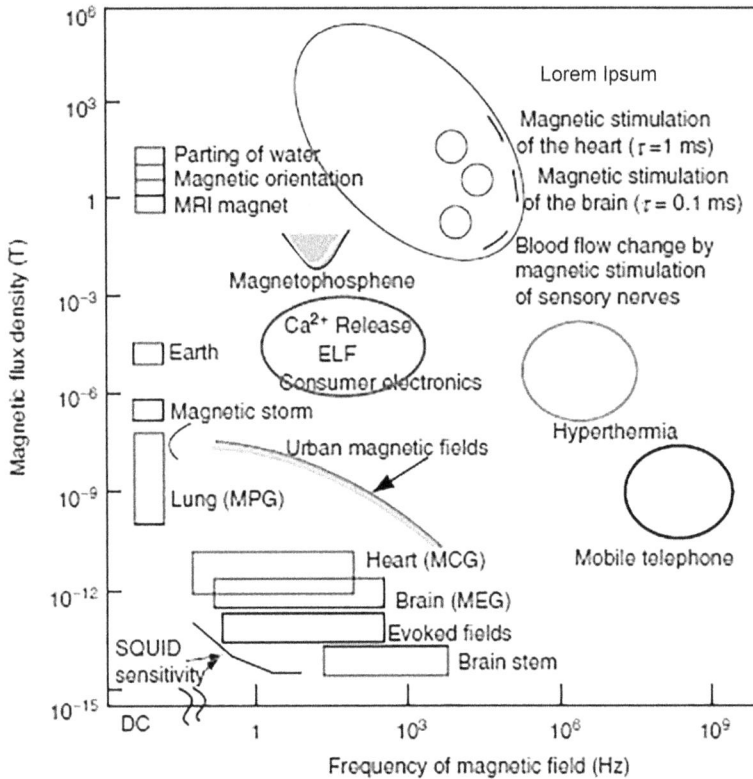

FIGURE 2.13 Various biomagnetic phenomena observed or used at different frequency (Hz) and magnetic flux density (T)

Metals), Japan. It is a powerful permanent magnet (Sagawa et al., 1984). At the same time, a new powerful permanent magnet was invented and reported by General Motor Corporation, USA (Croat et al., 1984). The typical research developments will be introduced as magnetic-related phenomena; levitation of diamagnetic materials and physical property changes of water and Moses effects.

In 1778, Sebald Justinus Brugmans (1763–1819), a Dutch botanist and physician, professor at Francker, Botany at Leyden, the Netherlands, found first a new form of magnetism. His observation was that bismuth and antimony are repelled by a single pole of a magnet. This is the foundation of the science of diamagnetism (Jackson, 2015; Mottelay, 1922). From other work source, his father, Anton Brugmans (1732–1789), a Dutch physicist, had discovered a new form of magnetism, called diamagnetism (Küstler, 2007). Later, after the observation of diamagnetism, Thomas Johan Seebeck (1770–1831), a German physicist, Coulomb, and Alexandre Edmond Becquerel (1820–1891), a French physicist, had indicated independently, in the first third of the nineteenth century, the existence of a repulsive force exerted by a magnet, which was later called diamagnetism. There were no further examinations. Without regard to which of the Brugmans discovered first diamagnetic materials, Michael Faraday presented at the Royal Society conference on September 13, 1845, the discovery of magneto-optical effects in his laboratory (Faraday, 1846). Furthermore, he investigated the action of different substances to magnets. The substance investigations were included apples, caffeine, dried blood, sulfates, minerals, acids, uranium, phosphorus, arsenic, different gases, etc. Faraday classified the substances depending on their reaction to the magnet. The substances that move towards the magnet were called "magnetic" and the substances that move away from the strong point of the magnet were called "diamagnetic" (Küstler, 2007). In 1847, William Thomson theorized the feasibility of diamagnetic levitation (Thomson, 1847).

Recently, it is well known that magnetic fields affect flame behavior and gas flows such as quenching flames by magnetic fields. The effect was first reported by Michael Faraday in 1847 (Faraday, 1847). Faraday applied a magnetic field to a flame of a wax taper. He observed that permanent magnets could cause the flames to deform into equatorial disks and that the flames were more luminous when placed in the magnetic field. Faraday theorized that the change in the flames shape were due to the mixed presence of paramagnetic and diamagnetic gases in the flame. The constituent gases are of parametric and diamagnetic nature. Oxygen is a paramagnetic molecule, and nitrogen and CO_2 are diamagnetic materials. After Faraday's report, there has been great interest in the impact of magnetic fields on combustion behavior.

After about 140 years of Faraday's experiment and report, Japanese scientists began to investigate the effect of magnetic fields on biological and chemical processes with more advanced technologies (Ueno and Harada, 1986). Ueno and his co-workers investigated the effects of gradient magnetic fields on the burning flames of candles and their quenching (Ueno and Harada, 1986, 1987; Ueno, 1989). They exposed a candle that burned in the airgap between magnetic poles, and its flame was exposed to gradient magnetic fields. During the magnetic field exposures, the flames were pressed down and the shape changed and looked like a mushroom (Ueno and Harada, 1987). The field strength in the center of the airgap in this experiment was 1.2 T. They expanded their experiments to study the flow of gases. They observed that the flow of gases such as CO_2, nitrogen, oxygen, and argon are blocked or disturbed by magnetic fields. They introduced a model called magnetic curtain to explain these phenomena. These phenomena can be explained by the action of magnetic forces on paramagnetic molecules. The magnetic curtain is a wall of oxygen or air. Furthermore, they examined the effect of magnetic fields on quenching candle flames using an electromagnet with a pair of columnar magnetic poles in which the inner side-pieces were hollow and a candle burned in the hollowed space. The flames were quenched a few seconds after the magnetic field was turned on. The interception of oxygen by magnetic curtain has a quenching effect on the flames (Ueno and Iwasaka, 1990). This phenomenon must be taken into consideration when a human subject is positioned inside the magnetic curtain such as inside an MRI. We have to be careful about the respiratory functions, evaporation of water molecules, body temperature, blood circulation, and other physiological functions being disturbed due to the above effect.

From the 1990s with development of high magnetic field magnets, the most drastic demonstrations of diamagnetic materials have been investigated. E. Beaugnon and R. Tournier, from the Centre National de la Recherche Scientifique (CNRS) in Grenoble, France, demonstrated the magnetic levitation of diamagnetic materials such as water, ethanol, acetone, bismuth, antimony, graphite, wood, and plastic at room temperature in a strong inhomogeneous static magnetic field of up to 27 T with field gradients as high as 2,000 T^2/m (Beaugnon and Tournier, 1991a, b). They pointed out that the levitation technique opens the way to provide possibilities of a contactless, microgravity environment for the elaboration of a wide range of materials. Andre Geim, a Russian-born Dutch-British physicist, from the University of Manchester, showed the levitation of diamagnetic substances in a high static gradient magnetic field (Berry and Geim, 1997). They levitated living frog in a magnetic field of about 16 T. Andre Geim won the Ig Nobel Prize in 2000, for levitating a live frog in a high magnetic field. In addition, Andre Geim shared with Konstantin Novoselov, a Russian-born British physicist, University of Manchester, the Nobel Prize in Physics in 2010, for groundbreaking experiments regarding the two-dimensional material graphene. Andre Gerim is the only person to have won both Nobel and Ig Nobel Prizes.

Water is diamagnetic. When a container of water is placed in a magnet, there is a phenomenon where splitting the water into two parts. In 1994, Ueno and Iwasaka discovered the surface deformation of diamagnetic fluids under high gradient magnetic fields (Ueno and Iwasaka, 1994a, b), which later was called the Moses effect. They used a horizontal type superconducting magnet of 70 cm long and a 10 cm diameter bore, by which at 8 T, the maximum product of the magnetic field gradient was 400 T^2/m at ±75 mm along to horizontal bore axis. This phenomenon is due to the diamagnetic property of water. Permanent magnets induced opposite magnetization in water. Two "frozen" cascades were formed; the surface of the water near the center of the magnet was parted, and the bottom of the water chamber could

FIGURE 2.14   Demonstration of the Moses effect using a super-conducting magnet (From Ueno and Iwasaka)

be seen. The water level at both ends of the chamber was raised (Ueno and Iwasaka, 1994a). Figure 2.14 shows the demonstration of the Moses effect with agarose gels which allow the formation of surface changes in magnetic fields. The phenomenon of Moses effect is closely in association with the Biblical story. It describes that Moses, the most important prophet, divided the Red Sea into two parts when the Israelites wandered through the wildness after their Exodus from Egypt. Ueno received the d'Arsonval Medal from the Bioelectromagnetics Society in 2010.

If a container of a paramagnetic liquid is placed in a strong magnet, a phenomenon occurs in which the liquid rises in the center of the container. This phenomenon is called the inverse (reversed) Moses effect. This inverse Moses effect was first presented by Noriyuki Hirota from the University of Tokyo and others in 1995 (Hirota et al., 1995). They showed the change of surface profile of aqueous solutions with different magnetic susceptibility. In order to compare, they used distilled water and saturated copper sulfate ($CuSO_4$) aqueous solutions, the former has diamagnetic character with $\chi = -9.031 \times 10^{-6}$ (SI) and the latter solution is paramagnetic character with $\chi = +8.397 \times 10^{-6}$ (SI). Under a 10 T superconducting magnet, the changes in the surfaces of diamagnetic and paramagnetic liquids are due to differences in their respective magnetic susceptibilities. In the case of distilled water, the fall of the surface was shown in the center of the magnet. On the other hand, the rise of the surface of saturated $CuSO_4$ aqueous solution was obtained.

2.5.2.2 Magnetic Resonance Imaging

MRI was introduced as the most advanced diagnostic imaging technology in medicine in the early 1980s. MRI is based on the Nuclear Magnetic Resonance (NMR) of water molecules with the use of electromagnetic fields. Historically, NMR was first described and measured in molecular beams by Isidor Isaac Rabi (1898–1988), Nobel Laureate in 1944, Columbia University, with his associates Norman Foster Ramsey (1915–2011), Nobel Laureate in 1989. In 1946, Edward Mills Purcell (1912–1997), an American physicist at Harvard University, and Felix Bloch (1905–1983), a Swiss-American physicist at Stanford University, invented the NMR. NMR is a physical observation method in which nuclei in a strong and constant magnetic field are perturbed by a weak oscillating magnetic field. Purcell and Bloch shared the Nobel Prize in Physics in 1952 for their development of new methods for nuclear magnetic precision measurements and other discoveries in connection.

In 1971, Raymond Vahan Damadian, an American physician and inventor, observed the difference of the relaxation times between tumors and normal tissues and proposed the use of NMR as imaging for the detection of cancer (Damadian, 1971). In Japan, during the 1970s, using NMR, Zenmon Abe (1914–1999), professor at Hokkaido University, Sapporo, and his co-workers proposed and measured non-invasively biological imaging information by a magnetic focusing method (Abe et al., 1974). The basic principle and the technical development of image formation using of NMR was proposed by Paul Christian Lauterbur (1929–2007), professor at the University of Illinois, USA, and Sir Peter Mansfield (1933–2017), professor at the University of Nottingham, UK (Lauterbur, 1973; Mansfield, 1977). Their fundamental concept of NMR is the combination of high static magnetic fields and a gradient (time-varying) magnetic field. After this invention, they won the Nobel Prize in Physiology or Medicine in 2003 for their discoveries concerning "Magnetic Resonance Imaging." The root of MRI is widely known as NMR. However, to avoid alarming medical experts and the public, the word "nuclear" was deleted from the term "Nuclear Magnetic Resonance," although it has nothing to do with radioactivity.

The actual operation of MRI utilizes basically three different types of electromagnetic fields: a high static magnetic field, a rapidly changing gradient magnetic field, and a RF electromagnetic field. The high static magnetic field is the main magnetic field in MRI, and is usually generated by a strong super-conducting magnet. The gradient magnetic field with frequencies in the kHz range is used to localize aligned protons in the image reconstruction process. The RF electromagnetic fields in the range from 10 to 400 MHz are used to excite the protons within the stable magnetic fields.

Magnetic Resonance Spectroscopy (MRS) is based on in vivo NMR and is also a non-invasive imaging technique that permits in vivo measurements and quantification of the concentration of neurochemical metabolites. Kurt Wüthrich, professor at the Swiss Federal Institute of Technology (ETH) in Zurich, Switzerland, won the Nobel Prize in Chemistry in 2002 for his development of NMR spectroscopy for determining the three-dimensional structure of biological macromolecules in solution. Koichi Tanaka of Shimazu Corporation shared with Wüthrich the 2002s Nobel Prize in Chemistry for his development of soft desorption ionization methods for mass spectrometric analyses of biological macromolecules. The clinical application of MRS is reviewed by Faghihi (2017). He emphasizes that the in vivo MRS is a key technique for the investigation of human metabolism. For the development of two-dimensional NMR spectroscopy, Richard Ernst (1933–2021), professor at ETH in Zurich, won the Nobel Prize in Chemistry in 1991 for his contributions to the development of the methodology of high resolution NMR. Interestingly, until now, the invention of NMR and its fundamental research and applications led to 13 Nobel Prizes in Physics, Chemistry, and Physiology or Medicine (Boesch, 2003; Kauffman, 2014).

The technique to measure brain activity by measuring the cerebral blood flow (CBF) goes back to the time when Charles Smart Roy (1854–1897), professor of Pathology at the University of Cambridge, and Charles Scott Sherrington (1857–1952), a British neurophysiologist, reported using dogs, cats, and rabbits that changes of CBF reflect the neural activity of the brain (Roy and Sherrington, 1890). They characterized the relationship between CBF and brain function. Their observation means that CBF changes reflect tight coupling between cellular energy requirement and vascular delivery of glucose and oxygen (Lin et al., 2010). The measurements of brain functions that are made today are based on this report. Later, in 1932, Sherrington along with Edgar Douglas Adrian (1889–1977), a British neurologist, won the Nobel Prize in Physiology or Medicine for their discoveries regarding the function of neurons. When the activity of various functions occurs in the brain, the CBF and metabolism increase in association with that activity. The changes are spatially coincident with the site where the neural activity occurred. By measuring these changes in CBF and metabolism, we can study the functions of the brain.

Tissues are lined with blood vessels that supply oxygen. The blood reaches the veins from the arteries through the capillaries. The red blood cells in blood have a lot of hemoglobin, which carries oxygen. This hemoglobin is diamagnetic when bound to oxygen molecules, but becomes paramagnetic (deoxyhemoglobin) after oxygen is released in the capillaries. This deoxyhemoglobin causes a slight distortion of the magnetic fields around intravenous blood vessels. This distortion weakens the proton signal in the vicinity. This phenomenon is called the blood oxygenation level dependent (BOLD) effect. Using this effect, it is possible to image dynamic changes in local oxygen metabolism and regional blood flow. As a more advanced imaging technology, functional fMRI was introduced in 1990, due to the blood-oxygenation level-dependent effect, which was discovered by Seiji Ogawa, now at Tohoku Fukushi University, and his co-workers at Bell Laboratory (Ogawa et al., 1990a, b, 1992). Hemoglobin is diamagnetic, and deoxyhemoglobin is paramagnetic which means that the magnetic susceptibility of these two materials is slightly different. Focusing on the NMR signal caused by this slight difference in magnetic susceptibility, the measurement using the BOLD effects was proposed. Since Ogawa's proposal, fMRI was widely spread interest and applications in neurocognitive and neurophysiological studies. fMRI depends also on the basis of NMR. Since fMRI can image brain activity with high resolution in a non-invasive manner, fMRI is being used to elucidate the mechanisms of functional brain activity in the fields of medicine, physiology, cognitive science, and education.

As mentioned above, fMRI focuses on the slight differences in magnetic susceptibility of materials in the blood. Historically, in 1846, Michael Faraday investigated the magnetic properties of dried blood and found that blood was not magnetic (Faraday, 1846). Pauling pointed out that if he determined the magnetic susceptibilities of atrial and venous blood, he would have found them to differ by a large amount (as much as 20 per cent) for completely oxygenated and completely deoxygenated blood (Pauling and Coryell, 1936a).

Afterward, Pauling focused on the oxygen saturation in the blood and the magnetism of the blood. Linus Carl Pauling (1901–1994), professor at California Institute of Technology, and his student, Charles Dubois Coryell (1912–1971), later professor at the Massachusetts Institute of Technology (MIT), reported that oxygenated as well as carbonmonoxy forms of hemoglobin were diamagnetic, but discovered that the deoxygenated protein was magnetic (Pauling and Coryell, 1936a, b). The magnetic susceptibility of blood hemoglobin changed as a function of whether it was bound to oxygen or not. Pauling and Coryell reported that ferrohemoglobin (hemoglobin) is strongly paramagnetic and contains four unpaired electrons per heme. The oxygen molecule, O_2, contains two unpaired electrons and is also paramagnetic. However, when O_2 is combined with ferromagnetic to form oxyferrohemoglobin (also called oxyhemoglobin) with no unpaired electrons and is diamagnetic. Pauling received in 1954 the Nobel Prize in Chemistry for his research into the nature of the chemical bonds and its application to the elucidation of the structure of complex substances. He was awarded the Nobel Peace Prize in 1962 for his opposition to weapons of mass destruction.

2.5.2.3 Bioelectric-Generated Magnetism

Biomagnetism is found in electric current changes. This means that biomagnetic fields can be produced by electric current flow in biological systems. Recording of biomagnetic activity of weak signals is different from the recording of bioelectric signals. The former needs the use of sensitive magnetometers. The technical development of sensitive magnetometers requires detecting very weak signals of electrical phenomena. ECG and EEG measures directly and non-invasively the endogenous oscillatory, electrical activity in the heart and brain. Magnetocardiography (MCG) and magnetoencephalography can measure indirectly and non-invasively the magnetic fields generated by the changes of electrical activity. The time course of ECG and MCG signals are basically similar.

In 1963, Gerhard Baule and Richard McFee, both at Syracuse University, USA, detected for the first time MCG of electrical activity in a human heart with an induction coil magnetometer outdoors in an open field (Baule and McFee, 1963). The magnetometer was made by winding two million turns of copper wire on a dumbbell-shaped ferrite core (about 30 cm in length and 9 cm in diameter). The two copper pick-up coils were connected in opposition so that they canceled the induced voltage from the uniform magnetic background fluctuations. Later, a group from the USSR repeated the Baule-McFee's experiment and confirmed their experimental results (Cohen, 1969). Cohen pointed out that the measurement by both groups was limited by the complexity of the gradient detection, imperfect coil cancellation, and field distortion due to ferrite which was necessary to raise the heart signal waveforms above the amplifier input noise (Cohen, 1969). To overcome these limitations, a magnetically shielded chamber with a 2.23 m inner size of the room was used. It consisted of two cubic shells of 1.5 mm thick ferromagnetic shielding and one layer of aluminum. In this chamber, David Cohen recorded the MCG by a magnetometer which is an air-core with a 200,000 turn coil of 5 cm in length and 8 cm in diameter (Cohen, 1967a, b). Then, Cohen described first the magnetic field of the brain's spontaneous α-rhythm with an induction coil magnetometer using this chamber (Cohen, 1968). Later, Cohen moved to MIT, and there he built a more conventional five-layer shielded room with a SQUID magnetometer. The introduction of a SQUID magnetometer gave the increase in sensitivity of biomagnetic measurements (Zimmermann et al., 1970). SQUID device works at the liquid helium temperature, $-270°C$, and detects extremely low magnetic fields $\sim 10^{-9}$ T. Cohen and his co-workers recorded first human MCG with SQUID in 1969 (Cohen et al., 1970). The introduction of SQUID and the work of Cohen and Zimmerman in MIT was a breakthrough in biomagnetism. Cohen and James Edward Zimmerman (1923–1999) recorded MCG with a SQUID magnetometer in this shielded room. Cohen is considered to be the pioneer in biomagnetism.

From these pioneering investigations with the newly developed SQUID in the 1970s, biomagnetism became a more popular research field. Many laboratories developed SQIUD-magnetometers for biomagnetic investigations. The first biomagnetic conference took place at MIT on August 8–11, 1976, as an International Workshop (Cohen, 2004). There were only 23 attendees: from USA 15, Finland 3, Canada 2, France 2, and Japan 1. The attendee from Japan was Makoto Kotani, professor and later the President of Tokyo Denki University, who is the pioneer in the research of biomagnetism in Japan (Kotani, 2011; Uchikawa, 2020). Topics in this workshop were MCG, magnetoencephalography, and SQUID. Although the first Workshop was small, the Workshop became a regular Conference. The International Conference on Biomagnetism (BIOMAG 2022) will he held in Melbourne, Australia in 2022.

During the 1980s, research groups in biomagnetism grew steadily all over the world and published a number of studies obtained with single-channel SQUID magnetometers. In this way, the first measurements were performed with a single-channel magnetometer. After that, the instrumentation gradually improved from single-channel to multi-channel magnetometers. All commercially available multi-channel systems are based on low-temperature SQUIDs. Now, 275-channel SQUID magnetometers are available for medical uses.

2.5.2.4 Non-Invasive Electric and Magnetic Stimulations

The early history of electromagnetic stimulation was reviewed in Section 2.4.2. The use of the Transcranial Electrical Stimulation for therapy goes back to the discovery of electrosensitive animals such as the torpedo fish. This electrotherapy continued until the end of the seventeenth century. More than 200 years ago, Luigi Galvani and his nephew, Giovanni Aldini, used electrical stimulation to improve melancholy in patients with invasive electric stimulation. Transcranial Direct Current Stimulation (tDCS) as the kind of transcranial electrical stimulation has been considered a neuro-modulatory intervention. In 1938, the use of electrical stimulus treatment (ECT) in epilepsy and in mental illnesses was introduced by two Italian physicians, Ugo Cerlett (1877–1963), neurologist, professor at University of Rome La Sapienza, Rome, and his student Lucio Bini (1908–1964) (Rowbottom and Susskind, 1984). ECT is a therapy in which electric current is used to provoke a seizure for a short duration. This newly developed ECT spread rapidly through Europe and North America in the 1950s and 1960s. ECT was cleared by the FDA for the treatment of depression as a preamendment device. ECT has been used to treat a variety of psychiatric disorders: depression (unipolar and bipolar), schizophrenia, bipolar manic (and mixed) states, catatonia, and schizoaffective disorder. The wave-forms for ECT are of high current, approximately 800 mA with trains (AC or pulsed bursts) lasting 1–6 Hz. The electrodes are placed either unilaterally or bilaterally on the cranium. The non-invasive techniques had been investigated only in a small number of studies. During the operation of tDCS, a weak-amplitude direct electric current is passed over the brain through two surface electrodes placed on the scalp. Now, non-invasive methods for measurement of brain activity through tDCS and TMS are effective methods. Both tDCS and TMS are now widely used for rehabilitation treatments in clinic. Furthermore, they can be used to evaluate neuronal activity in the brain, brain connectivity for diagnostic purpose.

Now, we introduce the earliest demonstrations of magnetic stimulations in animals. It has passed over half a century from D'Arsonval's demonstration of magnetic stimulation in human retina. In 1959, Alexander Kolin (1910–1997), a Russian-born Germany-US biophysicist, from the University of California, accomplished the first magnetic stimulation of a frog's sciatic nerve and muscles submerged in Ringer's solution by means of time-varying magnetic fields (Kolin et al., 1959). He presented that this effect is due to eddy current induced in conductive tissue and their surroundings. In the past, he invented in 1936, the electromagnetic blood flow meter and isoelectric focusing. For this invention, he was the candidate for the Nobel Prize.

In 1965, Reginald G. Bickford, electroencephalographer, and B. D. Fremming stimulated the peripheral nerves in animals and humans using a pulsed magnetic field lasting 40 ms (Bickford and Freming, 1965). The resulting long-lasting activation interval made it impossible to record nerve or muscle action potentials, and the work was not pursued further. In 1973, Per Åke Öberg of Linköping University investigated the magnetic stimulation on excitable tissues in frog nerve muscles which were placed in high-frequency magnetic fields of 1 kHz–1 MHz (Öberg, 1973). Shoogo Ueno, professor at Kyusyu University and later at the University of Tokyo, and his co-workers carried out experiments to measure action potentials of lobster giant axons placed on magnets with time-varying magnetic fields (Ueno et al., 1981, 1984, 1986). After the study of this single nerve axon, Ueno proposed a new type of magnetic nerve stimulation (Ueno et al., 1988). In 1988, in order to more precisely localize the stimulating region over the head, Ueno introduced the magnetic stimulation with a figure-of-eight coil as the stimulating coils (Ueno et al., 1988). Figure 2.15 shows a TMS of the primary motor cortex with the figure-of-eight coil. Then, they succeeded in stimulating the human cortex with a 5 mm resolution (Ueno et al., 1990). Ueno soon improved it by changing the coil arrangements. The figure-of-eight shaped coil has the merit to induce more concentric electric field (~mm in size) than that of a circular single coil, which results in better focusing and control of the spatial extent of the excitation.

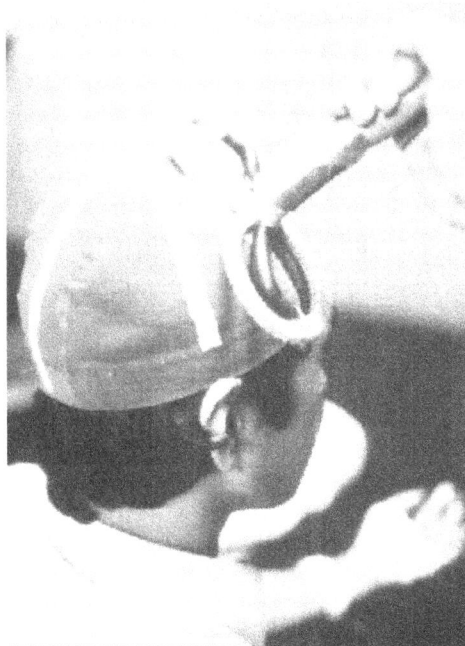

FIGURE 2.15   TMS of the primary motor cortex (From Ueno)

According to these fundamental approaches, TMS of the motor cortex gradually obtained interest. In parallel, Anthony Tony Barker, professor at the University of Sheffield, Teaching Hospital's Department of Medical Physics and Clinical Engineering, and his co-workers stimulated successfully superficial nerves via magnetic fields by placing the stimulation coils on the human head and proposed a prototype of a magnetic stimulator for peripheral nerve stimulation (Barker et al., 1985, 1987). They used 2 ms duration pulses of time-varying magnetic fields to induce electric currents in the brain and recorded for the first time motor-evoked potentials obtained by median nerve magnetic stimulation with circular single coils. As the first clinical examinations, Barker's group developed the first TMS device which produces cortical depolarization via magnetic stimulation in 1986 (Barker et al., 1986). This was the beginning of magnetic brain stimulation in clinical examinations. These early researches used circular single coils. Placing these circular single coils on the head and passing a current pulse through it. Eddy currents were induced in the head, which stimulates broad areas of the brain. Since then, the TMS has become a popular technique in the research of brain activity because it is easy to use as a non-invasive tool (Barker, 1991; Ueno and Sekino, 2014; Ueno et al., 2019). TMS is a non-invasive and effective method of cortical stimulation for medical diagnosis, in particular for studying the function of the human brain. This method is based on the utilization of time-varying magnetic fields. Wassermann described the basic principles of TMS as follows:

> TMS uses the principle of inductance to get electrical energy across the scalp and skull without the pain of direct percutaneous electrical stimulation. It involves placing a small coil of wire on the scalp and passing a powerful and rapidly changing current through it. This produces a magnetic field that passes unimpeded and relatively painlessly through the tissues of the head. The peak strength of the magnetic field is related to the magnitude of the current and the number of turns of wire in the coil. The magnetic field, in turn, induces a much weaker electrical current in the brain. The strength of the induced current is a function of the rate of change of the magnetic field, which is determined by the rate of change of the current in the coil. In order to produce enough current to excite neurons in the brain, the current passed through the coil must change within a few hundred microseconds.

Wassermann (1998)

Depending on the parameters of stimulation such as frequency, duration and magnetic field strength, TMS is subdivided into two repetitive TMS (rTMS) treatments: low-frequency (1 Hz or less) rTMS and high-frequency rTMS with frequencies from 5 to 20 Hz. The former leads to reduction of excitability, and the latter is considered to produce an excitatory effect. The rTMS was first produced by Cadwell Laboratories in 1988. In 2008, rTMS was first approved for clinical treatments of patients with major depression by the US Food and Drug Administration (FDA). The FDA guidelines for rTMS were updated in 2011 (Lin, 2016). Now, TMS has become a tool in basic neuroscience investigations, therapeutic and rehabilitation applications such as depression, Parkinson's disease (PD), Alzheimer's disease, stroke, pain, etc. (Kadosh, 2014). In 2017, Barker received "The First International Brain Stimulation Award" for his outstanding contributions in developing TMS at the Conference in Barcelona. This International Conference award was given by publisher Elsevier.

2.5.3 Others

After constructing his first large telescope in 1774, Frederick William Herschel (1738–1822), a German-born British astronomer and composer, while measuring the thermal effects of sunlight with a prism observed the infrared radiation from the spectrum of sunlight by means of a glass thermometer in 1800. André Marie Ampère in 1835 confirmed that this infrared radiation has shorter wavelengths of electromagnetic waves. Infrared radiation is often divided into three regions according to the ISO: 20473, which specifies the following: near-infrared (NIR) with wavelengths of 0.78–3 μm, mid-infrared with

3–50 μm, and far-infrared with 50–1,000 μm. Johann Wilhelm Ritter (1776–1810), a German physicist, found in 1801 that invisible ultraviolet (UV) radiation from sunlight through their action in darkening silver chloride. The first industrial application of near-infrared radiation began in the 1950s. In the first applications, near-infrared spectroscopy (NIRS) was used only as an add-on unit to other optical devices that used other wavelengths such as UV, visible, or mid-infrared spectrometers.

Several other medical applications of optical radiations will be introduced. Next, as historically well-known medical applications, the UV light of the electromagnetic spectrum was used. In Greek and Roman times, the exposure to the sun had been practically used for health and therapeutic reasons. In the eighteenth and early nineteenth century, sun therapy was recommended after the fact that UV light could kill various bacteria had been firmly established. In 1899, Erik Johan Widmark (1850–1909), professor in Ophtalmiatrics at the Karolinska Institute, had shown that UV light is responsible for erythema and pigmentation of the skin (Rowbottom and Susskind, 1984).

Niels Ryberg Finsen (1860–1904) developed the Finsen lamp based on electric carbon arcs. First, he designed and used a source of artificial light in therapy. Then he modified it. For a lens, common glass was initially used and later he replaced it with fused quartz to allow the separation of light and the formation of UV light (Grzybowsli and Pietrzak, 2012). Light treatment had two uses: natural sun therapy and artificial light therapy. In 1893, he observed that light had a beneficial effect on smallpox scars. Niels Ryberg Finsen, professor at the University of Copenhagen, Denmark, was the founder of light therapy (Figure 2.16) (Finsen, 1899, p. 67). He was born in the Faroe Islands and educated in Iceland and Copenhagen. The Nobel Prize in Physiology or Medicine in 1903 was awarded to Finsen in recognition of his work on the treatment of diseases, and in particular the treatment of *lupus vulgaris* by means of concentrated light, by which he has opened a new avenue of light therapy for medicine. Unfortunately, he died the following year. His Nobel Prize entered into some controversy. His supporters argued that Finsen had a great mind, and his research was innovative and beneficial all humanity. However, his opponents claimed that Finsen's contribution to medicine had too little theoretic foundation and was not academic enough. During the Nobel Prize decision process, the members of the Nobel Prize Committee paid a few

FIGURE 2.16 Treatment in the Finsen lamp (From Finsen, 1899.)

visits to the institute before the final decision was made (Grzybowsli and Pietrzak, 2012). Special attention is paid to that Finsen donated much of his Nobel Prize award to both his institute and the sanatorium for heart patients. Currently, it should be mentioned that light therapy is treated with much greater caution because over exposure to UV light can lead to melanoma and other skin cancer.

Next, we introduce another topic. The infrared region lies between the visible and microwave portions of the electromagnetic spectrum. Infrared radiation is invisible to the human eye but it can be focused, reflected, and polarized just like visible light. The NIR band has a wavelength range from 0.78 to 3 μm. NIRS uses near-infrared radiation. In the 1980s, a single-unit, stand-alone NIRS system was made available, but the application of NIRS was focused more on chemical analysis. With the introduction of optical-fiber in the mid-1980s and with the monochromator-detector developments in the early 1990s, NIRS became a more powerful tool for other than chemical analysis. The use of NIR for biomedical measurements dates back to 1977, when Frans Jöbsis (1929–2006), professor at Duke University, published the non-invasively measurement of the changes of oxygenation of the brain in an intact cat head (Jöbsis, 1977). In the 1990s, it became clear that NIRS could capture changes in hemoglobin concentration that were linked to changes in the blood flow in the brain, which was linked to neural activity. This attracted great attention as a new functional imaging method, fNIRS. The fNIRS is the term used in contrast to fMRI, which detects changes in the magnetic susceptibility of hemoglobin and images it to obtain BOLD information. On the other hand, fNIRS captures changes in the absorption of light as changes in the concentration of hemoglobin. The difference is that fNIRS can measure the cerebral cortex with surface optical sensors and fMRI is used to visualize deeper brain activity.

Under the suggestion of Jöbsis, Mamoru Tamura (1943–2009), professor at the Research Institute for Electronic Science, Hokkaido University, and David Thomas Delpy, professor at University College London, produced great advances in fNIRS (Delpy et al., 1987; Cope and Delpy, 1988). Tamura published with instrumental support from Hamamatsu Photonics, K.K., and Shimazu Corporation the first fNIRS human studies (Hoshi and Tamura, 1993a, b). They observed bilateral prefrontal cortex oxygenation changes in 14 volunteers during a mental task using CW instruments, each equipped with a single channel. Jöbsis is the founder of NIRS. Tamura was known as one of the great pioneers in biomedical optics and fNIRS. Japanese organizations such as Hamamatsu Photonics, K.K, and Shimazu Corporation support greatly to the development and the use of optical radiation in biology, chemistry, and medicine. In 2008, Delpy received the Rosalind Franklin Medal and Prize for his pioneering development of a range of novel techniques and instruments to monitor the health of patients in intensive care units and to image tissue physiology and metabolism. The Rosalind Franklin Medal and Prize is given for distinguished contributions to physics applied to life science including biological physics. This Medal and Prize was named after Rosalind Elsie Franklin (1920–1958), a British chemist and X-ray crystallographer. She performed X-ray analysis of DNA and helped the creation of the Watson-Crick Model.

In 1935, Karl Matthes (1905–1962), a German physician, made a device which measured continuously the oxyhemoglobin saturation of human blood, the in vivo transillumination of the ear (Severinghaus, 1986). In 1939, using two different wave lengths of light (red and infrared) he developed the first red and infrared ear oxygen saturation meter. Infrared wavelength is absorbed more by oxygenated hemoglobin and red wavelength is absorbed more by deoxygenated hemoglobin. Glenn Allan Millikan (1906–1947), an American physiologist, inventor, University of Pennsylvania, Philadelphia, was the son of Robert Andrews Millikan (1868–1953), an American experimental physicist. Father Millikan won the Nobel Prize in Physics in 1923 for the measurement of the elementary electric charge and for his work on the photoelectric effect. Glenn Millikan made an ear oximeter (Millikan and Taylor, 1942). Although the ear oximeter performed very poorly due to the fact that light absorption of ear is affected very little by arterial blood, he coined term oximeter. The pulse-oximeter would consist of a probe attached to the patient's ear lobe or finger and a display unit. Using NIR in the range of 0.6–1 μm, the pulse-oximeter was used as a non-invasive method for monitoring a person's oxygenation of the blood. The oxygenation of the blood was measured as a function of time by determining the absorption at two different wavelengths. In 1947, Glenn Millikan was killed by a falling rock during mountain climbing.

Oxygen concentration of erythrocyte in the artery is an important marker showing condition of respiratory system. One has to obtain some amount of blood from the artery for measuring the concentration with price-blood measuring device. Moreover, as oxygen concentration of circulating artery varies rapidly, one has to measure the concentration even in the night. This is heavy burden to the patient. Therefore, non-invasive method is critically important for clinical practice.

The most important work in pulse-oximeter was made by two Japanese: an engineer Takuho Aoyagi (1936–2020) of Nihon Kohden Co., and a surgeon Susumu Nakajima of Hokkaido University in the early 1970s. Aoyagi invented a new type of oximeter which uses pulsatile variation in optical density of tissues in the red and infrared wavelength to compute arterial oxygen saturation without need for calibration (Aoyagi et al., 1974). Nihon Koden Co. ordered Aoyagi to stop further investigation. However, Nakajima quickly noticed the idea of Aoyagi and ordered Nihon Kohden Co. to manufacture Aoyagi's prototype pulse-oximeter by the Government grand money. Nakajima brought the prototype into the laboratory of Professor Masaji Mochizuki (1922–2015), a world famed physiologist of oximeter of the RIAE of Hokkaido University. Then, Nakajima, joined by Aoyagi for early several sessions, performed animal experiments to compare the oximeter data obtained from platinum electrode inserted into femoral artery of the dog and the data obtained from Aoyagi's prototype pulse-oximeter. Platinum electrode method was invented by Mochizuki to measure oxygen concentration in blood. From the repeated animal experiments, Nakajima found and clearly confirmed that the device captured the subtle change of inspired SaO_2 without much of a delay. Based on these animal experiments, Nakajima compared those obtained from Elma (Wood's type)-oximeter. Then, Nakajima repeated for ten times on three healthy human subjects and two patients with low oxygen in blood if similar or the same results could be obtained, with successful results. Moreover, Nakajima found the same results of arterial blood and pulse-oximeter. From these preliminary investigations, Nakajima decided the Aoyagi's type pulse-oximeter can be useful for clinical practice (Nakajima et al., 1975, 1979). Figure 2.17 shows the world's first clinical application of a pulse-oximeter conducted by Nakajima in the patient room at Misumai National Sanatorium, Sapporo. The patient had respiratory failure after he underwent thoracoplasty for tuberculosis treatment. The pulse-oximeter is characterized by the fact that requires only the adjustment of the alarm position and volume, the setting of a low cut filter to prevent fluctuations due to

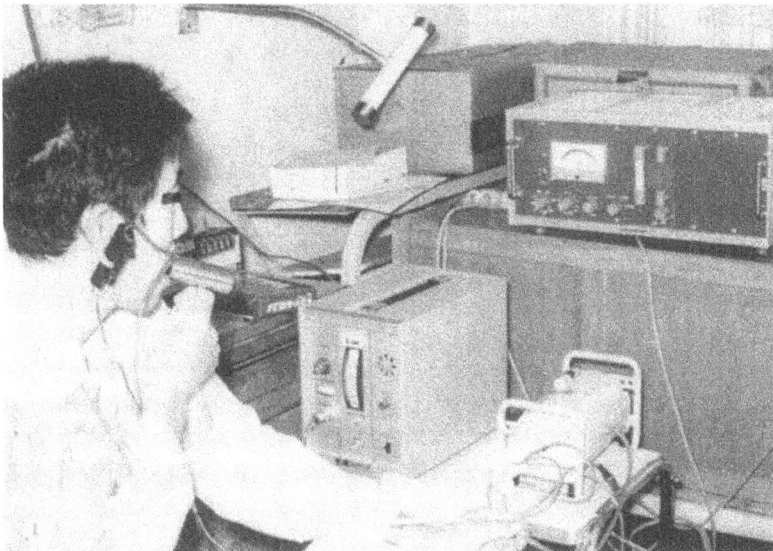

FIGURE 2.17 The world's first clinical application of ear-piece pulse-oximeter (Courtesy of Susumu Nakajima, Moriyama Memorial Hospital)

respiration, and the attachment of an earpiece, without the need for ischemic manipulation or calibration. Therefore, it can be easily operated by doctors and nurses. The Misumai National Sanatorium was reorganized into Hokkaido Medical Center of National Hospital Organization. Nakajima is now the director of Moriyama Memorial Center, Asahikawa.

Professor John Wendell Severinghaus (1922–2021), an anesthesiologist at the University of California, was searching for the origin of pulse-oximeter and reached to two Japanese (Severinghaus, 1987; Severinghaus and Honda, 1987).

The Aoyagi's type pulse-oximeter has been improved by many people at many stages, and at year 2021, the size of the pulse-oximeter is small enough can be held on one hand. It is not necessary to add that the pulse-oximeter is widely used in anesthesiology, ambulance, respiratory section, and many other sections in the hospital. It can be said that the finger-tip pulse-oximeter is one of the most significant advance of red and infrared wave length radiation in medicine and biology in the twentieth century.

2.5.4 New Instrumental Era

At the beginning of the twentieth century, the electronic inventions opened the new era in bioelectromagnetism. The year 1905 was the birth of the age of electronic, after Lee de Forest invented a three-electrode valve (vacuum tube) and the original was known as Audion valve. His invention was not only a detector of signals but also an amplifier (Rowbottom and Sussind, 1984). Historically, before de Forest's invention, Thomas Alva Edison (1847–1931), an American inventor, observed in 1883 the electron current flow in a vacuum, now called the Edison effect. Sir John Ambrose Fleming (1849–1945), a British engineer and physicist, professor at University of College London, utilized the Edison effect and invented the diode for rectifying high-frequency oscillation in 1904. Until the invention of the transistor, three-electrode valves were excellent devices for amplification. At the same time, Karl Ferdinand Braun invented the Braun tube. He won, with Guglielmo Marconi (1874–1937), an Italian inventor, the Nobel Prize in Physics in 1909 for their contributions to the development of wireless technology.

From the point of view of medical applications, the twentieth century began with the discovery of X-rays by Wilhelm Conrad Röntgen (1845–1923), professor of Physics at the University of Würzburg and its use in diagnosis. Röntgen won the first Nobel Prize in Physics 1901, for in recognition of the extraordinary services he has rendered by the discovery of the remarkable rays subsequently named after him. The clinical ECG recorder was commercially available in the 1920s, diathermy in the 1930s, discovery of EEG in the 1930s, and inventions of the pacemakers and defibrillators appeared in the 1950s and 1960s, respectively.

During the 1920s and 1930s, the use of electronic amplifiers and oscillators had increasingly gained popularity in electrophysiological research fields. In 1920, Alexander Forbes (1882–1965), professor of Physiology at Harvard Medical School, made the pioneering study of the nervous system. He invented with Catherine Thacher, the vacuum-tube amplifier and recorded nerve impulse from a single nerve fiber of frogs with it (Forbes and Thacher, 1920a, b). Joseph Erlanger and Herbert Spencer Gasser (1888–1963), an American physiologist, invented in 1921 the cathode-ray oscilloscope with a vacuum tube amplification device. Using this oscilloscope, they investigated the time course, transmission velocity of the nerve impulse from nerve fiber and action potential. Erlanger and Gasser received the Nobel Prize in Physiology or Medicine in 1944 for their discoveries relating to the highly differentiated functions of single nerve fibers. However, these instrumental achievements had their weaknesses such as being hard to handle, low Signal-to-Noise Ratio and drifting with time. To overcome these weaknesses, we had to wait until the introduction of solid-state electronic with transistors in the 1950s.

In 1947, John Bardeen (1908–1991), Walter Houser Brattain (1902–1987), and William Bradford Shockley (1910–1989) made an outstanding invention, the junction transistor (transfer+resistor). They worked together at the Bell Telephone Laboratory. The first n-p-n junction transistor appeared in 1950. This invention led to the outstanding development of electronic technology. This development in electronic technology allowed the miniaturization for the instruments of detection, recording, and monitoring of bio-signals generated from bioelectromagnetic phenomena. This transistor replaced the vacuum tube amplification device. The three inventors received the Nobel Prize in Physics in 1956 for their research on semiconductors and their discovery of the transistor effect. Bardeen has been the only person to be awarded the Nobel Prize in Physics twice. He won it with Leon Neil Cooper of Brown University and John Robert Schrieffer (1931–2019) of the University of Pennsylvania, in 1972 for their jointly developed theory of superconductivity usually called the Bardeen-Cooper-Schrieffer theory. As mentioned above, the great invention of the transistor affected the development of electronic technology and instrumentation in bioelectromagnetism. Using developed instruments, the detection, recording, and monitoring of weak bio-signals inside the body could easily be performed without weaknesses.

The history of the cardiac pacemaker and the cardiac defibrillator will be introduced shortly. Regarding direct cardiac pacing, in 1871, Franz Steiner, assistant of Christian Albert Theodor Billroth (1829–1894), surgeon of the University of Vienna, close friend of Johannes Brahms, pointed out first that cardiac arrest was a frequent complication due to chloroform anesthesia and reported that the heart could be made to beat by electrical stimulation with direct current (Geddes, 1994; Steiner, 1871). He stimulated rhythmically the heart of over-anesthetized animals to produce cardiac arrest. The animals included three horses, one donkey, ten dogs, 14 cats, and eight rabbits. After Steiner's suggestion, Hugo Wilhelm von Ziemssen (1829–1902), a German physician, in 1882 performed experiments with both mechanical and electrical heart stimulations and applied direct cardiac pacing to human subjects with normal chests, and found that periodic pulses of direct current applied at a rate slightly higher than the heart's own beat rate would accelerate the beat until it coincides with the external stimulus (Rowbottom and Susskind, 1984). He applied cardiac pacing to a 42-year-old woman.

Fifty years later, in 1932, Albert Salisbury Hyman (1893–1972), an American cardiologist, developed with his brother, the first clinical artificial pacemaker (not portable) for delivery of induction shock stimuli (60–120 minutes) to the atria. This was an electric machine device which was tested on experimental animals and at least one human patient. While he was still a student at Harvard University in 1918, he concluded that if the artificial stimulus could be delivered directly to the heart, and in particular to the right atrial area near the sinus node, a normal cycle would start up again and continue long enough to supply the coronary artery system with enough blood to reactivate the sinus node (Rowbottom and Susskind, 1984).

In 1952, Paul Maurice Zoll (1911–1999), an American cardiologist, performed cardiac pacing with modern equipment. The intensity and duration of the electrical pulses could be carefully controlled by electronic technology (Rowbottom and Susskind, 1984). So, he is the pioneer in the development of the artificial cardiac pacemaker and cardiac defibrillator. Back in 1899, Jean-Louis Prevost (1838–1927), a Swiss physiologist, from the University of Geneva, and Frédéric Battelli (1867–1941), an Italian physiologist, reported first about cardiac defibrillation. They showed that death from electric shock was caused by massive fibrillary contractions, and tested the induction of ventricular fibrillation on dogs using low voltage electric shocks. On the other hand, high-voltage electric shocks did not induce ventricular fibrillation. William Bennet Kouwenhoven (1886–1975), electrical engineer, professor at Johns Hopkins University, invented the original type of today's external defibrillator. During his student days, in the 1930s, his interest was in the relationship between electric shocks and its effects on the human heart. This led to the invention of the defibrillator. His defibrillator was first used by Claude Beck (1894–1971), professor of Surgery at Case Western Reserve University in 1947 and nominee for the Nobel Prize in 1952. He contributed to the innovation of various cardiac surgery techniques. During the 1960s–1970s, Kouwenhoven evaluated and documented the human health effect of linemen exposed to electric field generated from high voltage transmission lines.

In 1958, the engineer Rune Elmqvist (1906–1996) at the Karolinska Hospital in Stockholm and Åke Senning (1915–2000), a Swedish surgeon, developed the first implanted cardiac pacemaker with a nickel-cadmium battery, which lasted only a few days (Rowbottom and Susskind, 1984). This battery could be recharged from outside body. The invention of the transistor led to the charging of this battery successful. The development of portable pacemakers depended on transistor technology. Clarence Walton

Lillehei (1918–1999), an American surgeon, the University of Minnesota, and Earl Elmer Bakken (1924–2018), an American biomedical engineer and industrialist, built the first external battery-operated pacemaker which included two transistors in 1957. This gave transistorized and wearable pacemakers. This invention led to improve the pacemakers design and it spread quickly and widely over the world. Bakken with his brother-in-law Palmer Hermundslie (1918–1970) founded Medtronic, Inc on April 29, 1949. He founded in 1975 the Bakken Museum in Minneapolis, Minnesota, where the excellent exhibitions present the history and explanations of electricity, electromagnetism, and bioelectromagnetism. The cardiac pacemaker and cardiac defibrillator are the milestones in bioelectromagnetism.

2.6 Research on Biological and Health Effects of Electromagnetic Fields

The International Electro-Technical Exhibition was held in Frankfurt am Main, Germany, between May and October of 1891. At this exhibition, an important demonstration was given of long-distance high power transmission by means of a three-phase alternating current. The power was generated 175 km away at Lauffen am Neckar from Frankfurt where there was a small waterfall. Using this small waterfall, turbines were utilized to generate power of 300 horsepower (224 kW) which was taken up to drive two three-phase alternators. This generated three currents of 1,400 A at 55 V with 40 Hz. This voltage was raised to 8,500 V and was reduced at the Frankfurt end to 65 V which was then utilized for lighting incandescent lamps and driving motors. The transmission lines consisted of three copper wires, each 4 mm in diameter carried on porcelain insulators on 3,000 telegraph posts stretching over 175 km. This test showed that 74% of the energy given by the waterfall at Lauffen was transmitted electrically to Frankfurt. This test plant experiment produced great interest and surprised all the electrical enterprises due to the fact that the commercial practicality of transmitting water power for long distance by high voltage three-phase currents was demonstrated. Mikhail Dolivo-Dobrovolsky (1861–1919), a Polish-Russian engineer, created the three-phase transformer system which was used in the system used to transmit the electric power at this test plant.

The introduction of electric power transmission system brought the battle of the currents between Edison and Westinghouse. They fought a battle between Direct current (DC) and Alternating current (AC) transmission with Tesla who favored the latter. The AC transmission lines gradually gained the position with the use of Tesla's invention. However, with the technical development of easily handling AC-DC conversion at power stations, DC began to be used for the transmission. In the late nineteenth century, Thomas Edison promoted large-scale low voltage DC electricity supply indoor for incandescent lighting networks in New York. However, in 1888, Westinghouse opened an AC power plant in Buffalo, New York, and stepped down 1,000 V to 50 V to send to the consumers. George Westinghouse (1846–1914) introduced an AC-based power distribution network. William Stanley (1858–1916), an American physicist, developed the first AC transformer, and he helped building Westinghouse's first AC transmission systems.

The transmission of power from the Niagara Falls began to be considered around 1892. This Niagara Falls project would produce 5,000 horsepower (3,700 kW) with transmission as a 25 Hz AC current. This large-scale hydroelectric generator at Niagara Falls provided electricity to Buffalo, New York, via transmission lines. Nikola Tesla invented an AC induction motor which was used in the power industry. In 1895, George Westinghouse with Nikola Tesla constructed a large hydroelectric power plant at Niagara Falls and from there the electricity was supplied to the place of consumption by long-distance transmission of electricity. AC current was used due to its character for increasing or decreasing voltage and long-distance transmission with high voltages. This transmission system made it successful.

From around 1900s, people began to live in anthropogenic electromagnetic environments due to the successful development of electrification around the world (Fleming, 1921). With the expansion of the use of electricity, the levels of anthropogenic electromagnetic fields increased. Medical and clinical applications of electricity started and drew attention after the twentieth century began. This led to questions about the possible human health risks in relation to exposures to electromagnetic fields in bioelectromagnetism.

The following section discusses the history of the human health evaluations related to exposures to three different electromagnetic fields: static (DC; 0 Hz) electric and magnetic fields, extremely low frequency (ELF) fields focused on power frequency (50/60 Hz), and RF (100 kHz–300 GHz) electromagnetic field.

2.6.1 Static Fields

When large amounts of power need to be transmitted over long-distances, DC transmission line are cheaper to construct and are suitable for long-distance overhead and submarine cable transmission. In the past, DC systems were limited to shorter distances because it was difficult to convert voltages between high transmission lines and low distribution lines. The first DC transmission line system in the world was for the supply of electricity to the island of Godland in Sweden (HVDC Godland). At that time, the transmission distance was about 100 km. Over 600–900 km for overhead lines and 30–60 km for submarine cable transmission were considered economical. In the 1930s, economically available converters between DC and AC voltages were constructed. Thereafter, DC lines were constructed first in New York, and transmission of Extra High Voltage (EHV) DC transmission lines in the USA started in the 1970s.

The electrical environment of HVDC transmission line is more complex than that of the HVAC transmission line. The coupling of electric field to organism is entirely different. For the HVAC case, the coupling is capacitive and induced currents are generated inside organisms as a result of the changing electric fields. For the HVDC case, the electric field coupling is resistive. To characterize the DC electrical environment, three measures are used: (1) ion current density, (2) ion concentration, and (3) electric fields. In addition, a current produces a static magnetic field. It is generally accepted that people cannot perceived the electric field from the HVDC transmission line, with no harmful effects. Advantages of DC transmission lines include lower power losses and low cost of construction.

In 1973, the construction of a ± 400 kV HVDC transmission line was planned from the Coal Creek generation station near Underwood in North Dakota to the Dickinson converter station near the Twin Cities (Minneapolis and St. Paul) in Minnesota. This line extends 708 km. It is inverted to HVAC for further transmission to a local distribution system. In the planning stage and during construction processes, public opposition to high transmission lines and the use of new technology of DC transmission was voiced in different counties of Minnesota. At the time when the power line was constructed, the health and safety issues related to air ions, DC electric field, and ozone remained (Mains 1983; Wellstone and Casper, 2003). In 1975, the Minnesota Environmental Quality Board (MEQB) ordered environmental reports on the project, and formed a citizen's route-evaluation committee, which held 21 information meetings and 23 days of public hearings. A testimony report of over 3,000 pages was published, from which the farmers and landowners voiced the following concerns: (1) the use of eminent domain by the cooperatives to take over private land, (2) the financial compensation due to them for the land, (3) the loss of prime farmland for a large energy project, including interference with mechanical irrigation of the fields and with aerial application of agricultural chemicals, (4) the amount of ozone generated by the transmission lines and its effects on crops, animals, and people, (5) the hazards of working with large farming equipment under the lines, (6) the direct effects of the electric and magnetic fields caused by the lines on people and animals, (7) the relationship between this line and emerging controversies concerning exposure to the fields from AC high-voltage lines, (8) the lack of notice to the landowners from the counties, the state, and the cooperatives, (9) preemption of county authority by the state, (10) the fear that the project was mainly to provide power for a large metropolitan area and not for the farms, as they originally had been promised (Mains, 1983). The report presented that there were increasing concerns about compensation for private property, the effects of ozone on crops, animals, and people, the safety of farming under power lines, and the direct effects of electric and magnetic fields on animals and people.

For the adverse health effects from HVDC transmission lines, the Board asked the Minnesota Department of Health (MDH) to report on the health and safety effects of both HVDC and HVAC lines (Banks et al., 1977). The MDH completed the task with its conclusions as follows:

With regard to HVDC transmission, insufficient research and experience exists to propose any meaningful performance standards, whether empirically based or otherwise, that have an objective of protecting the public health. As a general observation, there does not seem to be as much concern among the scientific community over DC electric fields. However, this position appears to be based on extremely limited information. In particular, the whole issue of the space charge field needs investigation and clarification. For example, air ions have been suggested as responsible for both adverse and beneficial effects. The subject is controversial and considerable further research is unquestionably warranted.' (III-22-23), 'because of the unique importance of the HVDC transmission line environment to Minnesota, and the State must ensure that an adequate research program is undertaken at the proper time (IV-29).

Banks et al. (1977)

The State of Minnesota established the MEQB and approved a route of ±400 kV HVDC transmission line on June 3, 1976. To approve it, the State required the MEQB to study ozone levels and the effects of ozone on crops. But a movement against the construction of the power lines in Minnesota's Grant, Meeker, Stearns, and Traverse counties had been launched. The opposition grew outside the courthouse, bringing farmers and power company employees face-to-face (Figure 2.18).

In 1977, a science court of experts met to address the unresolved issues, including the health and safety issues related to transmission lines, the impact on agriculture, the necessity of transmission

FIGURE 2.18 Demonstration and protests in Lowry, Minnesota (Courtesy of Mike Knaak, from Wellstone and Casper, 2003.)

lines, the possibility of alternatives, and land security. That same year, a report was submitted by the MDH. The report stated that the study was not sufficiently experienced in HVDC transmission lines and that the extremely limited information did not address the concerns of the people living near the planned transmission line route. In the same year, the report stated the need to address the issue of air ions. Meanwhile, the campaign of those opposed to the construction became more militant and in 1978 the opposition escalated further, but by 1979 the transmission line started commercial operation.

The MDH established a program to monitor the electrical environment near transmission lines and to study the epidemiology of the people living near the line. In 1982, the Minnesota MEQB held a public hearing and reported following: (1) no adverse health effects of HVDC transmission lines, (2) the potential long-term effects of air ions is small, (3) there is no effect of transmission lines on Holstein cattle productivity, and (4) there is no electrical shock to humans in the Right of Way during normal operation. It also reported that (5) there was no effect of induced currents or voltages with proper grounding and (6) there was no effect of ozone on people or plants. After that, the number of protests against the operation of the transmission lines decreased. Although the intensity of static electric and magnetic fields generated from HVDC transmission lines are weak, there is public concern about possible biological and health-related problems of static electric and magnetic fields produced in the vicinity of HVDC transmission lines.

In 1983, the issue of the effects of HVDC transmission lines on humans was first addressed by Robert S. Banks and Roy C. Haupt. They conducted a human health survey on four HVDC transmission lines (Pacific Intertie, Nelson River, Square Butte line and CU line) operating in North America at that time (Banks and Williams, 1983). Agricultural landowners, people living near the power lines, and employees of the power company were surveyed to determine if there were suffering any health effects. The results showed no health effects. In addition, for a cross-sectional epidemiological study of HVDC transmission lines and their effects on health such as headaches, depression, allergies, and complaints, interviews with home visits were conducted (Haupt and Nolti, 1984). They found that at the time of the epidemiological study, the transmission lines had been in operation for 12 years. The exposure group consisted of 245 residents within 220 m of the power transmission lines (Right of Way) and the control group consisted of 193 residents more than 1 km away from the transmission lines, for a total of 438 residents. Although the population in this report was small, the data was carefully collected and analyzed so that there was no difference in the health status between the two groups and that no health effects could be linked to HVDC transmission lines. However, there were arguments against such report, such as that the results are worthless because the electrical environments were not identified (Bank and Williams, 1984; Haupt, 1984).

A Scientific Advisory Committee formed by the State of Minnesota concluded that health and safety evaluations of the ±400 kV HVDC transmission lines did not indicate that exposure to static electric fields and air ions were likely health hazards (Bailey et al., 1982, 1986, 1997). It was showed that no difference was found in milk production, calving intervals, rate of culling for reproductive problems, and incidence of abortions before and after the line was energized (Martin et al., 1986). The long-term exposure studies did not produce evidence for biological effects of HVDC transmission lines on people, growth, and reproduction of crops and beef cattle (Angell et al., 1990).

Over the history, there was large laboratory research on static electric fields and air ions (Charry and Kavet, 1987), but only few studies had been done for HVDC transmission lines. In the 1970s, the HVDC transmission line was introduced and this introduction triggered interest for the assessment of the biological effects to exposures to static electric fields and air ions. There are a wide variety of ions in the atmosphere (Reiter, 1992). These ions can originate from both natural sources such as cosmic rays, radioactivity, and water splashing and from artificial sources such as the HVDC transmission lines produced corona ions when the voltage of transmission lines is high enough to cause corona breakdown around cable. There are several reviewed papers concerning the effects of air ions on humans (Alexander et al., 2013; Perez et al., 2013).

As part of human health evaluation, the International Agency for Research on Cancer (IARC) assessed the potential carcinogenicity of static electric and magnetic fields and categorized them in group 3, *not classifiable as to their carcinogenicity to humans* (IARC, 2002). Many studies have been conducted on animals exposed to DC magnetic fields (Simon, 1992). The WHO concluded that no irreversible effects have been reported from magnetic fields up to 2 T (WHO, 1987). Based on the evaluation of biological effects research, the WHO carried out the human health assessment of static electric and magnetic fields and published a monograph of the Environmental Health Criteria (EHC) for them in 2006 (WHO, 2006), in which the health risk assessment was announced as follows:

> Static electric fields: There are no studies on exposure to static electric fields from which any conclusions on chronic or delayed effects can be made. IARC (IARC, 2002) noted there was insufficient evidence to determine the carcinogenicity of static electric fields. Static magnetic fields: The available evidence from epidemiological and laboratory studies is not sufficient to draw any conclusion with regard to chronic and delayed effects. IARC (IARC, 2002) concluded that there was inadequate evidence in humans for the carcinogenicity of static magnetic fields, and no relevant data available from experimental animals. Their carcinogenicity to humans is therefore not at present classifiable.
>
> *WHO (2006, p. 8)*

Today, the AC electricity from power stations through transmission and distribution lines to the end users is still used. Due to the development and incorporation of AC-DC converters, HVDC transmission lines are being challenged. The inverter used to transfer three-phase voltage is synchronized with the transmission line. DC electricity can be directly produced by solar/photovoltaic panels and offshore windfarms. HVDC systems are used for long-distance transmission of energy and for interconnection between HVAC and HVDC. HVDC transmission lines have been already operating for a long time in the United States, Canada, Great Britain, Sweden, Norway, China, Japan, etc. China has built 25,000 km Ultra-HVDC transmission lines 800 and 1,100 kV. In the past, although many HVDC transmission lines were built in the 1970s to 1980s, the study of human perception of DC electric field was rarely investigated. As new HVDC transmission lines are proposed, concern on the environmental and human health effects of static electric and magnetic fields associated the operation of these lines have again increased. For example, HVDC transmission lines are to be employed in Germany as part of new long-distance power transmission lines from the renewable energy sources, offshore windfarms, in the north to the western and southern metropolitan industrial areas. Recently, the research group of Aachen University started to investigate the human perception thresholds for DC electric fields under whole body exposure and well-controlled environments in order to determine statistically verified detection thresholds of the population (Jankowiak et al., 2021: Kursawe et al., 2021). The Commission on Radiological Protection (SSK) recommended further research projects will be conducted on perception, mainly in the form of human studies under well-controlled conditions (SSK, 2013).

Systematic reviews on the biological effects of static electric and magnetic fields on humans, vertebrates, invertebrates, and plants have been published (Driessen et al., 2020; Petri et al., 2017; Schmiedchen et al., 2018). It is necessary to further evaluate the biological and human health effects in relation to HVDC transmission lines, although the WHO already carried out the human health assessment of static electric and magnetic fields.

2.6.2 Extremely Low-Frequency Fields

In 1968, the U.S. Navy proposed a submarine communication system, called Project Sanguine. Project Sanguine was going to build and operate a system with a frequency at 76 Hz on a large tract of land from Clam Lake, Wisconsin to Michigan. However, the people living in Clam Lake near the facility campaigned against Project Sanguine. This Project was not successful because of its cost, protests, and environmental impact. After Project Sanguine, in 1975, the U.S. Navy proposed a new Project called Seafarer (Surface ELF Antenna for Addressing Remotely-deployed Receivers) Project. This new Project was smaller than Project Sanguine. The opposition did not abate. The U.S Navy established the Committee on Biosphere Effects of Extremely-Low-Frequency Radiation to study possible health effects of the signals emanating from the Seafarer system. A study of the environmental impact also started in 1975. On February 16, 1978, the 39th president James Earl Carter terminated the Seafarer Project. The committee concluded and recommended for biological effects the following:

> Beginning with a decision to build Seafarer and continuing into the period of its operation, research should be conducted to increase the knowledge of the basic effects of weak ELF fields associated with Seafarer. This should include fundamental research concerned with the biophysics and physiology of magnetic-and electric-field detection and use, and studies of the related behavior of birds, insects, bacteria, and electrosensitive fish. In addition, research on the underlying mechanism of cell division and on information processing and integration in complex nervous systems in relation to ELF environments should be conducted and evaluated as part of the requirement for continued monitoring of the operating Seafarer system for its possible effects on biologic systems.

> *NRC (1977, p. 53)*

On October 8, 1981, under the direction of the 40th president, Ronald Wilson Reagan (1911–2004), the U.S. Navy proposed a scaled-down ELF system with 76 Hz operation, called Project ELF. As a result, Project ELF (a smaller version of the Seafarer Project) started. It consisted of two transmission facilities, one at Clam Lake in northern Wisconsin and another at Republic, Michigan, with a total of 135 km of an above-ground transmission antenna. Although opposition to its construction developed, its movement gradually faded away and the ELF facilities at Clam Lake began operating in 1985 and in Michigan in 1989. During this period, monitoring of the impact on the ecosystem around the facility began in the 1970s, and since 1982 the distribution of arthropods, earthworms and amphibians in the soil was monitored in addition to electric and magnetic fields. The results were reported in 1997 (NRC, 1997b). The system operated from 1989 until 2004.

   During the Cold War in the 1960s, Soviet Union investigators presented a number of reports concerning the incidence of compliance in substation workers (Asanova and Rakov, 1966; Sazonova, 1967). In 1966, Asanova and Rakov reported neurological symptoms in workers at a high voltage power switchyard. It was shown that the analysis of the major causes was inadequate and that the health effects probably did not originate from induced current flow. Spark discharges, low-frequency noise, and ozone smell originating from inside substation and distribution equipment was shown to be more likely the agents responsible.

   The above-mentioned original papers from the Soviet Union were written in Russian. These papers did not attract the attention of Western people until a presentation at the 1972 International Conference on CIGRE in Paris, where an overview of the problem was revealed to researchers in Western countries. At the conference, Korobkova and others examined the health of 250 workers who work at 500 and 700 kV switchyards and reported that prolonged work without protective gear against electric shock impaired the central nervous system, the heart and the vascular system, and in young people there was a loss of sexual desire (Korobkova et al., 1972). It was also noted that this tended to increase with prolonged work in an electric field. In response to this report, researchers from Western countries questioned the lack of control over the study, and efforts were made in Germany, France, Canada, and the United States to reconfirm the results from the Soviet Union. As a result of these efforts, the health effects seen in the reports from the Soviet Union could not be replicated. In retrospect, the Soviet report was important and significant in the sense that it was the first to address the health effects of electric fields exposure.

   Up until the early 1970s, it was assumed that exposure to electromagnetic fields at environmentally relevant field strengths produced no harmful effect on humans. In the early 1970s, concerns about health hazards associated with electric field from transmission lines and from the workplace led to many studies and continued controversy about whether adverse health effects on humans occur. In particular, the construction of Extra High Voltage (EHV)-AC transmission lines gave public and scientific interest. The scientific research focused on the possible effects of electric fields, audible, and radio noises and ozone. The papers on the biological and health effects of high electric field were published (Hauf, 1974, 1976; Malboysson, 1976; Silney, 1976). Apart from the work of the Soviet Union investigators, there were almost no reports of harmful effects related to electric fields. These reports include medical studies of 10 energized line workers exposed to energized 350 kV transmission lines (Singewald, 1973), medical studies of 56 substation maintenance workers at 735 kV substations (Roberge, 1976), medical studies of 53 workers with over 5 years at 400 kV substations (Knave et al., 1979), and medical studies of residents near 200/400 kV transmission lines (Strumza, 1970), etc.

A study conducted to confirm the reports from the Soviet Union was made by professor Kouwenhoven and his group of John Hopkins University (Kouwenhoven et al., 1966, 1967). The group looked at the health effects on linemen working on live lines at 138 and 345 kV electric fields and added evidence to refute the hazard report from the Soviet Union. A total of 11 male workers, ranging from 30 to 47 years of age, were subjected to a total of 42 months of continuous health examinations from December 1962 to May 1966. During this period, five physical examinations were conducted at the John Hopkins University Hospital. The results of these periodic physical examinations showed no effect of the electric fields. The health of the workers engaged in these live line operations continued to be followed for a total of 9 years until 1972. Although one of the 11 workers who participated in the follow-up study did not participate in the study after 1967, Kouwenhoven and his colleagues reported that the electric field did not cause any health problems based on nearly 9 years of research. Kouwenhoven who led this study was born in Brooklyn, New York, and began investigating the effects of electric shocks on the human body and electrical stimulation of the heart in the 1920s, and is known as a pioneer in the study of the effects of electricity on human.

In another subject, In 1973 Louse B. Young (1919–2010), a physicist and science writer, who lived in a small village in Ohio, published the book *Power over People* (Young, 1973, 1992). Back cover of this book showed a photograph of a woman, presumably Mrs. Young, holding two fluorescent bulbs, while standing directly under a high voltage transmission line and the fluorescent bulbs glow brightly in a dim light. The book described how plans were being made to build a high voltage transmission line across the land where a family had lived for several generations, and how they joined the movement to oppose it from 1969 to 1972, when it was being built. Opposition to the line was strong in 1970s in Minnesota, New York, and other states. Until the first half of the 1970s, concerns for the biological effects of electromagnetic fields generated by high voltage power transmission lines had focused on induced voltages on objects and human. For example, concerns were on perceptible shocks experimentally by humans near metallic structures, near and under high voltage power transmission lines, and installation. Humans sensed clearly vibrations in the hair.

In the late 1970s and early 1980s, human health issues came from the epidemiological studies that established a link between an increase of cancer in child and adult due to exposure to extremely low-frequency electromagnetic fields (ELF-EMF). The issue of the influence of electromagnetic fields on human health was taken up widely. The research expanded in the 1980s to include the study of a possible association between ELF-magnetic fields and cancer in residents near power transmission lines. In 1979, Nancy Wertheimer (1927–2007), an epidemiologist at Medical Center, University of Colorado, and Edward A. Leeper, an engineer, compared the distribution of childhood cancer in the Denver area of Colorado with wire code, the index of degree of current flow of power lines. Actual exposure was not measured. They published its results as an epidemiological study in a journal that showed an increased risk of leukemia in children living near power transmission lines (Wertheimer and Leeper, 1979). They inferred that a possible association exists between childhood cancer and exposure to ELF-magnetic fields. The incidence was roughly doubled in the exposed cases compared to the control cases. Since the publication of the Wertheimer-Leeper study, many new research efforts related to the safety of

ELF-magnetic fields emerged in both epidemiological and biological areas. Wertheimer was the seventh recipient of the d'Arsonval Award of Bioelectromagnetics Society in 1999. Many reports on the possibility of a link between residential and occupational exposures to ELF-magnetic fields and cancer appeared. The topic focused on leukemia and brain tumor in children, and leukemia, brain tumor, and breast cancer in adults. The occupational study focused mainly on leukemia and brain tumor, although there were concerns with suicide, depression, neurological diseases like Alzheimer's disease, and amyotrophic lateral sclerosis. Laboratory studies were divided into two areas, in vitro and in vivo studies. The approach of in vitro studies has been used in an effort to find possible mechanisms for interaction between ELF-magnetic field exposure and biological systems. The in vivo studies provide the information on how ELF-magnetic fields interact with biological system, using whole organisms such as laboratory animals and humans. In 1982, Milham studied the relationship between the cause of death and occupation for 438,000 individuals between 1950 and 1979 (Milham, 1982). He found that for occupations where opportunities for exposure to ELF- electromagnetic fields were believed to be high, a high death rate from leukemia was observed. In 1986, Lennart Tomenius, a medical officer of Health, county of Stockholm, Sweden, published a replication of the Wertheimer and Leeper study with a 50 Hz electricity supply (Tomenius, 1988).

At the same time when the Seafarer Project started, the Power Authority of the State of New York (PASNY) announced a Project to build a 765 kV AC transmission line to transmit the electricity generated by a hydroelectric power at James Bay from Canadian border across 250 km in New York State (Young, 1973). At roughly the same time, Rochester Gas and Electric Corporation and Niagara Mohawk Power Corporation decided to build another 765 kV AC transmission line. Interestingly, professor Robert Otto Becker owned his property close to one of the planned transmission lines. Becker and his co-worker, Andrew A. Marino, an Orthopedic Surgery at the Veterans Administration Hospital, had been studying the effect of electricity on living systems. Later, Marino became a professor at Louisiana State University in Shreveport. They were concerned about the health effects due to exposure to ELF-electromagnetic fields produced by these transmission lines. Gradually, the campaign against the construction became more and more intense. In particular, concerns about the environmental impact of the construction of the transmission line and the health effects of ELF-electromagnetic fields generated from it, became so great that public hearings were held. Historically, Becker and Marino were known as the persons leading the early opposition to high voltage transmission lines.

The public hearing by the New York State concerning the construction of the PASNY 765 kV transmission line, which began in 1973, was marked by intense arguments concerning the biological effects of ELF-electromagnetic fields. Research efforts in this area were intensified after this public hearing. Especially, the Department of Energy (DOE), the Electric Power Research Institute (EPRI), and the New York State expanded and initiated research in this area. In Europe, the Central Electricity General Board (CEGB) (UK), the Ente Nazionale per l'energia ELettrica (ENEL) (Italy) and the Électricité de France (EDF) (France) also initiated researching.

The public hearings were held over a 3-year period and more than 10,000 pages of testimonies were submitted into the record. As a result though public hearings, it was agreed that ELF-electromagnetic field research project on the safety of transmission lines would be set up. PASNY and the seven electric companies financially supported the Project and the New York State Department of Health oversaw it. The 5-year five-million-dollar Project ran through 1987, and was initiated with the name of the "New York State Power Lines Project." The project was launched in 1981, and a report summarizing the Project was compiled and completed in 1987 (Ahlbom et al., 1987). The Project covered seven themes involved 16 research groups and the goal of the Project was to determine the effects of ELF-electromagnetic fields on human, animals and plants. The studies carried out in the Project did not report any results of concern for the health effects of ELF-electromagnetic fields. A major theme of research in this Project was to replicate and confirm the epidemiological results of Wertheimer and Leeper relating childhood leukemia to ELF-magnetic field exposure. This confirmation study was given to David A. Savitz, an epidemiologist, professor at School of Public Health at the University of North Carolina, and later Brown University.

Savitz and his co-workers concluded that prolonged exposure to low-level ELF-magnetic fields may increase the risk of developing cancer in children (Savitz et al., 1988). Other epidemiological studies have indicated that there is a possible association between ELF-electromagnetic field exposure and cancer in humans. After the publication of these reports, many research efforts focused to investigate possible associations between ELF-magnetic field exposure and cancer in humans. The final report of the New York State Power Lines Project appeared July 1, 1987, which announced the following as conclusions:

> In conclusion, results of the New York founded projects document biological effects of electric and magnetic fields in several systems. The variety of effects of magnetic fields have not been previously appreciated. Several areas of potential concern for public health have been identified, but more research must be done before final conclusions can be drawn. Of particular concern is the demonstration of possible association of residential magnetic fields with incidence of certain childhood cancers. Further study of this possible association and mechanism to explain it are important. The variety of behavioral and nervous system effects may not constitute a major hazard because most appear to be reversible, but they may impact temporarily on human function. Further research should also be done in this area.
>
> *Ahlbom (1987, p. 10)*

Later, epidemiological and human exposure studies were conducted on the safety of the ELF-magnetic field exposure including: in animal experiments, carcinogenesis, reproduction, behavior and perception, and neuroendocrine behavior. Animal experiments did not provide sufficient results to demonstrate the influence on human health.

In the 1990s, researchers in Sweden reported the results of a study on the adverse health effects of ELF-magnetic field exposure (Feychting and Ahlbom, 1993). Different researchers reported the same health concerns from ELF-magnetic field exposure. In order to clarify the credibility of the results of ELF-magnetic field exposure, the focus of the researchers shifted from the effect of electric fields to identifying the effects of ELF-magnetic fields on human health, and research of ELF-magnetic field exposure were performed around the world.

In 1992, the U.S Congress authorized the 5-year National EMF research and communication program known as Electric and Magnetic Fields Research and Public Information Dissemination (EMF RAPID) Program in the Energy Policy Act (1994–1998). This Program was funded by Federal and non-Federal funds. The U.S. DOE administered this Program, and the National Institute of Environmental Health Sciences (NIEHS) and the National Institute of Health supervised the health effects study and risk assessment. The main goal of this program was to determine whether or not exposure to power frequency electromagnetic field (ELF-EMF) from the generation, transmission, and use of electricity affects human health and if so, to reduce the exposure and to communicate electromagnetic field information to the public. The Project was completed after 5 years and a final report was submitted to the U.S. Congress. The final report of this Program said as follows:

> The scientific evidence suggesting that ELF-EMF exposures pose any health risk is weak. The strongest evidence for health effects comes from associations observed in human populations with two forms of cancer: childhood leukemia and chronic lymphocytic leukemia in occupationally exposed adults. While the support from individual studies is weak, the epidemiological studies demonstrate, for some methods of measuring exposure, a fairly consistent pattern of a small, increased risk with increasing exposure that is somewhat weaker for chronic lymphocytic leukemia than for childhood leukemia, In contrast, the mechanistic studies and the animal toxicology literature fail to demonstrate any consistent pattern across studies although sporadic findings of biological effects have been reported. No indication of increased leukemias in experimental animals has been observed.
>
> *NIEHS (1999)*

Research was conducted in Japan around the 1960s concerning the biological effect of electrostatic induction by a 500 kV transmission line. It was shown that perception of electrostatic induction could be prevented by application of technical standard limits. The intensity of the electric fields near road surfaces should be below 3 kV/m (Takagi and Muto, 1971). In Japan, the health and biological effects of power frequency electromagnetic fields are taken charge by the Ministry of International Trade and Industry (MITI, now METI). MITI set a technical standard for electric fields of 3 kV/m in 1976 to prevent the electrostatic induction. Since then, MITI organized the Electric Field Effect research group and the Electromagnetic Field Effect research group in the 1980s and 1990s, respectively. The former group concluded that electric fields produced in residential environments have no harmful effects on humans. The latter surveyed the scientific papers and concluded that

> It was said that at present, there is little need to regulate or standardize power frequency magnetic fields based on the effects on human health. However, it is important to scientifically investigate the effects of power frequency magnetic field on human health in order to provide accurate information.

The METI has the responsibility for the electric power facilities. After long research and discussion, METI decided the Japanese regulation to be 50/60 Hz for electric and magnetic fields. METI introduced the limit of magnetic field strength of 200 µT which is based on ICNIRP guidelines, in order to protect the general public from acute health effects. The Central Research Institute of Electric Power Industry (CRIEPI), funded by Japanese electric utility companies, has conducted the study concerning to biological effects of ELF-electromagnetic field exposure. CRIEPI initiated research studies in the 1970s, starting with the evaluation of the biological effects of electric fields. The exposure experiments were conducted to investigate the effects of AC and DC electric fields on the growth of trees and the harvest of crops and its effects on reproduction, blood components, and cellular activities using experimental animals such as mice and hamsters to clarify possible effects related to the planned Ultra High Voltage transmission lines. In addition to animal experiments, measuring and calculating methods of AC and DC electric fields have been developed. In the next steps, CRIEPI has conducted a study to investigate whether ELF-magnetic field effects on cellular function, reproduction, neuroendocrine system occur or not using cells and experimental animals. During the overall exposure study, CRIEPI has constructed animal care and animal exposure facilities and the in vitro exposure unit. In order to investigate the in vivo biological effect of electric fields, CRIEPI developed the electric field facility, which generated the maximum 2.5 kV/m at 50 Hz. The in vivo ELF-magnetic field exposure system produces both rotating polarized and linearly polarized ELF-magnetic fields up to maximum of 0.5 mT at 50 Hz. For the in vitro exposure, a facility producing up to 10 mT with rotating and linearly polarized ELF-magnetic fields was constructed (Shigemitsu et al., 1993; Yamazaki et al., 2000). During these times, co-operative activities between MITI of Japan, the Department of Energy (DOE) in USA, and CRIEPI investigated and supported the biological effects of ELF-electromagnetic fields on operant and the social behavior in baboons which was carried out by the South West Research Institute (SwRI), San Antonio, Texas (Greenebaum, 1995).

The evaluation of biological studies and human health effects of ELF-EMF and intermediate frequency (IF) magnetic field studies in Japan began in the 1980s. Yoshitaka Otaka has briefly reviewed Japanese research efforts on ELF-EMF (partly including the study of RF (RF-EMF)) (Otaka, 2001). Historically, from the end of the 1970s to the early 1980s, research on biological and engineering issues had begun in several universities and research organizations across Japan (Amemiya, 1994; Kato, 2006; Miyakoshi and Shigemitsu, 2014; Takebe et al., 2000; Taki 2016). Some research involved studies on the biological effects of electric fields on animals and the distribution of the electric fields and induced currents from a high voltage transmission line in humans. Models of the human body represented by simple structures such as cylinders were used to estimate induced currents and surface electric fields (Chiba et al., 1984; Shimizu et al., 1988 a, b). The results of these studies explained the electric field distribution inside and outside the human body exposed to a high voltage transmission line (Shigemitsu and Yamazaki, 2012). For biological studies using mice, rats, and cats, carefully constructed exposure facilities were successfully developed (Kobayashi et al., 1983; Negishi et al., 2008; Shigemitsu et al., 1981). These facilities helped the research group and made great contributions to animal-related research (Kato et al., 1986, 1989, 1993; Kato and Shigemitsu, 1997). So far, no basic research data on the biological effects of IF-magnetic field is available (Shigemitsu et al., 2007). Further studies on the biological effects and health risk assessment of IF-magnetic field ranging from tens of kHz to about 100 kHz are needed. Based on Japanese traditional research dedicated to the study of ELF-EMF, pioneering research on the biological effects of IF-magnetic field started in Japan in the early 2000. The CRIEPI developed in vivo and in vitro exposure facilities for the investigation of toxicological and reproductive effects in experimental animals (Nakasono et al., 2008; Shigemitsu et al., 2009). In cellular experiments, microorganisms and cells were used to study the mutagenic effect of IF-magnetic field, using the newly built in vitro exposure facilities (Fujita et al., 2007; Miyakoshi et al., 2007).

During the last four decades, research on the interaction between extremely low-frequency (50/60 Hz) ELF-EMF and biological systems has increased significantly (WHO, 1984, 1987). In addition to biological research, major research on the possible adverse effects of ELF-EMF on human health has been completed. During this time, there have been many review publications on the potential biological and human health effects of ELF-EMF (NRC, 1997a; NIEHS, 1998; IARC, 2002; ICNIRP, 2003). In 1996, the WHO established the International EMF Project which focused first on research to determine the link between ELF-magnetic field exposure to human and childhood leukemia. In 2001, the IARC assessed the carcinogenicity of ELF-magnetic field exposure, which resulted in a health risk assessment of ELF magnetic field exposure. IARC categorized ELF-magnetic field to be as *possible carcinogenic to humans* in the IARC carcinogenic category of 2B. Based on the evaluation of biological and human health effect research, the WHO carried out the human health assessment of ELF-EMF and published a monograph of the EHC for ELF-EMF (up to 100 kHz) in 2007 which states:

> Although a causal relationship between magnetic field exposure and childhood leukemia has not been established, the possible public health impact has been calculated assuming causality in order to provide a potentially useful input into policy.

> *WHO (2007, p. 12)*

This means that the possibility of causality was taken into consideration in the health risk assessment. Even after the IARC categorized ELF-magnetic field as Group 2B, the discussion about a causal relationship between ELF-magnetic field and childhood leukemia still has controversy. On the other hand, the ELF-electric field has no association with carcinogenicity: *not classifiable as to its carcinogenicity to humans* (Group 3) (IARC, 2002).

2.6.3 Radiofrequency Fields

After the experimental discovery of electromagnetic waves by Hertz, the medical and military applications using RF fields have been developed since roughly around the 1930s. In the late 1940s, reports on microwave exposure were related to the microwave diathermy in medical research. In the early 1950s, the military and industrial concern focused on the hazards related to microwave exposure. In particular, with the introduction of medical use for diathermy, the debate over non-thermal and thermal effects of microwaves began and the debate continued still even after World War II.

During and after World War II, microwave technology was studied not only for military use but also for civilian use. Communication technologies (e.g., wireless telephony) have advanced rapidly in recent years; hence, microwave energy now is ubiquitous in the atmosphere. In the late 1940s, it was reported that clicking sounds could be heard near a radar station. The RF hearing effect was systematically studied about 15 years later (Frey, 1961), concomitant with other studies of microwave effects on other organs and tissues, such as the eyes and the nervous system. Two shaped development of early research on the biological effects of microwave radiation was considered, medical applications of microwave diathermy and the health hazards (Cook et al., 1980).

During the 1950s, increased use of microwave energy in diathermy, radars, and industrial equipment stimulated interest in the biological effects of microwave energy. In the 1950s and 1960s, the research activity of RF fields and microwaves in bioelectromagnetism had shifted from medical applications to military use. The research changed from medical applications such as diathermy to hazard identification of microwaves. It plays deeply important roles in developing researches on the biological and health effect of microwave exposure. With the rapid expanse of military applications of microwave energy, in 1957, the "Tri-Service Program" was initiated to investigate the biological effects of microwaves in order to develop the safety standards of microwave exposure. The primary concern of the Tri-Service program which continued until 1960 was the hazard issue, not medical applications (Lin, 1994; Steneck, 1984).

Historically, the thermal effect of RF fields was important issue which is given by the temperature rise due to RF field energy absorption into the body. This thermal effect was the basis for setting safety standards to limit exposure to RF field.

Between 1953 and 1979, one of the Cold War affairs was the Moscow Signal episode which has acquired important discussion on the interactions between RF fields and human health (Steneck, 1984). During the Cold War, signals of low-level microwave radiations from nearby buildings were targeted to the floors of the US embassy in Moscow. Although the existence of this microwave signal had been kept secret, in 1976, a 2-year epidemiological study was conducted by the US Department of State and the Johns Hopkins University (Lin, 2017). The study involved the Moscow embassy's 1,827 person staffs and their families, 3,000 persons and 2,561 employees and their families, 5,000 persons as a control population from other Eastern US embassies including Belgrade, Budapest, Leningrad, and Prague. The purpose of this study was to assess any difference in morbidity and mortality between the Moscow embassy and the control population. After the investigation, in 1978, the report of this study appeared. The conclusion was that the Moscow embassy and the other groups did not significantly differ in overall and specific mortality, and no compelling evidence was observed to implicate the Moscow embassy microwave signals in any adverse health effects. In addition, the study pointed out that the study population was relatively young, and it might have been too early to detect long-term health and mortality outcomes (Lin, 2017). Forty years after the publication of the 1978s epidemiological study, Jose A. Martinez, researcher of the Technical University of Cartagena, Spain, pointed out, while reviewing the original data and applying several Fisher exact tests, that the results clearly show a significantly worse health status for the Moscow group (for both male and females), as well as for the overall sample (Martinez, 2019). The episode of the Moscow signal is still a controversy.

Under the asking of the U.S. Food and Drug Administration (FDA) of the Department of Health and Human Service, the National Academies organized a workshop to identify research needs and gaps in knowledge of the biological effects and adverse health outcomes due to exposure to RF energy from wireless communications devices. The first cellular phone was invented in 1973 by the engineer Martin Cooper. It weighted 1.1 kg and could be used for about 30 min. Ten years later, Motorola released commercially its first mobile phone. After the development of the first cellular phone, mobile phone technologies became widespread. The National Academy organized a seven-member committee to plan the workshop. The workshop was held on August 7–9, 2007, in Washington, DC. The report on the research needs and gaps identified by the committee was published by National Research Council (NAC, 2008). The important research needs were found by the committee, included the following ten issues:

1) Characterization of exposure to juveniles, children, pregnant women, and fetuses from personal wireless devices and RF fields from base station antennas, 2) Characterization of radiated electromagnetic fields for typical multiple-element base station antennas and exposure to affected individuals, 3) Characterization of the dosimetry of evolving antenna configurations for cell phones and text messaging devices, 4) Prospective epidemiologic cohort studies of children and pregnant women, 5) Epidemiologic case-control studies and childhood cancer, including brain cancer,

6) Prospective epidemiologic cohort studies of adults in a general population and retrospective cohorts with medium to high occupational exposures. 7) Human laboratory studies that focus on possible adverse effects on electroencephalography activity and that include a sufficient number of subjects. 8) Investigation of the effect of RF electromagnetic fields on neural networks, 9) Evaluation of doses occurring on the microscopic level, and 10) Additional experimental research focused on the identification of potential biophysical and biochemical/molecular mechanisms of RF action.

NRC (2008, p. 2)

After the IARC categorization, the INTERPHONE study was conducted by the IARC. It was based on a large-scale, case-control epidemiological study which was performed in 13 countries in the world during the years 2000–2012 (Interphone, 2011). The results did not prove any connection between the usage of mobile phone and the risk of developing glioma, meningioma, or acoustic neuroma. Eventually, an increased risk of glioma for the largest RF field exposure level was observed. However, the presence of biases and errors in the data prevented a causal interpretation of such results.

At roughly the same time when the workshop was held, the US National Program of Toxicology (NTP) conducted a long term (2 years) bioassay to evaluate the Carcinogenicity and genotoxicity of RF fields on rats and mice (NTP, 2018a, b, 2020). The toxicology studies in rats (Hsd: Sprague Dawley SD) and mice (B6C3F1/N) by the NTP with a $ 30 million budget conducted studies for more than 10 years to help clarify potential health hazards from exposure to RF fields used in 2G and 3G cellular phones. The rats (in the womb) and mice (5–6 weeks old) were exposed to RF fields in special whole-body chambers at 900 (Code Division Multiple Access; CDMA) and 1,900 MHz (Global System for Mobile Communications; GSM). The exposure was intermittent, 10 minutes on and 10 minutes off, 9 hours/day with 1.5, 3, or 6 W/kg of body weight in rats, and 2.5, 5 or 10 W/kg in mice. The results of these studies were associated with: (1) clear evidence of an association with tumors (malignant Schwannoma) in the hearts of male rats, (2) some evidence of an association with tumors (malignant gliomas) in the brains of male rats, (3) some evidence of an association with tumors (benign, malignant, or complex combined pheochromocytoma) in the adrenal glands of male rats, and (4) it was unclear for female rats and male and female mice whether the cancer observed in the studies were associated with exposure to RF fields (NTP, 2020). The NTP study used 2G and 3G cellular phones technologies and did not apply 4G or 5G technologies, and it also did not investigate frequencies and modulations used for Wi-Fi (NTP, 2020).

In 2011, an expert working group of the IARC reviewed the large amount of published literature and categorized RF field as *possibly carcinogenic to humans (Group 2B)*. From the evaluation:

1) there is limited evidence in humans for the carcinogenicity of radiofrequency radiation. Positive associations have been observed between exposure to radiofrequency radiation from wireless phones and glioma, and acoustic neuroma, 2) There is limited evidence in experimental animals for the carcinogenicity of radiofrequency radiation.

IARC (2013)

Finally, we introduce two important points for human health issues related to RF fields: (1) the WHO has launched a systematic review evaluation on the potential health effects of RF field exposure. The systematic review includes three topics: (a) cancer (animal studies), (b) adverse reproductive outcomes (animal and in vitro studies), and (c) effect of exposure to heat from any source on pain, burns, cataract, and heat-related illnesses (WHO, 2020). The review result will require the next few years to be available. The health outcomes required for the review have been prioritized by experts (Verbeek et al., 2021). (2) An Advisory Group of IARC employed of 29 scientists from 18 countries recommended a priorities evaluation to the IARC so that the inclusion of RF fields as a part of the agent lists of carcinogenicity is reassessed for the IARC Monograph program in the next 5-year period (IARC, 2019). RF field was again recommended with high priority for evaluation. This priority was assigned on the basis of evidence of human exposure and the extent of available evidence for evaluating carcinogenicity (IARC, 2019).

As the definition of ICNIRP, RF fields are in the frequency range from 100 kHz to 300 GHz (ICNIRP, 2009). The future technologies of wireless telecommunications, the development of 5 G mobile networks, Wi-Fi, Bluetooth, and others are included in this frequency range. Wireless connectivity through the internet (Internet of things) has also become increased. With the development and the application of these new technologies, exposure to RF fields of the general population will become increased. It is important that in the future the health assessment of the effects of exposure to RF fields should be more systematically reviewed and evaluated (Karipidis et al., 2021: Wood et al., 2021). The health outcomes include cancer, adverse birth and pregnancy outcomes, cognitive impairment, electromagnetic hypersensitivity, etc. As mentioned above, recent laboratory and epidemiological studies led to the conclusion that RF fields should be re-evaluated and re-categorized as a human carcinogen. In future, there is a need for more multidisciplinary studies.

2.7 Conclusion

Since ancient times, electricity and magnetism have been considered to be separate. In 1820, Oersted showed that electricity and magnetism are common mechanical phenomena. To explain this phenomena, he used for the first time the term "electromagnetism" in his paper (Oersted, 1820). Afterwards, Faraday immediately gave the explanation and meaning to this phenomena, and actively used the term "electromagnetism," which became established as a scientific term within a year (Faraday, 1821, 1822). After Oersted's discovery, the scientific knowledge in electromagnetism was improved. This improvement led to a better understanding of the basic electromagnetic phenomena in bioelectromagnetism.

While looking at the history of electromagnetism, the long history of bioelectromagnetism and its development from basic research to advanced technology was chronologically reviewed. It covered mainly from the consideration of the generation and detection of electromagnetic fields within biological systems, through the various applications of electromagnetic phenomena, to the debate about possible health hazards associated with exposure to static electric and magnetic fields, ELF-EMF, and RF fields. Research developments in electromagnetism provide a better understanding of the basic interactions between electromagnetic fields and biological systems in bioelectromagnetism. Although bioelectromagnetism has a long research history, the term "bioelectromagnetism" gained popularity around 1980. The time around 1980 was the time when the Bioelectromagnetics Society (BEMS) was founded and the *Journal of Bioelectromagnetics* was first published (BEMS, 2005). This Society covers research related to electromagnetic phenomena ranging from static fields through RF field to terahertz frequencies and acoustic energy with biological systems. It has passed over 40 years since the establishment of the BEMS. In 2021, from the merger between the BEMS and the EBEA (European Bioelectromagnetics Association), a new Society called the BioEM will arise and will expand the research and activity of bioelectromagnetism more internationally.

As mentioned in this chapter, the understanding of the nature of the electromagnetic phenomena in modern science was made gradually through the experiments made by Oersted, Faraday, Ampère, Maxwell, Hertz, and others. In particular, Maxwell proposed and unified the mathematical description of the electromagnetic wave with a set of equations, which was formulated by Oliver Heaviside (1850–1925), an English mathematician and physicist, later called the Maxwell equation, from which it is clear that the electromagnetic waves travel in free space with the speed of light. This theory was later confirmed experimentally by Hertz. After these scientific developments, we began to understand the character of the electromagnetic waves (radiation). Most importantly, the electromagnetic radiation behaves as both particles (photons) and waves, and that it is divided into several bands depending on wavelengths.

The electromagnetic radiation are of two types: non-ionizing radiation such as electric and magnetic fields, RF fields including microwaves, infrared, UV, and visible radiations. The other is ionizing radiation such as part of the UV radiation, X-rays, and gamma rays. Non-ionizing radiations are ubiquitous around us on our daily living and working environments. The ubiquity of our daily exposure to non-ionizing radiations has increased. For example, we use the telegraph, microwave oven, radio, television, wireless telecommunication such as WLAN, cellular phones, Bluetooth, and WiMax on a daily basis without thinking about the principle of their operations. MRI used in medicine has also advanced on the basis of the discoveries made by Maxwell and Hertz. People use many highly developed technologies conveniently without knowing or remembering who invented or discovered them. Due to the development and use of these technologies, the safety and health effects of non-ionizing radiations have also been studied and questioned.

As mentioned above, the technology we use employs mainly electromagnetic phenomena from static fields through low-frequency fields to RF fields. This broad band spectrum of electromagnetic radiation is widely employed for wireless power transfer systems, telecommunications, therapeutic and diagnostic applications in medicine, etc. Now, the applications are quickly expanding using higher frequency ranging from millimeter to Tera-Hertz (10^{12} Hz) such as in optical radiations for medicine and chemistry (Ueno, 2020).

Owing to the development of new technologies producing electromagnetic fields, the electromagnetic environment around us will become more ubiquitous and more complex in domestic, industrial, and medical environments, which will bring new environmental issues. With these rapid expansions, public concern regarding human health effects is likely to increase more than in the past. In bioelectromagnetism, the electromagnetic environment will be the center of attention as areas where the safety for humans and global systems are not fully understood. The research studies related to bioelectromagnetism have been changing over time and they are expected to change furthermore with the newly developing technologies. A further consideration for these researches is the need to bring researchers from other disciplinary fields such as physics, chemistry, biology, medicine, biophysics, engineering, and social science.

Acknowledgments

The authors would like to thank Professor Emeritus Masao Taki of Tokyo Metropolitan University, Dr. Soichi Watanabe of the National Institute of Information Technology, Dr. Koichiro Kobayashi, professor at Iwate University, Dr Susumu Nakajima, director of the Moriyama Memorial Hospital, Drs. Kenichi Yamazaki, Satoshi Nakasono, Masayuki Takahashi and Atsushi Saito of CRIEPI, Mr. Yoshinobu Kawahara of Tokyo Electric Company, Dr. Hiroaki Miyagi of HM Research & Consulting Co., Ltd, Dr. President Amane Hayashi and Dr. Carlos Ordonez of Forestic Co., Ltd, for their kindly help for preparing this chapter.

References

Abe Z, Tanaka K, Hotta K and Imai M (1974): Noninvasive measurements of biological information with application of nuclear magnetic resonance. In: *Biological and Clinical Effects of Low Magnetic and Electric Fields*. ed. Llaurado JG, Sances A and Battocletti JH. – – 295–317. Charles C. Thomas, Springfield.

Ahlbom A, Albert EN, Fraser-Smith AC, et al. (1987): *Biological Effects of Power Line Fields* New York State Power Lines Project Scientific Advisory Panel Final Report.1-154, New York State Power Lines Project.

Alexander DD, Bailey WH, Perez V, Mitchell ME and Su S (2013): Air ions and respiratory function outcomes: a comprehensive review. *J Nagat Results Biomed* 12:14.

Amemiya Y (eds) (1994): Special issue on biological effects of electromagnetic fields. *IEICE Trans Commu* E77–B:683–767.

Angell RF, Scott MR, Raleigh RJ and Bracken TD (1990): Effects of high voltage direct current transmission lines on beef cattle production. *Bioelectromagnetics* 11:273–283.

Aoyagi T (2003): Pulse oximetry: its invention, theory and future. *J Anesthesia* 17:259–266.

Aoyagi T, Kishi M, Yamaguchi K and Watanabe S (1974): Improvement of ear-piece oximeter (in Japanese). *Jap J Med Eletron Biol Eng* 12 (Suppl):90–91.

Aquilina O (2006): A brief history of cardiac pacing. *Images Paediatric Cardiol* 8:17–81.

Arnold WM and Zimmermann U (1982): Rotating-field induced rotation and measurement of the membrane capacitance of single mesophyll cells of *Avena Sativa*. *Z. Naturforsch Sect C. Biosci* 37c:908–915.

Asanova TP and Rakov AI (1966): The state of health of persons working in electric fields of outdoor 400 and 500 kV switch-yards. *Gig Tr Prof Zabol*, 10:50–52 (in Russian) (translated by G.Knickerbocker for IEEE Power Engineering Society, Piscataway, New Jersey, in Special Publication No.10).

Bachman CH and Reichmanis M (1973): Some effects of high electrical fields on barley growth. *Int J Biometeorol.* 17:253–262.

Bailey WH, Bissell M, Brambl RM, et al. (1982): *A Health and Safety Evaluation of the ±400 kV Powerline.* Science Advisors' Report to the Minnesota Environmental Quality Board. Minnesota Environmental Quality Board, St.Paul, MN.

Bailey WH, Bissell M, Dorn DR, et al. (1986): *Comments of the MBQB Science Advisors on Electrical Environmental Outside the Right of Way of CU-TR-1,* Report 5. Science Advisors Reports to ther Minnesota Envirobnmental Quality Board. Minnesota Environmental Quality Board, St.Paul, MN.

Bailey WH, Weil DE and Stewart JR (1997): *HVDC Power Transmission Environmental Issues Review.* Oak Ridge National Laboratory, Oak Ridge, TN. Doi: 10.2172/580576.

Banks RS, Kanniainen CM and Cleark RD (1977): *Public Health and Safety Effects of High-Voltage Overhead Transmission Lines- An Analysis for the Minnesota Environmental Quality Board.* Minnesota Department of Health. Divison of Environmental Health, Minneapolis, Minnesota.

Banks RS and Williams AN (1983): The public health implications of HVDC transmission lines: An assessment of the available evidence. *IEEE Trans PAS* 102:2640–2648.

Barker AT (1991): An introduction to the basic principles of magnetic nerve stimulation. *J Clin Neurophysiol* 8:26–37.

Barker AT, Dixon RA, Sharrard WJW and Sutcliffe ML (1984): Pulsed magnetic field therapy for tibia non-union. *Lancet* 1 (8384):994–996.

Barker AT, Freeston IL, Jalinous R and Jarratt JA (1986): Clinical evaluation of conduction time measurements in central motor pathways using magnetic stimulation of the human brain. *Lancet*(8493):1325–1326.

Barker AT, Jalinous R and Freeston I (1985): Non-invasive magnetic stimulation of the human motor cortex. *Lancet* 1(8437):106–1107.

Barlow HB, Kohn HL and Walsh EG (1947): Visual sensations aroused by magnetic fields. *Am J Physiol* 148:372–375.

Barron SL (1950): The development of the electrocardiograph in Great Britain. *British Med J* 1 (4655):720–725.

Bassett CAL and Becker RO (1962): Generation of electric potential in bone in response to mechanical stress. *Science* 137:1063–1064.

Bassett CAL, Mithell SN and Faston SR (1981): Treatment of ununited tibial diaphyseal fractures with pulsing electromagnetic fields. *J Bone Joint Surg Am* 63:511–523.

Bassett CAL, Pawluk RJ and Pills AA (1974): Augmentation of bone repair by inductively coupled electromagnetic fields. *Science* 184:575–577.

Baule GM and McFee R (1963): Detection of the magnetic field of the heart. *Am Heart J* 66:95–96.

Beaugnon E and Tournier R (1991a): Levitation of water and organic substances in high static magnetic fields. *Journal de Physique III France* 1:1423–1428.

Beaugnon E and Tournier R (1991b): Levitation of organic materials. *Nature* 349:470.

Beck A (1890): Die Bestimmung der Localisation der Gehirn-und Rückenmarkfuntionen vermittelst der elektrischen Erscheinungen. *Centralblatt für Physiologie* 4:473–476.

Beer B (1902): Über das Auftretten einer objective Lichtempfindung in magnetichen Felde. *Klinische Wochen- Zeitschrft* 15:108–109.

BEMS (2005): *The Bioelectromagnetics Society. History of the First 25 years.* The Bioelectromagnetics Society. 1–44.

Benjamin P (1898): *History of Eectricity-from antiquity to the days of Benjamin Franklin-.* John Wiley Sons, Inc., New York.

Berry MV and Geim AK (1997): Of flying frogs and levitrons. *Eur J Phys* 18:307–313.

Bickford RG and Fremming BD (1965): Neuronal stimulation by pulsed magnetic fields in animals and man. In: *Digest of the 6th International on Medical Electronics and Biological Engineering (Tokyo),* Japanese Society for Medical Electronics and Biological Engineering, August 22–27, Tokyo, Abstract 7–6.

Binder A, Parr G, Halzman B and Fitton-Jakson S (1984): Pulsed electromagnetic field therapy of persistent rotator cuff tendinitis. A double-blind controlled assessment. *Lancet* 1 (8379): 695–698.

Blackman VH (1924a): Field experiments in electro-culture. *J Agri Sci* 14:240–267.

Blackman VH and Legg AT (1924b): Pot culture experiments with an electric discharge. *J Agri Sci* 14:268–273.

Boesch C (2003): Nobel prizes for nuclear magnetic resonance: 2003 and historical perspectives. *J Mag Resonan Imag* 20:177–179.

Briggs LJ, Campbell AB, Heald RH and Flint LH (1926): *Electroculture.* United States Department of Agriculture. Department Bulletin No.1379:1–38. U.S.Gov.Print.Office, Washington, DC.

Burt JPH, Pethig R and Talary MS (1998): Microelectrode devices for manipulating and analyzing bioparticles. *Trans Inst Meas Control.* 20:82–90.

Busch W (1866): Über den Einfluss welche heftigere Erysipeln zuweilig auf organisierte Neubildungenausben. *Verhandlungen des Naturhistorischen Vereines der Preussichen Rheinlande und Westphalens* 23:28–30.

Cambiaghi M and Parent A (2018): From Aldini's galvanization of human bodies to the Modern Prometheus. *Medicina Historica* 2:27–37.

Cambiagihi M and Sandrone S (2014): Robert Bartholow (1831–1904). *J Neurol* 261:1649–1650.

Cambridge NA (1977): Electrical apparatus used in medicine before 1900. *Proceed Royal Soc Med* 70:635–641.

Cavendish H (1776): An account of some attempts to imitate the effects of the torpedo by electricity. *Philos Trans R Soc London* 66:196–225.

Charry JM and Kavet RI (ed) (1987): *Air Ions: Physical and Biological Aspects.* CRC Press, Boca Raton, FL.

Chiba A, Isaka K, Yokoi Y, et al. (1984): Application of finite element method to analysis of induced current densities inside human model exposed to 60 Hz electric fields. *IEEE Trans PAS* 103:1895–1902.

Cohen D (1967a): Magnetic fields around torso: production by electrical activity of the human heart. *Science* 156:652–654.

Cohen D (1967b): Shielded facilities for low-level magnetic measurements. *J Appl Phys* Suppl 38:1295–1296.

Cohen D (1968): Magnetoenephalography: evidence of magnetic fields produced by alpha-rhythm currents. *Science* 161:784–876.

Cohen D and Chandler L (1969): Measurement and a simplified interpretation of magnetocardiograms from humans. *Circulation* 39:395–402.

Cohen D, Edelsack A and Zimmerman JE (1970): Magnetocardiograms taken inside a shielded room with a superconducting point-contact magnetometer. *App Phys Lett* 16:278–280.

Cohen D (2004): Boston and the history of biomagnetism. *Neurol Clin Neurophysiol* 86 (Nov 30):1–4.

Cole KS (1928a): Electrical impedance of suspensions of spheres. *J Gen Physiol* 12: 29–36.

Cole KS (1928b): Electrical impedance of suspensions of arbacia eggs. *J Gen Physiol* 12:37–54.

Cole KS and Baker RF (1941): Longitudinal impedance of the squid giant axon. *J Gen Physiol* 24:771–788.

Cole KS and Curtis HJ (1939): Electric impedance of the squid giant axon during activity. *J Gen Physiol* 22:649–670.

Cole KS and Curtis HJ (1941): Membrane potential of the squid axon during current flow. *J Gen Physiol* 24:551–563.

Cole KS and Cole RH (1941): Dispersion and absorption in dielectrics. I. Alternating current characteristics. *J Chem Phys* 9:341–351.

Colwell HA (1922): *An Essay on the History of Electrotherapy and Diagnosis.* William Heinemann. Ltd, London.

Cook HJ, Steneck NH, Vander AJ and Kane GL (1980): Early research on the biological effects of microwave radiation: 1940–1960. *Ann Sci.* 37:323–351.

Cope M and Delpy DT (1988): System for long-term measurement of cerebral blood and tissue oxygenation on newborn infants by near infrared transillumination. *Med Biol Eng Comput* 26:289–294.

Croat JJ, Herbst JF, Lee RW and Pinkerton FE (1984): Pr-Fe and Nd-Fe-based materials: A new class of high-performance permanent magnets (invited). *J Appl Phys* 55:2078–2082.

Damadian R (1971): Tumor detetion by nuclear magnetic resonance. *Science* 171:1151–1153.

Deipolyi AR, Folberg A, Yarmush ML, et al (2014): Irreversible electroporation: evolution of a laboratory technique in interventional oncology. *Diagn Interv Radiol* 20:147–154.

Delbourgo J (2006): *A Most Amazing Scene of Wonders: Electricity and Enlightenment in Early America.* Harvard University Press, Cambridge, MA.

Della Porta G (1589): *Natural Magick* (Magiae Naturalis). Reproduction of original work with English translation in 1658.

Delpy DT, Cope MC, Cady EB, et al. (1987): Cerebral monitoring in newborn infants by magnetic resonance and near infrared spectroscopy. *Scand J Clin Lab Invest* 47 (Sup 188): 9–17.

Deschanel AP (1876): *Elementary Treatise on Natural Philosophy Part 3. Electricity and Magnetism.* The English revised edition (1884). D. Appleton and Company, New York.

Driessen S, Bodewein L, Dechent D, et al. (2020): Biological and health-related effects of weak static magnetic fields (≤1 mT) in humans and vertebrates: A systematic review. *PLoS One* 15(6):e0230038.

Faghihi R, Zeinali-Rafsanjani B, Mosleh-Shirazi MA, et al. (2017): Magnetic resonance spectroscopy and its clinical applications: a review. *J Med Imaging Radiation Sci* 48:233–253.

Faraday M (1821): Historical sketch of electro-magnetism. *Ann Philos* 2:195-200, 274–290.

Faraday M (1822): Historical sketch of electro-magnetism. *Ann Philos* 3:107–119.

Faraday M (1839): Experimental researches in electricity: fifteenth series: notice of the character and direction of the electric force of the gymnotus. *Philos Trans R Soc London* 129:1–12.

Faraday M (1846): II. Experimental reseaerches in electricity; twentieth series: on new magnetic actions and on the magnetic condition of all matter. *Philos Trans R Soc London* 136:21–40.

Faraday M (1847): LXIV. On the diamagnetic conditions of flame and gases. *The London, Edinburgh, and Dublin Philos Magaz J Sci* 31(210):401–421.

Feychting M and Ahlbom A (1993): Magnetic fields and cancer in children residing near Swedish high-voltage power line. *Am J Epidemiol* 138:467–481.

Figuier L (1865): *Les Grandes Inventions Anciennes et Modernes Dans les Sciences, l'Industrie et les Arts.* L.Hachett et Cie, Paris. p.262.

Finger S, Piccolino M and Stahnisch FW (2013a): Alexander von Humboldt: Galvanism, animal electricity, and self-experimentation Part 1: Formative years, Naturphilosophie, and Galvanism. *J History Neurosci: Basic Clin Perspect* 22:225–260.

Finger S, Piccolino M and Stahnisch FW (2013b): Alexander von Humboldt: Galvanism, animal electricity, and self-experimentation Part 2: The electric eel, animal electricity, and Later years. *J History Neurosci: Basic Clin Perspect* 22:327–352.

Finsen NR (1899): *Ueber die Anwendung von concentrirten Chemischen Lichtstrahlen in der Medicin.* Verlag von F.C.W.Vogel, Leipzig.

Fitzgerald PB (2014): Transcranial pulsed current stimulation: A new way forward? *Clin Neurophysiol.* 125:217–219.Fleming JA (1921): *Fifty Years of Electricity.* The Wireless Press, Ltd, London.

Forbes A and Thacher C (1920a): Electron tube amplification with the string galvanometer. *Am J Physiol* 51:177-178.

Forbes A and Thacher C (1920b): Amplification of action currents with the electron tuber in recording with the string galvanometer. *Am J Physiol* 52:409–471.

Foster KR and Schwan HP (1989): Dielectric properties of tissues and biological materials: a critical review. *CRC Crit.Rev.Biomed.Eng.* 17:25–104.

Franken Haeuser B and Widén L (1956): Anode break excitation in desheathed frog nerve. *J Physiol* 131:243–247.

Franklin B (1752): XCV. A letter of Benjamin Franklin, Esq; to Mr. Peter Collinson, F.R.S. concerning an electrical Kite. *Philos Trans R Soc Lond* :565–567.

Franklin B and other Commissioner (1784): *Report of the Commission Charged by the King of France with the Examination of Animal Magnetism as Now Practiced at Paris,* 1–129, Translated from French. London, Printed for J. Johnson.

Frey AH (1961): Auditory system response to radio-frequency energy. *Aerosp Med* 32:1140–1142.

Fricke H (1925): A mathemetial treatment of the electric conductivity and capacity of disperse systems. 2. The capacity of a suspension of conducting spheroids surrounded by a non-conducting membrane for a current of low frequency. *Phys Rev* 26:678–681.

Fricke H and Morse S (1924): A mathematical treatment of the electric conductivity and capacity of disperse system. 1. electric conductivity of a suspension of homogeneous spheroids. *Phys Rev* 24:575–587.

Fricke H and Morse S (1925a): The electric capacity of suspensions with special reference to blood. *J Gen Physiol* 9:137–152.

Fricke H and Morse S (1925b): The electric resistance and capacity of blood for frequencies between 800 and 800 and 4·1/2 million cycles. *J Gen Physiol* 9:153–167.

Fujita A, Hirota I, Kawahara Y and Omori H (2007): Development and evaluation of intermediate frequency magnetic field exposure system for studies of in vitro biological effects. *Bioelectromagnetics* 28:538–545.

Fukada E and Yasuda I (1957): On the piezoelectric effect of bone. *J Phys Soc Japan* 12:121–128.

Füredi AA and Ohad I (1964): Effects of high-frequency electric fields in the living cell: 1. Behavior of human erythrocytes in high-frequency electric fields and its relation to their age. *Biochim Biophys Acta* 79:1–8.

Gas P (2011): *Essential Facts on the History of Hyperthermia and Their Connections with Electromedicine.* AGH University of Science and Technology. ISSN 0033-2097:37–40.

Geddes LA (1984): The beginnings of electromedicine. *IEEE Eng Med Biol Magazine (December)*: 3:8–23.

Geddes LA (1994): The first stimulators-reviewing the history of electrical stimulation and the devices crucial to its development. *IEEE Eng Med Biol Magazine (August/ September)* 13:532–542.

Geddes LA (2008): The history of magnetophosphenes. *IEEE Eng in Medicine and Biology Magazine (July/August)* 4:101–102.

Green RM (1953): *Commentary on The Effect of Electricity on Muscular Motion.* Waverly Press, Inc, Baltimore, MD.

Greenebaum B (ed) (1995): *Bioeletromagnetics* 16 (Supplement 3):1–122. The Bioelectromagentisc Society.

Grimnes S and Martinsen OG (2000): *Bioimpedance and Bioelectricity Basics.* Academic Press, London, San Diego.

Grzybowsli A and Pietrzak K (2012): From patient to discover - Niels Ryberg Finsen (1860-1904) - the founder of phototherapy in dermatology. *Clin Dermatol* 30:451–455.

Hamilton J (2002): *A Life of Discovery-Michael Faraday, Giant of the Scientific Revolution-.* Random House, New York.

Hamilton WA and Sale AJH (1967): Effects of high electric fields on microorganisms: II. Mechanism of action of the lethal effects. *Biochimica et Biophysica Acta* 148:789–800.

Hart FX and Schottenfeld RS (1979): Evaporation and plant damage in electric fields. *Int J Biometeorol* 23:63–68.

Hashimoto U (1984): *Kyu-rigen (Japanese Text)*. Reprint of original by Unsai Hashimoto. Ohmsha.

Hauf R (1974): Effects of 50 Hz alternating fields on man. *Elektrotech. Z* B26:318–320.

Hauf R (1976): Influence of 50 Hz alternating electric and magnetic fields on human beings. *Rev Gen Elektr. Special Issue (July)*:31–49.

Haupt RC (1984): Letters from the editors: response from Haupt. *Am J Public Health* 74: 1042–1043.

Haupt RC and Nolfi JR (1984): The effects of high voltage transmission lines on the health of adjacent resident populations. *Am J Public Health* 74:76–78.

Haygarth J (1800): Of the imagination, as a cause and cure of disorders of the body, exemplified by fictitious tractors, and epidemical convulsions. *Ann Med (Edinb)* 5:133–145.

Hermann L (1872): Ueber eine Wirkung galvanischer Ströme auf Muskeln und Nerven. *Pflüger, Arch* 5:223–275.

Hirota N, Honma T, Sugawara H, et al. (1995): Rise and fall of surface level of water solutions under high magnetic field. *Jpn J Appl Phys* 34:L991–L993.

Hodgkin AL and Huxley AF (1952a): The components of membrane conductance in the giant axon of *Loligo. J Physiol (London)* 116:473–496.

Hodgkin AL and Huxley AF (1952b): Currents carried by sodium and potassium ions through the membrane of the giant axon of *Loligo. J Physiol (London)* 116:449–472.

Hodgkin AL and Huxley AF (1952c): The dual effect of membrane potential on sodium conductance in the giant of *Loligo. J Physiol (London)* 116:497–506.

Hodgkin AL and Huxley AF (1952d): A quantitative description of membrane current and its application to conduction and excitation in nerve. *J Physiol (London)* 117:500–544.

Hodgkin AL, Huxley AF and Katz B (1952): Measurement of current-voltage relations in the giant axon of *Loligo. J Physiol (London)* 116:424–448.

Höber R (1910): Eine Methode, die elektrische Leitfähigkeit im Innern von Zellen zu messen. *Pflüg Arch Phys* 133:237–253

Höber R (1912): Eine zweites Verfahren die Leitfähigkeit im Innern von Zellen zu messen. *Pflüg Arch Phys* 148:189–221.

Höber R (1913): Messungen der inner Leitfähigkeit von Zellen. *Pflüg Arch Phys* 150:15–45.

Hoshi Y and Tamura M (1993a): Detection of dynamic changes in cerebral oxygenation coupled to neuronal function during mental work in man. *Neurosci Lett* 150:5–8.

Hoshi Y and Tamura M (1993b): Dynamic multichannel near-infrared optical imaging of human brain activity. *J Appl Physiol* 75:1842–1846.

Humboldt A (1826): *Ansichten der Natur*. Nikol Verlag, Hamburg.

HUSCAP (Hokkaido University Collection of Scholarly and Academic Papers) (1980): Bukyokusi (in Japanese). Bukyokusi. 1207–1250. http:hdl.handle.net/2115/29991.1980-03-20.

Huxley AF and Stämpfli R (1949): Evidence for salutatory conduction in peripheral myelinated nerve fibers. *J Physiol (London)* 108:315–339.

IARC (2002): *Non-Ionizing Radiation, Part 1: Satic and Extremely Low-Frequency (ELF) Electric and Magnetic Fields*. Vol 80. IARC Monographs on the Evaluation of Carcinogenic Risks to Humans. International Agency for Research on Cancer, Lyon, France.

IARC (2013): *Non-Ionizing Radiation, Part 2: Radiofrequency Electromagnetic Fields*. Vol 102. IARC Monographs on the Evaluation of Carcinogenic Risks to Humans. International Agency for Research on Cancer, Lyon, France.

IARC Monographs Priorities Group (2019): Advisory group recommendations on priorities for the IARC monographs. *Lancet Oncol* 20: 763–764.

ICNIRP (2003): *Exposure to Static and Low Frequency Electromagnetic Fields, Biological Effects and Health Consequences (0-100 kHz)*. (eds). Matthes R, Vecchia P, McKinlay AF, Veyret B and Bernhardt JH. International Agency for Research on Non-Ionizing Radiation Protection, Munich.

ICNIRP (2009): *Exposure to High Frequency Electromagnetic Fields, Biological Effects and Health Consequences (100 kHz-300 GHz)*. (eds.) Vecchia P, Matthes R, Ziegelberger G, Lin J, Saunders R and Swerdlow A. International Agency for Research on Non-Ionizing Radiation Protection, Munich.

Interphone Study Group (2011): Acoustic neuroma risk in relation to mobile telephone use: results of the interphone international case-control study. *Cancer Epidemiol* 35: 453–464.

Jackson R (2015): John Tyndall and the early history of diamagnetism. *Ann Sci* 72:435–489.

Jankowiak K, Drieoen S, Kaifie A, Kimpeler S, et al (2021): Identification of environmental and experimental factors influencing human perception of DC and AC electric fields. *Bioelectromagnetics* 42: 341–356.

Jerrold WC (1891): *Michael Faraday-Man of Science*. Fleming H. Revell Company, New York.

Jöbsis FF (1977): Noninvasive, infrared monitoring of cerebral and myocardial oxygen sufficiency and circulatory parameters. *Science* 198:1264–1267.

Kadosh RC (2014): *The Stimulated Brain- Cognitive Enhancement Using Non-invasive Brain Stimulation*. Academic Press, Elsevier Inc., London, San Diego.

Karipidis K, Mate R, Urban D, Tinker R and Wood AW (2021): 5G mobile networks and health-a state-of-the-science review of the research into low-level RF fields above 6 GHz. *J Exposure Sci Environ Epidemiol* 31: 585–605.

Kato G (1970): The road a scientist followed: notes of Japanese physiology as I myself experienced it. *Annu Rev Physiol* 32:1–22.

Kato M (1960): The conduction velocity of the ulnar nerve and the spinal reflex time measured by means of the H wave in average adults and athletes. *Tohoku J Exper Med* 73:74–85.

Kato M (ed) (2006): *Electromagnetics in Biology*. Springer Verlag. Tokyo.

Kato M, Honma K, Shigemitsu T and Shiga Y (1993): Effects of exposure to a circularly polarized 50-Hz magnetic field on plasma and pineal melatonin levels in rats. *Bioelectromagnetics* 14:97–106.

Kato M, Ohta S, Kobayashi T and Matsumoto G (1986): Response of sensory receptors of the cat's hindlimb to a transient, step-function DC electric field. *Bioelectromagnetics* 7:395–404.

Kato M, Ohta S, Shimizu K and Matsumoto G (1989): Detection threshold of 50 Hz electric fields by human subjects. *Bioelectromagnetics* 10:319–327.

Kato M and Shigemitsu T (1997): Effects of 50 Hz magnetic fields on pineal function in the rat. In: *The Melatonin Hypothesis, Breast Cancer and Use of Electric Power*. (ed). Stevens RG, Wilson BW, Anderson LE, 337–376. Battelle Press, Richland.

Kauffman G (2014): Nobel Prize for MRI imaging denied to Raymond V. Damadian a decade ago. *Chem Educator* 19:73–90.

Kerr J (1875): A new relation between electricity and light: dielectrified media birefringent (second paper). *Philosophical Magazine Series* 4. 50:446–458.

Knave B, Gamberale F, Bergström S, et al. (1979): Long-term exposure to electric fields: a crosssectional epidemiologic investigation of occupationally exposed workers in high-voltage substations. *Scand J Work Environ Health* 5:115–125.

Kobayashi T, Shimizu K, Matsumoto G, et al. (1983): Uniform DC electric field exposure system for mice and cats. *Bioelectromagnetics* 4:303–314.

Kolin A, Brill NQ and Broberg PJ (1959): Stimulation of irritable tissues by means of an alternating magnetic field. *Proc Soc Exp Biol Med* 102:251–253.

Korobkova VP, Morozov UA, Stolyarov MD and Yakub YA (1972): Influence of the electric field in 500 and 750 kV switchyards on maintenance staff and means for its protection. In *International Conference on Large High-Voltage Electric Systems, Proceedings of a Meeting Held at Paris*, Aug 28–Sep 6, 1972, Paper 23–06.

Kotaka S and Krueger AP (1968): Studies on the air-ion-induced growth increase in higher plants. *Adv Front Pl Sci* 20:115–208.

Kotani M (2011): Researches on biomagnetism in Japan-Past and future (in Japanese). *Trans Japanese Soc Med Biol Eng* 49:316–321.

Kouwenhoven WB, Miller CJ, Barnes HC and Burgess TJ (1966): Body currents in live line working. *IEEE Trans PAS* 85:403–411.

Kouwenhoven WB, Langworthy OR, Singwald ML and Knickerkocker GG (1967): Medical evaluation of man working in AC electric fields. *IEEE Trans PAS* 86:506–511.

Krasny-Ergen W (1936): Nicht-thermische Wirkungen elektrischer Schwingungen auf Kolloide. *Hochfrequenz-technik und Elektroakust* 48:126–133.

Krasny-Ergen W (1937): Der Feldverlauf im Bereiche sehr kurzen Wellen: spontane Drefelder. *Hochfrequenz-technik und Elektroakust* 49 195–201.

Krueger AP, Kotaka S and Andriese PC (1962): Studies on the effects of gaseous ions on plant growths. I. The influence of positive and negative air ions on the growth of *Avena Sativa. J Gen Physiol* 45:879–895.

Kursawe M, Stunder D, Krampert T, Kaifie A, Driessen S, Kraus T and Jankowiak K (2021): Human detection thresholds of DC, AC, and hybrid electric fields: a double-blind study. *Environ Health* 20: 92.

Küstler G (2007): Dimgnetic levitation-historical milestones. *Rev. Roum. Sci. Techn-Electrotechn. et. Energ* 52:265–282.

Lauterbur PC (1973): Image formation by induced local interactions: Examples employing nuclear magnetic resonance. *Nature* 242:190–191.

Lemström KS (1904): *Electricity in Agriculture and Horticulture*. The Electrician Printing and Publishing Co, London.

Liebesny P (1938): Athermic short wave therapy. *Arch Phys Therapy* 19:736–740.

Lillie RS (1925): Factors affecting transmission and recovery in the passive iron nerve model. *J Gen Physiol* 7:473–507.

Lin AL, Gao JH, Duong TO and Fox PT (2010): Functional neuroimaging: a physiological perspective. *Front Neuroenerg* 2:17.

Lin JC (1994): Early contribution to electromagnetic field in living systems. In *Advances in Electromagnetic Fields in Living System*. (ed). Lin JC, pp. 1–25, Prenum Press, New York, London.

Lin JC (2016): Minimally invasive transcranial magnetic stimulation (TMS) treatment for major depression. *Radio Sci Bullet* 357:57–59.

Lin JC (2017): The Moscow embassy Microwave Signal. *Radio Sci Bullet* 363: 90–93.

Lodge O (1908): Electricity in agriculture. *Nature* 78:331–332.

Lund EJ (1947): *Bioelectric Fields and Growth*. The University of Texas Press, Austin, TX.

Magnusson CE and Stevens HC (1911): Visual sensations caused by the changes in the strength of a magnetic fields. *Am J Physiol* 29:124–136.

Magnusson CE and Stevens HC (1914): Visual sensations created by a magnetic fields. *Phil Mag* 28:188–207.

Mains S (1983): The Minnesota power-line wars. *IEEE Spect* 20:56–62.

Malboysson E (1976): Medical control of men working within eletromagnetic fields. *Rev Gen Electro: Special Issue (July)*:75–80.

Mansfield P and Maudsley AA (1977): Planar spin imaging by NMR. *J Magnet Resonance* 27:101–119.

Martin FB, Bender A, Steuernagel G, et al. (1986): Epidemiologic study of Holstein dairy cow performance and reproduction near a high-voltage direct-current powerlines. *J Toxicol Environ Health* 19:303–324.

Martinez JA (2019): The "Moscow signal" epidemiological study, 40 years on. *Rev Environ Health* 34:13–24.

Masuda S, Washizu M and Iwadera N (1987): Separation of small particles suspended in lipid by nonuniform traveling field. *IEEE Trans IA* 23: 474–481.

Masuda S, Washizu M and Kawabata I (1988): Movement of blood cells in liquid by nonuniform traveling field. *IEEE Trans IA* 24:217–222.

McComas AJ (2011): *Galvani's Spark: The Story of the Nerve Impulse*. Oxford University Press, Oxford.

Milham S (1982): Mortality from leukemia in workers exposed to electrical and magnetic fields. *New England J* 307:249–249.

Millett D (1998): Illustrating a revolution: an unrecognized contribution to the 'golden era' of cerebral localization. *Notes Rec R Soc London* 52:283–305.

Millikan GA and Taylor CB (1942): The oximeter, an instrument for measuring continuously the oxygen saturation of arterial blood in man. *Rev Sci. Instrum* 13:434–444.

Miyakoshi J, Horiuchi E, Nakahara T and Sakurai T (2007): Magnetic fields generated by an induction heating (IH) cook top do not cause genotoxicitiy in vitro. *Bioelectromagnetics* 28:29–37.

Miyakoshi J and Shigemitsu T (2014): Bioelectromagnetism in the living body. *Comp Biomed Phys* 10:55–70.

Mizuno F, Ishiya K, Matsushita M, et al. (2020): A biomolecular anthropological investigation of William Aams, the firt SAMURAI from England. *Sci Rep* 10:2651.

Mottelay PF (1893): *De magnete (English version translated from original), The Dover edition (1958)*, Dover Publications Inc., New York.

Mottelay PF (1922): *Bibliographical History of Electricity and Magnetism*. Charles Griffin & Company, Ltd, London.

Murr LE (1963): Plant growth response in a simulated electric field environment. *Nature* 200: 490–491.

Murr LE (1964a): Mechanism of plant-cell damage in an electrostatic field. *Nature* 201: 1305–1306.

Murr LE (1964b): The biophysics of plant electrotropism. *Trans NY Acad Sci* 27:759–771.

Murr LE (1965): Biophysics of plant growth in an electrostatic field. *Nature* 206:467–469.

Murr LE (1966): The biophysics of plants growth in a reversed electrostatic fields; a comparison with conventional electrostatic and electrokinetic field growth responses. *Int J Biometeorol* 19:135–146.

Muth E (1927): Über die Erscheinung der Perlschurkettenbildung von Emulsionspartikelschen unter Einwirkung eines Wechselfeldes. *Kolloid Z* 41:97–102.

Nakajima S, Hirai Y, Takase H et al., (1975): Experimental and clinical application of new pulsed type earpiece oximeter (English translation available). *Respiration Circul* 23:709–714.

Nakajima S, Ikeda K, Nishioka H et al. (1979): Clinical application of a new (finger type) pulse wave oximeter (in Japanese). *Jap J Sur*: 41–61. English version available from https://moriyamamemorialhp.com/director/

Nakasono S, Ikehata M, Dateki M, et al. (2008): Intermediate frequency magnetic fields do not have mutagenic, co-mutagenic or gene conversion potentials in microbial genotoxicity tests. *Mutat Res* 649:187–200.

NIEHS (National Institute of Environmental Health Sciences) (1999): *Health Effects from Exposure to Power-Line Frequency Electric and Magnetic Fields*. NIH Publication No.99-4493, Cincinnati, OH.

National Research Council (1977): *Biological Effects of Electric and Magnetic Fields Associated with Proposed Project Seafarer*. Report of the Committee on Biosphere Effects of Extreme- Low-Frequency Radiation, Division of Medical Sciences, Assembly of Life Sciences, National Research Council, National Academy of Sciences.

National Research Council (1997a): *Possible Health Effects of Exposure to Residential Electric and Magnetic Fields*. National Academy Press, Washington DC.

National Research Council (1997b): *Committee to evaluate the U.S.Navy's extremely low frequency Communications System Ecological Monitoring Pprogram: An Evaluation of the U.S. Navy's extremely low frequency Communications System Eological Monitoring Pogram*. National Academy Press, Washington, DC.

National Research Council (1999): *Research on POWER-FREQUENCY FIELDs Completed under the Energy Policy Act of 1992*. National Academy Press, Washington, DC.

National Research Council (2008): *Identification of Research Needs relating to Potential Biological or Adverse Health Effects of Wireless Communications Devices*. National Academies Press. Washington, DC.

National Toxicology Program (2018a): *Technical Report on the Toxicology and Carcinogenesis Studies in Sprague Dawley (Hsd: Sprague Dawley® SD®) Rats Exposed to Whole-Body Radio Frequency Radiation at a Frequency (900 MHz) and Modulations (GSM and CDMA) Used by Cell Phones.* NTP TR 595 U.S. Department of Health and Human Survice ISSN: 2378-8925.

National Toxicology Program (2018b): *Technical Report on the Toxicology and Carcinogenesis Studies in B6C3F1/N Mice Exposed to Whole-Body Radio Frequency Radiation at a Frequency (1,900 MHz) and Modulations (GSM and CDMA) Used by Cell Phones.* NTP TR 596 U.S Department of Health and Human Survices ISSN: 2378-8925.

National Toxicology Program (2020): Cellphone radio frequency radiation studies. https://www.niehs.nih.go/health/materials/cell_phone_radiofrequency_radiation_studies_508.pdf.

Negishi T, Imai S, Shibuya K, et al (2008): Lack of promotion effects of 50 Hz magnetic fields on 712-Dimethybenz (α) anthracene-induced malignant lymphoma/lymphatic leukemia in mice. *Bioelectromagnetics* 29:29–38.

Neumann E and Rosenheck K (1972): Permeability changes induced by electric impulses in vesicular membranes. *J Memb Biol* 10:279–290.

Neumann E, Schaeffer-Ridder M, Wang Y and Hofschneider PH (1982): Gene transfer into mouse lymphoma cells by electroporation in high electric fields. *EMBO J* 1:841–845.

New York Tribune (1912): Electricity aids children- experiments prove it promotes bodily growth and intelligence. *New York Tribune*, April 16th.

NIEHS (1998): *Assessment of Health Effects from Exposure to Power-Line Frequency Electric and Magnetic Fields* (NIEHS Working Group Report). National Institute of Environmental Health Sciences of the National Institutes of Health. NIH Publication No.98-3981. Research Triangle Park, North Carolina.

Nobili L (1825): Ueber einen neuen Galvanometer. *J Chemie und Physik* 45:249–254.

Öberg PÅ (1973): Magnetic stimulation of nerve tissue. *Med Biol Eng Comput* 11:55–64.

Oersted HC (1820): Nouvelles expériences electro-magnétique. *Journale de Physique.* 91:78–80.

Ogawa S, Lee TM, Kay AR and Tank DW (1990b): Brain magnetic resonance imaging with contrast dependent on blood oxygenation. *Proc Natl Acad Sci USA* 87:9868–9872.

Ogawa S, Lee TM, Nayak AS and Glynn P (1990a): Oxygenation-sensitive contrast in magnetic resonance image of rodent brain at high magnetic fields. *Magnet Reson Med* 14:68–78.

Ogawa S, Tank DW, Menon R, et al. (1992): Intrinsic signal changes accompanying sensory stimulation: Functional brain mapping with magnetic reso- nance imaging. *Proc Natl Acad Sci. USA* 89:5951–5955.

O'Konski CT (1981): A history of electro-optics. In: *Molecular Electro-Optics.* ed Krause DS, pp. 1–15, Plenum Press, New York.

Otaka Y (2001): An introduction to bioeffects studies of ELF and RF EMF in Japan. *Bioelectromagnet Newsletter* No. 158:9–10.

Parent A (2004): Giovanni Aldini: From animal electricity to human brain stimulation. *Can J Neurol Sci* 31:576–584.

Pauling L and Coryell CD (1936a): The magnetic properties and structure of the hemochromogens and related substances. *Proc. Natl. Acad. Sci. USA* 22:159–163.

Pauling L and Coryell CD (1936b): The magnetic properties and structure of hemoglobin, oxyhemoglobin and carbonmonoxyhemoglobin. *Proc. Natl. Acad. Sci. USA* 22:210–216.

Perez V, Alexander DD and Bailey WH (2013): Air ions and mood outcomes: a review and meta-analysis. *BMC Psychiatry* 13: 29. http://www.biomedcentral.com/1471-244X/13/29.

Pethig R and Kell DB (1987): The passive electrical properties of biological systems: their significance in physioilogy, biophysics and biotechnology. *Phys Med Biol* 32:933–970.

Pethig R and Schmueser I (2012): Marking 100 years since Rudolf Höber's disvcovery of the insulating envelope surrounding cells and of the β-dispersion exhibited by tissue. *J Electric Bioimpedance* 3:74–79.

Petri AK, Shmeidchen K, Stunder D, et al. (2017): Biological effects of exposure to static electric fields in humans and vertebrates: a systematic review. *Environ Health* 16:41.

Philippson M (1920): Sur la résistance électrique des cellules et des tissue. *Compl. Rend. Soc. Boil* 83:1399–1402.

Philippson M (1921): Les lois de la résistance électrique des tissus vivants. *Bull Acad. f. Belg., Cl. Sci* 7:387–403.

Pliny the Elder (77 AD) (1938): *Natural History.* English Translation by Rackham H, Jones WHS and Eichholz DE. https://en.wikisource.org/wiki/Natural_History_(Rackman,_Jones,_%26_Eichholz).

Pohl HA and Hawk I (1966): Separation of living and death cells by dielectrophoresis. *Science* 152 (No.3722):647–649.

Pohl HA and Todd GW (1981): Electroculture for crop enhancement by air anions. *Int J Biometeorol* 25:309–321.

Ponchon P, Delorme G and Pallandre B (2020): André-Marie Ampère and the two hundred years of electrodynamics. *Electra* No.311: 38–46.

Possenti L and Selleri S (2017): Leopoldo Nobili: His galvanometer and his connections to the Faraday-Neuman-Lenz law of induction. *Radio Sci Bulletin* 363:77–79.

Potamian R and Walsh JJ (1909): *Makers of Electricity.* Fordham University Press, New York.

Reischauer EO and Tsuru S (eds) (1983): *Encyclopedia of Japan.* Kodansha Ltd., Tokyo.

Reiter R (1992): *Phenomena in Atmospheric and Environmental Electricity.* Elsevier Science Publishers, Amsterdam, London, New York, Tokyo.

Roberge PF (1976): Study on the state of health of electrical maintenance workers on Hydro- Québec's 735 kV power transmission system. Health Department, Hydro-Québec, Montreal.

Rockwell AD (1903): *The Medical and Surgical Uses of Electricity Including the X-rays, Phototherapy, the Finsen Light, Vibratory Therapeutics, High Frequency Currents, and Radio- Activitys.* E.B.Treat & Company, New York.

Rohdenburg GL and Prime F (1921): The effect of combined radiation and heat on neoplasms. *Arch Surg* 2:161–190.

Rowbottom M and Susskind C (1984): *Electricity and Medicine: History of Their Interaction.* San Francisco Press, Inc, San Francisco, CA.

Roy CS and Sherrington CS (1890): On the regulation of the blood-supply of the brain. *J Physiology (London)* 11:85–108.

Sagawa M, Fujimura S, Yamamoto H, Matsuura Y, Hiraga K (1984): Permanent magnet materials based on the rare Earth-Iron-Boron tetragonal compounds. *IEEE Trans Mag* 20:1584–1589.

Saito M and Schwan HP (1961): The time constants of pearl-chain formatin, In: *Biological Effects of Microwave Radiation*, Vol.1, Plenum Press. 85-97. Plenum Press, New York.

Saito M, Schwan HP and Schwarz G (1966): Response of nonspherical biological particles to alternating electric fields. *Biophys J* 6:313–327.

Sale AJH and Hamilton WA (1967a): Effects of high electric fields on microorganisms: I. Killing of bacteria and yeasts. *Biochimica et Biophysica Acta* 148: 781–788.

Sale AJH and Hamilton WA (1967b): Effects of high electric fields on microorganisms: III. Lysis of erythrocytes and protoplasts. *Biochimica et Biophysica Acta* 163:37–43.

Sarkar TK, Mailloux RJ, Oliner AA, et al (2006): *History of Wireless.* John Wiley & Sons.

Sauer FA (1983): Forces on suspended particles in the electromagnetic field. In: *Coherent Excitations in Biological Systems.* (eds) Frohlich H and Kremer F. pp. 134–144. Springer, Berlin, Heidelberg.

Savitz DA, Wachtel H, Barnes FA, et al (1988): Case-control study of childhood cancer and exposure to 60-Hz magnetic fields. *Am J Epidemiol* 128:21–38.

Sazonova TE (1967): *Physiological and Hygienic Assessment of Labour Conditions at 400–500 kV Outdoor Switch-Yards*, Profizdat, Institute of Labour Protection of the All-Union Central of Trade Unions (Scientific Pubs. Issue 46); 1975, Piscataway, NJ, IEEE Power Engineering Soviety (Translation in Special Publication No.10).

Schereschewsky JW (1928): The action of currents of very high frequency upon tissue cells: A.Upon a transplantable mouse sarcoma. *Public Health Rep (Wash)* 43:927–939.

Schereschewsky JW and Andervont HB (1928b): The action of currents of very high frequency upon tissue cells. B. Upon a transplantable fowl sarcoma. *Public Health Rep (Wash)* 43: 940–945.

Schiffer MB (2003): *Draw the Lightning Down-Benjamin Franklin and Electrical Technology in the Age of Enlightenment.* University of California Press, Berkeley, CA.

Schoenbach KH, Beebe SJ and Buescher ES (2001): Intracellular effect of ultrashort electrical pulses. *Bioelectromagnetics* 22:440–448.

Schoenbach KH, Peterkin FE, Alden RW and Beebe SJ (1997): The effect of pulsed eletric fields on biological cells: experiments and applications. *IEEE Trans Plasma Sci* 25:284–292.

Shoenbach KH, Joshi RP, Kolb J et al (2004): Ultrashort electrical pulses open a new gateway into biological cells. *Proc IEEE* 92: 1122–1137.

Schmiedchen K, Petri AK, Driessen S and Bailey WH (2018): Systematic review of biological effects of exposure to static electric fields. Part II: Invertebrates and plants. *Environ Res* 160:60–76.

Schwan HP (1957): Electrical properties of tissue and ell suspensions. *Adv Biol Med Phys* 5: 147–209.

Schwan HP (1988): Biological effects of non-ionizing radiation: cellular properties and interactions. *Ann Biomed Eng* 16:245–263.

Schwan HP (1989): Dielectrophoresis and rotation of cells. In: *Eletroporation and Electrofusion in Cell Biology.* (eds). Neumann E, Sowers AE and Jordan CA, pp. 1–21, Springer, Boston, MA.

Schwan HP and Sher LD (1969): Alternating-current field-indued forces and their biological implications. *J Electrochem Soc* 116:22c–26c.

Schwarz G, Saito M and Schwan HP (1965): On the orientation of nonspherical particles in an alternating eletric field. *J Chem Phys* 43:3562–3569.

Seegenschmiedt MH and Vernon CC (1995): A historical perspective on hyperthermia in oncology. In: *Thermoradiotherapy and Thermochemotherapy.* (eds). Seegenschmiedt MH, Fessenden P and Vernon CC, pp. 3–44, Springer-Verlag, Berlin, Heidelberg, New York.

Senda M, Takeda J, Abe S and Nakamura T (1979): Induction of cell fusion of plant protoplasts by electrical stimulation. *Plant Cell Physiol* 20:1441–1443.

Severinghaus JW (1987): History, status and future of pulse oximetery. In: *Continuous Transcutaneous Monitoring. Advances in Experimental Medicine and Biology.* (ed). Huch A, Huch R and Rooth G, vol 7, pp. 3–8, Springer, Boston, MA.

Severinghaus JW and Honda Y (1987): History of blood gas analysis. VII. Pulse oximeter. *J Clin Monit* 3:135–138.

Shibusawa M (1963): *Denkai zuiso (Japanese Text).* Korona publisher, Tokyo.

Shibusawa M and Shibata K (1927): The effect of electric discharges on the rate of growth of plants (in Japanese). *J Inst Electric Eng Japan* 47 (473): 1259–1300.

Shigemitsu T, Negishi T, Yamazaki K, et al. (2009): A newly designed and constructed 20 kHz magnetic field exposure facility for in vivo study. *Bioelectromagnetics* 30:36–44.

Shigemitsu T, Tsuchida Y, Nishiyama F, Matsumoto G et al., (1981): Temporal variation of the static electric field inside an animal cage. *Bioelectromagnetics* 2:391–402.

Shigemitsu T, Takeshita K, Shiga Y and Kato M (1993): A 50 Hz magnetic field exposure system for small animals. *Bioelectromagnetics* 14:107–116.

Shigemitsu T and Yamazaki K (2012): Interaction of extremely low-frequency electromagnetic fields with biological systems. In: *Electromagnetic Fields in Biological Systems.* (ed). Lin JC, pp. 197–260, CRC Press, Boca Raton, FL.

Shigemitsu T, Yamazki K, Nakasono S and Kakikawa M (2007): A review of studies of the biological effects of electromagnetic fields in the intermediate frequency range. *IEEJ Trans* 2:405–412.

Shimizu K, Endo H and Matsumoto G (1988a): Visualization of electric fields around a biological body. *IEEE Trans BME* 35:296–302.

Shimizu K, Endo H and Matsumoto G (1988b): Fundamental study on measurement of ELF electric field at biologivcal body surfaces. *IEEE Trans IM* 37:779–784.

Silney J (1976): Effet d'influence d'un champ èlectrique à 50 Hz sur l'organisme. *Rev. Gen. [Elektr.]: Special Issue (July)*:81–90.

Simon NJ (1992): *Biological Effects of Static Magnetic Fields-A Review.* International Cryogenic Materials Commission, Inc. Boulder, CO.

Singewald ML, Langworthy OR and Kouwenhoven WB (1973): Medical follow-up study of high voltage linemen working in AC electric fields. *IEEE Trans PAS* 92:1307–1309.

Solly E (1845): The influence of electricity on vegetation. *J Horticultural Soc London* 1:81–109.

SSK (Strahlenschutzkommission) (2013). *Biologische Effekte der Emissionnen von Hoch- spannungs-Glei- chstrom-übertragungsleitungen (HGÜ), Empfehlungen der Strahlenschutz- kommission mit wissen- schaftenlicher Begründung* (In German with English excerpt). SSK, Bonn.

Stamm JS and Rosen SC (1969): Electrical stimulation and steady potential shifts in prefrontal cortex during delayed response performance by monkeys. *Acta Biol Exp (Warsz)* 29:385–399.

Stämpfli R (1954): A new method for measuring membrane potentials with external electrodes. *Experimentia* 10:508–509.

Stämpfli R (1957): Reversible electrical breakdown of the excitable membrane of a Ranvier node. *Ann Acad Brasil Ciens* 30:57–63.

Stämpfli R and Willi M (1957): Membrane potential of a Ranvier node measured after electrical destruc- tion of its membrane. *Experimentia* 13:297–298.

Steiner F (1871): Ueber die Elektropunctur des Herzens als Wiederbelebungsmittel in der Chloroformsyncope, zugleich eine Studie über Stichwunden des Herzens. *Archiv für kleinische Chirurgie* 12:741–790.

Steneck NH (1984): *The Microwave Debate.* The MIT Press. Cambridge, MA.

Stewart GN (1899): The relative volume or weight of corpuscles and plasma in blood. *J Physiol (London)* 24:356–373.

Strumza MV (1970): Influence on the human health of close electric conductors at high tension, medical inquiry result. *Arch Mal Prof. Med. Trav. Secur. Soc* 31:269–276.

Takagi T and Muto T (1971): *Influence Upon Human Bodies and Animals of Electrostatic Induction Caused by 500 kV Transmission Lines,* Report No.SC.36-WG/2/25B, Tokyo Electric Power Co., Inc., Tokyo.

Takebe H, Shiga T, Kato M and Masada E (eds) (2000): *Biological and Health Effects from Exposure to Power-line Frequency Electromagnetic Fields.* Ohmsha, Tokyo.

Taki M (2016): Bioeletromagnetics researches in Japan for human protection from electromagnetic field exposures. *IEEJ Trans* 11:683–695.

Talary MS, Burt JPH, Tame JA and Pethig R (1996): Electromanipulation and separation of cells using travelling electric fields. *J Phys D: Allp Phys.* 29:2198–2203

Tasaki I (1939a): Electric stimulation and the excitatory process in the nerve fiber. *Am. J. Physiol* 125:380–385.

Tasaki I (1939b): The electro-saltatory transmission of the nerve impulses and the effect of narcosis upon the nerve fiber. *Am J Physiol* 127:211–227.

Tatar MM (1978): *Spellbound-Studies on Mesmerism and Literature.* Princeton University Press, Prinston, NJ.

Teixeira-Pinto AA, Nejelski LL, Cutler JL and Heller JH (1960): The behavior of unicellular organisms in an electromagnetic field. *Exp. Cell Res* 20:548–564.

Tesla N (1891): Message with currents of high frequency. *Electric Eng* 12:679.

Thompson SP (1910): A physiological effect of an alternating magnetic field. *Proc. Roy. Soc. Lond B* 82:396–398.

Thomson W (1847): On the forces experienced by small spheres under magnetic influence; and on some of the phenomena presented by diamagnetic substances. *Cambridge Dublin Math J* 2:230–235.

Thomson W (1854–1855): On the theory of the electric telegraph. *Proc. Roy.Soc.London* 7: 382–399.

Tomenius L (1986): 50-Hz electromagnetic environment and the incidence of childhood tumors in Stockholm county. *Bioelectromagnetics* 7:191-207.

Uchikawa Y (2020): Regard and Memory from 3D biomagnetic measurement (in Japanese). *Magnetics Japan* 15: 58-63.

Ueno S (ed) (2020): *Bioimaging-Imaging by Light and Electromagnetics in Medicine and Biology.* CRC Press, Taylor and Francis Group, Boca Raton, FL, London, New York.

Ueno S (1989): Quenching of flames by magnetic fields. *J Appl Physics* 65:1243-1245.

Ueno S and Harada k (1987): Effects of magnetic fields on flames and gas flow. *IEEE Trans Mag* 23: 2752-2757.

Ueno S and Harada K (1986): Experimental difficulties in observing the effects of magnetic fields on biological and chemical processes. *IEEE Trans Magn* 22:868-873.

Ueno S, Harada K, Ji C and Oomura Y (1984): Magnetic nerve stimulation without interlinkage between nerve and magnetic flux. *IEEE Trans Magn* 20:1160-1662.

Ueno S and Iwasaka M (1994b): Parting of water by magnetic field. *IEEE Trans on Magn* 30: 4698-4700.

Ueno S and Iwasaka M (1994a): Properties of diamagnetic fluid in high gradient magnetic fields. *J. Appl. Phys* 75:7177-7179.

Ueno S and Iwasaka M (1990): Properties of magnetic curtain produced by magnetic fields. *J Appl Physics* 67:5901-5903.

Ueno S, Lövsund P and Öberg PA (1981): On the effect of alternating magnetic fields on action potential in lobster giant axon. *Proceedings of the 5th Nordic Meeting on Med. and Biol. Eng,* 262-264, Linköping, Sweden.

Ueno S, Lövsund P and Öberg PA (1986): Effects of time-varying magnetic fields on action potential in lobster giant axon. *Med Biol Eng Comput* 24:521-526.

Ueno S, Matsuda T and Fujiki M (1990): Functional mapping of the human motor cortex obtained by focal and vectorial magnetic stimulation of the brain. *IEEE Trans Magn* 26:1539-1544.

Ueno S, Tashiro T and Harada K (1988): Localized stimulation of neural tissue in the brain by means of a paired configuration of time-varying magnetic fields. *J. Appl. Phys* 64:5862-5864.

Ueno S and Sekino M (eds) (2014): *Biomagnetics-Principles and Application of Biomagnetic Stimulation and Imaging.* CRC Press, Taylor and Francis Group, Boca Raton, FL, London, New York.

Ueno S, Sekino M and Shigemitsu T (2019): Transcranial magnetic and electric stimulation. In: *Handbook of Biological Effects of Electromagnetic Fields.* ed. Greenebaum B and Barnes F, pp. 345-405, CRC Press, Taylor and Francis Group, Boca Raton, FL, London, New York.

Vanbever R and Prèat V (1999): In vivo efficancy and safety of skin electroporation. *Adv Drug Deliv Rev* 35:77-88.

Van Wert SL and Saunders JA (1992): Electrofusion and electroporation of plants. *Plant Physiol* 99:365-367.

Verbeek J, Oftedal G, Feyhting M et al (2021): Prioritizing health outomes when assessing the effects of exposure to radiofrequency electromagnetic fields: a survey among experts. *Environ Int* 146: 106300.

Verkhratsky A, Krishtal OA and Petersen OH (2006): From Galvani to patch clamp: the development of electrophysiology. *Pflugers Arch-Eur J Physiol* 453:233-247.

Verkhratsky A and Parpura V (2014): History of electrophysiology and the patch clamp. In: *Patch-Clamp Methods and Protocols, Methods in Molecular Biology.* ed. Martina M and Taverna S. vol.1183, 1-19. Humana Press, New York.

Walsh J (1773-1774): *Three tracts Concerning the Torpedo. Published in the Philosophical Transactions for the Years 1773 and 1774.* ECCO Print Editions. London.

Wassermann EM (1998): Risk and safety of repetitive transcranial magnetic stimulation: report and suggested guidelines from the International Workshop on the Safety of Repetitive Transcranial Magnetic Stimulation, June 5-7, 1996. *Electroencephalogr Clin Neurophysiol* 108:1-16.

Wellstone P and Casper BM (2003): *Powerline.* University of Minnesota Press, Minneapolis, MI.

Wertheimer N and Leeper ED (1979): Electrical wiring configurations and childhood cancer. *Am J Epidemiol* 109:273–284.

Westermark N (1927): Effect of heat upon rat-tumors. *Skandinavisches Archiv für Physiologie* 52:257–322.

WHO (1984): *Extremely Low Frequency (ELF) Fields* (Environmental Health Criteria No 35), World Health Organization, Geneva.

WHO (1987): *Magnetic Fields* (Environmental Health Criteria No 69), World Health Organization, Geneva.

WHO (2006): *Static Fields* (Environmental Health Criteria No 232), World Health Organization, Geneva.

WHO (2007): *Extremely Low Frequency Fields* (Environmental Health Criteria No 238), World Health Organization, Geneva

WHO (2020): Relaunch call for expressions of interest for systematic reviews (2020). https://www.who.int/news-room/artiles-detail/relaunch-call-for-expressions-of-interest-for-systematic-reviews-(2020)

Wood AW, Mate R and Karipidis K (2021): Meta-analysis of in vitro and in vivo studies of the biological effects of low-level millimeter waves. *J Expos Sci Environ Epidemiol* 31: 606–613.

Yamaguchi FM (1983): Electroculture of tomato plants in a commercial hydroponics greenhouse. *J Biological Physics* 11:5–10.

Yamazaki K, Fujinami H, Shigemitsu T and Nishimura I (2000): Low stray ELF magnetic field exposure for in vitro study. *Bioelectromagnetics* 21:75–83.

Yasuda I (1954): Piezoelectric activity of bone. *J Jap Orthoped Surg Soc* 28:267–269.

Young LB (1973): *Power over People*. Oxford University Press, Oxford.

Young LB (1992): *Power over People*. Reissued as an Oxford University Press, Oxford.

Zimmermann JE, Thiene P and Hardings J (1970): Design and operation of stable r-f biased superconducting point-contact quantum devices. *J. Appl. Phys* 41:1572–1580.

Zimmermann U (1982): Electric field-mediated fusion and related electrical phenomena. *Biochimica et Biophysica Acta* 694:227–277.

Zimmerman U, Pilwart G and Riemann F (1974): Dielectric breakdown of cell membranes. *Biophysical J* 14:881–899.

Zimmermann U and Scheurich P (1981): High frequency fusion of plant protoplasts by electric fields. *Planta* 151:26–32.

The Coupling of Atmospheric Electromagnetic Fields with Biological Systems

Tsukasa Shigemitsu

Shoogo Ueno

3.1 Introduction

People live in an environment where there is always electromagnetic radiation from natural sources. Natural sources include the sun which produces radiation in the range from visible light, infrared (IR) to ultraviolet (UV), the earth which generates a magnetic field in its core, the atmospheric related to lightning strikes, and cosmic phenomena. As a result of the development of civilization, people began to produce man-made electromagnetic fields about 150 years ago. These electromagnetic fields fit into the existing spectrum of the natural electromagnetic radiation.

Life on earth evolved in the sunlight and began to perceive light. Within the spectrum of the sun, people perceive light in the visible region of wavelengths (0.38–0.78 μm) as natural light. After Newton first decomposed light into seven colors (red, orange, yellow, green, blue, indigo, and violet) in 1672, the existence of infrared and UV light also became known. In the case of spectroscopy of light, the thermal action of red is greater than that of blue. Furthermore, infrared light, which has a strong thermal action, was discovered by Herschel in 1800. Ritter found in 1801 that UV light has a chemical effect.

As mentioned above, the sun produces electromagnetic radiations in the range from the UV, through visible light to infrared region (IR), called optical radiation. About 45% of radiation falls within visible region of wavelengths. The UV region is from below 0.10 to around 0.40 μm which is subdivided into UVC (0.10–0.28 μm), UVB (0.28–0.32 μm), and UVA (0.32–0.40 μm). The IR is divided into IRA (0.78–1.40 μm), IRB (1.40–3.00 μm), and IRC (3.00 μm–1 mm).

There are a variety of electromagnetic phenomena on earth. Since prehistoric times, biological systems have been exposed to the electromagnetic fields produced by the earth itself, atmospheric ions, and electromagnetic fields associated with lightning discharge and solar activity. Therefore, it is presumed that the evolution of life has been affected by these electromagnetic phenomena in various ways. In other words, the question of what relationship exists between electromagnetic phenomena on earth and the evolution of life has been a matter of great interest to both scientists and the general public. For example, it has been reported that animals use actively the geomagnetic field. These include magnetite-based magnetotactic bacteria, the behavior of honeybees, and the migration of bird. It has also been reported that different kinds of fish use the Lorenzini organ to sense electric field and control their behavior. Electromagnetic fields are ubiquitous in earth's atmosphere. Natural electromagnetic fields are considered to be one of the most important environmental factors.

Biological systems and their surrounding environment are closely related to each other. The specificity of this relationship is said to be a major characteristic that sustains life. Among these environments, there has already been a great deal of research on the relationship between biological systems and atmospheric pressure, temperature, humidity, visible light, UV, IR, radiation, and gravity. A number of papers and books have been published. However, research on the relationship between the electromagnetic environment and biological systems has not progressed as much. This may be because we are not considered to have electromagnetic receptors for light as we do for vision and sound as we do for hearing.

In this chapter, we will discuss mainly how electromagnetic phenomena generated from atmospheric activity relate to biological systems. In addition, we will focus on electroreception and magnetoreception. They have a connection with the change of electromagnetic fields in nature. Electroreception is the ability to detect external electric field. Magnetoreception is the ability to sense changes in a magnetic field to perceive direction, position, and navigation.

3.2 Atmospheric and Cosmic Environments

Earth was born as one planet in the solar system 4.6 billion years ago. The primordial atmosphere consisted of volcanic gases that erupted slowly from earth's crust, and its main components are thought to have been carbon dioxide, nitrogen, hydrogen, and water vapor. Nitrogen, which makes up 80% of air, is contained in the amino acids that form proteins and the nucleic acids that form genes. Nitrogen is also important for our respiration, making it a necessary element for life on earth.

The synthesis of amino acids, the precursors for making proteins, is well known from Miller's experiments (Miller, 1953). A flask containing methane, ammonia, and water was heated to boiling, and the resulting vapor was repeatedly discharged. As a result, organic substances including amino acids such as glycine and alanine were produced. Miller's experiment was based on the assumption of electrical discharge in nature. Later, it was revealed that UV from the sun, radiation, and thermal energy can also produce amino acids.

After hundreds of millions of years, the water vapor coagulated into water, which creates oceans that covered about 70% of the earth's surface. It transforms the earth into a watery planet. The carbon dioxide in the atmosphere was gradually replaced by oxygen through the photosynthetic action of cyanobacteria and green algae. Thus, it is believed that a water- and oxygen-rich global environment was formed that enabled life to live.

On the ancient earth, intense UV poured down from the sky. The UV light's energy prevented the emergence of life, so the first life was born in the ocean, where the UV light could not reach. It has been found in rocks about two billion years ago as fossils of cyanobacteria. The presence of polymeric hydrocarbons, which are also thought to be the decomposition products of chloroplasts, have been found in the rocks. It suggests that photosynthesis was already present at that time. This increase in atmospheric oxygen due to photosynthesis led to a second important change for the organism. This increase in atmospheric oxygen was influenced by UV radiation from the sun to become ozone and form an ozone layer around the earth. The ozone layer shields out UV radiations, thus creating a mild global environment in which life can survive on land.

The earth is covered by an atmospheric layer that is divided into several layers. Up to an altitude of about 12 km from the earth's surface is called the troposphere, where meteorological phenomena prevail, and where the atmospheric gains heat from the oceans and the earth, rises, expands and cools, and then condenses into clouds and snow. Next, the altitude of about 10–50 km is called the stratosphere, where temperatures rise. The stratosphere is formed from the presence of the ozone layer inside the stratosphere, which absorbs the UV radiation in solar and superheats the atmosphere. The space above the stratosphere, 50–100 km, is called the mesosphere. From an altitude of about 80 km, ionization phenomenon prevails and is called the ionosphere, where oxygen atoms absorb the sun's UV radiation and X-rays. Oxygen atoms collide with protons and electrons, resulting in ionization phenomena. Depending on the distribution of electron density, the space below about 60–90 km is called the D layer, called also D region. D layer is the bottom side region of the ionosphere with small electron densities. The space 90–160 km is called the E layer, called also E region. This layer has ionization maximum at 110 km. The space 140–400 km is called the F layer, called also F region. In the ionosphere, the movement of electrons and ions causes electric currents to flow, and diurnal variations in the geomagnetic fields are caused by currents flowing in the tidal movement of the upper atmosphere and the action of the Earth's magnetic field.

Every object emits light, which is determined by its temperature. The need to treat this light quantitatively arose, and the ideal object, called the black body, was conceived. Based on this concept, the energy spectrum of sunlight can be approximated by the energy spectrum of light emitted by a black body at about 5,780 K. Radiation is transfer of energy by electromagnetic waves; solar and terrestrial radiation are considered to be two types of heat in and out of the earth. The sun provides the earth with 1.95 cal/cm^2 per minute from about 150 million km away, by means of a wide range of electromagnetic waves with wavelength ranging from a few hundred nm to several µm (the sun's surface temperature is assumed 5,780 K). This is equivalent to a 1.36 kW/m^2 which is called solar constant. In solar radiation, it has a maximum at a wavelength of about 0.475 µm. The earth is warmed by its thermal energy and electromagnetic waves (from a few µm to several hundred µm), corresponding to its temperature (255 K), emitted from the earth's surface, sea surface, and atmosphere into space. The radiant energy from the earth's radiation and the incident energy from the sun are in equilibrium, and the temperature of the earth's surface is maintained at a constant temperature suitable for the support of life. The earth's radiation has a maximum intensity of about 11 µm. It is also weakly absorbed by the earth's atmosphere at 8–12 µm and reaches outside the earth's atmosphere. For this reason, we call the wavelength range, "atmospheric window." On the other hand, the carbon dioxide has a strong absorption band at the wavelength of 2.5–3 µm and 4–5 µm. The existence of this absorption band has led to the discussion of the increase in carbon dioxide as a problem of global warming.

The attenuation of UV light is due to absorption and scattering of ozone. Water and carbon dioxide in atmosphere reduce the irradiance of infrared light. Visible light is also attenuated by the atmosphere. Since the spectrum of solar radiation and the spectrum of earth radiation are almost separated after 5 µm, sunlight up to 5 µm which is longer than visible light, is called near-infrared radiation and earth radiation is called infrared radiation. By separating them in this way, we can avoid overlapping the classification of wavelengths and the classification of radiation sources. Solar radiation consists of optical radiation from the UV, visible, and IRs of the electromagnetic spectrum, while earth radiation covers the IR.

However, this sun-blessed, prehistoric environment of the earth was never an easy one for living systems. The reason is that there was intense radiation around the earth. This is eloquently illustrated by the discovery of a natural fission reactor at the Oklo deposit, in the Republic of Gabon, Africa, discovered by the French nuclear agency in 1972 (Kuroda, 1983), which showed that the nuclear "fire" had existed on the earth and large-scale transmutations of the elements. The Oklo had high concentrations of uranium ore, about the same as those used in nuclear power plant today. This natural fission reactor is estimated to have reached criticality about two billion years ago and burned the equivalent of about 30 kW of atomic energy over a long period of 150,000 years. This natural fission reactor shows that there were large amounts of highly radioactive rocks on the surface of the ancient earth. Therefore, the radioactive environment on earth must have been severe. However, with the passage of more than a billion years, various types of radiation have gradually disappeared, and the present calm earth is thought to have settled down.

All life on earth has also been exposed to cosmic rays pouring down from the far heavens. Fortunately, cosmic rays are weakened by the influence of the earth's magnetic field and ionosphere, but a certain amount of cosmic rays always plow through the atmosphere and into biosphere. Biological systems have survived and evolved in both the radiation generated by Earth's rocks and radiation coming from space. Radiation threatens life and generates mutations that have become the driving force of biological evolution.

In addition to electromagnetic wave from the sun and radiation from the space and earth, living organisms have been exposed to electromagnetic fields created by the earth itself, air ions, and electromagnetic fields of frequencies running around earth (ELF-electromagnetic fields: ELF-EMF) since the beginning of life. In ancient times, when natural radiation was strong, the generation of air ions due to ionization was much more intense than it is now. This may have had a significant impact on life. In the era of tropical conditions and active geology, thunderclouds cause violent updrafts and volcanic explosions were common, and the phenomena of electricity generation by lightning must have been much more intense than today. Therefore, it is presumed that there were always strong static electric field and ELF-EMF on the earth. They must have had various effects on the evolution of life.

The energy by electric, magnetic, and ELF-EMF is incomparably weaker than that of radiation. With such low energy natural phenomena, it is assumed that organisms have been actively using that energy to sustain life, rather than as a means of defense, such as an adaptive response. For example, geomagnetism is used in the migration of birds and fish. Also, the static electric fields, air ions, seem to be closely related to the growth and health of organisms. It is also speculated that ELF-EMF are closely related to the formation of biological clocks that create the rhythm of life.

As factors of natural origin, various radiations such as cosmic rays, solar wind, X-rays, UV, visible light, and IR that fall to the earth from outer space are known. These are caused by electromagnetic phenomena in space. On the other hand, most of the phenomena on earth are lightning and electromagnetic phenomena in the ionosphere, and weather-related phenomena such as atmospheric pressure, temperature, and various forms of water have important interactions with biological systems and ecology. There are electric fields generated by electrical phenomena, magnetic fields generated by magnetic phenomena, and electromagnetic waves that propagate in pairs in space. The static electricity caused by lightning is also an electromagnetic wave. In this way, there are many naturally occurring electromagnetic waves around in our living space. Here, the relationship between these electromagnetic phenomena and biological systems is briefly discussed.

3.3 Natural Background Fields

The natural origins of the electromagnetic waves include the earth, the sun, atmosphere, and cosmic phenomena. The earth generates a magnetic field. Lightning strikes occur in the atmosphere which produces the electric field variation. Static electric fields are derived from the earth's atmosphere as a part of global electric circuit. The naturally originated electric field on earth is highest near the surface from about 0.1 to 0.15 kV/m during fair weather conditions to several thousand V/m under thunderclouds.

The earth is a huge magnet. Its magnetic field is called the geomagnetic field. The intensity of geomagnetic field ranges from about 70 μT at the North and South Poles to about 30 μT on the equator.

The total field intensity in Japan is around 50 μT. Geomagnetism is believed to be caused by convection movements in the liquid outer core of the earth. The huge current of hundreds of millions to billions of amperes of conductive fluid flows through it. This is called self-excited magneto-hydrodynamic dynamo. The geomagnetic field is described by three components: (1) total magnetic intensity, (2) declination, and (3) inclination. There are solar, lunar, and diurnal variations of the geomagnetic field. The diurnal variations may be pronounced during the day. Irregular pulsations and magnetic storms can be recorded in addition to these periodic variations. On the time scale of hundreds of years, the earth magnetic dipole is decreasing and the magnetic field strength decreased. It is believed that the inversion of the earth magnetic dipole occurs. This inversion could cause appearance and disappearance of biological species and life. Unlike the magnetic field which depends on the location on the earth, the electric field strength depends on the latitude to a small extent.

Electromagnetic fields have been present throughout the evolution of life on earth. In the marine environment, the geomagnetic field is the dominant of electromagnetic fields. Due to Faraday's induction law, the electric field is induced in marine species, resulting from its moving through seawater in the geomagnetic field.

The earth forms the global atmospheric electric circuit between earth surface and ionosphere as shown in Figure 3.1 of the paper by Rycroft et al. (2012). It shows various important processes in the global atmospheric electric circuit. Charge separation in thunderstorms creates a potential difference between the highly conductive regions of the ionosphere and the earth's surface. A global atmospheric electric circuit is said to form between the negatively charged surface of earth and positively charged surface of the ionosphere. This structure has a resemblance to the spherical capacitor.

On a dry winter day, we have had the experience of receiving a violent electric shock to our fingertips when crossing a carpet, touching a metal doorknob or the body of a car, causing us to jump. We have also heard the sound of small sparks when taking off a synthetic fiber shirt. These are caused by static electricity generated by friction and discharged between your fingertips and the knob or car body, or between you and your synthetic fiber shirt. This kind of charging phenomena can also be observed in the open air. If the metal supported by insulating rods is placed at a hight of one meter above the ground, and measured its electric potential ranging from several tens of volts to several hundreds of volts can be observed on a sunny day. The higher the potential rises above the ground surface, the higher the value. When there are thunderclouds nearby, the potential is several times higher, ranging from several thousand volts to several tens of thousands of volts. This is due to the presence of electric fields, and positive and negative electrified molecules and particles, or air ions, in the atmosphere, which charge the metal plate. This kind of electrical phenomenon in the natural atmosphere is called atmospheric electricity.

In the air at about 50 km, some of the gases are ionized by UV and X-ray from the sun and become positive ions. These ions are stratified in order to weigh, starting from the bottom, and each layer is thought to have an amount of electrons equal to the total amount of positive ions.

The higher the altitude, the stronger the UV radiation and X-rays which increase ionization and ion density. Also, the mean free path of the ions increases because the density of air decreases. Therefore, the electrical conductivity also increases, and there is no difference in electrical potential in the horizontal direction. For example, the electrical conductivity at 1, 5, and 10 km from the surface is about 2.5, 10, 25×10^{-14} S/m, respectively. The number of small ions at an altitude of 15 km is about 5,000 per cc, producing an electrical conductivity of 100×10^{-14} S/m. After 50 km above the ground, the conductivity increases rapidly, and at 100 km it reaches about 10^{-2} S/m, about the same as the conductivity of the earth can be assumed that there is a good conductor called the ionosphere 100 km above the ground. On a global scale, electromagnetic fields are ubiquitous, and their links to biology have been studied more extensively over the last century (König, 1977; König et al., 1981). More recently, evidence has emerged that biology is linked with static electric fields throughout the earth's atmosphere (Clarke et al., 2013; Morley and Robert, 2018; Hunting et al., 2019).

(a)

(b)

FIGURE 3.1   Photographic examples of a typical lightning stroke. (a) The lightning stroke in the top photo was captured during summer season at Akagi Test Center, CRIEPI. (b) The bottom photo shows the upward lightning stroke occurred in the winter season, Hokuriku, Japan. (Courtesy of CRIEPI, Tokyo.)

3.3.1 Lightning

The lightning that streaks across the sky with deafening thunder and lightning bolts, accompanied by a violent evening shower, is a wonderful summer tradition in Japan and the most important source of electric fields generated near the earth's surface.

Summer thunderclouds occur at a height of 0.5–1 km above the ground and often consist of several parallel cloud masses that are about 10 km wide and 10–13 km high. Thunderclouds go through three stages of development, maturity, and weakening. Each individual cloud mass is called a thundercloud cell, and within each cell, convection occurs. These convection currents create violent updrafts with wind speeds of up to 30 m/s and these currents carry water vapor up to 10–15 km in the air. The water vapor is condensed by the cold air above, but is quickly cooled to super-cooled water droplets. In addition, if there is a nucleus of dust or other particles, it will freeze into ice flakes. The super-cooled water droplets collide with the ice flakes and crystallize, growing into large powdery ice particles. As the ice particles grow and become heavier and cannot be supported by the updraft, they begin to fall, dragging the surrounding air with them and causing downward airflow. This is how violent convection occurs. In this convection storm, water droplets and ice fragments collide with each other, splitting and becoming electrically charged. Thus, positive charges are distributed above the cell and negative charges are distributed below. As a result, a large potential difference of tens of volts to 100 million volts is generated between the cloud base and the ground surface. A lightning strike is when the electric charges accumulated in the thundercloud discharge toward the earth, and the current of a single lightning strike reaches several thousand amperes to several tens of thousands of amperes.

Lightning discharges are considered as a source of electric currents, and it is believed that lightning discharges can generate from several kA to several hundred kA. The path of a lightning stroke (Figure 3.1), which can be of various lengths, acts as a huge antenna. Electromagnetic waves, of frequencies determined by the length of the lightning stroke path, are emitted. The length of the discharge path can exceed several kilometers, and the frequencies range from several Hz to the GHz band. These are naturally emitted electromagnetic fields.

When observing the occurrence of thunderclouds on the earth with satellites, we can see that there are always 1,000–2,000 thunderclouds of various sizes developing simultaneously on the earth. The average frequency of lightning strikes is about 160 times per second, and the current flowing into the earth due to lightning strikes is estimated to be about 200,000 A/s.

Electrical potential is negative on earth and positive in the upper atmosphere. Thunderclouds are the origin of a large global electric field, which is created by pumping electricity near the earth's surface far up into the sky. The earth's electric field is about 0.1 kV/m when it is highest and calmest near the surface, 0.03 kV/m at an altitude of 1 km, and about 0.01 kV/m at an altitude of 10 km. As a result, the potential difference between the earth and the ionosphere is 200–400 kV. When a well-developed thundercloud is overhead, the electric field at the earth's surface is about 3–20 kV/m. Thus, a large current circuit is thought to be formed between the earth and the ionosphere. The potential at the earth's surface and ionosphere is 20–50 kV (or 0.1–0.15 kV/m), and the return current in good weather is about 1–1.5 kA (or 3×10^{-12} A/m^2). The electric field at the ground surface is about 1.5 kV/m in the presence of atmospheric disturbances and 3–20 kV/m during thunderstorms, depending on meteorological conditions such as humidity, temperature, wind, and fog, and on the ion concentration in the atmosphere.

Lightning strikes have been known to affect biological systems directly by causing injury, or death in trees, cattle, and humans. In this way, direct effects of lightning on biological systems are obvious. Indirect effects have also been observed.

The natural sources of electromagnetic processes are associated with lightning discharges, and the resultant signals are called "atmospherics" or "sferics." About 100 lightning discharges per second occur globally. For example, one cloud-to-ground flash occurs about every second, averaged over the year in the USA (Home page of GHRC: https://ghrc.nsstc.nasa.gov/home/). They vary with time and location, and they have waves in ELF and very low frequency (VLF) ranges (ending at 300 kHz). Atmospheric

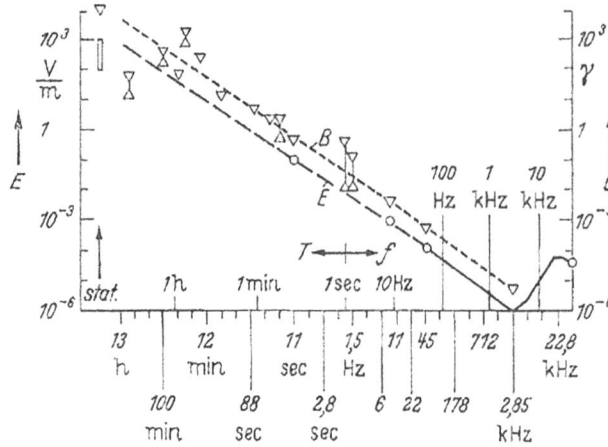

FIGURE 3.2 Intensity of natural electric and magnetic fields in the frequency range 0–50 kHz. Circle: electric field (*E*) in V/m, Triangles: magnetic field (*B*) in 10^{-9} T. (From Oehrl and König, 1968.)

consists primarily of electromagnetic waves in the ELF and VLF ranges. In ELF atmospherics, the electromagnetic phenomena measured over the earth's surface result from the excitation of the earth-ionosphere cavity resonator by thunderstorms. The earth's surface and the lower ionosphere are good conductors. The air between them is an insulating layer. This forms a waveguide for electromagnetic waves which depends on frequency. This waveguide forms earth-ionosphere cavity for low-frequency electromagnetic waves.

When ELF electromagnetic fields are produced by lightning discharge, the electric field causes electric current flow in the atmosphere. Magnetic fields of the same frequency as the result of current flow are produced. The higher the frequency of the electromagnetic waves, the greater the attenuation and the lesser the propagation range. As described by Oehrl and König (1968), the electric field and the magnetic field intensity and frequency range 0–50 kHz measured at one point are inversely proportional in log-log plot (Figure 3.2). The amplitude of the electric field is averaged 10^{-2} V/m at 1 Hz and 10^{-6} V/m at 3 kHz. This shows a relationship between electric and magnetic fields and frequency. Natural electric and magnetic field strengths decrease rapidly with increasing frequency. The strength of ultra-low frequency electric and magnetic fields over a broad frequency range is about 10^{-3}–10^{-5} V/m and 10^{-12}–10^{-14} T, respectively. The E/H ratio ($\approx$370 Ω) corresponds to that of electromagnetic waves. The E/H is the wave impedance.

The observed wave forms vary with the distance—which can be from 50 to 15,000 km—from the lightning strike (Figure 3.3) (Al'pert and Fligel, 1995). When the distance is short, the waveform is a single pulse. However, at distances greater than 1,000 km, the waveform approaches an oscillating form with a definite periodicity. The changing of the waveform originates from the electromagnetic wave emitted by lightning, which propagates by reflecting between the ionosphere and the earth's surface, which acts as a perfect conductor. This phenomenon exhibits resonance at specific frequencies. The space between the earth and ionosphere serves as a large waveguide for atmospherics. Of the signals propagating from lightning discharges, low-frequency components have low attenuation and can circle the earth several times. Standing waves develop from the excitation of the spherical, surface-cavity resonator between the earth's surface and the lower boundary of the ionosphere. The fundamental frequency of this resonance is near 7.5 Hz, which is determined by dividing the propagation speed of the electromagnetic wave with light velocity (3×10^5 km/s) by the diameter of the earth (4×10^4 km).

The quasi-static (i.e., relatively constant) field consists of a negatively charged earth and a positively charged atmosphere. The ground-level electric field is about 0.1 kV/m during fair weather, but electric fields above 100 kV/m have been observed during thunderstorms. Atmospherics exhibit considerable

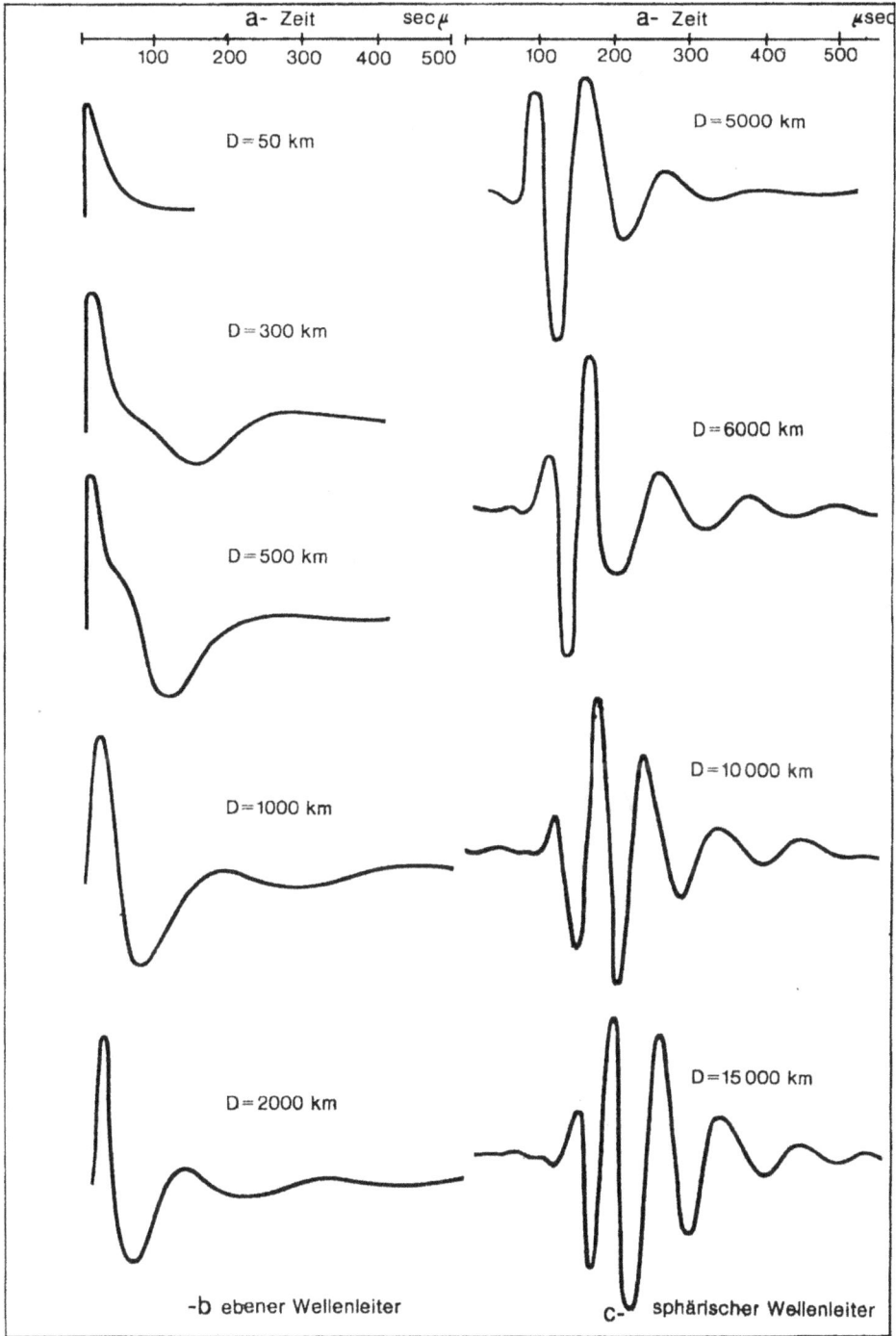

FIGURE 3.3 Waveforms of atmospheric signals (VLF) at various distances, between 50 and 15,000 km, from a lightning contact with the earth. (a) Time; (b) plane waveguide; (c) spherical waveguide. (From König et al., 1981.)

amplitude over a wide frequency band. Ground lightning with a duration of few microseconds can generate global-scale, alternating-current electric fields.

The atmospheric contains a wide variety of ions with difference in size and charge. These ions tend to attach to aerosols. They originate from natural sources such as cosmic rays, radioactivity, splashing water, and dust storms and anthropogenic sources such as high voltage power lines. Corona ions are produced when the voltage is high enough to cause corona breakdown around cable.

3.3.2 Electromagnetic Fields and Resonance

In 1952, Schumann theorized that the space between the earth and ionosphere forms a spherical capacitor. Assuming the ionosphere to be a perfect reflector, and based on the circumference of the earth-ionosphere cavity and the velocity of light, he predicted that its first mode of resonant frequency is 10 Hz (Schumann 1952). This resonance phenomenon is now called Schumann resonance (SR). In 1954, König first reported the measurement of the resonance phenomena (König, 1977). The circumference of this cavity resonator is approximately equal to the wavelength of 7.8 Hz electromagnetic waves. More rigorously, as the electrical conductivity of ionosphere's boundary layer are finite, the fundamental resonant frequency is about 7.8 Hz. The harmonic resonant frequency can be obtained from following equation.

$$f = 7.8\sqrt{\frac{n(n+1)}{2}}\,\mathrm{Hz}$$

This formula predicts a fundamental mode resonant frequency of $f_1 = 7.8$ Hz and harmonic resonant frequency, $n = 1, 2, 3…$ Power spectrum of measured electric fields of natural signals has been described (König et al., 1981). The resonances can be clearly seen in the measurement of the power spectrum as shown in Figure 3.4 (König et al., 1981; Toomey and Polk, 1970). There is a fundamental resonant frequency at about 8 Hz, and harmonic resonant frequencies are observed around 14, 20, and 26 Hz.

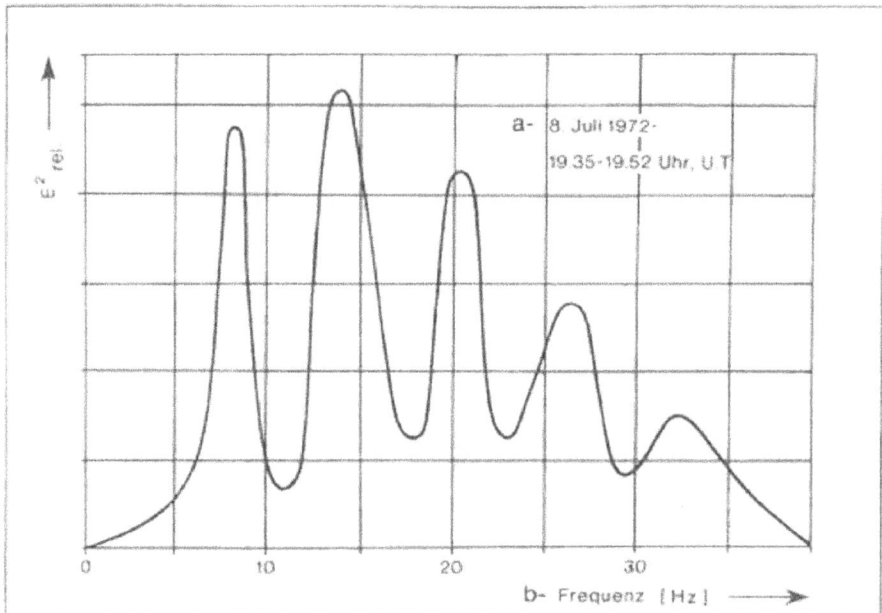

FIGURE 3.4 Power spectra of the naturally generated electric field in the ELF ranges (SR). (From König et al., 1981.)

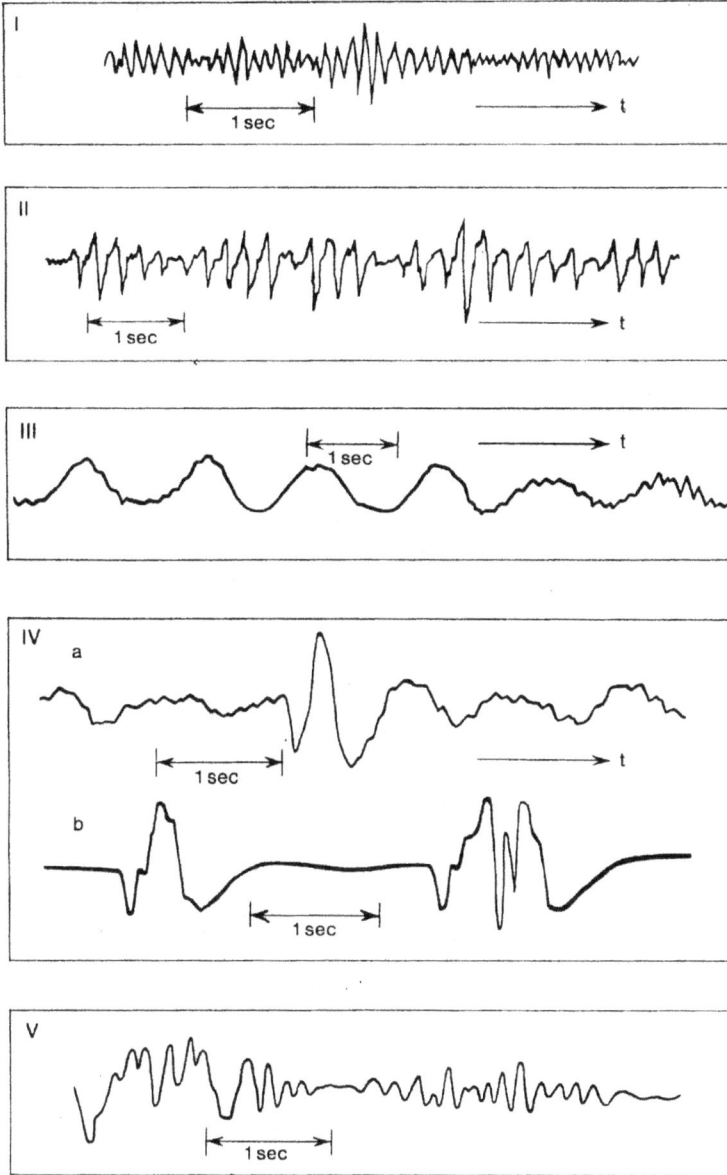

FIGURE 3.5 Electromagnetic processes of natural origin in the ELF range. (From König et al., 1981.)

As shown in Figure 3.5, in nature, there are mainly three types of naturally occurring ELF electric fields. Type I shows the SR waveform at 8 Hz. Type II is the wave form of naturally occurred local variation of electric fields at 3–6 Hz, and type III shows local variation at about 0.5–2 Hz. Type IV is field fluctuation as a result of thunderstorm activity. a is the field, thunderstorm not yet visible on the horizon, b is thunderstorm on the horizon. Type V is the signals of sunrise.

The earth frequency was first mentioned by Fitzgerald in 1893 (Fitzgerald, 1893). Tesla also mentioned this frequency in 1905 (Tesla, 1905, 2015; Jackson, 1998). He presented the idea of the earth as resonating circuit as the lowest resonant frequency as 6 Hz. He described generation and detection of these low frequency waves. He was quoted as follows:

Alpha waves in the human brain are between 6 and 8 hertz. The wave frequency of the human cavity resonates between 6 and 8 hertz. All biological systems operate in the same frequency range. The human brain's alpha waves function in this range and the electrical resonance of the earth is between 6 and 8 hertz. Thus, our entire biological system- the brain and the earth itself- work on the same frequencies. If we can control that resonance system electronically, we can direct control the entire mental system of humankind.

Danho et al. (2019)

Schumann resonance is the resonance mode that exist in the cavity established between the earth's surface and the lower edge of ionosphere. These signals remain topics of today's study.

3.4 Natural Electromagnetic Fields and Biological Systems

We are exposed daily to electromagnetic fields, both from natural and man-made sources. The naturally occurring electromagnetic fields originate from properties of the earth's molten liquid core, electric discharges in the atmosphere (terrestrial sources), and solar and lunar activities (extraterrestrial sources). Man-made electromagnetic fields come mainly from 50 to 60 Hz power transmission and distribution lines and from the electrical appliances driven at the 50/60 Hz and thus are classified as extremely low frequency (ELF) fields. The relationship between natural occurred ELF fields and bioactivity in human and vertebrates is briefly introduced.

3.4.1 Electromagnetic Fields and Humans

Many researchers have speculated that the frequencies of SR in the earth-ionosphere cavity have played a role in the evolution of all living organisms' brainwave patterns (König, 1977; Hainsworth, 1984; Price, 2016; Price et al., 2021). According to Hainsworth, the human central nervous system has the synchronization with biorhythms through the use of SR. For example, the frequencies of SR lie in the brain wave spectrum. These electromagnetic wave signals may be expected to have biological effects. Lightning discharge generates electromagnetic resonances within earth-ionosphere cavity. In the 1950s and 1960s, many experiments were being carried out to investigate the similarities between natural ELF electric fields that appeared in the atmosphere and the activity in human. Most interesting experiments were performed on human biorhythms.

Schumann and König were interested in the biological effects of atmospherics. They researched relationships between SRs and life on earth. Their research covered, from an influence on yeast cells, bacteria, animals to humans (König, 1977; König et al., 1981). Among their researches, they demonstrated a link between SR and brain rhythms. König showed that all of the fundamental frequencies and patterns of human brain wave (EEG) activity are measured within the Schumann cavity (earth-ionosphere) as local field variations during thunderstorms. König compared EEG recordings with natural electromagnetic fields in the environment and found that they lie in the same frequency range. He speculated that this is possibly no coincidence, but a human adaptation to the electromagnetic environment over the long course of evolution. As these frequencies are in the same region as those of human brain waves, the possibility of correlation between these ELF electromagnetic phenomena and human activity has been considered. König conducted an interesting experiment during an automobile traffic exhibition in Munich in 1953 (König and Ankermüller, 1960; König, 1977). Using visitors as subjects, the subject was seated in a chair, and as soon as the light in front of the seat was turned on, the hand-held reactor was measured. The correlation between naturally occurred ELF electric field and reaction time to light signal was investigated. The results showed that when type I signals increase, reaction time became shorter. When type II signals were present, reaction times became longer. These results were confirmed in experiments using artificially produced 8–10 and 3 Hz (1 V/m) (König, 1977). Although there were issues related to statistical analyses of the results, this was the first experiment to show some type of relationship between atmospheric signals and human response.

SRs are the background signal in the ELF regions of electromagnetic spectrum. It has peak at around 7.8, 14.3, 20.8, 27.3, and 33.8 Hz as shown in Figure 3.4. As shown in Figure 3.5, ELF electromagnetic fields which are electromagnetic phenomena in nature include SR waves with a frequency of about 8 Hz (type I), local field fluctuations with a frequency of 3–6 Hz (type II), and field fluctuations with a frequency of 0.5–2 Hz (type III). Students and office workers were exposed to 10 Hz electric fields to study its effect. One hundred and sixty-two students were exposed to a 30 V/m, 10 Hz electric field with the static electric field of 50 V/m on the head. The workers were repeatedly exposed to a 10 Hz electric field every 3 weeks for 6 months. Exposure assessment included attention, concentration, and effects on learning. Positive effects such as improved attention were found as a result of exposure. At the same time, the subjects reported mentally stable results (Altmann, 1976).

Behavioral effects of ELF electric fields have been sought with natural and artificial ELF fields. Reaction times were measured in man and monkeys. The subjects were required to press or release a button when given a simple light or tone stimulus. Reaction times in humans have been reported slow response times at times of high natural activity between 3 and 6 Hz, with a converse effect at times of 10 Hz peaks (König, 1977). These natural oscillations are believed to arise in the SR with typical amplitudes of 1–2 V/m. Further studies from the same group with artificial fields at frequencies of 5–10 Hz and strengths of 0.3–5.0 V/m again showed trends consistent with effects of natural fields, but these trends were not amenable to statistical analysis. Other tests of human reaction times in ELF fields of 1–20 Hz have suggested that reaction time is inversely related to field frequency (Hamer, 1968), but the statistical significance of these results has been questioned (The National Research Council, 1977).

There has been the speculation of the similarity between EEG rhythms and SR. The most common frequencies of human brain waves include α, β, δ, and θ. The α wave is the major rhythm in a normal relaxed condition. The β wave reflects active processing. The δ wave is the rhythm that occurs in a deep dreamless sleep or unconsciousness. The θ wave is associated with drowsiness. When healthy adults relax with their eyes closed, brain waves of 8–12 Hz frequency and about 5–100 μV can be measured (α waves). The α wave is the main component of brain waves of humans with β waves (13–30 Hz, 5–30 μV) being another component, representing normal alert mental state. The δ wave activity declines during deep sleepiness, and 4–7 Hz low voltage slow waves (θ waves) appear, representing dreaming states. The human electrical activity occurs in a frequency range below 50 Hz. It has been noted that the form of brain waves is similar to the SR waves. If one compares α and δ waves with the record obtained from the electric field in ELF range, there are similarities between α wave and type I signal, and between δ waves and type II signal (König et al., 1981). Under similar conditions, brain waves in the same frequency ranges are spontaneously observed for all vertebrates.

The remarkable similarity of the brain waves (EEG) with the SR waves was recognized in the 1950s (König et al., 1981; Schumann and König, 1954). They first showed the measured frequencies were consistent with a predicted mathematical model. The frequencies of the SR waves are closely related to α, β, and γ brain waves. Persinger and co-workers have studied EEG activity and the SR in real time (Persinger and Saroka, 2015; Saroka et al., 2016).

As mentioned above, the human EEG wave is very similar to the SR wave and the fluctuating waves of the local electric field. This suggests that ELF electromagnetic fields, which have existed in nature since the earliest days of biological development and have traveled around the earth day and night, may have had a significant impact on the formation of biological brain waves. If the formation of brain waves is closely related to the natural electromagnetic fields, then it is not just human, but all animals on earth should have the same frequency. Based on this assumption, the waves like EEG frequency measurements of animals, such as dogs, cats, guinea pigs, rabbits, and fish such as salmon, show that both animals and fish have brain waves of several Hz to several tens of Hz, similar to human's brain waves.

The brain waves of vertebrates, from humans to fish, are rhythmic, with a waveform close to a sine wave, and can be roughly divided into waves with a frequency of 10 Hz or lower and waves with a

(a)   (b)

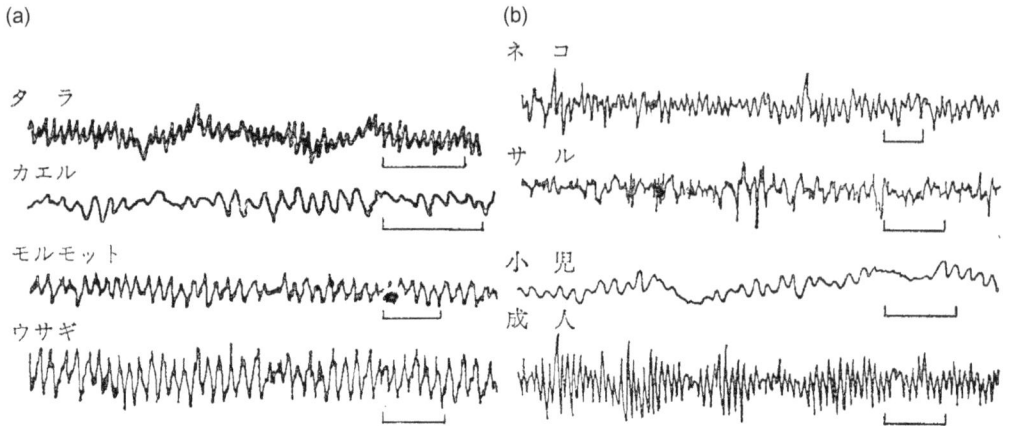

タ　ラ

カ エ ル

モ ル モ ッ ト

ウ サ ギ

ネ　コ

サ　ル

小　児

成　人

FIGURE 3.6 Brain waves of various animals. (a) Left from top; Codfish, frog, guinea pig, and rabbit and (b) Right from top; Cat, monkey, infant, and adult (From Iwase, 1970 and Morrell, 1967)

frequency of 50 Hz or lower. The former corresponds to α wave in humans, and the latter to β wave. There is no intrinsic difference between the brain waves of different species (Figure 3.6 and Table 3.1). In frog brains (odontoid and optic lobes), EEG can be seen even when the brain is removed, and EEG can be seen even when the brains is cut into pieces the size of a grain of rice. Invertebrate brains, or ganglia as they are properly called, can also record rhythmic waves like EEG. Such rhythmic waves can also be seen in simple nerve centers. For example, the optic ganglia of grasshoppers (*Melanoplus*) and genus damselfly (*Dysticus*) show rhythmic waves of 7–10 Hz. Light stimulation increases the frequency to 20–40 Hz. In the grasshopper (*Schistocerca*), the rhythmic waves are mixed with spini-form waves (Iwase, 1970).

After König's speculation, Bullock suggested that the origin of the frequency of electrical activity in many biological systems on our planet and many species exhibit electrical activity in the ELF range below 50 Hz, which is strongly similar to the natural ELF spectrum in the atmosphere produced by lightning (Bullock, 2002). It is surprising that many different types of species exhibit similar low-frequency electrical wave activity. Most vertebrates from fish to *Homo sapiens* show very similar electrical of activity (Bullock, 2002).

Recently, it was once again reported that there is a link between the constantly occurring atmospheric lightning-derived electromagnetic phenomena and the electrical activity in many biological systems acquired during evolution on our planet (Price et al., 2021). An example of the results, Figure 3.7 shows

TABLE 3.1 The Brain Waves of Humans and Animals

		Frequency (Hz)	Amplitude (μV)
Human (Adult)	α wave	8–12	5–100
	β wave	13–30	5–30
	δ wave	0.5–3	5–100
	θ wave	4–7	5–100
Animal	Dog·Cat	10–16	
	Guinea pig	6–7	
	Rabbit	3–6	
Fish	Salmon	14–16	
	Goldfish	14–16	

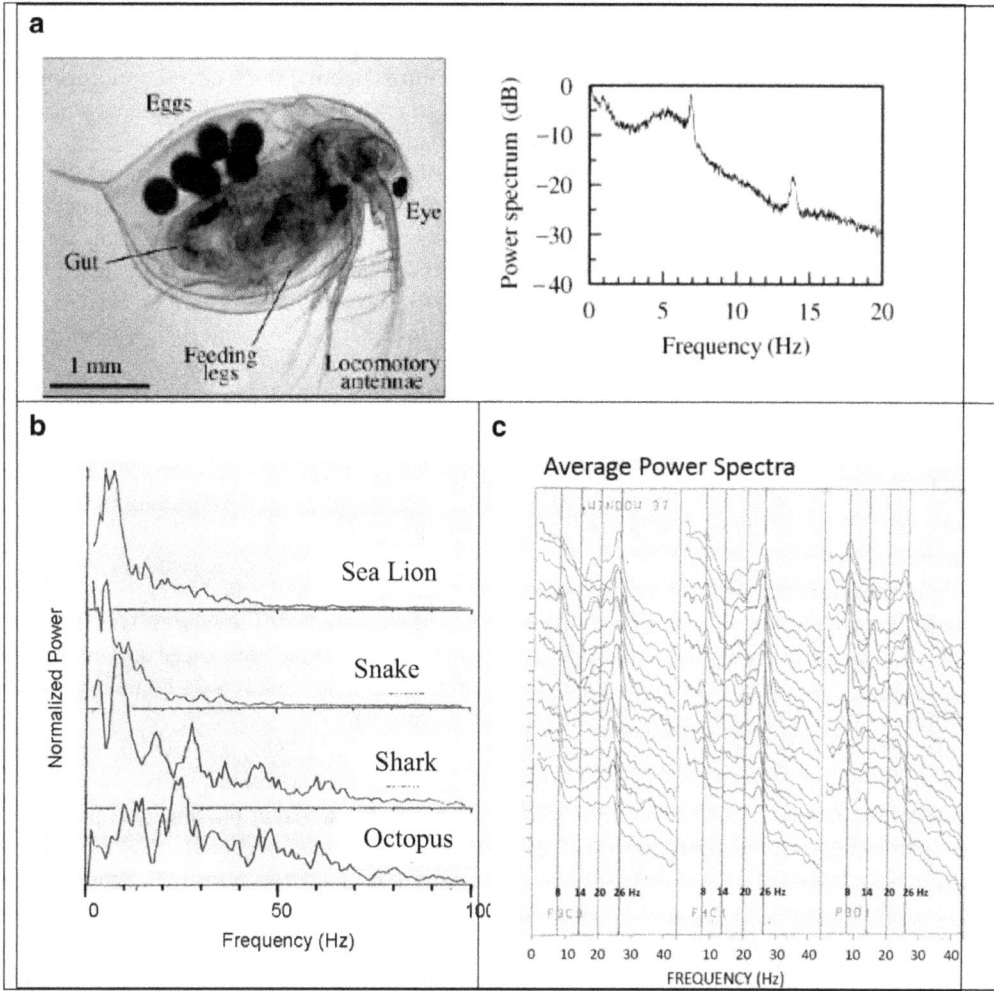

FIGURE 3.7 Power spectra of the electrical activity. (a) Zooplankton. (b) Vertebrates (sea lion, snake, shark) and invertebrates (octopus) and (c) human measured at three different locations on the head showing 1-minute averages. (From Price et al., 2021.)

the differences in the brain waves of different species across vertebrates and invertebrates, with peaks of electrical activity in zooplankton at 7 and 14 Hz, and frequencies below 50 Hz in sea lion, snake, shark and octopus, as well as humans. The spectra are almost the same and the dominant peak close to 8 Hz in all examples. As they indicated, many species from zooplankton to humans exhibit electrical activity in the ELF range below 50 Hz with significant maximum in their power spectra focused at specific frequencies. These frequencies show the similarity to the natural ELF spectra in the atmosphere produced by lightning from the comparison of Figures 3.5 and 3.6, Table 3.1. It is hypothesized that the background natural ELF fields in the atmosphere had a major role in the evolution of biological systems (Price et al., 2021).

3.4.2 Circadian Rhythm in Human and Vertebrates

In 1962, the French apeleologist, Michel Siffre attempted to live alone for 2 months in an underground cave, isolated from the rest of the world, without any clocks, calendars, or any other means of telling time. The purpose of this experiment was to find out how human rhythms are affected by the absence of the sense of time. The daily rhythms of sleeping and waking were recorded, and cycles that deviated by 0.8 hours from 24 hours were detected. This meant that Siffre's internal clock, or free-running period, was 24.8 hours in the cave, cut off from the outside world. After this pilot study, the research studies on human circadian rhythms were performed in artificial isolation facilities.

Various periodicities in biological processes are coupled, to a certain extent, to geophysical cycles. Animals have 23–25 hours periods, and plants have 23–28 hours periods. However, the period lengths and phases of these internally generated (endogenous) daily rhythms of biological organisms are readily controlled by external environmental factors, such as the 24 hours day-night cycle produced by the earth's rotation. Biological organisms can adjust their rhythms (*Entrainment*) based on external environmental conditions, which are called synchronizing factors (a *Zeitgeber*): visible light is a very important *Zeitgeber*.

Among the natural conditions that could serve as a *Zeitgeber*, lightning and other electromagnetic phenomena are possible synchronizing factors. Human body temperature and activity have 24-hours periods and are synchronized by external environmental factors. If external environmental stimuli are removed, the free running cycle rhythm becomes 25.3 hours. During the 1960s, research using underground rooms or caves, to isolate subjects from external information to investigate human circadian rhythms has been active.

First, the free-running circadian rhythms of activity such sleep and awake, body temperature, urination, and other rhythms of subjects living in the two rooms were studied. The subjects were informed about neither the shielding nor the artificial field. In both rooms, volunteer was placed for 1 month with no information from outside without letters. These parameters have usually a 24-hours cycle which synchronizes with day-night cycle produced by the earth's rotation (Wever, 1979).

Human circadian rhythms arise endogenously (Aschoff and Wever, 1962). Even without time cues in a constant environment, they deviate only slightly from 24 hours, with most autonomous human rhythms approximating 25 hours. Typically, oscillators governing different physiological parameters such as activity and rectal temperature are coupled and run synchronously (Wever, 1975) but may run asynchronously in "internal desynchronization." The rhythms of all different measured parameters run synchronously to each other. This is internal synchronization which is the case in most experiments. In about 20% of the experiments, the internal desynchronization occurs as shown in Figure 27 of Wever's book (1979, p. 48). It is the examples of internal desynchronization. Subject lives under constant conditions without time cues. Temporal course of the rhythms is shown. Presented are the successive periods, one beneath the other. The experiment is divided into two sections (A and B). The activity rhythm is represented by bars (black=activity; white=rest). The rectal temperature rhythm is represented by triangles, maximum temporal position (Δ) and minimum values (∇). The circadian rhythm of activity was "split" into two rhythms, with periods of 25.7 and 33.4 hours. The lengthening of the period for subjects in the shielded room was statistically significant. During first 14 days, measured variables run synchronously to each other with equal periods of 25.7 hours. After that time, the two rhythms ran separately, which means that internal desynchronization occurred spontaneously. On the 14th subjective day, the period of the activity rhythm lengthened to a mean of 33.4 hours without any environmental factor being changed and the period of the rectal temperature rhythm shortened to 25.1 hours. In about 20% of the experiments, internal desynchronization occurred. This phenomenon was not observed for subjects living in the unshielded room.

As early as the 1960s, studies on the effects of SR electric field on the circadian rhythm were done. To investigate the effect of natural electric fields on the circadian rhythm of human activity, Wever conducted an experiment focusing on the effects of 10 Hz electric fields. The reason why 10 Hz electric field is used comes from that the original suggestion of Schumann prediction should be at 10 Hz. To provide isolation from external sound, light, and other cues, two underground rooms were installed. Two identical rooms, one shielded and the other unshielded, allowed simultaneous testing of field-exposed and

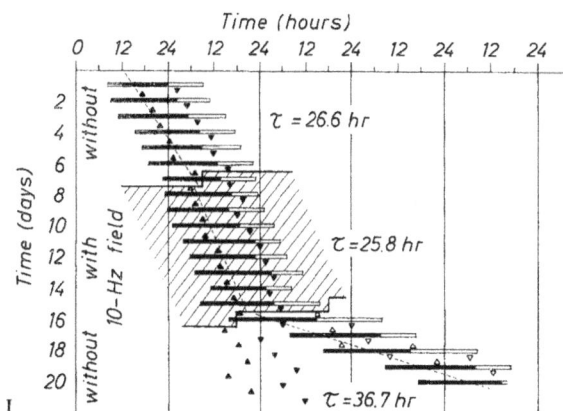

FIGURE 3.8 Effects of presence or absence of 10 Hz electric field on changes in free-running circadian rhythms in subjects under constant conditions. The hatched section is the period of 10 Hz electric field exposure. Activity rhythm is shown by bars (black filled = active period, white filled = resting period), body temperature rhythm, (▲) represent maxima and (▽) minima. Open triangles represent the temporal repetition of the maximum and minimum. Period (τ) represents the various phases of the experiment. (From Wever, 1968.)

control subjects who were not aware of differences between two rooms. One room was shielded from electromagnetic fields (reduction of natural static electric field) and the other is installed with a vertical 10 Hz, 2.5 V/m electric fields, which was almost equivalent to that in the natural environment (SR signal). Human circadian rhythm has been studied in an underground room shielded from natural electric and magnetic fields. It also appears sensitive to 10 Hz, 2.5 V/m electric field imposed in this shielded environment (Wever, 1968, 1973, 1974, 1975, 1977).

The period of the free-running circadian rhythm became shorter and returned to its original length, when the electric field was terminated. As an example, Figure 3.8 demonstrated the change of free-running circadian rhythm. Subjects were then exposed to a 10 Hz, 2.5 V/m electric field in the shielded room. The 10 Hz field was off during first and third periods in Figure 3.8. When the field was turned on during the second period, the period of free-running rhythm shortened. During the third period with turn off, the period of rhythm lengthened and internal desynchronization occurred.

Totally ten experiments show that the period was shorter with the 10 Hz field than without it by an average of 1.3 ± 0.7 hours with highly significant level ($p < 0.001$) (Wever, 1974). The shortening was greater for those subjects with the longest circadian cycles in the absence of the field. The internal desynchronization was not observed when the electric field was applied. In some subjects internal desynchronization occurred immediately after the field was switched off; in others it ceased immediately after the field was switched on. This showed that a 10 Hz electric field can affect circadian rhythms, including shortening the period and minimizing internal desynchronization. Reducing field exposure to 12 hours on and 12 hours off did not eliminate its *Zeitgeber* effect in temporarily restoring a 24.0 hour rhythm in ten subjects with free-running rhythms between 23.5 and 26 hours. Wever interpreted these results as indicating a significant *Zeitgeber* capacity of the 10 Hz field even when imposed intermittently (Wever, 1970, 1974).

Free-running period was measured by the activity and rectal temperature periodicities in initial experiments. In the unshielded room, subjects had free-running period shorter on the average by 20 minutes, and internal desynchronization was less likely than in the shielded room (unshielded room: $n = 57$, mean free-running period 24.87 ± 0.45 hours, internal desynchronization in four subjects; shielded room: $n = 80$, mean free-running period 25.21 ± 0.80 hours, internal desynchronization in 28 subjects). All these differences were significant at $P < 0.01$. Wever hypothesized that the differences were due to natural electromagnetic field present only in the unshielded room (Wever, 1970).

In view of Wever's emphasis on the role of natural electric fields in entrainment of the normal circadian rhythms, further information is needed on residual natural and artificial fields in his test rooms. Wever concluded that electromagnetic fields in the ELF range influence human circadian rhythms. However, replicable data are not shown.

In the past, it has been postulated that animal are sensitive to electric fields occurring in the atmosphere and may obtain time cues from the daily variations. It also suggested that the electric field may be responsible for the 24 hours day-night rhythm.

Several results of the change of circadian rhythms of animals are summarized. These experiments used 10 Hz electric fields with 10,000 times stronger than natural generating field in the 8–14 Hz range. This provided support for the concept that electric fields can affect circadian rhythms and act as a weak *Zeitgeber*. The effect of static electric fields on circadian rhythm of mice (*Mus musculus*) was investigated (Dowse and Palmer, 1969). The continuous locomotor activity of mice was recorded to determine the effects of electric field on circadian rhythm. After acclimatization to a light-dark cycle (200 fc, illumination, light period: 05:00–17:00 hours) for 10 days, electric fields were applied only in the light period for 10–14 days, and then the light-dark cycle was reversed. Under dim light (1 fc), the mice were left with the electric field applied for 25–35 days and then placed in constant light. The results suggested a possible effect of electric field on locomotor activity rhythm. Mouse activity was entrained with on-off cycle of the field. It was proposed that the mouse detected electric field extending into the cage because of imperfect Faraday shielding and distortion by the field wheel (Dowse and Palmer, 1969). However, Roberts added a comment to this report (Roberts, 1969). When laboratory animals such as mice are exposed to an atmospheric electric field level of about 100 V/m, it is unreasonable to assume that there is an effect considering the magnitude of the electric current in the animal body and the heat generated. It is necessary to consider the effects of corona discharge, noise, etc. during the experiment, and fluctuations in the electric field due to positive and negative air inside the cage. This can hardly be regarded as conclusive evidence for a static electric field interaction. The mice may have been responding to noise generated by the electrical equipment, or to corona discharges from the cage which the mice could hear.

The activity rhythms of *Drosophila melanogaster* were studied when a phase shift was added to the 24-hours cycle in an electric field of 10 Hz, 0.15 kV/m (Dowse, 1982). This activity rhythm was examined with the infrared beam blocked; a 6-hour phase shift in a 24-hour cycle disrupted the recorded activity rhythms.

Shortened circadian rhythms were seen in the green finches (*Carduelis chloris L.*) exposed to 10 Hz, 2.5 V/m electric field (Wever, 1973). The continuous field exposure for 10–20 days sharply shortened the circadian period from 24.8 to 23.9 hours. Nocturnal restlessness in the European brambling finch associated with seasonal migration was enhanced by a 10 Hz electric field that illuminated certain perches for which the birds exhibited a preference over unexposed perches in the same row. No differences were detected during the day (Wever, 1977).

As mentioned above, changes in free-running rhythms have been reported with exposure of 10 Hz, 2.5 V/m for human and fore green finches (*Carduelis chloris L.*). The field gave a shortening of the spontaneous period, averaged 1.3 hours in human, and 0.8 hours in green finches. Wever pointed that the field reduced internal desynchronization in humans and acted as a *Zeitgeber*. This experiment was repeated with green finches (Lintzen et al., 1989, 1992). Effects of the electric fields on activity of green finches (*Carduelis chloris L.*) and other experimental results have been reported (Lintzen et al., 1989). Green finches were exposed to an electric field of 10 Hz, 2.5 V/m, and its effect on the free running period was investigated. The free running period was 23.66±0.80 hours for the electric field exposure and 23.64±0.77 hours for the control, which is opposite to the result pointed out by Wever, and no effect of the electric field exposure was observed. There was also no effect at 8.7 and 65.2 V/m. The effect of exposure of fruit flies to 10 Hz, 1 and 10 kV/m electric fields on circadian locomotor activity was reported (Engelmann et al., 1996).

Recently, rat cardiac muscle cells were observed to be influenced by weak magnetic field (90 nT) in 7.8 Hz (Elhalel et al., 2019). The 3–4 days cultures are exposed to 7.8 Hz magnetic field and three experiments are performed: (1) the observation of the spontaneous mechanical contractions of the cardiac cell, (2) the observation of the spontaneous calcium (Ca^+) transients with and without magnetic field exposure, (3) the observation of the damage caused to the cardiac cells following stress due to hypoxia or the addition of H_2O_2 induced with or without magnetic field exposure.

Issues raised with regard to these results include (1) the lack of a clear cause and effect of the periodicity relationship between electromagnetic fields and organisms, (2) uncertainty regarding the mechanisms, (3) absence of measurements of electric field, and (4) inadequate explanation of the data analysis. However, it concluded that electromagnetic fields in the ELF range influence human circadian rhythms (Wever, 1974). Replicable data are not shown.

3.5 Atmospheric Electricity and Biological Systems

The static electric field that exists in nature due to electrical phenomena in atmosphere such as lightning seems to have an effect on living organisms. This is natural for organisms that have been born and evolved in the natural static electric field. The static electric field arises naturally in the environment such as with the approach of storm clouds or through triboelectric charge separation on clothing. All living organisms such as humans and animals are exposed to atmospheric static electric field. The atmospheric static electric field is generated between positively charged ionosphere and the negative earth.

In this section, we introduce the studies that have been conducted so far on the relationship between natural occurring and man-made static electric fields and living organisms, dividing into humans, vertebrates, and invertebrates. In general, vertebrates can be classified into five groups, based on their skin, their reproduction, the maintenance of their body temperature and characteristics of their arms, legs, wings, and fins. Mammals and birds are endothermic vertebrates. Ectothermic vertebrates include reptiles, fish, and amphibians. The main invertebrates used in the study of the electromagnetic field derived from atmospheric phenomena are insects such as honeybees, Drosophila, cockroaches, etc.

3.5.1 Static Electric Field in Humans and Others

When we stand directly under the high voltage, the soft hair of our head stands upright, and when we hold up our hand, the hair on our arm flutters as if pulled by the device. This is due to the electrostatic force (Coulomb force) of the electric charge induced on the tips of the hair. The stronger the electric field, the stronger this attractive force becomes. Although we cannot perceive the electric field directly, we can perceive it through the stimulation of the skin caused by the suction of head and body hair. The static electric field perception experiments provided evidence that detection thresholds for static electric field are much lower for whole-body exposure (Blodin et al., 1996; Chapman et al., 2005; Clairmont et al., 1989; Odagiri and Shimizu, 1999) than partial exposure (arm and forehand). Blodin et al. found that with whole-body exposure under static electric field strength up to 50 kV/m with 7–11 seconds per trial, the detection threshold of seated and grounded male and female subjects was 45.1 kV/m. Co-exposure to air ions with ion current densities of 60 nA/m^2 did not affect detection thresholds. With high concentrations (120 nA/m^2) of air ions, the subjects detected the lower static electric field, the detection threshold was 36.9 kV/m with some participants being able to perceive weaker field of 10 kV/m or less. Odagiri and Shimizu conducted the partial-body exposure experiments where only the participants' arm was exposed to static electric field with strengths up to 450 kV/m (1999). They presented that the subjects were able to perceive static electric field above 250 kV/m on their forearm when the relative humidity was 90%. The detection threshold increased to about 375 kV/m when the humidity was set to 50%. When the arm was shaved, the participants were no longer able to perceive a static electric field at intensities up to 450 kV/m. It means that the perceived sensation is dependent on body hair. Chapman et al. conducted the similar experiments which only the forearm of the subjects was exposed to static electric fields between 30 and 65 kV/m (2005). None of the subjects was able to perceive the fields. The authors indicate that applied field strengths were too low to be detected under partial-body exposure. From both experiments, it was concluded that exposure of body surface area with hair could play a crucial role in the detection of static electric fields.

Field perception experiments in human presented that detection thresholds for static electric field were significantly lower when the whole body was exposed compared to when only the subject's arm was exposed as partial body exposure. The most reasonable reason for this difference is that whole-body exposure increases the field strength at the top of the body such as head or shoulders. In addition, perception of static electric field appears to be influenced by several factors such as the relative humidity, awareness, and simultaneous presence of air ions.

To identify the mechanisms of static electric field, by recording action potentials from afferent fibers innervating various sensory receptors in the anesthetized cat's hindlimb, Kato et al. confirmed that body hair is involved in the perception of static electric field (1986). They recorded afferent impulse discharges of hair receptors of cats exposed to static electric field with 180–310 kV/m (both polarities). The stronger the electric field, the wider was the angle of the hair movement. Action potentials were evoked in the afferent fibers innervating G_1 hair receptor, G_2 hair receptor, and down hair receptor. No action potentials were evoked in afferent fibers innervating type I, type II, field receptors, muscle spindles, or joint receptors. Their results indicated that a strong static electric field induced movement of the hairs, eventually evoking excitation of the hair receptors.

In outdoor test under HVDC lines, most subjects do not detect electric fields below about 25 kV/m. Clairmont et al. investigated human's electric field perception and thresholds for DC and AC electric fields when HVAC and HVDC lines were operated in a hybrid fashion (1989). Each person rated various sensations at measurement locations along the lateral profile of the test line, while DC field, AC field, and ion current density were simultaneously monitored at each of the locations. When the DC and AC electric fields overlapped, the threshold of the sensed electric field was higher than when person was exposed to each field separately. However, it was pointed out that the number of subjects, their selection methods, and characteristics have not been fully described. There are some shortcomings such as the lack of sufficient description of the exposure conditions and the fact that it was not done in a double-blind manner.

In long-term experiment, mice were continuously exposed to static electric field (2 kV/m) during a period of 2 years (Kellogg et al., 1985a, b, 1986). They found no difference in the lifespan of mice exposed to 2 kV/m static electric field (both polarity) and environmental static electric fields and that increased values in serum glucose, and decreased urea nitrogen levels. The authors saw a connection between serum glucose levels and lifespan which lent support to their hypothesis that bioelectric processes are involved in mortality and aging rate. On the other hand, the researchers from the Rockefeller University found no effect on behavior or neurotransmitter activity in the brain of rats exposed to 3 and 12 kV/m fields for 2, 18, and 66 hours (Bailey and Charry, 1986, 1987; Charry et al., 1985).

Exposure associated with DC transmission line includes static electric field, static magnetic fields, and air ion (charged aerosols). There are a wide variety of ions in the atmosphere (Reiter, 1992). These ions originate from both natural such as cosmic rays, radioactivity, water splashing, etc., and high voltage transmission lines produced corona ions when the voltage of transmission lines are high enough to cause corona breakdown around cable. In the past, as mentioned above, there were several investigations concerning the human perception of static electric field. Subsequently, research on the effects of air ions and static electric fields began in 1980, focusing on the problem of health effects of the electromagnetic environment under DC transmission lines.

The new investigation will just start to identify the influence factors to human perception of static electric fields (Jankowiak et al., 2021). This research is focused on the influence of experimental and environmental factors such as the ramp slope, exposure duration, presence of air ions, and relative humidity on human perception. It will provide preliminary insights into the studies of whole-body static electric field with a sufficient number of participants of all ages under highly controlled conditions, which lead to determine statistically verified detection thresholds. This will give the insights for safety level of electric field due to high voltage transmission lines. Using newly developed of DC and AC electric field facility, a total of 203 healthy subjects were exposed to static electric field (DC), alternating electric field (AC), and hybrid electric field (EFs) in order to determine the detection thresholds of electric field (Kursawe et al., 2021). The results indicated that detection thresholds of human perception of DC, AC, and hybrid EL were lower compared to single EF presentation of DC or AC. Ion current exposure enhanced EF perception. High relative humidity facilitated DC EF perception, whereas low relative humidity reinforced the perception of AC EFs (Kursawe et al., 2021).

The first to point out the biological effects of air electricity was Beccaria of the Turin University, who said that nature apparently utilizes air electricity on a large scale for the plants growth (Mottelay, 1922). Toward the end of the nineteenth century, the discovery of air ions by Sir Joseph John Thomson, John Ludwig Julius Elster, and Hans Friedrich Geitel led to develop the study of the biological effects of air electricity. Elster and Geitel discovered the small ions in the air to be products of the natural radioactivity of the air (Fricke, 1992).

From the 1950s to 1980s, Albert Paul Krueger became strongly interested in the effects of electrically charged air particles on living organisms. He was the pioneer in the research field of ions. In the context of investigations of atmospheric phenomenon such as Foehn, heat stress such as headache and tension due to hot dry desert winds (Sharav), Felix Gad Sulman engaged in research and was a pioneer in biometeorology and bioclimatology (Sulman, 1980).

It is well known that static electric fields influence the air ionization. This has never been taken into consideration in reported experiments. So, it is difficult to determine whether the reported biological effects are due to the direct action of the static electric fields or associated changes in air ionization.

Reinhold Reiter made an important point in his book about the study of the biological effects of atmospheric electricity, especially air ions and static electric fields (Reiter, 1992, pp. 447–450). He emphasized very importantly that it must be pointed out that certain foolish assertions concerning the source and behavior of small ions have often been repeated in the scientific literature up to the present day, and in consideration of certain laws in physics and atmospheric physics its assertions are wrong.

3.5.2 Static Electric Field in Vertebrates and Invertebrates

In 1918, Jean-Henri Fabre suggested that invertebrate, *Geotrupes* respond to atmospheric electric fields. He described that

> They seem to be influenced above all by the electric tension of the atmosphere. On hot and sultry evenings, when a storm is brewing, I see them moving about even more than usual. The morrow is always marked by violent claps of thunder.
>
> *Fabre (1918, p. 289)*

There have been so great interests in the influence of static electric field on insect's behavior. In 1895, Sigmund Exner, uncle of Karl von Firsch, discovered that the feature of birds and the hairs of mammals store the electric charge created by friction with other materials (Exner, 1895). Heuschmann observed that insect cuticle also accumulates electric charge by friction. The positive charge build-up on flying insects such as bee has been appreciated for decades (Greggers et al., 2013; Heuschmann, 1929).

After the existence of the SR was suggested, research on the effects of atmospheric-derived static electric fields on living organisms was conducted mainly in Germany. In the 1960s and 1970s, Altmann studied the effects of static electric fields on vertebrate and invertebrate animals such as house fly, fruit fly, honey bee, wasps, cockroach, grasshopper, frog, budgerigar mice, and guinea pig. The effects of static electric fields on oxygen consumption and protein metabolism in typical representatives of animals were planned. It was shown that the effects of static electric fields accelerated the metabolism in the animal body, resulting in a significant increase in activity and oxygen consumption (Altmann, 1959, 1969; Altmann and Lang, 1974; Altmann and Soltau, 1974; Altmann and Warnke, 1978, 1979). The effects of electric fields of 0, 1.4, and 2.8 kV/m on oxygen consumption, food intake as a swarm, and survival of honey bees were compared in electric field and in Faraday cages. The electric field changes oxygen consumption and food intake. Increased oxygen consumption and increased food intake was found in bees, cockroaches, Indian stick insects, and wasps during exposure to weak static electric field. Altmann observed an increase in metabolic rate was accompanied by an increase in food consumption (sugar water) in various animal species and higher mortality under the influence of the field. For example, bees showed substantially higher oxygen consumption in the static electric field than that of control bees in a Faraday cage. It was reported that the static electric field has a stimulating effect on metabolism in guinea pigs (Altmann, 1962). In 1969, static electric fields in the studies by Altmann were 0.42 and 1 kV/m. The control was the Faraday cage condition. The metabolic activity such as oxygen consumption was compared to that of the Faraday cage condition. In guinea pigs exposed to the static electric field of 0.42 kV/m, the increase in activity was related to higher oxygen consumption, and free amino acid levels increased in all animals compared to Faraday shielding. It was shown that the static electric field accelerated the metabolism in the animals and greatly increased their activity level and oxygen consumption. Guinea pigs were exposed to the normal condition, the Faraday cage condition, and an electric field of 0.24 kV/m at 10 Hz for 13 days each. Blood, hematocrit, and protein were examined. The results showed that there were differences in the composition of the proteins (Altmann and Soltau, 1974). The three conditions were Faraday cage shield, 3.5 kV/m with 10 Hz and 3.5 kV/m were set up to see which of these the mice would choose (Altmann and Lang, 1974). The mouse nest location was in a shielded condition, and 10 Hz was the preferred play location. The same experiment was conducted with normal electric fields, where the nest was in the normal location and the play area was in the 10 Hz field.

Furthermore, comparisons were made with AC electric fields of 2–10 Hz, and it was founded that there was no difference in AC electric fields from those where the fields were shielded. Focusing on the physiological aspects of animals, Altmann investigated the effects of the static electric field and reported that static electric fields increased the metabolic rate and other parameters in various animals. For example, oxygen consumption was clearly higher in fruit flies than in Faraday cage controls. When looking at the activity and metabolic rate as indicators of the electrical state of honey bee hives made of wood, PVC, and metal, the activity state was higher in the PVC hive, which is in a highly electrified state, than in the grounded metal hive (Altmann and Warnke, 1979). A clear increase in metabolic rate was found in animals exposed to the static, 10 Hz electric fields. A rise in the oxygen consumption parallel to the increase in activity was also measured. Animals exposed to the 1.75 and 5 Hz AC field showed no changes related to the Faraday cage controls. Based on the oxygen consumption of goldfish, Altman showed that aquatic animals are not immune to the effects of electric fields. Goldfish exposed to static electric fields had higher oxygen consumption than Faraday shielded animals. The same result was seen for oxygen consumption in frogs. This increase was seen only when the animals were exposed to an electric field, and then the oxygen consumption immediately recovered to normal.

The systematic experiments about the effects of static electric fields on living organisms were also conducted at the University of Graz, Austria (Möse et al., 1969a, b, 1970, 1971, 1972, 1973, 1977). They first, examined the effects of electric fields on smooth muscle sensitivity to stimulant drugs. The ileum of guinea pigs and the uterus of rats exposed for several days showed decreased excitability to histamine, acetylcholine, bradoxine, and serotonin. Möse et al. reported on experiments with white mice exposed to about 24 kV/m of static electric field (1972). Based on control comparisons, statistically significant increases were found: running activity increased 55%; food consumption increased 19%; drinking water consumption increased 15%. The body temperature rose by 0.3°. Furthermore, the effect of static electric field on immune response was investigated. Mice were implanted with sheep leukocytes and exposed to electric fields of 0.04–24 kV/m. Immune responses were examined in terms of plaque formation, and it was reported that the maximum plaque formation occurred at 7–15 kV/m. In summary, they noted that the static electric field increased the activity level, oxygen consumption, water and food consumption, and immunity of the mice. In addition, the shielding of the electric field by the Faraday cage had a negative effect on the growth and activity of the mice. Based on these findings, they hypothesized that the natural electric field on earth has the effect of maintaining and promoting immunity in living organisms, and proposed the establishment of an electric field treatment room in a hospital, in which a static electric field and a 10 Hz electric field are superimposed. It was speculated that altered metabolic functions may be the result of direct effects of static electric field and air ions. Möse et al. (1971) suggested a mechanism through which static electric field act on cell functions by modifying bioelectrical potentials which in turn lead to increased cellular respiration. Möse and colleagues discussed that absorbed air ions may induce a serotonin release in the brain (1969b) or a shift in the metabolic activity of organs (1969a). However, they did not consider the possibility that the reported responses also could have been indirect effects resulting from external sensory stimulation by the static field. Additionally, one of these studies reported that mice which were kept in a Faraday cage had a low oxygen consumption compared to the control group under ambient conditions. According to the authors, lowered oxygen consumption and decreased metabolic activity of rodents held in a Faraday cage indicate that these animals were disadvantaged by the absence of both static electric field and air ions. The authors cited these results supported their hypothesis that exposure to static electric field was beneficial. On the other hand, the shielding of animals from static electric field had negative effects (Möse et al., 1971). The oxygen consumption by liver cells has been measured in mice exposed to different environmental conditions (static electric field 24 kV/m, normal room condition, and Faraday cage condition). The oxygen consumption had increased under the static electric field compared with that in the animals kept in a normal room conditions; the shielded condition caused an opposite effect. The results showed that the different electro-climatic conditions of the environment influence the oxygen consumption of the liver cells. The authors speculated that shielding from the natural static electric field may have adverse effects on health.

By the direct plaque-technique, Möse et al. determined the degree of immunization of mice under the influence of various environmental condition of bio-climate: static electric field (field strength: 0.04, 0.2, 1, 5 and 24 kV/m; control: room condition and Faraday cage condition), for 15 days' exposure (1973). Mice were pretreated with ovine erythrocytes. The highest plaque formation was found in the spleens of animals exposed to the static electric field. The same was found for spleen weight, splenocyte count, and hemagglutination titer. Interestingly, the field had the greatest effect on plaque values at field between 1 and 5 kV/m. But even at 0.2 kV/m, a considerable increase of immunization was observed in comparison with the controls. Based on these findings, it is suggested that the natural static electric field on the earth is an important factor in promoting and maintaining the immune system.

Further, comparative analyses of the development rate of a slow tumor (Methylcholanthrene: 0.1 mg per animal) in mice (weights are 25–30 g with 8 weeks old) were undertaken (Möse and Fischer, 1977). The experiments were under three conditions: (1) A static electric field with 0.2 kV/m, (2) a Faraday cage, and (3) a laboratory condition. Faraday cage has the shielding effectivity on atmospheric electric disturbances: 99%. The tumor was initiated following a 6-week acclimatization period. Next, the appearance rates over a period of 8 months at 14-days intervals were observed. Under laboratory conditions, these were perceptibly higher than in the static electric field or in Faraday cage. No difference was apparent between the latter two conditions. The results of the neoplastic activity for both in static electric field and in Faraday cage were reduced compared to a laboratory condition. However, the authors mentioned that an explanation for the results was difficult to interpret.

There had been so many investigations on the effect of static electric field in invertebrates, honey bee, cockroach, fruit fly, housefly, etc. The electric field perception and behavior, reproduction and development, metabolism, brain, and nervous system were included as endpoints of laboratory studies (Altmann, 1959; Edwards, 1961; Jackson et al., 2011; Maw, 1961a, b; Newland et al., 2008, 2015; Perumpral et al., 1978; Schuà, 1954; Watson, 1984).

Maw found that mosquitoes would aggregate in high atmospheric field and the fruit fly (*Drosophila melanogaster*) and the blow fly (*Calliphora vicina*) show decreased locomotion when exposed to fields of 0.5 kV/m (Maw, 1961a). Maw investigated stimulating effects of a weak static electric field (0.12 kV/m) on the oviposition rate of the ichneumon wasp (*Scambus buolianae*) (Maw, 1961b). Edwards investigated that after fruit fly (*Drosophila melanogaster*) and blow fly (*Calliphora vicina*) were exposed to a static electric field, they stopped moving. Blow fly needs a stronger field to elicit a response than fruit fly. Further, in order to investigate the effects of the static electric field caused by thunderclouds, Edwards studied the pupal period of the larvae of the pyralid beetle, the egg-laying sites of its adults, and the number of egg laid, etc. in the static electric field of 19 kV/m (Edwards, 1961). It was reported that the pupal period was longer and the mortality rate was slightly higher, and that 65% of the eggs were laid in the fields compared to the controls. Negative effects such as a decreased oviposition rate and delayed hatching in hemlock loopers (18 kV/m) were reported.

Static electric field influences on locomotor activity were reflected in reduced rates of locomotion proportional to electric field strength in blow fly (*Calliphora vicina*) and fruit fly (Drosophila melanogaster) (Watson, 1984) and in the cockroach (Jackson et al., 2011). As results of perception behavior experiments with blow fly and fruit fly (Watson, 1984), agitation occurred at about 200 kV/m. Mean values for paralysis of fruit flies at 416 kV/m, for blow flies at 359 kV/m, and insects were killed at field of 350–400 kV/m. In this report, it is unclear whether the effect is statistically significant or not because there is not enough data, and the effect size cannot be calculated. The animal becomes paralyzed in a high static electric field that does not produce flashover and recovers when the field is turned off. This occurs when the animal touches the static electric field during flight, and it is thought that the central nervous system is affected by the current from the corona discharge. The perception experiments in cockroaches (Periplaneta Americana; third and fourth instar; $n = 126$ in total) was reported (Jackson et al., 2011). Exposure to electric fields of 66–166 kV/m was used to examine detection and avoidance behavior. Decrease of activity an increased field avoidance behavior at higer field strengths; less distance covered in treated zone for field strength above 99 kV/m. animals spent less time in the treated zone at 132 kV/m.

It was reported that the static electric fields were 20–150 kV/m for 3 minutes for housefly (*Musca domestia L.*) and 25–150 kV/m for cabbage loopers (*Trichoplusia ni*) (Perumpral et al., 1978). Exposure to electric fields was used to examine detection and avoidance behavior. In cabbage loopers, the wing beat frequency was also examined. The flight behavior of the cabbage looper exposed to a static electric field of 20 kV/m was disturbed (Perumpral et al., 1978). When house flies were taught to choose between areas with and without electric field of 100 kV/m, they avoided areas with electric fields. They preferred static electric field of 75 kV/m and avoidance behavior by house fly occurred in the range of 100–150 kV/m. The effect of static electric fields on wingbeat frequency in male cabbage loopers was, in all cases, affected inconsistently although between 20 and 150 kV/m. On the other hand, females were not significantly affected.

Newland et al. examined the mechanism underlying the ability to perceive static electric field and scrutinized sensory structures for the detection and avoidance behaviors of such field in cockroaches (*Periplaneta Americana*; third and fourth instar and imagines with $n = 5$–40 per group) to static electric fields of 4–30 kV/m (2008). As a result, they found insect behavioral changes in response to static electric fields as tested using a Y-choice chamber an electric field generated in one arm of the chamber. Locomotor and avoidance were affected by the magnitude of the electric fields with up to 85% of individuals avoiding the charged arm when the static electric field at the entrance to the arm was above 8–10 kV/m. In order to determine mechanisms of perception, they ablated the antennae of cockroaches and found that the ability of cockroaches to avoid static electric fields was abolished. The authors concluded that the antennae are very crucial organs for the detection of static electric fields in cockroaches. Newland reported the experiments on fruit fly (*Drosophila melanogaster*; male and female) with different wing forms or excised wings $n = 8$–160 per group (2015). The animals were exposed to the static electric field of 28–183 kV/m for 5 minutes in avoidance experiments and to static electric fields of 70 kV/m between 4 and 72 hours in measurements of biogenic amine levels in the brain. It was shown that threshold for avoidance behavior at 34–43 kV/m, statistically significant in response index compared to that of controls. Intact wings played an important role in avoidance behavior. They suggested that static electric field generates an uneven charge distribution on the body surface and causes mechanical detection of negatively charged body attachments, such as antennae in cockroaches and wings in fruit flies, toward the positive electrode. This mechanical deflection was recognized by sensory receptors at the base of the antennae. In addition, chronic exposure induced change in biogenic amine levels such as serotonin, dopamine, and octopamine in the brain of fruit flies exposed to the static electric field of 70 kV/m (Newland et al., 2015).

Jackson et al. go further to analyze the locomotory behavior of cockroaches (*Periplaneta americana*) in response to static electric fields of 66–166 kV/m (2011). To analyze the effects of static electric fields on cockroach behavior, individual cockroaches were placed into the test arena and different electric fields were applied. Walking behavior (such as velocity, distance moved, turn angle, and time spent walking) were analyzed. It was obtained that the cockroaches turn away or were repulsed when they encountered an electric field and if continuously exposed to one, walked more slowly, turned more often, and covered less distance (Jackson et al., 2011). This study demonstrated that the behavior of free-moving cockroaches is significantly influenced by static electric fields and their responses are related to field strength. The walking activity in cockroaches is also reduced by static electric fields.

Studies of electric field detection in invertebrates have focused on the direct effects of electrical forces on sensory limbs such as touch and wings (Bindokas et al., 1989; Newland et al., 2008; Watson, 1984). Such tactile sensations and wings are related to the sensing of electric field. The sensory limbs of insects are concerned with mechanoreceptors which respond to mechanical disturbances in the insect's external environment, such as contact, extension, and pressure. The movement of sensory wings by electrical forces is the same as stimulation by the environment, which leads to activation of mechanoreceptors and elicits behavioral changers. For example, Bindokas (1989) reported that the wings and sense of touch of honeybees respond to low-frequency electric fields of 150 kV/m, and wing oscillations in *Drosophila melanogaster* occur when exposed to static electric and low-frequency electric fields of 500 kV/m.

There are many research studies on the biological effects of the static electric fields, but many papers do not clarify the electrical conditions. Therefore, we cannot take the results as they are. The static electric field experiments are greatly affected by the material of the used animal cage, the contamination of the cage, and the relative humidity and temperature in the air, and the magnitude of the electric fields to which the animals are exposed during experiments differs greatly from the calculated value (Mühleisen, 1966; Shigemitsu et al., 1981). The high relative humidity prevents charge build-up acting like a Faraday cage. Therefore, the reliability of the results is considered to be lacking. In the future, the static electric field exposure method should be established and the exposure conditions should be standardized.

3.6 Detection of Electromagnetic Fields

In nature, there are dozens of examples of electro- and magneto-receptions. Electro-sensitive species are able to detect electric fields in order to detect prey and predators, to communicate, and/or locally orientate. These species are also able to respond to magnetic fields, using electro-sensory organs through the induction laws of Faraday, and some species may have both electro- and magneto-sensing organs.

Electroreception is defined as the ability of an organism to detect weak electric forces. It has long been known animals living in aquatic conductive environment. Electroreception has been found in sharks, rays, amphibians, teleost fish, dolphins, platypuses, etc. They use electrosensory organs in their snout to detect prey in soil.

In mammals, including the human, there exist many types of specifically developed sensory receptors, such as the rods and cones of the retina that signal light stimuli, the hair cells in the inner ear that sense sound stimuli, the Pacinian corpuscles in the limbs that detect vibration, the baroceptors in the aorta that are sensitive to blood pressure changes, and the receptors on the tongue that provide taste sensation and others. Besides the natural stimuli that elicit activity of each receptor type, electrical stimulation also can activate effectively all of these receptors. Some kinds of fish, e.g., eel, skate and shark, possess electroreceptors that are called ampullary lateral line organs. These electroreceptors sense disturbance of electric field surrounding the fish (Chichibu, 1970, see Figure 3.12). However, there is no evidence indicating the existence of specific receptors in mammalian body that are sensitive solely to electricity.

3.6.1 Electroreception

Electroreception is common in aquatic animals. It has been found in a number of vertebrate species including the fish, amphibians, and platypus (Kalmijn, 1971; Hurd et al., 1984; Gregory et al., 1987). In aquatic animals living in electrically conductive environment such as seawater, electroreception relies on direct transmission of stimulus from the water (conductive medium) to the nervous system though Lorenzini (conductive receptor channels). Ampullary electroreception needs the presence of an electrically conductive medium.

In the past, electroreception in an aerial environment had been hypothesized. Recently, for terrestrial animals in air, Sutton et al. concluded that sensory hairs are a site of electroreception in the bumblebee (2016). In fair weather, a static electric field is the order of amplitude of 0.1–0.2 kV/m as a consequence of the global atmospheric electric circuit. The electric field is a very important key factor for the electrobiology. The relevance of static electric fields for electrobiology has been considered (Reiter, 1992). Human and terrestrial animals have no use of electroreception due to high electrical resistive medium of air (insulator) environment. They can detect electric discharge through sensory and motor nerve fibers stimulation from direct or indirect contact with conducting objects.

In animals living in terrestrial environment, the detection of electric field must operate differently using different sensory mechanisms (Clarke et al., 2017). Recent studies provided behavioral evidence in bees for electroreception. The electroreception of insects such as honey bees can detect very weak static electric field as the electrobiology will be introduced. The relevance of static atmospheric electric field for biology has been considered (König, 1979, König et al., 1981). The research focused on the relationship between insect pollinators and plants (Clarke et al., 2013). Bumblebees (*Bombus terrestris*), a solitary species and honey bees (*Apis mellifera*), and social hive species have the detection of weak static electric fields using different sensory mechanisms (Clarke et al., 2013; Greggers et al., 2013). In Bumblebees, the detection of static electric field is through mechanosensory hairs, which are mechanically deflected by an applied electric stimulation (Sutton et al., 2016). Flowers generate weak electric field and bumblebees can sense those electric fields using the tiny hairs on their fuzzy bodies. Bumblebees beat its wing up to 200 times per second through the air. Bumblebees can sense the presence of weak electric field surrounding flowers and discriminate between static electric fields with different radial geometries (Clarke et al., 2013). Bumblebees detect electric fields around plants and learn to use them to decide whether or not to visit flowers. The sensory basis for electroreception in honey bees was hypothesized to be the antennae, electromechanically coupled to the surrounding electric field in virtue of bees being electrically charged (Greggers et al., 2013). Honey bee uses their antennae. The antennae oscillate under static electric field. This stimulation can elicit activity in the antennal nerve. Honey bees with removed or fixed antennae are less able to associate food reward with static electric field stimulation (Greggers et al., 2013).

It was shown that spiders can use electric field in fair weather to balloon upwards (Morley and Robert, 2018). Spiders can detect static electric field in natural atmospheric electricity. The ballooning behavior of spiders is triggered by these static electric fields. In the experiment, a static electric field was generated with electrodes that simulated the atmospheric electric environment, and spiders (*Linyphild, Engone*) were placed in the field to observe their behavior. When the electric field was turned on, the spider's abdomen stood up, and it exhibited ballooning, a preemptive behavior in which the spider blew out its silk threads. On the other hand, when the electric field was turned off, the spider moved downward. Since spiders have trichobothria, which are mechanical receptors, it is possible that these act as receptors for sensing electric fields. The authors found that hairs respond to a small amount of air flow, or stimulation from an electric field. The spider, which does not have wings, uses silk threads blown out of its body to perform ballooning using atmospheric electricity. It was concluded that atmospheric electricity could provide forces sufficient for dispersal by ballooning in spiders and hair-shaped sensors are putative electroreceptors (Morley and Robert, 2018).

In 1917, Parker and van Heusen published historically well-known paper on the nibbling response of the catfish (*Amiurus nebulosus*) to metallic and non-metallic rods (1917). They found that a blindfolded catfish to be remarkably sensitive to metallic rod. They have a regular response to metallic rod even at a distance of some centimeters, whereas a glass rod did not elicit a reaction until it actually touched the skin of the animal (Kalmijn, 1971). They demonstrated that nibbling responses were due to the galvanic currents generated at the interface between metal and aquarium water. These responses were elicited by a current of approximately 1.0 μA between two electrodes about 2.0 cm apart. The avoidance reactions were by currents of 1 μA or more. These results led to the concept of electroreception.

Electroreception in aquatic animals such as ray fishes is the ability to detect weak static electric fields. It facilitates the detection of prey. The electroreception is obtained through direct transmission of stimulus via electrical sensitive organs, called "ampullae of Lorenzini." This name is from Stefano Lorenzini. He discovered special electric organs of the elasmobranchii (shark and ray). In the 1960s Dutch scientists Dijkgraaf and Kalmijn established that sharks and rays, which have dermal sense organs, ampullae of Lorenzini, could sense weak electric currents from their prey organisms such as flatfishes even when they were buried under sand. Kalmijn (1974) showed that fish maintains their orientation while swimming in the geomagnetic field. Sharks and rays have the ampullae of Lorenzini, which are located near the front of their brains that detect the extremely weak electric field induced by the geomagnetic field, i.e., earth currents. These behaviors can be summarized as shown in Figure 3.9 which gives various mechanisms for detecting electromagnetic fields (Kalmijn, 2000). In Figure 3.9a, the situation when a shark approaches the vicinity of a dipole field (0.2–0.5 μV/m) used to simulate prey is shown. As the shark swims through the geomagnetic field, in accordance with Faraday's induction law, a vertical electromotive force is induced. This induced electric field allows selection of direction relative to the direction of the geomagnetic field to be obtained. It has been conjectured that this is used to judge the direction that the fish is swimming. In Figure 3.9b, the vector product of the flow of velocity (v) of the ocean stream (i.e., water current) and the geomagnetic field's vertical component (Bv) is equivalent to the electrical gradient created: current flow (ion flow) occurs, and detection of this current flow allows perception of the direction (up vs. downstream) of the flowing water; this provides a means of passive electro-orientation. In a slow ocean current, surface electric fields are 0.05–0.5 μV/cm. In a tidal current level, using a cross section of the Gulf Stream an example, total electric fields up to 0.5 μV/cm were predicted (Rommel and McCleave, 1973).

Michael Faraday discovered a changing magnetic field induces electric currents in nearby conductive structures. This induction law predicts that the movement of animal in magnetic field would induce in electromotive force. This is the basis of electroreception in the aquatic animals. In Figure 3.9c, the shark is moving through the geomagnetic field: the electric field resulting from motion of the shark through the geomagnetic field gives it a magnetic compass heading; this is active electro-orientation. For a fish swimming at velocity (v) through the horizontal geomagnetic field component (Bh), an electric gradient is induced by the vector product. For example, a fish swimming at the speed of 1 m/s through the geomagnetic field horizontal component of 25 μT will induce an electric field of 0.25 μV/cm. This electrical gradient passes through the ampullae of Lorenzini. Because sharks and rays can detect electric fields of 0.01 μV/cm, they can readily detect this field. Thus, the aquatic animal might perceive an electric field induced by water current or by its own motion in the geomagnetic fields.

Behavior of a shark near a dipole imitating prey buried in the sand is shown in Figure 3.9a. In 1971, Kalmijn published the key paper in the discovery of electroreception in fishes. The behavioral experiment is shown in Figure 2 of his paper (Kalmijn, 1971, see p. 377). The results of these behavioral experiments are suggested as follows: (1) When a shark was eagerly searching for food and passed a plaice at a distance of 15 cm or less, he generally exhibited a very clear feeding response, although the prey was almost entirely hidden from view by a thin layer of sand. (2) When the sharks, eagerly swimming about, passed the agar-screened plaice at a distance of 15 cm or less, they still showed their characteristic, well-aimed turnings toward the animal. (3) The plaice in the agar chamber was exchanged for a small bag of loose Nylon tissue, containing pieces of whiting. In this situation the sharks and rays eagerly tried to find their food at the end of the outlet tube and did not show even the slightest response when swimming just over the agar roof. (4) After the test animals had been motivated with whiting juice, they searched

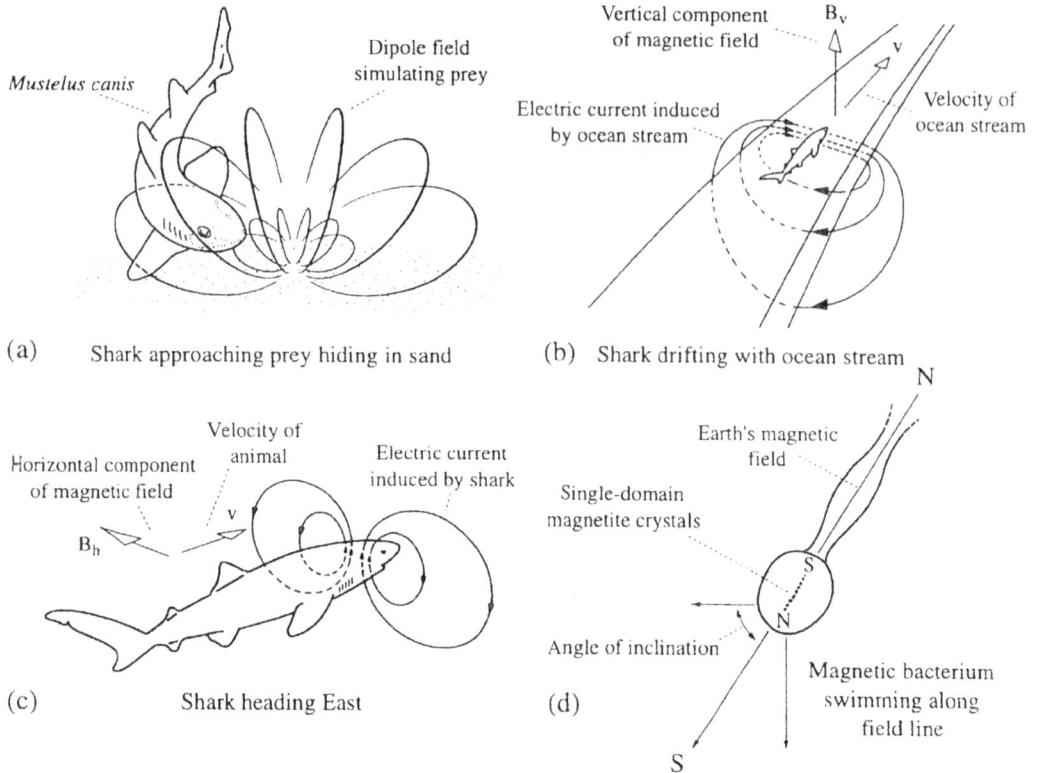

(a)   Shark approaching prey hiding in sand

(b)   Shark drifting with ocean stream

(c)   Shark heading East

(d)

FIGURE 3.9 Behavioral responses of organisms to electromagnetic fields in the aquatic environment. (a) Behavior of a shark near a dipole imitating prey buried in the sand. (b) Sharks in a situation like a tidal current with the geomagnetic field. Electric field induction in fish-oriented upstream and downstream in the flow is shown. (c) Sharks swimming in geomagnetic field. (d) As magnetic bacteria are anaerobic, movement is toward the bottom, where conditions are most anaerobic, as directed by geomagnetic field. (From Kalmijn, 2000.)

intensively all over the bottom of the pool, but did not notice the plaice any longer, although they often passed it very closely. Kalmijn proved directly that the sharks and rays are actually able to detect the source of such an electric field as produced by the plaice. (5) By stimulating the bioelectric fields of the plaice with a direct current of 4 μA, apparently natural feeding responses could be elicited. (6) Sharks and rays swam just over the piece of whiting to reach the site of the electrodes; often they swam away, after digging at the electrodes, without finding the piece of whiting. Kalmijn commented further that the electric fields acted as a much stronger directive force or the sharks and rays than did the visual and chemical stimuli of the piece of whiting (1971).

3.6.2 Magnetoreception

The electrophysiological in vitro responses to magnetic stimulation were reported from the pineal gland. Semm and his co-worker provided the first evidence that the cells in the pineal gland can respond to stimuli other than light (1980). These researchers demonstrated a significant diminution of in vivo electrical activity of single pinealocytes of guinea pig following acute inversion of the vertical component of the geomagnetic field by means of a Helmholtz coil. This finding suggested that pineal melatonin synthesis might be affected by magnetic fields. Welker demonstrated that rats exposed to a 15-minutes inversion of the horizontal component of the geomagnetic field during the nighttime hours in the presence of dim red light had a significantly decreased pineal melatonin synthesis as compared to unexposed control animals (Welker et al., 1983). Then, James Olcese reported that, in both albino and pigmented rats, melatonin synthesis was markedly inhibited following a single, 30-minutes exposure to a static magnetic field stimulus consisting of a 50° rotation of the earth's horizontal geomagnetic field (Olcese and Reuss, 1986).

Sensitivity to the geomagnetic field is widespread among reptiles and birds. Both behavioral and neurophysiological evidence suggests that there might be two magnetoreception mechanisms present in some vertebrates, one system functioning as a source of directional compass information and the other providing a geographic position ("map") information based on geomagnetic cues (Phillips and Deutschlander, 1997).

Contrary to the perception of electric field, no literature has been reported that human can detect magnetic field. For example, Schmitt and Tucker could not verify that human perceived magnetic field sensation with 0.7–1.5 mT, 60 Hz magnetic field exposure (Tucker and Schmitt, 1978).

In the late 1970s, the human homing experiments have been carried out to test magnetic field perception. Baker first performed human homing experiments (1980, 1987). He made three experiments: walk-abouts, bus experiments, and chair experiments. In walkabouts and bus experiments, subject students with helmet (magnetic strength between 0.014 and 0.030 T) had been blindfolded, and transported by bus over a route to a point 6–52 km from "a home site (Manchester University)" with changes of direction. Subjects were asked to indicate the direction of starting point. Baker found an overall ability to name the compass courses to the starting point and to point in the respective direction. Although many researchers tried to replicate the bus experiments, negative results were reported. In chair experiments, the blindfolded subjects sat on a rotatable chair in the center of the large coil sets. This experiment also showed a weak yet significant trend to name direction correctly (Wiltschko and Wiltschko, 1995). Two claims gave the Baker's experiment: One is in solving problems of orientation and navigation, humans have access to some non-visual ability, and the second is at least, this non-visual ability involves magnetoreception (Baker, 1987). Although many researchers attempted to replicate Baker's findings, they were unsuccessful. The controversy still continues. It is not easy to perform the perception of human. In 2019, Kirschvink's research group from California Institute of Technology co-operated with researcher of the University of Tokyo opening new research result on human sense to magnetic field equal to the strength of geomagnetic fields (Wang et al., 2019). His group recorded brain activity on the scalp to search for some response to changes in magnetic field. The 34 participants sat in testing chamber with recording of EEG. Following magnetic stimulation, a drop in amplitude of EEG alpha-oscillations (8–13 Hz) occurred in a repeatable manner. Alpha-ERD (alpha-event-related desynchronization) in response to the geomagnetic filed was triggered only by horizontal rotations when the static vertical magnetic field was directed downwards. No brain responses were elicited by the same horizontal rotations when the static vertical component was directed upwards. This test showed that the neural response was sensitive to static components of the magnetic fields. The neural response was also sensitive to the polarity of the magnetic field.

More interesting, apart from science, it has been suggested that very small special subsets of human population show a significant ability to sense magnetic fields (König et al., 1981; Tromp, 1968). Dowser belongs to this small special subset. Dowsing is known as diving rod (known as Wünschelrute, in Germany). Dowsing may have originated in Germany in the sixteenth century. It was used in attempts to find metals. The point is that dowser responds to the variation of geomagnetic fields, the site of anomaly. The original form of the dowsing was a forked wooden twig (Tromp, 1968). It was used originally to find metals (Agricola, 1556). The dowser walks over land with a forked wooden twig in his hands and experiences at certain places a contraction of his forearm muscles, which forces the forked wooden twig to turn either upwards or downwards. Now, it is said that dowsing is a pseudoscience.

The other document to magnetic field is the case of dances of honeybees on a horizontal comb. When honey bees communicate the direction and distance to food to their companions, and when food source is very close (within about 50 m), they make a simple circular movement of the whole body while they also make a figure-eight movement with their posteriors. The direction of the wagging dance of honey bee shows the angle between the location of the food and the sun, and the angle is transposed with respect to gravity. The speed of turning is said to represent the distance to the food. Martin Lindauer and Hermann Martin showed that this honeybee dance was affected by geomagnetic field (Lindauer and Martin, 1968). When the geomagnetic field is compensated to ±4%, misdirection disappears. The visible iron-containing granule cells are in the form of magnetite. Later, it was found that there was magnetite in the abdomen of honey bees that was strongly magnetized laterally. They are formed in hydrous iron oxides. It is thought that the magnetite is formed during the growth from larva to adult, with the magnetism being oriented with the geomagnetic field. Honeybees use this magnetite as a sensor for geomagnetic field, and they are thought to use this to determine direction (angle) even when it is cloudy and the sun cannot be seen. However, this angle is affected by the magnetic field of 0.1–0.001 µT.

The other case is the migratory birds. Alexander Theodor von Middendorff suggested that migrating birds orient along magnetic meridians (1859). Today it means that the migrating birds use a magnetic compass. Middendorff recorded the places and date of arrival of migrating birds from 1841 to 1854. Using these data, he drew a number of migratory routes of birds on a map. He referred to it as *Isepipteses*. Isepipteses means lines of simultaneous arrival (Al-Khalili and McFadden, 2014). In his report published in 1859, he proposed that the migratory birds orientate themselves by the geomagnetic fields. However, most nineteenth century zoologists had no interest and remained skeptical. It passed over hundred years since von Middendorff's suggestion that the migratory birds may solve their orientation by reference to the geomagnetic field. From the middle of the twentieth century, the supported experiments about effects of the geomagnetic field on bird migration have been investigated.

Friedrich Wilhelm Merkel observed the night-migrating birds can orient their orientation without help of celestial cues and identified the geomagnetic field as a key factor controlling bird orientation in the night. Merkel and Wolfgang Wiltschko published two papers on the discovery of magnetic compass in the European robin (Merkel and Wiltschko, 1965; Wiltschko and Merkel, 1966). Although magnetic compass remained still skepticism in the scientific community, Wiltschko's discovery that the birds used geomagnetic field as orientation compass was the breakthrough and expands the experimental and theoretical studies of magnetic compass sense of animals including vertebrate, invertebrate, bacteria, fishes, and birds. The ability to sense and respond to magnetic stimuli is known for a number of animals and bacteria. A number of experiments have reported that geomagnetic field was being used for migration and orientation (Wiltschko and Wiltschko, 1995).

There are three theories as to how animals including birds sense geomagnetism. The first one is electromagnetic induction, which is thought to occur in fishes, but the receptors needed for this mechanism are not found in birds or other animals. The second theory has to do with magnetized magnetite, which was discovered in the 1970s by a magnetotactic bacterium that uses magnetite in its body to move with the direction of geomagnetic field. Honeybee, fish, and birds also have fine crystals of magnetite. The third theory is that magnetic sensation in birds and other animals is mediated by biochemical reactions. In 2002, Ritz et al. have proposed a theory that birds can sense geomagnetism because a light-induced biochemical reaction occurs in the bird's eye and cryptochromes, a protein in the retina, act as a magnetic sensor (Ritz et al., 2000; Timmel, 2004). See other chapters for more discussions on magnetoreception.

Magnetoreception is widespread in the animal kingdom. There are lots of behavioral evidence from many outdoor experiments. A research group from the University of Duisburg-Essen, Germany, has repeated experiments on how wild animals perceive their surrounding environment and reported interesting results. In response to the question of what cues waterbirds such as ducks and geese flying in flocks at high altitudes, use when landing on the surface of a lake, researcher hypothesized that waterbirds are sensitive to magnetism and that their landing direction indicator follows magnetic field lines, and confirmed that they use a magnetic compass for landing (Hart et al., 2013a). The study showed that the landing direction indicator of a total of 15,000 birds from 14 species of waterbirds and suggested that they landed facing north-south in absence of any other common denominator determining the landing direction. More objective and reproducible data are needed. The relationship between predatory behavior and magnetoreception has also been reported for red fox (*Vulpes vulpes*), which are widespread in the northern hemisphere (Červený et al., 2011). Observations and records were taken of 592 jumping behaviors and the direction they were facing during predation in 84 red foxes, mainly in the Czech Republic. In particular, when capturing prey hiding under the snow, jumping in a certain direction increases the success rate. The specific direction is related to the earth's geomagnetic fields. The red fox's behavior incorporates a magnetic compass. The same research group reported that domestic dogs were found to be sensitive to small variations of the earth's magnetic field. Dogs align their body axis with north-south axis direction of the geomagnetic field for defecation and urination habits (Hart et al., 2013b). They reported also that domestic cattle as well as grazing resting wild deer, red deer (*Cervus alphas*), and roe deer (*Capreolus capreolus*) with the bodies aligned in the north-south direction, but that this aligned behavior is randomized in areas directly under or near power transmission lines which generates ELF-EMF (Begall et al., 2008).

Many animals have the ability to use the geomagnetic compass, including the long-distance migrating monarch butterfly, salmon returning to their home rivers; cetaceans and sea turtle are also known to use geomagnetism. Monarch butterfly migrates thousands of kilometers. They have substantially more magnetic material than other butterflies. The American monarch butterfly (*Danaus plexippuis*) inhabiting the North American continent migrates from Mexico to Southern Canada. It has been reported that they may use a magnetic compass as well as a solar compass for nearly 4,000 km migration (Guerra et al., 2014). Monarchs have light-sensitive magnetoreceptors which serve as an inclination compass.

It has been hypothesized that migratory cetaceans use magnetic anomalies as indicators of direction determination, navigation, and course determination. Some studies have confirmed this hypothesis with erroneous landward and stranding behavior during seasonal long-distance migrations. The distribution of geomagnetic fields is greatly affected by local topographic changes and is observed as a magnetic anomaly. Conversely, the point where the magnetic anomaly becomes weak can also be considered a magnetic anomaly. Margaret Klinowska, biologist at the University of Cambridge, has compiled a record of 3,000 cases of stranding over past 70 years that are preserved in the British Museum. Klinowska analyzed all reports on passive stranding (dead body) and active stranding (live animal) along the British coastlines (1985, 1986). She found that the difference between passive stranding and active stranding. Live stranding is associated with geomagnetic disturbances. The pattern of magnetic disturbance, not the absolute level of disturbance is the key factor for live stranding at all latitudes investigated. Further, Kirschvink professor of California Institute of Technology and his co-workers tested the hypothesis that cetaceans use weak anomalies in the geomagnetic field as cues for orientation, navigation, and/or piloting (1986). Live stranding of whales has also been correlated with local geomagnetic anomalies. Kirschvink employed 212 stranding events of live animals (active stranding) recorded and preserved at the Smithsonian Institution and data on stranding locations along the US East Coast to examine the correspondence between magnetic topographic changes and stranding locations (1986). They found that there are highly significant tendencies for cetaceans to beach themselves near coastal locations with local magnetic minima. After confirming these significant effects by Monte-Carlo simulations, he suggested that cetaceans may be sensing local geomagnetic fluctuations. This suggested that cetaceans have a magnetic sensing system comparable to those of other migratory and homing animals, which suggests that magnetic topography and the magnetic lineation of the ocean play an important role in guiding long distance migration.

Ferromagnetic materials in biological organisms interact with the earth's magnetic field to produce torque that can be sensed by some organisms and can provide a mechanism for sensing direction (Blakemore, 1975; Blakemore et al., 1981). The magnetic properties of magnetite crystal are related to the size and configuration of the material. If the balance between static magnetic energy and the wall energy in the magnetic domain of large crystals is considered, the situation becomes that of a multiple, magnetic-domain structure. However, for small crystals of less than 1 μm, a single-domain structure is more stable. If the crystals are much smaller, the anisotropic energy becomes about the level of thermal energy (kT) and the crystal becomes super-paramagnetic because of Brownian movement.

Magnetic bacteria are anaerobic and their movement is toward the bottom, where conditions are most anaerobic, as directed by geomagnetic field. The magnetic force lines of the geomagnetic field are horizontal near the equator. However, as the North and South Magnetic Poles are approached, the vertical components become larger, and magnetic dip, the downward slant toward the earth's surface, increases. If the magnetic force lines are followed, northward in the Northern Hemisphere and southward in the Southern Hemisphere, it is possible to move in a controlled direction within the low-oxygen environment of sediments. Interestingly, the discovery history of magnetotactic bacteria was introduced (Frankel, 2009). There are two unpublished manuscripts dated 1963 reported on "magneto-sensitivity" in aquatic bacteria. The original papers were written in Italian. The English translations of two manuscripts have been submitted for publications (Bellini, 2009a, b). In Bellini's manuscripts, he observed that bacteria oriented evidently themselves in a unique direction, and these microorganisms moved to the direction of the North Magnetic Pole, which he called "magnetosensitive bacteria." On the basis of these observations, he hypothesized that: (1) the bacteria contain an initial, biomagnetic dipole, (2) the magnetic dipole caused cells to migrate downward in the Northern Hemisphere toward the less oxygenated bottom sediments, (3) magnetosensitive bacteria in the Southern Hemisphere should swim toward the South Magnetic Pole, and (4) the biomagnetic dipole was fashioned from iron compounds (Frankel, 2009). Frankel mentioned, *The experimental observations presented by Bellini in the two manuscripts correspond very well with the observations of Blakemore and others* and *Bellini did make a valid, first discovery and present hypotheses that were later verified.* The finding of Bellini remained obscure until the Blakemore's finding. Frankel proposed that both discoveries be cited in future publications concerning magnetotactic bacteria. However, his proposal suggests strongly that Blakemore's discovery led to the development of today's scientific community in magnetobiology and bioelectromagnetism.

3.7 Discussion

How has the atmosphere-derived electromagnetic environment affected life on earth? There is no question that the earth is not a closed system but an open system that extends into the atmosphere and into space. For this reason, solar radiation and cosmic rays are also important sources of extraterrestrial natural electromagnetic fields.

Natural electromagnetic fields are environmental factors for all biological systems. Their origins are basically the atmosphere, the terrestrial electromagnetic fields, the radiation from the sun, the cosmic rays which are divided into non-ionizing radiation and ionizing radiation. All biological systems on earth live within and in harmony with these naturally originated electromagnetic fields and radiation. In particular, the atmosphere is host to a electromagnetic environment generating from a global electric circuit, atmospheric electric fields to local lightning strikes. Many environmental processes may interact with atmospheric electricity including lightning, thunderstorm, climate changes, etc. For a period of millions to a billion years, life on earth had the time in the evolution of the species to adapt to the natural originated electromagnetic environment. On the other hand, ever since it became possible to create man-made electricity, life has been also exposed to this man-made electromagnetic environment. The earth is covered with electromagnetic radiations not only from natural origin but also from man-made origin. These electromagnetic radiations are closely related to our daily activity and life.

In this chapter, the intimate connectivity of atmospheric electricity with biological systems was presented briefly. It shows how atmospheric electricity can govern human well-being as well as a whole animal kingdom. In particular, the effects of the earth's electric field on living organisms and how they are utilized by living organisms are discussed, as well as the relationship between the low-frequency electromagnetic fields generated by lightning discharges that occur constantly on a global scale and physiological changes in humans and animals and the relationship between natural electromagnetic fields and living things. It is also said that negative air ions produced by atmospheric electrical phenomena make us feel refreshed, while we feel deep under the presence of positive air ions. It is also believed that positive air ions increase near the ground surface when thunderclouds approach and that the discomfort during bad weather is caused by positive air ions. Globally, the effects of air ions charged aerosols on human health remain controversial. However, there have not been enough experiments to conclude these things. Insect behavior has been shown to be affected by high electric field. Flying insects avoid electric fields and cockroaches avoid static electric fields of 66–166 kV/m when walking. Bees may use electric fields as sensory cues in their interaction with flowers. It showed that bees learn themselves to acquire charge and learn to differentiate between artificial flowers when they were held at different voltages.

In the middle of twentieth century, the research focused on the effects of both climate changes, static electric field, and air ions on human health and plant growth (Baily et al., 2018; Charry and Kavet, 1987; Petri et al., 2017; Schmiedchen et al., 2018; Sulman, 1980). In Israel in the 1970s, Sulman conducted researches on atmospheric electricity, air ion and static electric field, and the relationship between static electric field and human (1980). Forty years later, Price and his research group in the same country took up again the issue of the relationship between atmospheric electrical phenomena and biological systems (Elhalel et al., 2019; Price et al., 2021). In particular, they began to investigate the connection between the SR in nature and the electromagnetic fields found in biological systems. Research on the relationship between the SR and the maintenance of life was conducted in the 1960s and 1970s by Professor Schumann, who predicted the SR phenomenon in 1952, and his student König. Recently, researchers again began to consider the atmospheric electricity as a meteorological parameter which is linked to the driving biological process.

There are many examples of human activities and technologies which cause environmental changes and impacts on ecosystems, climate change, air pollution, etc. The technologies are now found commonly. The electromagnetic fields generated from human daily activity is one of a critically important to research its impact on our environment. In particular, the radiofrequency electromagnetic fields increased dramatically from the 1990s and are ubiquitous around our environment. The effects of radiofrequency electromagnetic fields on people and the global systems need to be clarified (Cucurachi et al., 2013; Romanenko et al., 2017; Karipidis et al., 2021; Wood et al., 2021).

The world has been flooded with various electrical devices, equipment, and telecommunication devices that use electromagnetic fields, mainly non-ionizing radiation. In the future, electromagnetic fields will be more prevalent in our surrounding than ever before. Therefore, there is a need to clarify the effects of electromagnetic fields on human health as well as on vertebrates and invertebrates. These species make up the earth's natural global environment. In order to protect the earth's natural environment and maintain our sustainable society in harmony with nature and humanity in the future, we need to know what kind of effect these man-made electromagnetic fields will have on the environment, and whether the resulting flooding of the electromagnetic fields will interfere with the behavior of wildlife. The clarification of the effects of electromagnetic fields on them will help to solve the global environmental problems that we are facing today.

In addition to the problems of effects on biological systems and ecology, the effects of magnetic storms caused by solar activity and solar wind on power systems, broadcasting, and communication satellite systems have been discussed. It is imagined that as the sun's activity reaches it maximum, the solar wind will have a greater impact, and the energy released at that time will upset satellites, GPS, etc., causing electric currents in power lines and blackouts. Since such problems affecting space-based systems are attracting attention, efforts are being made to develop space weather forecasts to predict disturbances in the space environment. Space Weather Forecast Center of National Institute of Information and Communications Technology, Japan, provides information on solar activity, geomagnetic activity forecasts and warnings, solar flare, solar wind, solar proton, geomagnetic storms, sunspot number, geomagnetic indices, etc. (https://swc.nict.go.jp/en/). In connection with such accurate information, research is being conducted to clarify the interaction between electromagnetic phenomena of natural origin and biological systems and ecology, etc.

References

Agricola G (1556): *De Re Metallica*. English translated version published 1950. Dover Publications, New York.

Al-Khalili J and McFadden J (2014): *Life on the Edge-The Coming of Age of Quantum Biology*. Bantam Press, London.

Al'pert Ya L and Fligel DS (1995): *Propagation of ELF and VLF Waves Neat the Earth*. Springer Verlag, New York, London.

Altmann G (1959): Der Einfluss statischer elektrischer Felder auf den Stoffwechsel der Insekter. *Z.Bienenforsch* 4:199–201.

Altmann G (1962): Die physiologische Wirkung elektrischer Felder auf Tiere. *Verh Dtsch Zool Ges* 11:360–366.

Altmann G (1969): Die physiologische Wirkung elektrischer Felder auf Organismen. *Arch Met Geoph Biokl* B 17:269–290.

Altmann G and Lang S (1974): Die revierauftielung bei weissen Mausen unter naturilchen bedingungen, im Faradayschen raum and in kunstlichen luftelektrischen feldereichen. *Z Tierpsyschol* 34:337–344.

Altmann G, Lang S and Lehmaier M (1976): Psychotrope wirkungen des Wettergeschehens und eines kunstlichen elektrischen Rechteckimpuls-feldes der Frequenz 10 Hz. *Z Angew Bäeder Klimaheilkd* 23:407–420.

Altmann G and Soltau G (1974): Einfluss luftelektischer Felder auf des Blut von Meeschweichen. *Z Angew Bad u Klimaheilk* 21:28–32.

Altmann G and Warnke U (1978): Unspezifische analoge Reaktion von Mausen auf starke elektrisch Felder und elektrtishe Anschirmung. *Zeit Phys Med* 7:137–143.

Altmann G and Warnke U (1979): Über den Einfluß der elektrischen Umwelt auf das verhalten gekäfigter Honingbienen. *Anzeiger fur Schadlingskunde, Pflanzschtz, Umweltschtz* 52:17–19.

Aschoff J and Wever R (1962): Spontanperiodik des Menschen bei Ausschluss aller Zeitgeber. *Naturwissenschaften* 49:337–342.

Bailey WH and Charry JM (1986): Behavioral monitoring of rats during exposure to air ions and DC electric fields. *Bioelectromagnetics* 7:329–339.

Bailey WH and Charry JM (1987): Acute exposure of rats to air ions: effects in the regional concentration and utilizatyion of serotonin in brain. *Bioelectromagnetics* 8:173–181.

Bailey WH, Williams AL and Leonhard MJ (2018): Exposure of laboratory animals to small air ions: a systematic review of biological and behavioral studies. *BioMed Eng Online* 17:72.

Baker RR (1980): Goal orientation by blindfolded humans after long-distance displacement; possible involvement of a magnetic sense. *Science* 210:555–557.

Baker RR (1987): Human navigation and magnetoreception: the Manchester experiments do replicate. *Animal Behav* 35:691–704.

Begall S, Červený J, Neef J, Vojtech O and Burda H (2008): Magnetic alignment in grazing and resting cattle and deer. *Proc Natl Acad Sci USA* 105:13451–13455.

Bellini S (2009a): On a unique behavior of freshwater bacteria. *Chin J Oceanol Limnol* 27:3–5.

Bellini S (2009b): Further studies on "magnetosensitive bacteria". *Chin J Oceanol Limnol* 27:6–12.

Bindokas VP, gauger JR and Greenberg B (1989): Laboratory investigations of the electrical chracteristics of honey bees and their exposure to intense electric fields. *Bioelectromagnetics* 10:1–12.

Blakemore RP (1975): Magnetotactic bacteria. *Science* 19:377–379.

Blakemore RP, Frankel RB and Kalmijn AJ (1981): South-seeking magnetotactic bacteria in the southern hemisphere. *Nature* 286:384–385.

Blodin JP, Nguyen DH, Sbeghen J, et al. (1996): Human perception of electric fields and air ion currents associated with high-voltage DC transmission line. *Bioelectromagnetics* 17:230–241.

Bullock TH (2002): Biology of brain waves: natural history and evolution of an information-rich sign of activity. (eds). Arikan K and Moore N. *Advances in Electrophysiology in Clinical Practice and Research*. Kjellberg, Wheaton, IL.

Červený J, Begall S, Koubek P, Nováková P and Burda H (2011): Directional preference may enhance hunting accuracy in foraging foxes. *Biol Lett* 7:355–357.

Chapman CE, Blodin GB, Lapierre AM, et al. (2005): Perception of local DC and AC electric fields in humans. *Bioelectromagnetics* 26:357–366.

Charry JM and Bailey WH (1985): Regional turnover of norepinephrine and dopamine in rat brain following acute exposure to air ions. *Bioelectromagentics* 6:415–425.

Charry JM and Kavet RI (Ed) (1987): *Air Ions: Physical and Biological Aspects*. CRC Press, Boca Raton, FL.

Chichibu S (1970): Bioelectricity of the fish (Japanese text). In *Bioelectricity*. ed. Iwase Y, Tamashige M and Furukawa T., pp. 347–374, Nane-do Co, Tokyo.

Clairmont BA, Johnson GB, Zaffanerlla LE and Zelingher S (1989): The effect of HVAC-HVDC line separation in a hybrid corridor. *IEEE Trans PWRD* 4:1338–1350.

Clarke D, Morley E and Robert D (2017): The bee, the flower, and electric fields: electric ecology and aerial electroreception. *J Comp Physiol A* 203:737–748.

Clarke D, Whitney H, Sutton G and Robert D (2013): Detection and learning of floral electric fields by bumblebees. *Science* 340:66–70.

Cucurachi S, Tamis WLM, Vijver MG, Peijnenburg WJGM, Bolte JFB and de Snoo GR (2013): A review of the ecological effects of radiofrequency electromagnetic fields (RF-EMF). *Environ Int* 51:116–140.

Danho S, Schoellhorn W and Aclan M (2019): Innovative technical implementation of the Schumann resonances and its influence on organism and biological cells. *Mater Sci Eng* 564:012081.

Dowse HB (1982): The effects of phase shifts in a 10 Hz electric field cycle on locomoter activity rhythm of Drosophila melanogaster. *J Interdiscipl Cycle Res* 13:257–364.

Dowse HB and Palmer JD (1969): Entrainment of circadian activity rhythmus in mice by electrostatic fields. *Nature* 222:564–566.

Edwards DK (1961): Influence of electrical field on pupation and oviposition in Nepytia phantasmaria Strk (*Lepidoptera: geometridae*). *Nature* 191:976–993.

Elhalel G, Price C, Fixler D and Shainberg A (2019): Cardioprotection from stress conditions by weak magnetic fields in the Schumann resonance band. *Sci Rep* 9:1645.

Engelmann W, Hellrung W and Johnsson A (1996): Circadian locomotor activity of Musca flies: Recordingf method and effects of 10 Hz square-wave electric fields. *Bioelectromagnetics* 17:100–110.

Exner S (1895): Über die elektrischen Eigenschaften der Haare und Federn. *Pflügers Archiv* 61:427–449.

Fabre JH (1918): *The Sacred Beetle and Others (Translated by A T de Mattos)*. Dodd, Mead and Company, New York.

Fitzgerald GF (1893): The period of vibration of disturbances of electrification of the earth. *Nature* 48(1248):526.

Frankel RB (2009): The discovery of magnetotactic/magnetosensitive bacteria. *Chin J Oceanol Limnol* 21:1–2.

Fricke RGA (1992): *J.Elster & H.Geitel-Jugendfreude, Gymnasiallehrer, Wissenshaftler aus Passion*. Doring Druck, Druckerei und Verlag GmbH, Braunschweig.

Greggers MC, Koch G, Schmidt V, et al. (2013): Reception and learning of electric fields in bees. *Proc R Soc B* 280:20130528.

Gregory JE, Iggo A, McIntyre AK and Proske U (1987): Electroreceptors in the platypus. *Nature* 326: 386–387.

Guerra PA, Gegear RJ and Reppert SM (2014): A magnetic compass aids monarch butterfly migration. *Nat Commun* 5:4164 Doi: 10.1038/ncomms5164.

Hainsworth LB (1984): The effect of geophysical phenomena in human health. *Speculat Sci Technol* 6:439–444.

Hamer JR (1968): Effects of low-level, low-frequency electric fields on human reaction time. *Behav Biol* 2:217–222.

Hart V, Malkemper EP, Kušta T, Begall S, et al. (2013a): Directional compass preference for landing in water birds. *Fron Zool* 10:38.

Hart V, Nováková P, Malkemper EP, Begall S, Hanzal V, Ježek M, et al. (2013b): Dogs are sensitive to small variations of the Earth's magnetic field. *Fron Zool* 10:80.

Heuschmann O (1929): Über die elektrischen Eigenschaften der Insekten Haare. *Z vergl Physiol* 10:594–664.

Hunting ER, Harrison RG, Bruder A, et al. (2019): Atmospheric electricity influencing biogeochemical processes in soils and sediments. *Front Physiol* 10. 373.

Hurd CD, Fritzsch B, Wake MH et al. (1984): Electroreception in amphibians. *Am Sci* 72:228–232.

Iwase Y (1970): Electrical phenomena of the brain (Japanese text). In: *Bioelectricity*, Iwase Y, Tamashige M and Furukawa T (eds), pp. 271–315, Nane-do' Co, Tokyo.

Jackson J (1998): *Classical Electrodynamics*. 3rd Edition, p. 376, Wiley, New York.

Jackson CW, Hunt E, Sharkh S and Newland PL (2011): Static electric fields modify the locomotory behavior of cockroaches. *J Exp Biol* 214:2020–2026.

Jankowiak K, Drieôen S, Kaifie A, et al. (2021): Identification of environmental and experimental factors influencing human perception of DC and AC electric fields. *Bioelectromagnetics* Doi: 10.1002/bem.22347.

Kalmijn AJ (1971): The electric sense of sharks and rays. *J Exp Biol* 55:371–383.

Kalmijn AJ (1974): The detection of electric fields from inanimate and animate sources other than electric organs. In: *Handbook of Sensory Physiology Vol. III/3: Electroreceptors and other Specialized Receptors in Lower Vertebrates*, Fessard A (ed), pp. 147–200, Springer Verlag, Berlin, Heidelberg, New York.

Kalmijn AJ (2000): Detection and biological significance of electric and magnetic fields in microorganisms and fish. In: *Effects of Electromagnetic Fields on the Living Environment*, Matthes R, Bernhardt JH and Repacholi MH (eds), pp. 97–112. https://www.icnirp.org/en/publications/article/ICNIRP-emf-living-environment-2000.html.

Karipidis K, Mate R, Urban D, Tinker R and Wood AW (2021): 5G mobile networks and health- a state-of-the-science review of the research into low-level RF fields above 6 GHz. *J Expos Sci Environ Epidemiol* 31:585–605.

Kato M, Ohta S, Kobayashi T and G Matsumoto (1986): Response of sensory receptors of the cat's hindlimb to a transient step-function DC electric field. *Bioelectromagnetics* 7:395–404.

Kellogg EW and Yost MG (1986): The effects of long-term air ion and direct current electric field exposures on survival characteristics in female NAMRU mice. *J Geronotology* 41:147–153.

Kellogg EW, Yost MG, Reed EJ and Kruger AP (1985a): Long-term biological effects of air ions and dc electric fields on NAMRU mice: first year report. *Inter J Biometeorol* 29:253–268.

Kellogg EW, Yost MG, Reed EJ and Madin SH (1985b): Long-term biological effects of air ions and dc electric fields on NAMRU mice: second year report. *Inter J Biometeorol* 29:269–283.

Kirschvink JL, Dizon AE and Westphal JA (1986): Evidence from strandings for geomagnetic sensitivity in cetaceans. *J Exp Biol* 120:1–24.

Klinowska M (1985): Cetacean live stranding sites related to geomagnetic topography. *Aquatic Mammals* 11:27–32.

Klinowska M (1986): Cetacean live stranding dates related to geomagnetic disturbances. *Aquatic Mammals* 11:109–119.

König HL (1977): *Unsichtbare Umwelt*. Eigenverlag Herbert L. König, München.

König HL and Ankermüller F (1960): Über den Einfluss besonders niederfrequenter elektrischer Vorgänge in der Atmosphäre auf den Menschen. *Naturwissenschaften* 47:486–490.

König HL, Kruger AP, Lang S and Sönning W (1981): *Biologic Effects of Environmental Electromagnetism*. Springer Verlag, New Yok, Heidelberg, Berlin.

Kuroda PK (1983): The Oklo phenomenon. *Naturwissenschaften* 70:536–539.

Kursawa M, Stunder D, Krampert T, Kaifie A, Driessen S, Kraus T and Jankowiak K (2021): Human detection thresholds of DC, AC, and hybrid electric fields: a double-blind study. *Environ Health* 20:92. Doi: 10.1186/s12940-021-00781-4.

Lindauer M and Martin H (1968): Die Schwereorientierung der Bienen unter dem Einfluss der Erdmagnetfeldes. *Z Vergl Physiol* 60:219–243.

Lintzen T, Böse G, Müller M, Eichmeier J, et al. (1989): The stability of the circadian rhythms of green finches (*Cardeulis chloris*) under the influence of a week electric field. *J Biol Rhythm* 4:371–476.

Lintzen T, Böse G, Müller M, Falk M, et al. (1992): The stability of the circadian rhythm of green finches (*Carduelis chloris*) with in a weak 10 Hz electric fields or negatively ionizied atmosphere. In: *Electromagnetic Fields and Circadian Rhythmicity*, Moore-Ede MC, Campbell SS and Reiter RJ (eds), pp. 141–150, Birkhäuser, Boston, Basel, Berlin.

Maw MG (1961a): Suppression of oviposition rate of Scambus buolianae (Htg.) (*Hymenoptera: Ichneumonidae*) in fluctuating electrical fields. *Can Entomol* 93:602–604.

Maw MG (1961b): Behaviour of an insect on an electrically surface. *Can Entomol* 93:391–393.

Merkel FW and Wiltschko W (1965): Magnetismus und Richtungsfinden zugunruhiger Rotkehlchen (*Erithacus rubecula*). *Vogelwarte* 23:71–77.

Miller SL (1953): A production of amino acids under possible primitive Earth conditions. *Science* 117:528–529.

Morley EL and Robert D (2018): Electric fields elicit ballooning in spiders. *Curr Biol* 28:2324–2330.

Morrell F (1967): Electrical signs of sensory coding. In the *Neurosciences*, Quarton GC, Melnechuk T and Schmitt FO (eds), pp. 452–469. The Rockefeller University Press, New York.

Möse JR and Fischer G (1970): Zur Wirkung elektrostatisher Gleichfelder, weitere tierexperimentelle Ergebnisse. *Arch Hyg* 154:378–386.

Möse JR and Fischer G (1977): Die Entwicklung des Methylcholanthren-Tumors der Maus unter verschiedenen elektrobioklimatologischen Umwelteinflüssen. *Zbl Bakt Hyg I Abt Orig B* 164:447–454.

Möse JR, Fischer G and Fischer M (1969a): Einfluss des elektrischen Gleichfeldes auf die Wirkung einiger die glatte Muskulatur stimulierender Pharmaka. *Z Biol* 116:354–363.

Möse JR, Fischer G and Fischer M (1969b): Beeinflussung des serotoningehaltes von gehirn, darm und uterus durch das elektrische gleichfeld. *Z Biol* 116:363–370.

Möse JR, Fischer G and Porta J (1971): Wirkung des elektrostatischen Gleichfeldes auf den Sauerstoffverbrauch der Mäusleber. *Arch Hyg Bakteriol* 154:549–552.

Möse JR, Fischer G and Strampfer H (1973): Immunbiologische Reaktionen im elektrostatischen Gleichfeld und Faraday-Käfig. *Z Immunitatsforsch Exp Klin Immunol* 145:404–412.

Möse JR, Schuy S and Fischer G (1972): Versuchsanlage zum Studium der Wirkungen von elektrostatischen Gleichfeldern an kleinen Laboratoriumstieren und die damit erzielten Ergebnisse. *Biomedizinische Technik.* 17:65–70.

Mottelay PF (1922): *Bibliographical History of Electricity and Magnetism*. Charles Griffin & Company, Ltd, London.

Mühleisen R (1966): Messungen elektrischer Felder innerhalb von Tierkäfigen. *Z Vergl Physiologie* 54: 20–25.

The National Research Council (1977): *Biological Effects of Electric and Magnetic Fields Associated with Proposed Project Seafarer*, p. 257, Committee on Biosphere Effects of Etremely-Low-Frequency Radiation, National Academy of Sciences, Washington, DC.

Newland PL, Hunt E, Sharkh SM, et al. (2008): Static electric field detection and behavioural avoidance in cockroaches. *J Exp Biol* 211:3682–3690.

Newland PL, Ghamdi MSA, Sharkh SM, et al. (2015): Exposure to static electric fields leads to changes in biogenic amine levels in the brains of Drosophila. *Proc Biol Sci* 282:20151198.

Odagiri SH and Shimizu K (1999): Experimental analysis of the human perception threshold of a DC electric field. *Med Biol Eng Comput* 37:727–732.

Oehrl W and König HL (1968): Messung und Deutung elektromagnetischer Oszillationen natürlichen Ursprungs im Frequenzbereich unter 1 Hz. *Z Angew Phys* 25:6–14.

Olcese J and Reuss S (1986): Magnetic field effects on pineal gland melatonin synthesis: comparative studies on albino and pigmented rodents. *Brain Res* 369:365–368.

Parker GH and van Heusen AP (1917): The response of the catfish, *Amiurus Nebulosus*, to metallic and non-metallic rods. *Amer J Physiol* 44:405–420.

Persinger MA and Saroka KS (2015): Human quantitative electroencephalographic and Schumann resonance exhibit real-time coherence of spectral power densities: implications for interactive information processing. *J Sig Inform Proces* 6:153–164.

Perumpral JV, Earp UF and Stanley JM (1978): Effect of electrostatic field on locational preference of house flies and flight activities of cabbage loopers. *Environ Entomol* 7:482–486.

Petri AK, Shmeidchen K, Stunder D, et al. (2017): Biological bioeffects of exposure to static electric fields in humans and vertebrates: a systematic review. *Environ Health* 16:41.

Phillips JB and Borland SC (1992): Behavioural evidence for use of a light-dependent magnetoreception mechanism by a vertebrates. *Nature* 359:142–144.

Price C (2016): ELF electromagnetic waves from lightning: the Schumann resonance. *Atmosphere* 7:116.

Price C, Williams E, Elhalel G and Sentman D (2021): Natural ELF fields in the atmosphere and in living organisms. *Inter J Biometeorol* 65:85–92.

Reiter R (1992): *Phenomena in Atmospheric and Environmental Electricity*. Elsevier Science Publishers, Amsterdam, London, New York, Tokyo.

Ritz T, Adem S and Schulten K (2000): A model for vison-based magnetoreception in birds. *Biophys J* 78:707–718.

Roberts AM (1969): Effect of electric fields on mice. *Nature* 223:639.

Romanenko S, Begley R, Harvey AR, Hool L and Wallace VP (2017): The interaction between electromagnetic fields at megahertz, gigahertz and terahertz frequencies with cells, tissues and organisms: risks and potential. *J R Soc Interf* 14:20170585.

Rycroft MJ, Nicoll KA, Aplin KL, et al. (2012): Recent advances in global electric circuit coupling between the space environment and the troposphere. *J Atmosp Solar-Terrestrial Phys* 90–91:198–211.

Saroka KS, Vares DE and Persinger MA (2016): Similar spectral power densities within the Schumann resonance and a large population of quantitative electroencephalographic profiles: supportive evidence for Köning and Pobachenko. *PLoS One* 11 (1):e0146595. Doi: 10.1371/journal.pone.0146595.

Schmiedchen K, Petri AK, Driessen S and Bailey WH (2018): Systematic review of biological effects of exposure to static electric fields. Part II: Invertebrates and plants. *Environ Res* 160:60–76.

Schuà L (1954): Die Wirkung von luftelektrischen Feldern auf Tiere. *Verh Dtsch Zool Ges* 18:435–440.

Schumann WO (1952): Über die strahlungslosern Eigenschwingungen einer leitenden Kugel, die von einer Luftschicht und einer Ionensphärenhülle umgeben ist. *Z Naturforsch* 7(A):149–154.

Schumann WO and König HL (1954): Über die Beobachtung von "atmospherics" bei geringsten Frequenzen. *Naturwissenschaften* 41:183–184.

Semm P, Schneider T and Vollrath L (1980): Effects of an earth-strength magnetic field on electric activity of pineal cells. *Nature* 288:607–608.

Shigemitsu T, Tsuchida Y, Nishiyama F et al. (1981): Temporal variation of the static electric field inside an animal cage. *Bioelectromagnetics* 2:1259–1300.

Sulman FG (1980): *The Effect of Air Ionization, Electric Fields, Atmospherics and Other Electric Phenomena on Man and Animal*. Charles C Thomas Publisher, Springfield, IL.

Sutton GP, Clarke D, Morley EL and Robert D (2016): Mechanosensory hairs in bumblebees (*Bombus terrestris*) detect weak electric fields. *PNAS* 113:7261–7265.

Tesla N (1905): The transmission of electrical energy without wires as a means of furthering world peace. *Electr World Eng* 7:21–24.

Tesla N (2015): *Tesla Patent 787,412 Art of Transmitting Electrical Energy through the Natural Medium*. In Nikola Tesla Lectures & Patient, 655–660. Discovery Publisher, New York.

Timmel CR and Henbest KB (2004): A study of spin chemistry in weak magnetic fields. *Phil Trans R. Soc. Lond. A* 362:2573–2589.

Toomey J and Polk C (1970): *Research on Extremely Low Frequency Propagation with Particular Emphasis on Schumann Resonance and Related Phenomena. Contract Nr AF 19 (628)-4950*. University of Rhode Island, Kingston, RI.

Tromp SW (1968): Review of the possible physiological causes of dowsing. *Int J Parapsychol* 10:363–391.

Tucker RD and Schmitt OH (1978): Tests for human perception of 60 Hz moderate strength magnetic fields. *IEEE BME* 25:509–518.

von Middendorff A (1859): Die Isepiptesen Rußland. *Mem Acad Sci St Petersbourg VI Se* 8:1–143.

Wang CX, Hilburn IA, Wu DA, Mizuhara Y, et al. (2019): Transduction of the geomagnetic fields as evidenced from alpha-band activity in the human brain. *eNEURO* 6:e.0483–18.

Watson DB (1984): Effect of an electric field on insects. *N Z J Sci* 27:139–140.

Welker HA, Semm P, Willing RP, et al. (1983): Effects of an artificial magnetic field on serotonin N-acetyltransferase activity and melatonin content of the rat pineal gland. *Exp Brain Res* 50:426–432.

Wever R (1968): Einfluss schwacher elektromagnetischer Felder auf die circadiane Periodik des Menschen. *Naturwissenschaften* 55:29–33.

Wever R (1973): Human circadian rhythms under the influence of weak electric fields and the different aspects of these studies. *Int J Biometeorol* 17:227–232.

Wever R (1974): ELF-effects on human circadian rhythms. In *ELF and VLF Electromagnetic Field Effects*, Persinger MA (ed), pp. 101–144. Plenum, New York.

Wever R (1975): The circadian multi-oscillatory system of man. *Int J Chronobiol* 3:19–55.

Wever R (1977): Effects of low-level, low-frequency fields on human circadian rhythms. *Neurosci Res Program Bull* 15:39–45.

Wever RA (1979): *The Circadian System of Man-Results of Experiments Under Temporal Isolation.* Springer Verlag, New York.

Wiltschko W and Merkel FW (1966): Orientierung zugunruhiger Rotkehlchen im statischen Magnetfeld. *Verh Dtsch Zool Ges* 59:362–367.

Wiltschko R and Wiltschko W (1995): *Magnetic Orientation in Animals.* Springer Verlag, Berlin, Heidelberg.

Wiltschko R and Wiltschko W (2019): Magnetoreception in birds. *J R Soc Interf* 16:20190295.

Wood AW, Mate R and Karipidis K (2021): Meta-analysis of in vitro and in vivo studies of the biological effects of low-level millimeter waves. *J Expos Sci Environ Epidemiol* 31:606–613.

4

Spin Control Related to Chemical Compass of Migratory Birds

Hideyuki Okano

Tsukasa Shigemitsu

Shoogo Ueno

4.1 Introduction

For all living organisms, sensing stimuli from the external environment such as light, sound, temperature, pressure, and chemical substances and taking actions in response to stimuli are important to survive. These living organisms have developed highly sensitive and specific sensors in order to adapt to their own environment. In particular, it is known that many bacteria, plants, and animals can perceive the geomagnetic field (GMF), which is a weak magnetic field of about 50 μT, by using a certain kind of highly sensitive magnetic sensor in their own bodies.

It is currently widely accepted that birds possess at least two, maybe even three (Wu and Dickman, 2011, 2012; see for a review, Mouritsen and Hore, 2012), magnetosensory systems: (1) a visually based magnetosensor assumingly based on radical pair forming molecules in the retina (Ritz et al., 2000; Mouritsen et al., 2004, 2005; Heyers et al., 2007; Zapka, 2009), and (2) a magnetoreceptor innervated by the ophthalmic branch of the trigeminal nerve (V1) assumingly magnetite iron oxide nanoparticles based and located in the upper beak (Beason and Semm, 1987; Semm and Beason, 1990; Fleissner et al., 2003, 2007; Mora et al., 2004; Falkenberg et al., 2010; Heyers et al., 2010).

In the case of the radical pair forming molecules in the retina, it is strongly suggested that a blue-light photoreceptor protein, "cryptochrome (CRY)," may play a crucial role as one of the highly sensitive magnetic sensors (Ritz et al., 2000, 2004, 2009; Mouritsen and Ritz, 2005; Mouritsen and Hore, 2012; Lau et al., 2012; Wiltschko and Wiltschko, 2014, Bolte et al., 2016; Kerpal et al., 2019; Wiltschko et al., 2021). Challenging research studies have been conducted especially on migratory birds which may have highly sensitive magnetic sensors (Ritz et al., 2000, 2004, 2009; Mouritsen and Ritz, 2005; Mouritsen and Hore,

DOI: 10.1201/9781003181354-4

2012; Lau et al., 2012; Wiltschko and Wiltschko, 2014, Bolte et al., 2016; Kerpal et al., 2019; Wiltschko et al., 2021). In this CRY, a flavoprotein, "flavin adenine dinucleotide (FAD)" is a photofunctional molecule, and it is bound to the site buried in the helix domain by intermolecular interaction. When FAD is excited by blue light, electron transfer occurs from nearby tryptophan (Trp) that is charge-separated, and the consequent radical pair induces the efficiency of reaction to be detected, albeit with a weak magnetic field. Using such highly sensitive magnetic receptors (magnetoreceptors), migratory birds are assumed to be able to migrate in their intended proper direction.

Here, the historical background study of the "radical pair (recombination) mechanism (RPM)" models for quantum-assisted magnetic sensing and recent studies on both natural and artificial in vitro systems related to flavoproteins such as FAD are introduced. The forming processes of radical pairs differed between natural flavoproteins and artificial systems, which were focused in this review. The latter system was expected to provide advantages for precisely controlling the experimental condition. The avian magnetic compass is currently being actively investigated in the field of "Quantum Biology" (Ball, 2011; Al-Khalili and McFadden, 2014; Solov'yov et al., 2014).

4.2 Historical Background Study of the Radical Pair Mechanism Models

As historical background studies of spin chemistry, it has been generally considered difficult to control chemical and biological reactions using magnetic fields. The reason is that the electronic energy involved in the magnetic fields is extremely smaller than that in the chemical reactions. For example, the Zeeman splitting of electron spins by 1 T magnetic field is 0.935 cm^{-1}, while the thermal energy is about 200 cm^{-1}, and the activation energy of chemical reactions is usually 3.000 cm^{-1} or more (Hayashi, 1982). In spite of this historical background, however, it has been mainly published by Japanese researchers that when radical pairs are included in the reaction process, even relatively weak magnetic fields in the field range below 1 T could affect the chemical reaction rate and yield (Hata, 1976, 1978, 1985, 1986; Tanimoto et al., 1976; Hayashi and Nagakura, 1978; Hata et al., 1979; Sakaguchi et al., 1980a, b, 1981; Sakaguchi and Hayashi, 1982; Hata and Yagi, 1983; Hata and Nishida, 1985).

4.2.1 Magnetic Field Effects on Chemical Reactions via Radical Pair Recombination

Even in the 1 T magnetic field, it is difficult to thermodynamically change the chemical reaction at room temperature for the above-mentioned reasons, i.e., the Zeeman splitting energy in the magnetic field. Regarding the chemical magnetic field effects, from the 1970s to the 1980s, chemical reactions through unstable radical pairs or biradicals have been proved to be influenced by an external magnetic field (Kaptein, 1972; Sagdeev et al., 1973; Hata, 1976, 1978, 1985, 1986; Tanimoto et al., 1976; Hayashi and Nagakura, 1978; Hata et al., 1979; Turro and Chow, 1979; Sakaguchi et al., 1980a, b, 1981; Hayashi, 1982; Sakaguchi and Hayashi, 1982; Hata and Yagi, 1983; Hata and Nishida, 1985). The magnetic field effect on chemical reactions was interpreted in terms of the fact that a magnetic field enhances or reduces the singlet-triplet (S-T) conversion of intermediate radical pairs or biradicals (Kaptein, 1972; Hayashi and Nagakura, 1978) through the electronic Zeeman and electron-nuclear hyperfine coupling (HFC) interactions (Hayashi et al., 1966; Itoh et al., 1969). In particular, as one of the pioneer studies, Japanese researchers observed and found the magnetic field effects on the reaction yield and rate of the chemical reaction proceeding via a radical-pair intermediate in the solution from both experimental and theoretical perspectives (Hata, 1976, 1978, 1985, 1986; Tanimoto et al., 1976; Hayashi and Nagakura, 1978; Hata et al., 1979; Sakaguchi et al., 1980a, b, 1981; Hayashi, 1982; Sakaguchi and Hayashi, 1982; Hata and Yagi, 1983; Hata and Nishida, 1985).

Photochemical reaction proceeding via a radical-pair intermediate in the solution can generally be expected to show an external magnetic field effect which arises from an electric Zeeman interaction

(Δg mechanism), or hyperfine interaction (HFI) mechanism including an electron-exchange interaction in the singlet hydrogen-bonded radical-ion pair (HFI-J mechanism) in a radical-pair intermediate (Salikhov et al., 1984). The magnetic field effect due to the HFI-J mechanism is considered to be particularly interesting and important from the viewpoint of mechanistic photochemistry because it is expected when hydrogen or electron transfer between a photoexcited molecule and the hydrogen-bonded species occurs to form an appropriate hydrogen-bonded radical pair or radical ion-pair intermediate in a solvent cage (Hata, 1985).

In 1976, for the first time, Hata found this type of a photochemical magnetic field effect due to the HFI-J mechanism in the case of the photochemical isomerization of isoquinoline *N*-oxide in ethanol (Hata, 1976). Hata (1976) investigated the photochemical reaction of isoquinoline *N*-oxide in ethanol with or without magnetic field up to 17 kg (1.7 T) and measured the chemical yield of lactam (isocarbostyril). The magnetic field effects on the yield of lactam (isocarbostyril) in the photochemical reaction of isoquinoline *N*-oxide in ethanol is shown by Hata (1976). Here, the chemical yield of lactam was 65%–68% below 5 kg (0.5 T), and decreased drastically to be ca. 52% at about 10 kg (1 T). Further increase in the field strength resulted in the recovery in the chemical yield to reach a constant value of ca. 65%. Thus, the yield of lactam indicated a minimum value at about 10 kg (1 T). These results suggested that magnetic field could enhance intersystem crossing from the excited singlet spin state of isoquinoline *N*-oxide at about 10 kg (1 T). In 1978, this new phenomenon was successfully interpreted in terms of HFI-J mechanism assumed to be a transient intermediate of this reaction (Hata, 1978). These studies were partially reported in preliminary form (Hata, 1976, 1978; Hata et al., 1979, 1983).

Hata (1985) presented a further detailed mechanism of the photochemical isomerization of isoquinoline *N*-oxide. The magnetic field dependence of the chemical yield of lactam **2** in the photochemical reaction of isoquinoline *N*-oxide **1** is shown by Hata (1985). When the chemical reaction was carried out with or without magnetic fields up to 1.6 T, the chemical yield of lactam **2** was measured. The chemical yield of lactam was ~67% below 0.8 T and decreased drastically to be ~52% at about 1 T. Further increase in the field strength resulted in the recovery in the chemical yield to reach a constant value of ~67%. The conversion remained almost constant at ~17%.

The chemical yield of oxazepine **5** vs. magnetic field strength in the photochemical reaction of 1 cyanoisoquinoline *N*-oxide **4** is presented by Hata (1985). The results of the chemical yield of oxazepine **5** against the field strength proved to be independent of an external magnetic field. Here, also, the conversion remained almost constant at ~30%.

As for the second example of the photochemical magnetic field effects, Hata and Nishida (1985) reported the photoinduced substitution reaction of 4-methyl-2-quinolinecarbonitrile in ethanol and cyclohexane. External magnetic field effects on the photosubstitution reaction (1→2) in ethanol are shown by Hata and Nishida (1985). Here, **1**, 4-methyl-2-quinolinecarbonitrile; **2**, 2-(1-hydroxyethyl)-4-methylquinoline. Chemical yield of **2** vs. magnetic field strength. [**1**]=4.01×10⁻³ mol/dm³. (1) open circles: [C_5H_8]=0, (2) closed circles: [C_5H_8]=3.0×10⁻¹ mol/dm³. The chemical yield of 2-(1-hydroxyethyl)-4-methylquinoline **2** is plotted as a function of the field strength in the absence or presence of 1,3-pentadiene, where the conversion remained almost constant (20%–22%). In the absence of 1,3-pentadiene, as shown by curve (**a**), the chemical yield of **2** was ca. 48% at the zero field. However, it increased quadratically with an increase in the field strength to be ca. 58% at about 1.5 T (the magnetic field effect due to Δg mechanism). The chemical yield of **2** also showed a minimum (ca. 49%) at about 1.1 T (the magnetic field effect due to the HFI-J mechanism). The Δg magnetic field effect, as shown by curve (**b**), disappeared completely upon the addition of 1,3-pentadiene, although the magnetic field effect due to the HFI-J mechanism was still observed. Thus, the chemical yield of **2** was ca. 58% at a magnetic field below 0.8 T, but it decreased steeply with an increase in the field strength to become ca. 48% at about 1.1 T. Further increase in the field strength resulted in the quadratic recovery in the chemical yield to reach a constant value of ca. 58%. The results explicitly indicate that the Δg or the HFI-J magnetic field effect observed in a photochemical reaction can be assigned to the field dependence of the chemical yield of the T_1- or S_1-born cage product.

External magnetic field effects on the photosubstitution reaction (**1→3**) in cyclohexane are shown by Hata and Nishida (1985). Here, **1**, 4-methyl-2-quinolinecarbonitrile; **3**, 2-cyclohexyl-4-methylquinoline. In the case of the chemical yield of 2-cyclohexyl-4-methylquinoline **3**, the results are plotted against the field strength in either the absence or presence of 1,3-pentadiene. In either case, the conversion was almost independent of the field strength (19%–21%). In the absence of 1,3-pentadiene, as shown by curve (**a**), the chemical yield of **3** was ca. 65% at the zero field. However, it decreased steeply upon the application of a magnetic field of 40 mT to be ca. 54% (the magnetic field effect due to HFI mechanism). Further increase in the field strength resulted in the quadratic recovery in the chemical yield to reach a constant value of ca. 63% (the magnetic field effect due to Δg mechanism). However, the addition of 1,3-pentadiene as shown by curve (**b**) caused a complete disappearance of these magnetic field effects. Consequently, the chemical yield of **3** became independent of the field strength to show ca. 67%. This means that both HFI and Δg magnetic field effects observed in this reaction can be assigned to the field dependence of the chemical yield of the T_1-born cage product. Also, the fact that product **3** was obtained in a high field (ca. 67%), even in the of 1,3-pentadiene, suggests strongly that the photosubstitution reaction proceeds from the S_1 state as well as the T_1 state.

Tanimoto et al. (1976) examined the photodegradation reaction at room temperature under magnetic fields of up to 4.2 T to determine the relative magnetic field change of the cage product. This study is the first study to confirm the magnetic field effect on the reaction yield of the chemical reaction in the solution. Dissipative product yields and chemical reactions from triplet precursors can be treated in the same way (Hayashi and Nagakura, 1978; Sakaguchi et al., 1980a, b).

Furthermore, regarding the rate of the chemical reaction proceeding, it is theoretically estimated that the singlet-triplet (*S-T*) conversion rate (k_{ST}) (in the coherent mixing between singlet and triplet spin states) may be decreased by the magnetic field (Hayashi and Nagakura, 1978; Sakaguchi et al., 1980a, b). In order to measure k_{ST} experimentally, the research team devised a method of measuring the time change of the absorption intensity of the electron spectrum in the radical pair (Sakaguchi et al., 1980a, b, 1981; Sakaguchi and Hayashi, 1982). They used a pulsed laser as the excitation light source to examine the magnetic field effects on k_{ST} in photochemical reactions in a micelle for the first time (Sakaguchi et al., 1980a, b, 1981; Hayashi, 1982; Sakaguchi and Hayashi, 1982).

When benzophenone (BP) in a sodium dodecyl sulfate (SDS) micelle is laser excited, the triplet excited state ($^3BP^*$) abstracts hydrogen from the micelle molecule (RH) to form a triplet radical pair (Sakaguchi et al., 1980a, b).

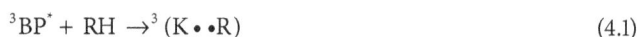

$$^3BP^* + RH \rightarrow {}^3 (K \bullet \bullet R) \tag{4.1}$$

Here, $BP = (C_6H_5)_2 CO$, $K\bullet = (C_6H_5)_2 \dot{C}OH$, $R\bullet = \bullet C_{12}H_{24}SO_4^-$. k_{ST} can be obtained from the time change of the absorption intensity of K•. Since these k_{ST} values are similar to T_1^{-1} of electrons in radicals, k_{ST} could be affected by T_1. Magnetic field strength dependence of the *S-T* conversion constant (k_{ST}) at remote pairs of 4,4'-difluorobenzophenone ketyl radical, $(C_6H_4F)_2 \dot{C}OH$, and alkyl radical, $C_{12}H_{24}SO_4^-$, in an SDS micelle is shown by Hayashi (1982), modified from Sakaguchi and Hayashi (1982). The experimental results suggested magnetic field strength dependence of the *S-T* conversion constant. Surprisingly, the k_{ST} value at 13.4 kg (1.34 T) is about 1/10 of the k_{ST} value at 0 kg, and such a large magnetic field effect on k_{ST} has not been reported so far (Hayashi, 1982). The reason is thought to be that radical-pair intermediates can exist for a long time in micelles (Hayashi, 1982). Since the reaction in the micelle is similar to the reaction in the living systems, the magnetic field effects on the biological reaction can be greatly expected (Hayashi, 1982).

4.2.2 Magnetic Field Effects on Biological Systems via Radical Pair Recombination

With regard to magnetic field effects on radical pair recombination in biological systems, the yield of excited triplet electronic states in the photosynthetic reaction center was first shown to be magnetic

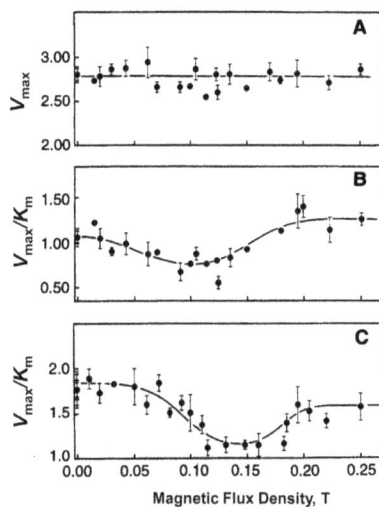

FIGURE 4.1 Ethanolamine ammonia-lyase (Harkins and Grissom, 1994). Magnetic field dependence of (a) V_{max} with unlabeled ethanolamine, (b) V_{max}/K_m with unlabeled ethanolamine, and (c) V_{max}/K_m with 1, 1, 2,2-D$_4$-ethanolamine. Each assay contained 100 mM N-(2-hydroxyethyl)piperazine-N'-2-ethane sulfonic acid (Hepes) pH 7.48, 5 µM adenosylcob(III)alamin, and ethanolamine ammonia-lyase at 25°C. Each data point represents the kinetic parameter derived by fitting observed $d[P]/dt$ vs [ethanolamine] data to $d[P]/dt = V_{max} - S]^n/K_m + [S]^n$ by nonlinear methods. The Hill number, n, varied only slightly between 0.75 and 0.85. In order to keep the measured rates with deuterated and unlabeled substrates similar, 8.59-fold more EAL enzyme was used in assays with deuterated ethanolamine than in assays with unlabeled ethanolamine. This yields an observed kinetic isotope effect of $^DV_{max} = 6.8 \pm 0.2$ and $^DV_{max}/K_m = 5.4 \pm 0.4$ at 0 T (Reproduced with permission from Harkins and Grissom, 1994, Copyright 1994, AAAS.)

field sensitive in 1977 (Blankenship et al., 1977; Hoff et al., 1977). Only one other biological system, an enzyme with radical pair intermediates, has been shown to exhibit any magnetic field-dependent parameters (Harkins and Grissom, 1994). Harkins and Grissom (1994) presented the magnetic field dependence of V_{max} and V_{max}/K_m for ethanolamine ammonia-lyase (Figure 4.1). The kinetic parameter V_{max} is independent of applied magnetic field up to 250 mT, whereas V_{max}/K_m exhibits a decrease of 25% at 100 mT (Harkins and Grissom, 1994). Under conditions of saturating substrate (expressed by the kinetic parameter V_{max}), product dissociation is followed immediately by the binding of another substrate molecule that is poised for the next round of catalysis (Harkins and Grissom, 1994). The net effect is no dependence of V_{max} on the magnetic field (Harkins and Grissom, 1994). In contrast, under conditions of a less-than-saturating substrate (expressed by the kinetic parameter V_{max}/K_m), product dissociation occurs and the "resting" state of the enzyme-cofactor complex is restored (Harkins and Grissom, 1994). Deuteration of ethanolamine produces a larger 60% decrease in V_{max}/K_m at 150 mT (Harkins and Grissom, 1994).

Grissom (1995) presented the rate of Cbl^{II} formation vs magnetic field (Figure 4.2, modified from Harkins and Grissom, 1995). This result unambiguously identifies {AdoCH$_2$ • Cbl^{II}} recombination as the magnetic field-sensitive step in ethanolamine ammonia-lyase (Harkins and Grissom, 1995). No deuterium isotope effect on the rate of Cbl^{II} formation is observed, and the magnetic field dependence of Cbl^{II} formation is independent of the isotopic composition of ethanolamine (Harkins and Grissom, 1995).

More recently, Chen and Ke (2018) reported that application of an external magnetic field in the range of 650–850 mT triggers intersystem crossing to the singlet {cob(II)alamin – substrate} radical-pair state. Spin-conserved H back-transfer from deoxyadenosine to the substrate radical yields a singlet {cob(II)

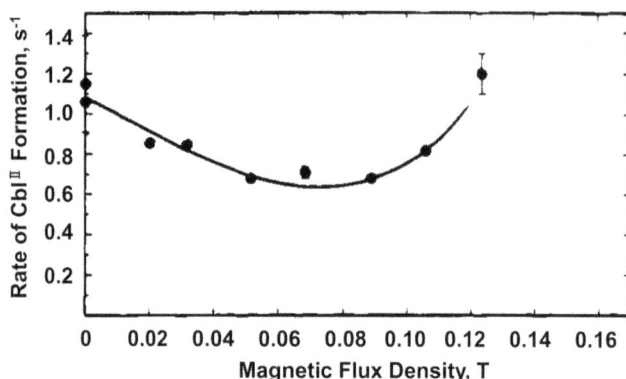

FIGURE 4.2 Ethanolamine ammonia-lyase stopped-flow kinetic study (Grissom, 1995, modified from Harkins and Grissom, 1995). The magnetic field dependence of the first-order rate of appearance of CbI[II] with unlabeled ethanolamine is shown. Standard error bars may be smaller than the plotted symbol. Identical rates are observed with unlabeled and deuterated ethanolamine. (Reproduced with permission from Grissom, 1995; Harkins and Grissom, 1995, Copyright 1995, American Chemical Society.)

alamin-5′-deoxyadenosyl} radical pair. Spin-selective recombination to adenosylcobalamin decreased the enzyme catalytic efficiency kcat/K_m by 16% at 760 mT.

Klevanik (1996) examined the magnetic field effects on primary reactions of Photosystem (PS) I by measuring changes of the fluorescence yield in preparations isolated from the cyanobacterium *Synechococcus elongatus*. Photoaccumulation of the reduced phylloquinone A_1 (the electron acceptor of PS I) in the presence of dithionite under anaerobic conditions led to an increase of chlorophyll (Chl) fluorescence yield and appearance of magnetic field effects (Klevanik, 1996). A magnetic field effect has been found which is characterized by half-saturation values of the order of ~2.5 mT (Klevanik, 1996). The addition of salts of monovalent and divalent cations to the suspension significantly modified the profile of the magnetic field dependence (Klevanik, 1996). The modification of the magnetic field effect by salts is ascribed to conformational changes in the PS I reaction center that affect the dipolar and/or spin exchange coupling in the radical pair P700$^+$ A_0^- (P700, the primary electron donor of PS I that best absorbs light at a wavelength of 700 nm; A_0, the primary electron acceptor of PS I) (Klevanik, 1996). Injection of neutral red into suspensions of samples containing reaction centers in the state P700 A_0 A_1^{red} gives rise to light-induced fluorescence quenching and disappearance of the magnetic field effect (Klevanik, 1996). In the dark, the high fluorescence and the magnetic field effect are restored (Klevanik, 1996). These phenomena are ascribed to photoaccumulation of state P700 A_0^- A_1^{red} and a strong quenching effect of the Chl an anion radical A_0^- that closely resembles that of Pheo⁻ in PS II (Klevanik, 1996).

The magnetic fields have been shown to change radical concentrations (Batchelor et al., 1993; Grissom, 1995; Hayashi, 2004; Timmel and Henbest, 2004). Spin modulation in the RPM is considered a plausible explanation of how the static magnetic fields including the geomagnetic field (GMF) can alter the chemistry of reactions and cause biological effects (Barnes and Greenebaum, 2015). The RPM is used to exemplify how applied magnetic fields weaker in strength than typical hyperfine interactions can influence the yields and kinetics of recombination reactions of free radicals in the solution (Timmel et al., 1998). "Spin-correlated radical pairs" can undergo coherent mixing between singlet (antiparallel) and triplet (parallel) spin states, which have different reactive fates, and this mixing process can be modulated by magnetic fields (Steiner and Ulrich, 1989; Woodward et al., 2009). Here triplet products are chemically different from the singlet products and thus may play a role in magnetoreception.

The RPM is a promising hypothesis to explain the mystery of the navigation of birds (Zhang et al., 2015). This theoretical study has demonstrated the role of weak magnetic fields play in the product yields of the radical pairs. In addition, this type of study has inspired scientists to design highly effective devices to detect weak magnetic fields and to use the GMF to navigate. The anisotropic hyperfine coupling (HPC) between the electron spins and the surrounding nuclear spins can play a crucial role in avian magnetoreception. The HPC can affect not only the product yields but also the entanglement of the electron spin states. By involving more nuclear spins one can greatly enhance the quantum entanglement (Sadiek et al., 2008). Additionally, mimicking this anisotropic magnetic environment can be very useful for creating detectors of weak magnetic fields. By studying the role of the intensity of the magnetic field in avian navigation, for instance, Zhang et al. (2015) found that birds could be able to detect the change of the GMF intensity and the approximate direction of parallels instead of sensing the exact direction. The plausible mechanism in which birds can utilize the signal has been investigated.

While the strong magnetic field effects have been investigated and understood in depth, Kerpal et al. (2019) made progress to obtain a more complete picture of the typically much less pronounced sensitivity to weak fields. Experimentally, only one study on a model system has succeeded in providing proof for an isotropic Earth strength effect, while an orientation dependence of the magnetic field effect was only observed for fields >3 mT (Maeda et al., 2008). The orientation dependence in this high field region is caused by anisotropic HPCs in the radical pair, the anisotropic dipolar coupling being negligible compared to HPCs or indeed their anisotropies (D=0.06 mT for the center-to-center distance of 3.6 nm in this pair) (Di Valentin et al., 2005). Previously only founded in theoretical simulations, it is speculated that these anisotropic HPCs in a radical pair with restricted motion may result in an orientation-dependent magnetic field response even in extremely weak magnetic fields including the GMF (Ritz et al., 2000).

Using a custom-designed transient absorption spectrometer, Kerpal et al. (2019) verified this hypothesis by testing if a quantum compass can function in fields as weak as the GMF. This is not only crucial regarding the discussion of the magnitude of any expected effects, but, importantly, the quantum dynamics in high- and low-field regimes are dominated by different processes (Lewis et al., 2018). The previously demonstrated existence of a chemical compass response of certain radical pair-based reactions in high fields (Maeda et al., 2008) is, therefore, a necessary condition but by no means sufficient to explain the avian compass sense within the quantum system's low field regime (Kerpal et al., 2019).

As shown in Figure 4.3, chemical structure, photocycle, and time dependence of the magnetic field effect (MFE) of carotenoid-porphyrin-fullerene (**CPF**) moieties are presented by Kerpal et al. (2019). Figure 4.3a shows the structure of the investigated model chemical compass, a molecular triad consisting of covalently linked carotenoid (**C**), porphyrin (**P**), and fullerene (**F**) moieties. Its photophysical behavior and response to high fields, in the absence and presence of resonant radiofrequency fields, have been studied previously (Kodis et al., 2004; Maeda et al., 2008, 2011, 2015). As shown in Figure 4.3b, photoexcitation of the porphyrin at 532 nm is followed by rapid intramolecular electron transfer, first generating a primary radical pair $C-P^{+}-F^{\bullet-}$ of picosecond lifetime, before subsequent electron transfer leads to the formation of the secondary radical pair $C^{\bullet+}-P-F^{\bullet-}$, which lives for up to roughly a microsecond. Previous work, in similar solvent and temperature conditions to those employed here, demonstrated that this secondary radical pair is formed predominantly in the singlet state, $^{S}[C^{\bullet+} \pm P-F^{\bullet-}]$, with just 7% of radical pairs being created in the triplet state (Maeda et al., 2011). While each radical pair is born in a spin-correlated state (either singlet or triplet), the magnetic field characteristics of the radical pair ensemble are complex.

The measurements were carried out at 120 K, where the solvent, 2-methyltetrahydrofuran (MTHF), forms an optically transparent glass. Recombination of $C^{\bullet+} \pm P-F^{\bullet-}$ is possible from either the singlet or triplet states and occurs with rate constants k_S and k_T, respectively. The rates are strongly dependent on the solvent properties, notably its dielectric constant. Under similar conditions, k_S has been shown to be some three orders of magnitude faster than k_T, and consequently, a significant change in the recombination kinetics is observed upon application of a magnetic field (Maeda et al., 2011).

Most experimental investigations of magnetic field effects have relied on optical methods in which either the concentration of the radicals themselves or of one of their recombination products is determined as a function of field. Here nanosecond transient absorption spectroscopy was used to obtain

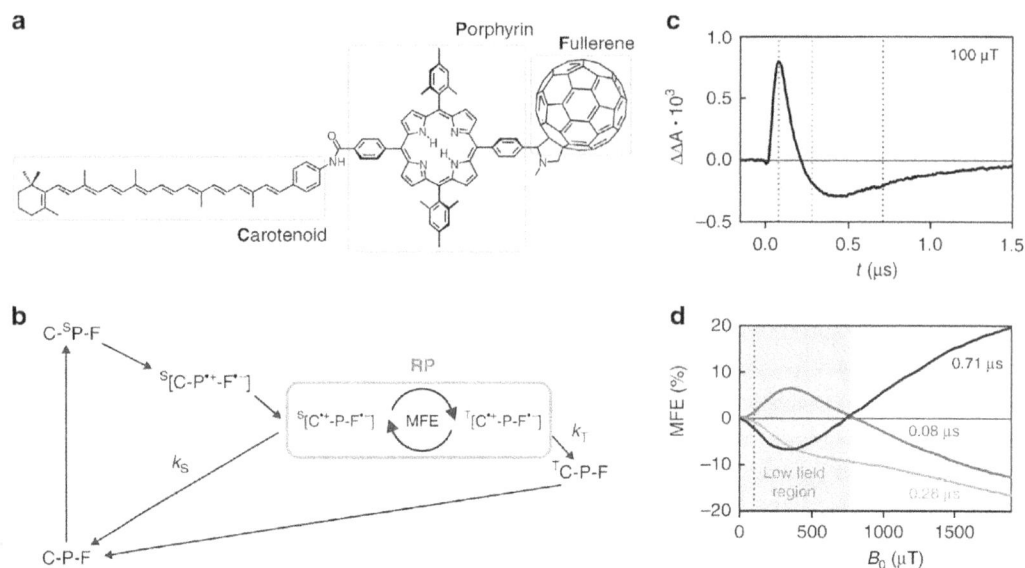

FIGURE 4.3 Chemical structure, photocycle, and time dependence of the magnetic field effect (MFE) of **CPF** moieties (Kerpal et al., 2019). (a) Structure of the molecular **CPF** triad. (b) Simplified photo scheme including all processes of relevance for this study. For simplicity, the secondary radical pair $\mathbf{C^{\cdot+} - P - F^{\cdot-}}$ is shown as created in a pure singlet state. For more detailed photochemical information (Kodis et al., 2004; Maeda et al., 2011). (c) Transient absorption subtraction signal $\Delta\Delta A$ of the radical pair, probed at 980 nm, obtained upon application of a magnetic field of $B_0 = 100$ µT. The dotted lines indicate delay times of 0.08, 0.28, and 0.71 µs after laser excitation, respectively. (d) Magnetic field dependence of the MFE averaged for a time window of 20 ns centered around the indicated delay times. The dotted black line indicates the field position of 100 µT and the field region approximately corresponding to the low field region is highlighted in gray. (Reproduced with permission from Kerpal et al., 2019, Copyright 2019, Springer Nature.) It is licensed under the Creative Commons Attribution 4.0 International.

the concentration profile of the carotenoid radical cation $\mathbf{C^+}$ via its absorbance in the near-infrared, following radical pair creation by a 532 nm laser pulse. Application of a magnetic field, B_0, is expected to change the rate of singlet-triplet (S-T) interconversion and consequently the overall radical pair kinetics, concentration, and absorbance. The effect of the field is typically quantified via $\Delta\Delta A(t, B_0) = \Delta A(t, B_0) - \Delta A(t, B_0 = 0)$, where $\Delta A(t, B_0)$ and $\Delta A(t, B_0 = 0)$ refer to the absorbance of the transient species at 980 nm (predominantly $\mathbf{C^{\cdot+}}$), in the presence and absence of the field, respectively. t defines the time after the 532 nm pump laser pulse. It can also be instructive to calculate the so-called percentage field effect, defined as magnetic field effect $(t, B_0) = \Delta\Delta A(t, B_0)/\Delta A(t, B_0 = 0) \times 100\%$.

Figure 4.3c demonstrates that a 100 µT field confers a pronounced effect on the recombination kinetics of the radical pair. In the presence of the magnetic field, the concentration of radicals immediately following the laser pulse is enhanced, $\Delta\Delta A(t < 220$ ns, 100 µT$) > 0$, but fewer radicals survive to microsecond timescales, $\Delta\Delta A(t > 220$ ns, 100 µT$) < 0$. This biphasic behavior has been noted before and seems to be characteristic of singlet-born radical pairs with $k_S > k_T$ undergoing spin-lattice relaxation at a rate comparable to recombination (van Dijk et al., 1998; Maeda et al., 2008). The mixed initial spin state (93% singlet: 7% triplet) in $\mathbf{C^{\cdot+} - P - F^{\cdot-}}$ further enhances this effect. Moreover, semiclassical spin dynamics simulations have recently reproduced some of the complex field- and time-dependent transient absorption characteristics of $\mathbf{C^{\cdot+} - P - F^{\cdot-}}$ without implicit consideration of relaxation processes or mixed initial spin states (Lewis et al., 2018).

As shown in Figure 4.3d, the percentage of magnetic field effect (t, B_0) is presented at different times (t) after laser excitation. The initial discussion will concentrate on the data obtained at early and late times, i.e., $t=0.08$ µs and $t=0.71$ µs. Both magnetic field effect traces follow the expected behavior, with the application of both low and high fields affecting changes in radical concentration of opposite signs. Following the discussion by Lewis et al. (2018), weak magnetic fields mainly enhance the S–T_0 interconversion efficiency, which results in an increase in radical concentration at early times after laser excitation (the initially predominantly singlet population is driven by radical concentration at late times when formed triplet radical pairs can return more efficiently to singlet radical pairs, which might subsequently recombine).

In contrast, higher fields affect the radical recombination via the Zeeman effect, energetically isolating the S/T_0 manifold from the T_+/T_- levels, therefore impeding efficient singlet-triplet (S-T) mixing. Following the arguments above, this results in a decrease in radical concentration at early times after the laser pulse and a corresponding increase on longer timescales.

It is, at first sight, perhaps surprising that the magnetic field effect data obtained at intermediate times, namely, 0.28 µs after the laser pulse, do not exhibit a sign inversion of the magnetic field effect. This finding is, however, in agreement with the results of Lewis et al. (2018), in which it was demonstrated that the field effects on the populations of T_0 and T_+/T_- are not only in opposite directions but evolve at different timescales. While the initially positive low field effect has, at 0.28 µs, already changed sign, the high field effect lags about 0.2 µs behind in its evolution.

Barnes and Greenebaum (2015, 2016) made some theoretical considerations on the weak magnetic field effects on radical pair reactions as follows. Radicals are produced during many biological reactions, including the mitochondrial metabolic processes. The vector representations of the components of the electron spin, electron angular momentum, and the nuclear spin with respect to the applied magnetic field are shown by Barnes and Greenebaum (2015). Here, a radical pair forms either a singlet state, where the spins are aligned with electron spins with opposite spins, or a triplet state, with the spins parallel (Barnes and Greenebaum, 2015).

In the singlet state, these pairs recombine with typical lifetimes between 10^{-6} and 10^{-10} s. In the triplet state, they are not allowed to recombine, and the opportunity for them to diffuse away increases so that they can react with other molecules (Barnes and Greenebaum, 2016). The coupling between the unpaired electrons and the nuclei in each fragment of the radical pair is different and, typically, can be described by magnetic fields in the range 10 µT–3 mT (Brocklehurst and McLauchlan, 1996). Barnes and Greenebaum (2016) proposed a schematic diagram of the evolution of spins of two members of a radical pair, one with only an electron spin and the other with both an electron and a nonzero nuclear spin, illustrating changes between relative S and T states under two sets of conditions. Here, (1) Precession of spins in an external magnetic field. (2) Stimulated transition by absorption of a photon of energy corresponding to the energy difference between levels in one radical. A photon must also carry angular momentum corresponding to the difference between levels.

For many radicals, this is stronger than the GMF (~50 µT) so that the quantum numbers describing the state of each fragment are determined by the sum F of the electron angular momentum and electron spin J and the nuclear spin I (Barnes and Greenebaum, 2016). The unpaired electrons in the outer orbit of each of the radical pair fragments can be thought of as rotating about their nuclei at different rates, so the net magnetic moments for the two fragments switch from an S to a T state and back (Brocklehurst and McLauchlan, 1996). The rate at which this happens is perturbed by the external magnetic field. The energy levels in each fragment are shifted by different amounts by the external magnetic fields (Barnes and Greenebaum, 2016).

Changes in the applied magnetic field shift the size of the energy barrier for the recombination and the recombination rate (Barnes and Greenebaum, 2015, 2016). Nuclear magnetic spectra may have very narrow absorption lines with bandwidths of a few cycles with corresponding lifetimes for excited states of seconds or longer (Barnes and Greenebaum, 2015, 2016). The energies of D_2 molecule states as a function of the magnetic field with low field (F, m) and high field (J, m_J, I, m_I) are estimated by Barnes and Greenebaum (2015). Here, quantum number labels m_J and m_I are the projections of the electron angular moment and nuclear spin on the external magnetic fields. Note the linearity of curves in the low field region, where $F = J + I$ is a good quantum number, and curvature as well as crossovers as field increases (Ramsey, 1956). Vertical lines (left diagram) indicate allowed transitions. Relative orientations of one transition's upper and lower state angular momenta are shown (right upper and lower diagrams). In the left diagram, circles indicate the examples of possible level-crossing transition points, and box on the horizontal axis indicates the region of possible zero-field transitions. Magnetic fields at the frequency corresponding to differences in the energy levels can drive molecules between energy levels of different nuclear spin states and change the concentration in these energy levels, which, in turn, can change the recombination lifetimes for radial pairs (Barnes and Greenebaum, 2015, 2015).

These narrow line widths can lead to saturation effects with magnetic fields in the range $10^{-8} - 10^{-9}$ T (Bovey, 1988). With large molecules that contain many atoms with nuclear spins, the calculations of the recombination rates are very complex as the contributions to the magnetic field seen by the active electron that is dependent on the nuclear spin of each atom, its distance from the electron, and the shielding by other electrons in different orbits (Batchelor et al., 1993; Brocklehurst and McLauchlan, 1996; Woodward et al., 2001; Wang and Ritz, 2006; Rodgers et al., 2007). Barnes and Greenebaum (2015, 2016) assumed that the sum of these fields is large enough so that coupling can lead to relatively sharp resonances, and the nuclear spin states are important in determining the recombination rates for the radical pairs. Nuclear resonance spectroscopy at radio frequencies (RFs) showed that nuclear spin states may have lifetimes of seconds or longer and corresponding resonant line widths of a few cycles (Bovey, 1988). In weak magnetic fields, where the magnetic coupling between the active electrons and the nuclei in the radicals is stronger than the perturbing external field, Barnes and Greenebaum (2015) postulate that they will also see shifts in radical concentrations that are frequency- and amplitude-dependent with relatively narrow line widths. Prato et al. (2013) also gave an explanation for effects seen when the ambient magnetic is shielded. For then level energy differences are below the natural line widths and spontaneous transitions can occur (Barnes and Greenebaum, 2015).

From these theoretical considerations, Barnes and Greenebaum (2015) concluded that the application of magnetic fields at frequencies ranging from a few Hertz to microwaves at the absorption frequencies observed in electron and nuclear resonance spectroscopy for radicals can lead to changes in free radical concentrations, and these effects have the potential to lead to biologically significant changes. Moreover, Barnes and Greenebaum (2016) supposed that there are now both the theoretical bases and sufficient experimental results for further consideration of the possibility that long-term exposures to magnetic fields can lead to both useful applications in treating diseases and to undesired health effects. It is expected that these effects are frequency, amplitude, and time-dependent (Barnes and Greenebaum, 2016). These effects will also be dependent on other biological conditions that can lead to changes in radical concentrations (Barnes and Greenebaum, 2016).

The RPM explains how a pair of reactive oxygen species with distinct chemical fates can be influenced by a low-level external magnetic field through Zeeman and hyperfine interactions. So far, a study of the effects of complex spatiotemporal signals within the context of the RPM has not been performed. Recently, Castello et al. (2021) presented a computational investigation of such effects by utilizing a generic pulsed electromagnetic field (PEMF) test signal and RPM models of different complexity. Their theoretical simulations showed how substantially different chemical results can be obtained within ranges that depend on the specific orientation of the PEMF test signal with respect to the background static magnetic field, its waveform, and both of their amplitudes (Castello et al., 2021). These results provide a basis for explaining the distinctive biological relevance of PEMF signals on radical pair chemical reactions (Castello et al., 2021). Their study establishes the role of PEMF as a diagnostic tool that may indicate the involvement of magnetosensitive radical pair reactions in biological systems (Castello et al., 2021). They speculated that extending this tool to determine orientation and amplitude dependence in which the input PEMF waveforms affect the reaction products can reveal the chemical nature of the radical pairs involved (Castello et al., 2021). They proposed that using the oscillating magnetic field or PEMF input waveform as a diagnostic tool to modify singlet quantum yields can easily be transferred to finding the optimal control to maximize the singlet yield (Castello et al., 2021).

4.3 Magnetic Sense via Radical Pair Mechanism

Surprisingly, it has been reported that migratory birds seem to determine the direction of migration using the GMF through the above-mentioned similar reaction process in their certain physiological sensors. In brief, it has been reported that the GMF would affect the singlet-triplet (S-T) interconversion in an orientation-dependent manner relative to the sensor molecule, leading to a change in the S-T yield that would, in turn, trigger a physiological and behavioral response in migratory birds (Rodgers and Hore, 2009; Hore and Mouritsen, 2016). We review and describe the history of several distinguished studies on the RPM models below.

For the magnetic compass of migratory birds, the RPM model for quantum-assisted magnetic sensing was first proposed by Schulten et al. in 1978. It is suggested that this hardcore theoretical physics paper formulated the RPM hypothesis of magnetoreception for the first time, and it is now clear that it was decades ahead of its time (Mouritsen, 2018). At the same time, since the 1970s, Mr. and Mrs. Wiltschko have begun to investigate the magnetic sensation of a migratory bird, European robin (*Erithacus rubecula*) in terms of ethology, which is the study of animal behavior (Wiltschko and Wiltschko, 1972). Later, they also found the magnetic compass of non-migratory birds, such as homing pigeons *Columba livia* (Wiltschko et al., 1981; Wiltschko and Wiltschko, 1998; Fleissner et al., 2003, 2007; Mora et al., 2004, 2014; Falkenberg et al., 2010; Wilzeck et al., 2010; Alexander et al., 2020; Rotov et al., 2020), and domestic chickens *Gallus gallus* (Freire et al., 2005, 2008; Wiltschko et al., 2007; Denzau et al., 2013a, b).

By altering the magnetic field, it has been experimentally demonstrated that it is possible to change the direction of flight of birds within a cage. Wiltschko and Wiltschko (1972) discovered the so-called "inclination compass" or "axial compass" in European robin. Wiltschko et al. (2011) presented a schematic section through the GMF from the west to illustrate the functional mode of the inclination compass (Figure 4.4). In principle it is possible to detect the inclination of the magnetic field lines, but which is north polarity or south polarity. It is impossible to obtain information on the direction itself. The inclination compass in the magnetic sense that only the compass information on the dip angle, which is the orientation component of the magnetic vector, can be obtained. Thus, the avian magnetic compass does not distinguish between magnetic "north" and "south" as indicated by polarity, but between "poleward" where the field lines point to the ground, and "equatorward" where they point upward (Wiltschko and Wiltschko, 2005; Wiltschko et al., 2011).

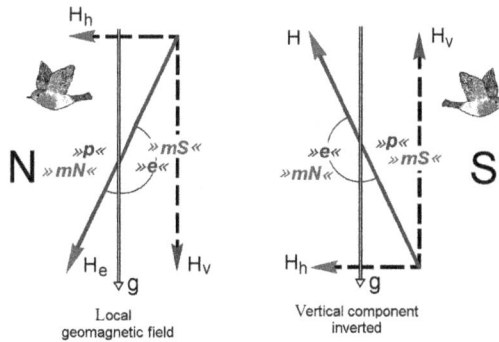

FIGURE 4.4 Schematic section through the GMF from the west to illustrate the functional mode of the inclination compass (Wiltschko et al., 2011). N, S, North and South; H_e, vector of the GMF; H, vector of the experimental field; H_h, H_v, horizontal and vertical components of the magnetic fields; g, gravity vector. The arrowheads indicate the polarity of the fields, with mN, mS, indicating magnetic North and magnetic South, respectively. The axial direction of the vector and its inclination, i.e., its relation to gravity is crucial for the inclination compass, with p, e indicating "poleward" and "equatorward," the readings of the inclination compass. The birds fly towards the directions that they assume to be their spring migratory direction. (Reproduced with permission from Wiltschko et al., 2011, Copyright 2011, Elsevier.)

When robins were tested in experimental fields with different intensities, it became evident that their magnetic compass is narrowly tuned to the ambient magnetic field (Wiltschko et al., 2011). At the test site in Frankfurt am Main, Germany (50°08′N, 8°40′E), the local GMF has an intensity of ~46 µT (Wiltschko et al., 2011). Robins caught and kept at this intensity were disoriented when the total intensities were decreased or increased by about 30%, indicating a narrow functional window (Wiltschko and Wiltschko, 1995, 2007). The disorientation in higher fields was especially surprising, because it clearly showed that the loss of orientation was not caused by the intensity getting below threshold (Wiltschko et al., 2011). Further tests showed that the functional window is flexible and can be adjusted to intensities outside the normal functional range (Wiltschko et al., 2011). Robins regained their ability to orient when they are exposed to lower or higher intensities, with an exposure of about 1 hour at 92 µT sufficient to enable them to orient at this intensity (Wiltschko et al., 2006a). At the same time, the birds did not lose their ability to orient in the local GMF (Wiltschko et al., 2011). This adjustment to new intensities is neither a shift nor a simple enlargement of the functional range; rather, experiencing an intensity outside the normal functional range seems to establish a new functional window around the respective intensity (Wiltschko and Wiltschko, 1995, 2007).

Moreover, Wiltschko and Wiltschko (1972) speculated with great insight that on the whole, this magnetic compass represents a highly flexible direction-finding system. They further estimated that its ability to adjust to a varying intensity range makes it independent of any secular variation in total intensity, and the fact that it does not use the polarity of the magnetic fields so-called "polarity compass" means that it is not affected by the geomagnetic reversals that have taken place several times since the phylogenetic origin of birds (Runcorn, 1969). The published data of inclination compass in European robin are shown by Wiltschko and Wiltschko (2005), compiled from Wiltschko and Wiltschko (1999) and Wiltschko et al. (2001). Orientation behavior of migrating European robins in spring was tested in the local GMF and in two experimental fields (Wiltschko and Wiltschko, 1999, 2005; Wiltschko et al., 2001). The triangles at the periphery of the circle mark mean headings of individual birds, the arrows represent the grand mean vectors with their lengths proportional to the radius of the circle. The two inner circles are the 5% and the 1% significance border of the Rayleigh test.

Subsequent research has revealed when the birds are blindfolded, they cannot perceive the magnetic field, and therefore, the magnetic perception requires light radiation or stimulation (Wiltschko and Wiltschko, 1999, 2001; Wiltschko et al., 2000a, b, 2001). More specially, this behavior was recorded under 565 nm green light at an intensity of 2.1 mW/m^2 as shown by Wiltschko and Wiltschko (2005), compiled from Wiltschko et al. (2000b, 2001) and Wiltschko and Wiltschko (2002). The orientation behavior of European robins in spring was monitored under monochromatic lights of different wavelengths.

Wiltschko et al. (2006b) subjected migratory Australian silvereyes, *Zosterops lateralis*, to a short, strong magnetic pulse and tested their subsequent response under different magnetic conditions. In the local GMF, the birds preferred easterly headings as before, and when the horizontal component of the magnetic field was shifted 90° anticlockwise, they altered their headings accordingly northwards (Wiltschko et al., 2006b). In a field with the vertical component inverted, the birds reversed their headings westwards, indicating that their directional orientation was controlled by the normal inclination compass (Wiltschko et al., 2006b).

Thus, in addition to the inclination compass, another important characteristic of the avian magnetic compass is its "light-dependency." Wiltschko et al. (2011) presented wavelength dependency of the avian magnetic compass (Figure 4.5, data from Wiltschko et al., 1993, 2007, 2010; Wiltschko and Wiltschko, 1998; Rappl et al., 2000; Muheim et al., 2002). Normal compass orientation requires light from the short-wavelength part of the spectrum. European robins and Australian silvereyes are well oriented in their migratory directions under 373 nm ultraviolet (UV), 424 nm blue, 502 nm turquoise, and 565 nm green light. Under 590 nm yellow and beyond, they were disoriented, indicating that their magnetoreception system works no longer properly under longer wavelength (Wiltschko et al., 1993; Muheim et al., 2002; Wiltschko and Wiltschko, 2007). Experiments using interference filters with a half-band width of only 10 nm could narrow down the onset of disorientation in robins even further to between 561 and

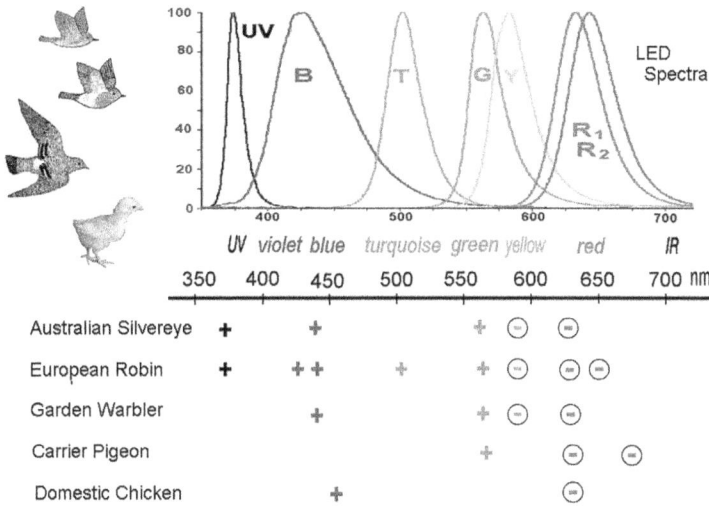

FIGURE 4.5 Wavelength dependency of the avian magnetic compass (Wiltschko et al., 2011, data from Wiltschko et al., 1993, 2007, 2010; Wiltschko and Wiltschko, 1998; Rappl et al., 2000; Muheim et al., 2002 and unpublished). Above: spectra of the light-emitting diodes used in the tests; below: orientation of five bird species tested, with + indication oriented behavior and Θ indicating disorientation. (Reproduced with permission from Wiltschko et al., 2011, Copyright 2011, Elsevier.)

568 nm (Muheim et al., 2002). Thus, it requires short-wavelength light from UV to green (Wiltschko and Wiltschko, 2014). This pattern seems to be common to passerine species (Rappl et al., 2000), homing pigeons (Wiltschko and Wiltschko, 1998), and domestic chickens (Wiltschko et al., 2007).

The experiments mentioned above used low-intensity monochromatic light of a quantal flux of about 7×10^{15} quanta/s m² as found 45 minutes before sunrise and after sunset (only with UV, the intensity was about 0.7×10^{15} quanta/s m², i.e., 1/10). Under monochromatic light of higher intensities and under bichromatic light, migratory birds no longer prefer their migratory direction (Wiltschko et al., 2010). However, birds are able to use their magnetic compass also under high light levels, provided the light is "white," i.e., composed of a wide variety of different wavelengths—the magnetic compass can be used, e.g., by homing pigeons in bright daylight.

Amphibians and reptiles were found to use an inclination compass-like birds, but that of amphibians shows a different wavelength dependency (Phillips, 1986a, b), and that of marine turtles does not require light at all (Light et al., 1993; Lohmann and Lohmann, 1993). In contrast to the amphibians, the spectral range where birds obtain normal magnetic compass information includes the larger part of the visual spectrum. At the same time, this wavelength dependency of magnetoreception shows no relationship to the peaks of the four color cones of the birds' visual system (Maier, 1992), and thus speaks against their involvement in mediating magnetic directions, suggesting the existence of another type of receptor. The birds' response looked like an "all-or-none"-response that could be attributed to one receptor, yet the rather abrupt transition to disorientation, which persisted under the increased intensity of the yellow or red light (Wiltschko and Wiltschko, 2001; Wiltschko et al., 2004), seems to suggest an antagonistic interaction with a second receptor (Wiltschko and Wiltschko, 2005). When the eyes were illuminated with monochromatic light of various wavelengths, units with a peak of responsiveness around 503 nm and others with a peak beyond 580 nm were identified, thus suggesting the two types of receptors with different absorption maxima, a finding that is in agreement with the behavioral studies likewise indicating two types of receptors with absorption peaks in the blue-to-green and in the long-wavelength range (e.g., Möller et al., 2001; Wiltschko et al., 2004).

In the case of RPM in oscillating magnetic fields, the singlet-triplet (*S-T*) interconversion rate can be significantly affected by oscillating magnetic fields of specific frequencies in the "MegaHertz range" (1–100 MHz) (Ritz et al., 2000). The radical pair model predicts that an oscillating magnetic field in the megahertz range can disrupt the magnetic compass due to the electron paramagnetic resonance (EPR) effect (Timmel and Hore, 1996). The intensities required for the resonance effect are so low that they would not affect any of the magnetite-based mechanisms currently considered as explained below, so that disruption of magnetic orientation would be diagnostic for the involvement of a radical pair mechanism (RPM) underlying the magnetic compass (Ritz et al., 2000). Such fields with frequencies of 0.64 MHz and above led to disorientation (Ritz et al., 2004, 2009; Kavokin et al., 2014).

Theoretical considerations and in vitro studies indicate that they are to be expected in the 0.1–10-MHz range. The effect of the oscillating magnetic fields should depend on their orientation with respect to the static background field (Cranfield et al., 1994). These resonances are generally very broad and might therefore lead to disturbing effects at virtually all frequencies within this range, provided the intensity of the oscillating magnetic field is sufficiently strong (Henbest et al., 2004). However, a special resonance occurs when the frequency of the oscillating magnetic field matches the energetic splitting induced by the static GMF; here, one expects a marked effect regardless of the structure of the molecules forming the radical pairs.

Wiltschko et al. (2011) presented the orientation of European robins in the GMF (Control, C), and in high-frequency fields added to the GMF in two different orientations (Figure 4.6, compiled from Wiltschko and Wiltschko, 2005; data from Thalau et al., 2005). First tests with a weak broadband noise field of frequencies from 0.1 to 10 MHz added to the GMF indeed showed that this disrupted the orientation of migratory birds (Ritz et al., 2004). Further tests used the single frequencies of 1.315 and 7.0 MHz with an intensity of about 480 nT. When these fields were presented parallel to the geomagnetic vector, the birds were oriented in their migratory direction, whereas they were disoriented when the same fields were presented at an angle of 24° or 48° to the GMF (Wiltschko et al., 2011, compiled from Wiltschko and Wiltschko, 2005, data from Thalau et al., 2005). This is in agreement with the radical pair model and clearly shows that the observed effect of the high-frequency field is a specific one. Together, these findings indicate that the primary process of magnetoreception in birds involves an RPM.

From the perspective of biophysical theory, Ritz et al. (2000) firstly explained that the birds can perceive even a weak magnetic field at the GMF level on the basis of the RPM model. Schematics of the light-dependent RPM model for quantum-assisted magnetic sensing are shown by Wiltschko and Wiltschko (2006), modified from Ritz et al. (2000). Here, a donor molecule (D) exists in an inactive ground state. A donor molecule (D) absorbs a photon. By electron transfer to an acceptor molecule (A),

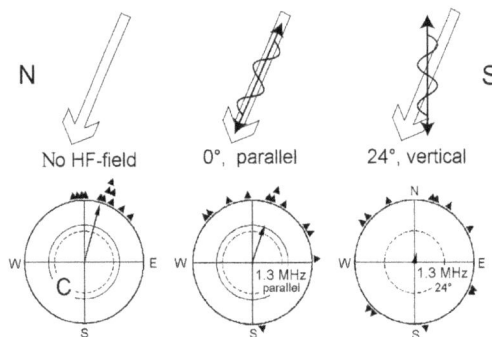

FIGURE 4.6 Orientation of European robins in the GMF (Control, C), and in high-frequency fields added to the GMF in two different orientations (Wiltschko et al., 2011, compiled from Wiltschko and Wiltschko, 2005, data from Thalau et al., 2005). The upper part of the diagram illustrates the orientation of the GMF and the high-frequency field in the three test conditions. (Reproduced with permission from Wiltschko et al., 2011, Copyright 2011, Elsevier.)

a singlet radical pair ($\bullet$D$^+$ and $\bullet$A$^-$)^S is formed. The absorption of blue light promotes electron transfer to an acceptor molecule (A), resulting in a singlet radical pair. An external magnetic field affects the *S-T* conversion. The inset gives a possible dependency of the triplet yield from the angle of the radical pair to the external magnetic field. Thus, the *S-T* conversion leads to a triplet radical pair ($\bullet$D$^+$ and $\bullet$A$^-$)^T, with the triplet yield depending on the alignment of the molecules in the external magnetic field. Here triplet products are chemically different from the singlet products and thus may play a role in magnetoreception. That is, when a photoreceptor, flavin ($\bullet$D$^+$) is excited by blue light, radical pair is generated when electron transfer occurs from nearby Trp ($\bullet$A$^-$). The *S-T* conversion is affected by the external magnetic fields (Zeeman interaction), and the hyperfine coupling (HPC). This *S-T* conversion is magnetic field strength-dependent (Brocklehurst, 1976; Werner et al., 1978, 1983; Lewis et al., 2018; Hore, 2019). Theoretically, it has been clarified that the efficiency of the mixing process of a singlet-born radical pair into a triplet spin state is maximized in a relatively weak magnetic field (Timmel et al., 1998; Timmel and Henbest, 2004; Evans et al., 2013; Lewis et al., 2018; Kerpal et al., 2019).

In chemical terms, the minimum requirement for a radical pair reaction to be sensitive to an external magnetic field is that at least one of the *S* and *T* states undergoes a reaction that is not open to the other, usually as a consequence of the imperative to conserve spin angular momentum (Rodgers and Hore, 2009). A simple reaction scheme that could form the basis of a compass magnetoreceptor is shown by Rodgers and Hore (2009). Here, (**A**) and (**B**) could be portions of the same molecule or distinct molecules held in close proximity by their surroundings (e.g., two cofactors or a cofactor and an amino acid residue in a protein) (Rodgers and Hore, 2009). (**C**) is either the signaling state or leads to the signaling state via subsequent chemical transformations (which are not shown) (Rodgers and Hore, 2009). An applied magnetic field can alter the yield of (**C**) by regulating the competition between its formation (from the *S* and *T* states, step 4), and the regeneration of (**A**) (**B**) (exclusively from the *S* state, step 2) (Rodgers and Hore, 2009). If the *S-T* interconversion is hindered by the external magnetic field, then less (**C**) will be produced and correspondingly more radical pairs recombine directly to (**A**) (**B**) (Rodgers and Hore, 2009). The opposite follows if the field enhances the *S-T* interconversion (Rodgers and Hore, 2009). It is important to note that the external magnetic field is far too weak to initiate new radical reactions (Rodgers and Hore, 2009). Variants on the reaction scheme are possible, and although the details of the chemistry may differ, the principles remain the same (Rodgers and Hore, 2009). Some of the more likely alternatives include an excited triplet precursor state; electron transfer in the reverse direction to form A$^{\bullet-}$B$^{\bullet+}$; and formation of the radical pair via sequential electron transfer steps (Solov'yov et al., 2007).

Spin dynamics simulations of anisotropic reaction yields for the RPM model are shown by Rodgers and Hore (2009), compiled from Rodgers (2007) and Efimova and Hore (2008). Here, upper images indicate polar plots; lower images show the corresponding signal modulation patterns for a bird looking directly along the GMF vector. The heights of the vertical scale bars in the upper images correspond to singlet yields of 2% or 0.2%. The simulations demonstrate that a relatively simple orientation dependence of the reaction yield (A) can be obtained from radical pairs containing a small number of hyperfine interactions or from more complex radicals when a few symmetry-related hyperfine interactions dominate (B). More intricate anisotropy patterns (C) can be dramatically simplified if the radical pairs are axially rotationally disordered (D). The signal modulation pattern for C is identical to that for D and is only shown once. Note that in all cases the reaction yield is invariant to the exact reversal of the magnetic field vector, i.e., the response is that of an inclination compass rather than a polarity compass. Unless the radicals contain very few magnetic nuclei or possess some degree of molecular symmetry or are favorably disordered, the shape of the anisotropy can be complex (Rodgers and Hore, 2009). There is no clear picture of what would constitute the optimum sensory input for the bird; however, it seems reasonable to suppose that strongly anisotropic but relatively simple directional information would be favored (Rodgers and Hore, 2009). Simulations and experiments on solution-phase reactions suggest a few simple design features discussed in the section on CRYs (Cintolesi et al., 2003; Efimova and Hore, 2008; Rodgers et al., 2007).

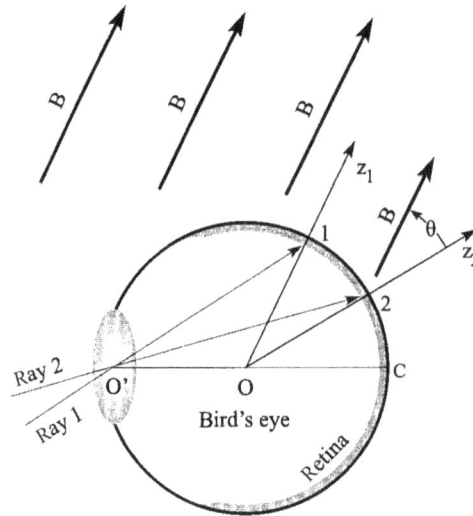

FIGURE 4.7   A bird's eye model used for the calculation of visual modulation patterns via plausible radical pair (Ritz et al., 2000). (Reproduced with permission from Ritz et al., 2000, Copyright 2000, Elsevier.)

CRY was immuno-chemically identified in the outer segments of the UV/V cones (SWS1 receptors) in the retinae of chickens, robins and several other species (Nießner et al., 2011; Bolte et al., 2021). This indicates the eyes as the site of the reception for magnetic directions, even if details are not yet entirely clear (Wiltschko and Wiltschko, 2019; Bolte et al., 2021).

In the case of the bird's eye model, the yield of the triplet spin state depends on the angle formed by the external magnetic field and the radical pair. Ritz et al. (2000) proposed a bird's eye model used for the calculation of visual modulation patterns via plausible radical pair (Figure 4.7).

Rays 1 and 2 enter through an infinitesimal hole at O′ and are projected onto a spherical retina (Figure 4.7). The receptor molecules are assumed to be oriented normally to the retina surface (directions z_1 and z_2), thus forming different angles with the direction of the magnetic field vector. This angle dependence has important implications. This is because, when the radical pair exists perpendicular to the retina sphere, the angle between the radical pair and the external magnetic field (GMF) is different depending on the retina surface point, and the concentration of the product is different at each point. It is thought that this concentration difference is replaced with a signal transmitted to nerves, and finally, the configuration image of the magnetic field is acquired in the migratory bird's brain.

Behavioral studies in European robins (Zapka et al., 2009) strongly suggest that a forebrain region named "Cluster N" (Mouritsen et al., 2005), which receives input from the eyes via the thalamofugal visual pathway (Heyers et al., 2007), is involved in processing magnetic compass information. Several studies on the migratory birds' brains showed that bilateral lesions of Cluster N disables magnetic orientation, and therefore, Cluster N is assumed to be a light-processing forebrain region (Möller et al., 2004; Mouritsen et al., 2004; Zapka et al., 2009). Moreover, it is presumed that it is the cryptochrome (CRY) of the retina that meets the conditions under which the electron transfer reaction occurs at the photoreceptors on the retina sphere (Ritz et al., 2000). According to the model proposed by Ritz et al. (2000), this would allow birds to perceive the GMF as a pattern superimposed on the visual image. In a sense, they could literally "see" the magnetic field (Ritz et al., 2000).

The RPM of the avian magnetic compass deals with the quantum evolution of highly non-equilibrium electron spin states of pairs of transient spin-correlated radicals residing inside a bird's retina as illustrated in Figure 4.8 (Pedersen et al., 2016).

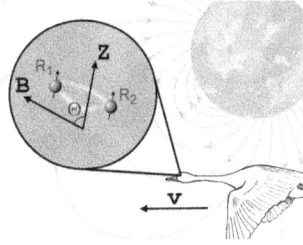

FIGURE 4.8 Schematic illustration of the avian radical pair-based compass (Pedersen et al., 2016). Magnetoreceptive molecules in the bird's eyes host a pair of radicals (R_1, R_2), and endow the bird with capabilities to sense the GMF. In the most simplified case, each radical pair is associated with a coordinate frame such that internal magnetic interactions are considered isotropic in the xy-plane, while the anisotropy defines the z-axis. The radical pairs participate in spin-dependent chemical reactions that are sensitive to the angle Θ between this z-axis and the direction of the GMF B, which, in turn, could be related to the direction of bird motion, denoted by v. (Reproduced with permission from Pedersen et al., 2016, Copyright 2016, Springer Nature.) It is licensed under the Creative Commons Attribution 4.0 International.

These spin-correlated radicals could form electronically entangled singlet and triplet states, which are respectively characterized by an anti-parallel and parallel alignment of the unpaired electron spins of the radicals (Pedersen et al., 2016). The core of the radical pair mechanism (RPM) of avian magnetoreception relies on the possible engagement of the radicals in biochemical reactions that could be affected by magnetic fields even though the Zeeman interaction of an unpaired electron spin with the GMF is more than six orders of magnitude smaller than the thermal energy available inside the biological surroundings (Pedersen et al., 2016). Hence, from a classical perspective, a magnetic sensitivity should not arise from the biochemical reactions, but arise from the quantum reactions (Pedersen et al., 2016). Such quantum reactions enter the stage through the RPM, which so far is the only known way an external magnetic field can influence a chemical reaction (Steiner and Ulrich, 1989; Brocklehurst, 2002; Timmel and Henbest, 2004; Rodgers, 2009; Solov'yov et al., 2014; Hore and Mouritsen, 2016). The RPM has been studied for about half a century by now and has been successfully applied to various phenomena such as spin polarization (Muus et al., 1977) and magnetic isotope effects (Salikhov, 1996). The anisotropy of the internal magnetic interactions in the radical pair, i.e., the hyperfine interactions, defines a molecular coordinate system, that, in turn, determines the orientation between the radical pair and the magnetic field using just a single angle Θ (Pedersen et al., 2016, Figure 4.8).

Here, taking more advanced parameters into account, Pedersen et al. (2016) linked the microscopic proposition of the chemical compass to the macroscopic scale through the angular probability distribution $R(\Theta)$, obtained from the spin dynamics of a radical pair. Using a simple model, they have simulated flight trajectories of a large number of birds assuming their navigation to rely on chemical compass distributions, $R(\Theta)$, of different precision (Pedersen et al., 2016). It was revealed that the precision of the chemical compass has a great impact on not just the spread of birds but also largely influences the time it takes to make the trip (Pedersen et al., 2016). Hence a precise chemical compass is of great importance, and it should once again be emphasized that a spin chemical mechanism with a fast spin-dependent reaction would be able to provide the needed precision (Pedersen et al., 2016).

4.4 Radical Pair-Based Magnetoreceptor and Cryptochrome

Perhaps most importantly, Ritz et al. (2000) hypothetically proposed that cryptochrome (CRY) in the bird retina is a strong candidate for function as a radical pair-based magnetoreceptor as a "CRY-based RPM theory." The CRY is a class of photopigments known from plants and related to photolyases (Sancar, 2003), and they are assumed to possess chemical properties crucial for the model, including the ability to form radical pairs (Giovani et al., 2003). Actually, an in vitro experimental study firstly suggested that bird CRY1a is excited by blue light (420–470 nm), and forms long-lived radical pairs blue light irradiation (Liedvogel et al., 2007; Muheim and Liedvogel, 2015).

Maeda et al. (2008) demonstrated that a radical-pair-based chemical reaction could respond to a weak magnetic field in the strength of the GMF in a chain-like molecular triad (CPF triad) as a completely artificial system using transient absorption measurements. Furthermore, Maeda et al. (2012) found that a radical-pair-based chemical reaction could also respond to magnetic fields of several tens of mT, which magnitudes are larger than that of the GMF in an experimental system that reconstructs a natural protein CRY. This CRY was derived not from the migratory birds but from a plant *Arabidopsis thaliana*. From these results, the RPM models became realistic and pioneered that CRY actually functions as a high-sensitivity magnetic sensor (Maeda et al., 2012).

In fact, concerning the question of whether the RPM models really function as a magnetic sensor in the retina of migratory birds, the following reports have been made: The first is a report on the localization of CRY and neural activity in the retina of migratory birds under magnetic fields. The second, though not a report on magnetic field effects, was the detection of a relatively long-lived radical pair formation in the order of milliseconds at room temperature in the CRY of a migratory bird called garden warbler (*Sylvia borin*) using transient absorption measurements. From the experimental results, in which the radical pair formation and the spin dynamics of radical pairs were detected by these biophysical methods and numerical simulations, the RPM models are regarded as promising candidates for the magnetic sensor of migratory birds.

4.4.1 Chemical Characteristics of Cryptochrome

FAD, which is a functional molecule and chromophore of cryptochrome (CRY), is known to have various electronic states depending on the external environment and to be involved in many redox reactions in the living systems. From *Arabidopsis thaliana* cryptochrome 1 (*At*CRY1), an anion radical is obtained after 450 nm photoexcitation of FAD via an electron transfer from a tryptophan (Trp) residue found nearby the protein cavity (Maeda et al., 2012). Giovani et al. (2003) proposed a scheme of light-induced reactions in purified *At*CRY1 (Figure 4.9).

Light-induced intra- or interprotein electron transfer has been proposed to trigger conformational changes of CRY that allow binding of a signaling partner for further signal transduction (Yang et al., 2000). Giovani et al. (2003) consider the following products of the electron transfer reactions as potential

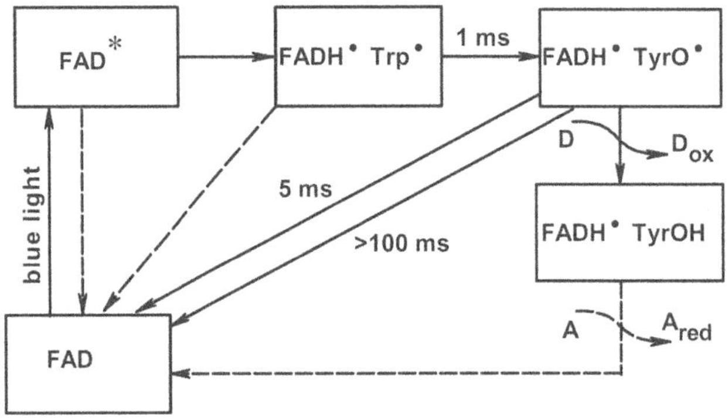

FIGURE 4.9 Scheme of light-induced reactions in *Arabidopsis thaliana* cryptochrome 1 (*At*CRY1) (Giovani et al., 2003). Solid arrows indicate reactions observed directly. Broken arrows indicate additional reactions. Protonation-deprotonation accompanying electron transfer is not explicit because it was not kinetically resolved. FAD*, singlet excited FAD after absorption of blue light; D, external electron donor; A, external electron acceptor. (Reproduced with permission from Giovani et al., 2003, Copyright 2003, Springer Nature.)

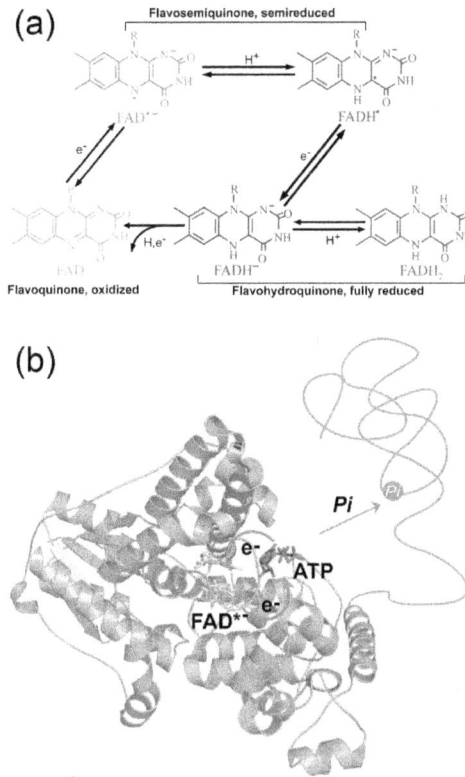

FIGURE 4.10 Photoexcitation of *Arabidopsis thaliana* cryptochrome 1 (*At*CRY1) (Liu et al., 2011). (a) Five possible redox forms of flavins. The two different forms of semiquinone radicals: anion radical (e.g., FAD$^-$), neutral radical (e.g., FADH$^\cdot$), two forms of reduced flavins: protonated hydroquinone (e.g., FADH$_2$), and anionic hydroquinone (e.g., FADH$^-$) are shown. R: side groups of flavins. (b) The photolyase-like cyclic electron shuttle model of CRY photoexcitation. In this model, the resting state of a CRY contains the anion radical semiquinone (FAD$^-$). Upon photon absorption, the excited FAD$^-$ transfers an electron to ATP, triggering phosphotransfer and autophosphorylation of the CRY. The electron is subsequently transferred back to flavin to complete the cycle. The putative locations of phosphorous group and electron transfer path are indicated. (Reproduced with permission from Liu et al., 2011, Copyright 2011, Elsevier.) It is licensed under the Creative Commons Attribution 4.0 International.

trigger states: the flavin radical FADH$^\cdot$, the tyrosyl radical TyrO$^\cdot$, the oxidized form D_{ox} of an external electron donor that reduced TyrO$^\cdot$, and the reduced form A_{red} of an external electron acceptor that oxidized FADH$^\cdot$ (Figure 4.9).

The FAD is a two electron carrier that can exist in one of the three different redox states or five different protonated forms: oxidized flavin (FAD), semireduced flavin (anion radical FAD$^{\cdot-}$ or neutral radical FADH$^\cdot$), and fully reduced flavin (FADH$^-$ or FADH$_2$) in *At*CRY1 (Liu et al., 2011, Figure 4.10a). Here, FADH$^\cdot$ is the semiquinone form of FAD and a radical species in which proton transfer occurs following electron transfer. Among the different redox forms, only the oxidized flavin, FAD, and anion radical semiquinone flavin, FAD$^{\cdot-}$ (FAD* in Figure 4.9), absorb significant amounts of blue light (~400–500 nm). It has been proposed that the oxidized flavin, FAD may be the ground state chromophore of *At*CRY1, because it absorbs blue light most effectively.

The CRY contains two domains, the N-terminal PHR (Photolyase-Homologous Region) domain of about 500 residues, and the CRY C-terminal extension (CCE) domain of various lengths and sequences (Liu et al., 2011, Figure 4.10b). PHR is the chromophore-binding domain of CRYs that bind non-covalently to the chromophore, FAD and possibly the second chromophore, 5,10-methenyltetrahydrofolate (MTHF) (Lin et al., 1995; Malhotra et al., 1995; Banerjee et al., 2007; Klar et al., 2007).

For some animal CRYs, it can even be the primary photoproduct after blue light irradiation (Song et al., 2006; Liu et al., 2010). Here, FAD$^{\bullet-}$ (FAD* in Figure 4.9) is a semi-reduced anion radical species in which electron transfer alone occurs. When FAD is irradiated with blue light, tryptophan (Trp) is able to act as an electron donor due to relatively higher occupied molecular orbital (HOMO) energy level. In addition to Trp, tyrosine (Tyr) is an amino acid that causes photoexcited electron transfer with respect to FAD. The highest HOMO energy level of these amino acids are higher than the HOMO of FAD. Therefore, when FAD is photoexcited, electron transfer occurs from the amino acid HOMO to the vacant orbit of FAD, and a charge-separated state and a radical pair are formed.

Recent studies have shown that these differences depend not only on Trp but also on what amino acids are located near the FAD (Giovani et al., 2003; Liedvogel et al., 2007; Berndt et al., 2007; Damiani et al., 2010; Iwata et al., 2010; Kattnig and Hore, 2017; Hammad et al., 2020). The structure has been investigated in the known CRY, especially for amino acids that affect the protonation of FAD. It has been reported that mutating asparagine (Asn), near the FAD of *Synechocystis* CRY-DASH (*Scy*CRY-DASH) (Brudler et al., 2003), to cysteine (Cys) stabilizes the flavin radical state FAD$^{\bullet-}$ (Iwata et al., 2010). Here, CRY-DASHs (*Drosophila, Arabidopsis, Synechocystis*, human-type cryptochromes) form one subclade of the cryptochrome/photolyase family (CPF), and CPF members are flavoproteins that act as DNA-repair enzymes (DNA-photolyases), or as UV-A/blue-light photoreceptors (cryptochromes) (Kiontke et al., 2020). The name CRY-DASH was invented by Brudler et al. (2003), together with the discovery of another photolyase homolog in *Arabidopsis thaliana* (*At*CRY3; AT5G24850.1) that contained FAD and did bind to undamaged dsDNA and single-stranded (ss) DNA, but was unable to complement photo-reactivation in photolyase-deficient *E. coli* cells (Kleine et al., 2003). It has also been reported that the original natural protein stabilizes the flavin radical state FADH$^{\bullet}$ more than the mutant protein in which Asn, in the vicinity of the FAD of photolyase (flavoprotein having a function of repairing DNA), was replaced by Cys (Damiani et al., 2010).

In the case of the amino acid sequences of natural CRY and photolyase in the *Drosophila melanogaster* CRY (*Dm*CRY), Cys is located in the vicinity of FAD (Iwata et al., 2010). In this case, the flavin radical state FAD$^{\bullet-}$ is stabilized and has the function of a circadian clock (Iwata et al., 2010). The form of FAD in the circadian clock for *Drosophila melanogaster* is regarded as FAD$^{\bullet-}$ (Kattnig and Hore, 2017, Figure 4.11).

The FAD cofactor leads to the formation of radical pairs via sequential electron transfers along the "tryptophan (Trp)-triad," a chain of three conserved Trp residues within the protein (Zeugner, et al., 2005; Biskup et al., 2009; Giovani et al., 2003). This process reduces the photoexcited singlet state of the FAD to the anion radical, FAD$^{\bullet-}$, and oxidizes the terminal, surface-exposed, Trp (Trp$_C$H) to give the cation radical, Trp$_C$H$^{\bullet+}$ (Kattnig and Hore, 2017). Formed with conservation of spin angular momentum,

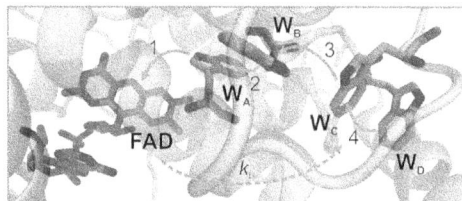

FIGURE 4.11 Electron transfer pathway in cryptochromes in the *Drosophila melanogaster* (*Dm*CRY) (Kattnig and Hore, 2017). After photoexcitation of the FAD cofactor, three or four rapid sequential electron transfers along a triad or tetrad of tryptophan (Trp) residues (W_A, W_B, W_C, W_D) generate a spin-correlated radical pair [FAD$^-$ Trp$_C$H$^+$] or [FAD$^-$ Trp$_D$H$^+$]. The figure is based on the crystal structure of *Dm*CRY (PDB ID: 4GU5) (Zoltowski et al., 2011; Levy et al. 2013). (Reproduced with permission from Kattnig and Hore, 2017, Copyright 2017, Springer Nature.)

the radical pair is initially in an electronic singlet state, [FAD$^{\bullet-}$ Trp$_C$H$^{\bullet+}$] (Henbest et al., 2008; Weber et al., 2010; Maeda et al., 2012). This form of the protein is a coherent superposition of the eigenstates of the spin Hamiltonian which comprises the Zeeman, hyperfine, exchange, and dipolar interactions of the electron spins. As a consequence, the radical pairs oscillate coherently between the singlet and triplet states, a process that manifests itself in the yields of subsequent spin-selective reactions of the radicals. In particular, when the protein is immobilized, the anisotropy of the electron-nuclear hyperfine coupling (HFC) interactions causes the reaction product yields to depend on the orientation of the protein with respect to an external magnetic field.

Sheppard et al. (2017) reported spectroscopic measurements of photo-induced FAD and Trp radicals in recombinantly expressed, purified *Dm*CRY. In brief, a combination of transient absorption and broadband cavity-enhanced absorption spectroscopy has been employed to explore the effects of external magnetic fields (of up to 22 mT) on the key species involved in the photocycle of *Dm*CRY (Sheppard et al., 2017). Details of these techniques can be found elsewhere (Maeda et al.. 2012; Neil et al., 2014). The protein concentration (~50 μM), temperature (267–278 K) and glycerol content (~50% for transient absorption measurements and 20% for the cavity-enhanced absorbance experiments) of the solutions were chosen to optimize the magnetic responses.

In the case of the *At*CRY1, the distance of the finally generated radical pair [FADH$^{\bullet}$ TyrO$^{\bullet}$] is 16 Å or more, and the lifetime of the radical pair is extended by suppressing charge recombination (Giovani et al., 2003). In the case of the migratory bird CRY, its structure and function might be similar or almost the same as those of the *At*CRY, and it is estimated that Try could contribute to the stability of radical state FADH$^{\bullet}$ (Liedvogel et al., 2007). The lifetime of the final radical pair [FADH$^{\bullet}$ TyrO$^{\bullet}$] is reported to be 14 ms at room temperature (Liedvogel et al., 2007).

In the case of the migratory bird CRY (*gw*CRY1a), it is uncertain because the exact structure has not been clarified, but when estimated from the amino acid sequences of the CRY, there is a possibility that aspartic acid (Asp) is located in the vicinity of FAD, and the flavin radical state FADH$^{\bullet}$ is stabilized (Liedvogel et al., 2007). It is speculated that it may have acquired the function of a magnetic compass (Liedvogel et al., 2007). Furthermore, in the migratory bird *gw*CRY1a (Liedvogel et al., 2007) and *At*CRY1 (Giovani et al., 2003), where FADH$^{\bullet}$ is a key radical, electron transfer from Try to a tryptophanyl radical, Trp$^{\bullet}$ (generated by electron transfer from Trp to FAD) has been reported. Here, a flavin radical, FADH$^{\bullet}$ has broad absorption from 500 to 650 nm, 5 ms half-life in *At*CRY1, 14 ms lifetime in *gw*CRY1a, and Trp$^{\bullet}$ has a narrow component from 500 to 550 nm, 1 ms half-life in *At*CRY1, 4 ms lifetime in *gw*CRY1a (Liedvogel et al., 2007). Focusing on the function of CRY, the form of FAD in the magnetic compass for avian bird CRY is estimated to be FADH$^{\bullet}$ (Liedvogel et al., 2007).

4.4.2 Magnetic Sense and Cryptochrome

It is said that a blue light-sensing protein called "cryptochrome (CRY)" in the retina of the eye plays an important role as a magnetic sensor or magnetoreceptor (Ritz et al., 2000, 2004, 2009; Möller et al., 2004; Mouritsen et al., 2004; Mouritsen and Ritz, 2005; Zapka et al., 2009; Mouritsen and Hore, 2012; Lau et al., 2012; Wiltschko and Wiltschko, 2014, 2021; Bolte et al., 2016; Kerpal et al., 2019; Wiltschko et al., 2021). CRY proteins are also components of the central circadian clockwork (Yuan et al., 2007) and are closely related to the light-dependent DNA repair enzymes, the photolyases (Cashmore, 2003). It is reported that the CRY is fingered as the smoking gun in the exquisite magnetic reception of birds (Roberts, 2016). As a putative magnetoreceptor, five different isoform CRYs (CRY1a, CRY1b, CRY2, CRY4a, and CRY4b) have been identified in the retinae of several bird species. CRY1a (Liedvogel et al., 2007; Nießner et al., 2011) is found from garden warblers (*Sylvia borin*) and European robins. CRY1b (Bolte et al., 2016; Nießner et al., 2016) is from European robins, migratory northern wheatears (*Oenanthe oenanthe*), and homing pigeons. CRY2 (Mouritsen et al., 2004) is from migratory garden warblers. CRY4 (Günther et al., 2018; Pinzon-Rodriguez et al., 2018; Hochstoeger et al., 2020; Wu et al., 2020) is from European robins.

CRY4 has been found in non-mammalian vertebrates, but it lacks the circadian transcriptional regulatory function (Kobayashi et al., 2000; Kubo et al., 2006; Takeuchi et al., 2016) or the photorepair activity (Kobayashi et al., 2000). Though the function of CRY4 is not well understood, chicken CRY4 (cCRY4) may be a magnetoreceptor because of its high level of expression in the retina and light-dependent structural changes in retinal homogenates. To further characterize the photosensitive nature of cCRY4, Mitsui et al. (2015) developed an expression system using budding yeast and purified cCRY4 at yields of submilligrams of protein per liter with binding of the FAD chromophore.

By short durations of cCRY4 irradiation with blue light, Mitsui et al. (2015) detected reduction of the FADox chromophore to FADH•, which was not observed in the previous study by Ozturk et al. (2009). Extended durations of irradiation reduced FADH• to the FADH⁻ form. Mitsui et al. (2015) detected for the first time that the FADH⁻ form returned to FADox in the dark via FADH• formation, and they could depict the putative photocycle of cCRY4 (Figure 4.12a). Although the dark oxidation of FADH⁻ to the FADH• form was observed in the previous study by Ozturk et al. (2009), their observations are not fully congruent with the photocycle (Figure 4.12a), probably because there are many differences in experimental conditions such as the different irradiation wavelengths (UV-A or blue light), the light intensity, the expression system used, the presence of the FLAG tag, and perhaps the concentration of dissolved oxygen.

The reduction from FADox to FADH is likely composed of two steps: (1) electron transfer to FADox to generate FAD•⁻ anion radical and (2) its protonation to generate the neutral FADH• form (Liu et al., 2010). Mitsui et al. (2015) did not observe temperature dependency in this reduction process, implying that both of these steps might be composed of temperature-independent mechanisms under the experimental conditions.

Mitsui et al. (2015) further estimated the absolute absorbance spectra of FADox, FADH, and FADH•⁻ forms (Figure 4.12b) using the difference absorption spectra obtained from our present measurements. Then, by using the photon fluence rate, the absorbance of the sample, and the rate of photoreaction of FADox to FADH• (blue light irradiation), Mitsui et al. (2015) could roughly estimate the quantum yield for the photoreduction ($\Phi 1$) to be $\approx 3\%$. Similarly, the quantum yield for the photoreduction of FADH• to FADH⁻ ($\Phi 2$) (red light irradiation) was estimated to be $\approx 2\%$. These values are lower than the quantum yields for the photoreduction of the other CRYs such as *Chlamydomonas* aCRY (animal-like CRY) ($\Phi 1 \approx 7\%$) (Spexard et al., 2014), but these values may be enough to receive an external light or magnetic signals accounting for the wide and relatively strong expression of cCRY4 in the retina.

The speculated photocycle of cCRY4 (Figure 4.12a) is similar to that of *At*CRY1 (Lin et al., 1995), but in the case of *At*CRY1, FADH• was not detected spectroscopically when FADH⁻ was incubated in the dark for reoxidation to FADox (Müller and Ahmad, 2011). This may be due to the rapid oxidation of FADH• to FADox in the two-electron reoxidation process of *At*CRY1. In this study, both the FADH• and FADH⁻ forms of cCRY4 were oxidized in vitro under dark conditions. A recent study of chicken CRY1a (cCRY1a) (Nießner et al., 2013), another CRY identified in the chicken retina, implied that cCRY1a may absorb not only blue light but also that of longer wavelengths (e.g., green and yellow) utilizing the FADH• state and changing its CRY C-terminal extension (CCE) conformation in the FADH⁻ state. Nießner et al. (2013) analyzed cCRY1a activation using the chicken retina in vivo; therefore, future investigation extending our novel yeast expression system to other CRY proteins both in vivo and in vitro may be beneficial to further analyses.

Concerning the biological functions of CRY4, avian CRYs are thought to work as light-driven magnetoreceptors (Ritz et al., 2000) based on their localization in the retina (Mouritsen et al., 2001; Nießner et al., 2011; Watari et al., 2012), and the photosensitivity of the purified protein (Liedvogel et al., 2007; Du et al., 2014). On the other hand, in the Western clawed frog (*Xenopus tropicalis*), CRY4 is highly expressed in the ovary and testis rather than the retina (Takeuchi et al., 2016), and hence more likely to be implicated in unknown photic function(s) in the gonadal tissues instead of magnetoreception. Considering that CRY involves multiple functions such as nonvisual photoreception (Emery et al., 1998; Tu et al., 2004), magnetoreception (Gegear et al., 2008), and vision (Mazzotta et al., 2013), CRY4 may play multiple roles in different cells and/or organs.

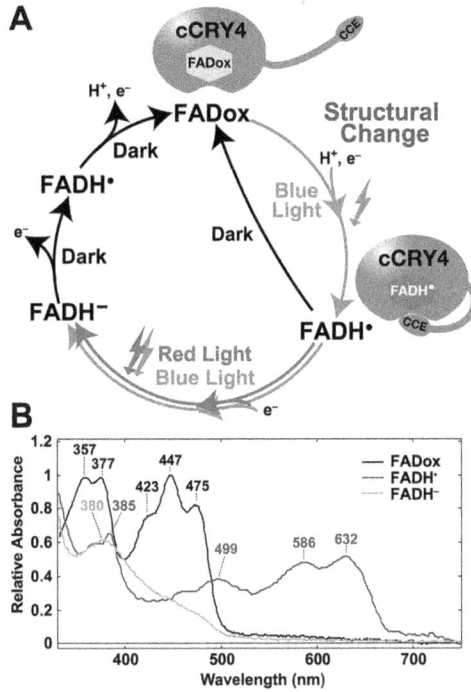

FIGURE 4.12 Model for the cCRY4 photocycle and calculated absolute absorbance spectra of photointermediates (Mitsui et al., 2015). (a) Short duration (0.5 minutes) irradiation with blue light (453 nm, 1 mW/cm²) reduced the FADox chromophore to the FADH• form. Extended blue light or red light irradiation further reduced FADH• to FADH⁻. FADH• is likely oxidized to FADox directly in dark, while FADH⁻ is likely oxidized to FADox via FADH• during dark incubation. (b) Absolute absorbance spectra of cCRY4 photointermediates estimated by calculation. The dark-adapted samples are considered to be mainly composed of FADox with a small amount of residual FADH•. The sample, after long irradiation with white light (128 minutes, 1 mW/cm²), is presumed to be composed of only FADH⁻ forms. Spectral changes induced by the initial 0.5 minutes blue light and the following 8 minutes red light irradiations are assumed to correspond to conversions from FADox to FADH• and FADH• to FADH⁻, respectively. The absorbance spectra were estimated as follows. (i) All spectra were normalized by their optical density and presumed to have absorbances at the peak of the dark-adapted spectrum such that the FADox form (447 nm) was 1.0. Absorbances at putative isosbestic points between the FADH• and FADH⁻ forms or the FADox and FADH⁻ forms were used. (ii) When the putative photobleaching rates of FADox were changed from 20% to 50%, putative absolute absorption spectra for FADH• were calculated and the photobleaching rate was determined to be 31%, placing the isosbestic point between FADH• and FADH⁻ at 446 nm. (iii) The determined absolute absorption spectrum of FADH• was diminished from "dark" or "white 128 min" in the spectrum with changing putative contents of the contaminating FADH• form. Mitsui et al. (2015) inferred the FADH• content in the FADox and FADH⁻ forms to be 3%, assuming that FADox had no absorbance in the longer wavelength region (> 520 nm). (Reproduced with permission from Mitsui et al., 2015, Copyright 2015, American Chemical Society.)

Moreover, using spectroscopic methods, Hochstoeger et al. (2020) showed that pigeon (*Columba livia*) cryptochrome *Cl*CRY4 (Zoltowski et al., 2019) is photoreduced efficiently and forms long-lived spin-correlated radical pairs via a tetrad of tryptophan residues. Hochstoeger et al. (2020) reported that *Cl*CRY4 is broadly and stably expressed within the retina but enriched at synapses in the outer plexiform layer in a repetitive manner. A proteomic survey for retinal-specific *Cl*CRY4 interactors identified molecules that are involved in receptor signaling, including glutamate receptor-interacting protein 2, which colocalizes with *Cl*CRY4. Their data support a model whereby *Cl*CRY4 acts as a UV-blue photoreceptor and/or a light-dependent magnetosensor by modulating glutamatergic synapses between horizontal cells and cones (Hochstoeger et al., 2020).

In particular, more recently, Wiltschko et al. (2021) reviewed and speculated that CRY1a appears to be the most likely receptor molecule for magnetic compass information due to its location in the outer segments of the UV cones with their clear oil droplets. These CRYs could generate free radical pairs, which play a key role as a kind of "quantum compass" (Hiscock et al., 2016b), and are deeply involved in the magnetic sense. That is, the CRY-dependent magnetoreception is currently proposed to be a result of light-initiated electron transfer chemistry in the protein, which is magnetically sensitive by virtue of the RPM (Rodgers and Hore, 2009; Dodson et al., 2013). In the principle of magnetoreception mechanism according to the RPM models, it is not possible to distinguish whether the directions of electron spins are opposite or the same, so in principle, it is possible to detect the inclination of the magnetic field lines, but which is north polarity or south polarity. It is impossible to obtain information on the direction itself. Therefore, as described above, the quantum compass is an inclination compass that only the information on the dip angle can be obtained.

In a follow-up study, Bradlaugh et al. (2021) reported that coupling blue light exposure with a 100 mT static magnetic field is sufficient to potentiate the effect of activated DmCRY on increasing the firing rate of action potentials in an identified motoneuron, termed the anterior corner cell, aCC. Again, no effects of either blue light or magnetic field were observed without prior expression of DmCRY or to orange light (590 nm) (Giachello et al., 2016). The effect was also abolished under conditions where K_v channels were blocked, and a similar mechanism was proposed for clock cells requiring HYPERKINETIC (Fogle et al., 2015). It seems clear that exposure to blue light is a key factor in all assays that have measured magnetosensitivity (Bradlaugh et al., 2021). This raises the exciting possibility that free FAD (blue light-sensitive flavin) may form RPs with available Trp residues in other intracellular proteins (Bradlaugh et al., 2021). However, a necessity for CRY, or at least the C-terminal, to transduce this effect suggests that CRY is required for downstream signaling following RP formation, whether this is due to an RP formed with the Trp of the C-terminal or with those of other unidentified proteins (Bradlaugh et al., 2021). In this respect, it is possible that the previously discussed protein-protein interaction domain located in the C-terminal fragment may provide a nucleation point for further components of this signal transduction cascade to assemble into a functioning magnetosensory complex (Bradlaugh et al., 2021). Thus, it is conceivable that CRY functions as an amplifier to boost weak magnetic field effects that occur within free FAD (Bradlaugh et al., 2021). A recent study reports that FAD-RPs are magnetic field-sensitive at physiological pH (Antill and Woodward, 2018).

As another CRY-related mechanism, Zaporozhan and Ponomarenko (2010) hypothesized that CRY is a transcriptional repressor of the major circadian complex CLOCK/BMAL1 (reviewed by Langmesser et al., 2008), and therefore, magnetic fields via some modulation of CRY function can influence circadian gene expression and modify the activity of the transcription factor nuclear factor-κB (NF-κB)- and glucocorticoids-dependent signaling pathways. In addition, Zaporozhan and Ponomarenko (2010) proposed a theory that magnetic fields induce definite genetic effects due to the existence of magnetic field–sensitive transcription factor repressors capable of regulating the biological activity of organisms through epigenetic mechanisms. These substances are proteins of the "CRY-photolyase family." Radical pair intermediates were observed in a number of proteins from the CRY-photolyase family using time-resolved EPR (Gindt et al., 1999; Weber et al., 2002; Biskup et al., 2009; Nohr et al., 2016).

Putative magnetoreceptor, CRY was first found in a flowering plant *Arabidopsis thaliana* in 1993 (Ahmad and Cashmore, 1993) termed as AtCRY1, which played an important role in photomorphogenic responses (Lin and Todo, 2005). Recently, the role of AtCRY1 has been investigated as a candidate for magnetoreceptor. For example, the growth-inhibiting influence of blue light on the *Arabidopsis thaliana* is moderated by magnetic fields in a way that may use the RPM (Ahmad et al., 2007). Moreover, the removal of the local GMF negatively affects the reproductive growth of *Arabidopsis*, which thus affects the yield and harvest index (Xu et al., 2012, 2013, 2014b), and delays the flowering time through down-regulation of flower-related genes (Xu et al., 2012; Agliassa et al., 2018a). The expression changes of three AtCRY1-signaling related genes, PHYB, CO, and FT suggest that the effects of a near-null magnetic field are CRY-related, which may be revealed by a modification of the active state of CRY and the subsequent signaling cascade plant CRY has been suggested to act as a magnetoreceptor (Xu et al., 2012). Artificial reversal of the GMF has confirmed that *Arabidopsis* can respond not only to magnetic field intensity but also to magnetic field direction and polarity (Bertea et al., 2015). Moreover, the GMF was found to impact photomorphogenic-promoting gene expression in etiolated seedlings of *Arabidopsis*, indicating the existence of a light-independent magnetoreception mechanism (Agliassa et al., 2018b). With regard to exposure of plants to magnetic fields higher than the GMF, the magnetic fields ranging ~1–30 mT have been reported to produce changes in quantum yield of flavin semiquinone radicals in *At*CRY1 (Maeda et al., 2012).

In the context of animal magnetoreception, one can well imagine that relaxation and recombination rates for the CRY radical pair might have been optimized for function by evolution, e.g., through interaction with binding partners, slight variations of protein structure as well as solution accessibility (protonation/deprotonation) of the radicals, especially the terminal tryptophan (Hiscock et al., 2016b). The second condition of strong axiality of the hyperfine couplings is, at least to some degree, fulfilled by the flavin radical (Lee et al., 2013). However, a hyperfine-coupling-free second radical is harder to imagine (Kerpal et al., 2019). The frequently evoked hypothesis of the involvement of a superoxide radical $O_2^{\cdot-}$ fails as demonstrated in the literature (Hogben et al., 2009), owing to the large spin-orbit coupling in $O_2^{\cdot-}$ leading to extremely fast electron spin relaxation (Kerpal et al., 2019). As a result, all spin coherence is lost on a nanosecond timescale and with it any magnetic field sensitivity (Kerpal et al., 2019).

As mentioned above, the magnetically sensitive species is commonly assumed to be [FAD$^{\cdot-}$ TrpH$^{\cdot+}$], formed by sequential light-induced intraprotein electron transfers from a chain of tryptophan residues to the FAD chromophore (Henbest et al., 2008; Weber et al., 2010; Maeda et al., 2012). However, some evidence points to superoxide, $O_2^{\cdot-}$, as an alternative partner for the flavin radical. The absence of hyperfine interactions in $O_2^{\cdot-}$ could lead to a more sensitive magnetic compass, but only if the electron spin relaxation of the $O_2^{\cdot-}$ radical is much slower than normally expected for a small mobile radical with an orbitally degenerate electronic ground state. Player and Hore (2019) used spin dynamics simulations to model the sensitivity of a flavin-superoxide radical pair to the direction of a 50 μT magnetic field. By varying parameters that characterize the local environment and molecular dynamics of the radicals, Player and Hore (2019) identified the highly restrictive conditions under which a $O_2^{\cdot-}$-containing radical pair could form the basis of a geomagnetic compass sensor. Player and Hore (2019) concluded that the involvement of superoxide in compass magnetoreception must remain highly speculative until further experimental evidence is forthcoming.

One further hypothesis regards the involvement of the ascorbyl radical characterized by few and small isotropic hyperfine couplings (Evans et al., 2016). In the solution, the flavin/ascorbyl pair demonstrated sensitivity to weak fields much exceeding previously reported effects in other flavin-containing radical pairs, including CRYs (Evans et al., 2016). However, recent molecular dynamics simulations suggest that the brief and infrequent encounters of the ascorbyl radical with CRY make this also an unlikely candidate in the search for an Earth strength field sensor (Nielsen et al., 2017).

4.5 Discussion and Conclusions

A great deal of experimental work on the structure and function of migratory bird CRY (as described above) has been carried out so far. Simply speaking, a photo-pigment CRY could function as a magnetic receptor molecule with a chromophore FAD in the retina (Rodgers and Hore, 2009; Hore and Mouritsen, 2016). However, the exact details are still unknown, because using the above-mentioned RPM-based magnetoreception, there remain unsolved mechanisms/pathways to determine the direction indicating "compass information," and more specifically, the location indicating "map information" of the latitude and longitude coordinates. Here, the plausible basic reaction mechanisms of biological systems and conventional artificial systems are reviewed and discussed. Light-dependent magnetic field effects in vitro have been reported for CRY1 from the plant *Arabidopsis thaliana* (*At*CRY1), and the closely related DNA photolyase from *Escherichia coli* (*Ec*PL) (Goez et al., 2009; Maeda et al., 2012). The magnetic responses of both molecules are explained by the RPM (Hore and Mouritsen, 2016; Rodgers and Hore, 2009), and the *Ec*PL photocycle which also provides a framework for the discussion of *Dm*CRY (Maeda et al., 2012). In the natural flavin system, e.g., *At*CRY1, *Ec*PL, and *Dm*CRY, the singlet radical pair was generated from the flavin-excited singlet state, and then the spin transition from the singlet to triplet states occurred at a rate depending on the strength and angle of the magnetic field (Maeda et al., 2012).

Meanwhile, in the artificial flavin system, the magnetic field effects on the FAD in an aqueous solution have been reported in the process, in which the isoalloxazine ring (flavin ring) and adenine form a stacking structure, and the intramolecular electron transfer occurs from adenine to flavin under acidic conditions (Murakami et al., 2005). That is, one of flavin semiquinone radicals FADH$^\bullet$ has a broad absorption band at 500–730 nm (somewhat environment-dependent), and in particular, the magnetic field effects on the FADH$^\bullet$ showed the transient absorption at 580 nm during photoexcitation of FAD under acidic conditions (Murakami et al., 2005). From this result and the transient absorption of the cation form of flavin^{3}FH$^+$ (exited triplet state) at 650 mn, it is estimated that the triplet-born radical pair formation occurred from the triplet state of flavin by exposure to a magnetic field (Murakami et al., 2005).

Horiuchi et al. (2003) studied the magnetic field effect on the electron transfer reactions from indole derivatives to flavin derivatives in micellar solutions. In this study, the hydrophobic nature of the flavin derivatives and the dynamics of the radical pair were studied by observing the influence of the magnetic field effect on the transient absorption (Horiuchi et al., 2003). The magnetic field effect on the transient absorption is sensitive to the restricted diffusion process (Horiuchi et al., 2003). It is a nice tool for the analysis of the incorporation and the diffusion process of the radical pair generated by the photochemical reaction of flavin and indole derivatives (Horiuchi et al., 2003).

Horiuchi et al. (2003) presented that the magnetic field effects on the transient absorption spectra were observed in the riboflavin-indole system (Figure 4.13). Initially, the T-T absorption band was observed around 650 nm (Horiuchi et al., 2003). Synchronized with the decay of this band, the absorption observed around 500–600 nm is recognized (Horiuchi et al., 2003). This absorption band has been assigned to the neutral radical of riboflavin (Dudley et al., 1964; Müller et al., 1972). When the external magnetic field ($B=0.2$ T) was applied, the absorption changed (Horiuchi et al., 2003). Action spectra of the magnetic field effect (Ali et al., 1997; Murakami et al., 2002) were obtained at various delay times after pulsed laser irradiation by plotting the change of the transient absorption versus monitoring wavelength and are shown in Figure 4.13b (Horiuchi et al., 2003). The action spectra of the magnetic field effect clearly distinguish between the contribution of the radical species and the overlapped spectrum of the triplet excited states (Horiuchi et al., 2003). The observed action spectra of the magnetic field effect are also assigned to the neutral radical of riboflavin (Dudley et al., 1964; Müller et al., 1972).

The magnetic field effect on the free radical yields observed by transient absorption reflected effectively the association of the donor and acceptor molecules with the micelles (Horiuchi et al., 2003). In the riboflavin-indole system, the magnetic field effect increased rapidly with an increasing concentration of sodium dodecyl sulfate (SDS) higher than the critical micellar concentration (Horiuchi et al., 2003). In contrast, in the flavin mononucleotide-indole system, the increase of magnetic field effect was very slow even at higher concentrations of SDS (Horiuchi et al., 2003). This result showed that riboflavin was well associated with the SDS micelle and the diffusion process of the radical pair was restricted by the micellar cages (Horiuchi et al., 2003).

Horiuchi et al. (2003) presented the magnetic field effects on the time profiles observed at 510 nm in the riboflavin-indole system (Figure 4.14). It is difficult to analyze the rising kinetics of the absorption band of the radical species because of the overlap with strong fluorescence and the contribution of T-T absorption. However, subtraction of the time profiles of the magnetic field effect, $A(B=0.2\,\text{T})-A(B=0\,\text{T})$, showed that the magnetic field effect of flavin radical is positive and grows with a similar timescale as the decay of the T-T absorption observed at 690 nm.

The difference in the magnetic field effect is not only due to the association with micelles in the ground state but also due to the dynamic process of the radical pair. The comparison of the rapidly decaying components (within 1 μs) between the riboflavin-indole system (Figure 4.14a) and the

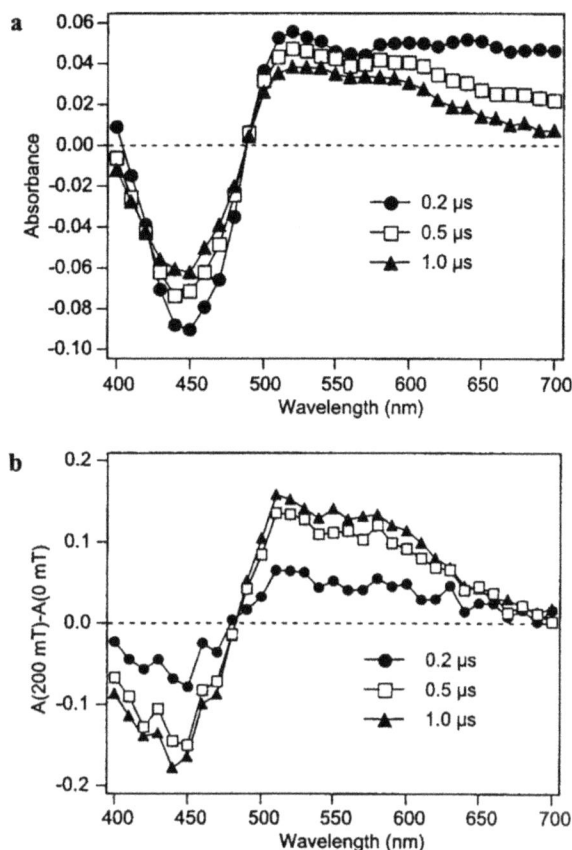

FIGURE 4.13 The magnetic field effects on the transient absorption spectra observed in the riboflavin-indole system (Horiuchi et al., 2003). (a) Time-resolved transient absorption spectra in an SDS micellar solution of ribo-flavin and indole. The concentration of SDS was 50 mM. Other conditions of the experiment are given in the text. (b) Time-resolved action spectra of the magnetic field effect obtained by subtraction. (Reproduced with permission from Horiuchi et al., 2003, Copyright 2003, Springer Nature.)

riboflavin-tryptophan (Trp) system (Figure 4.14b) apparently showed that the efficiency of the geminate recombination is much higher in the riboflavin-indole system. Additionally, the difference in the subtraction of the time profiles is important. In the riboflavin-indole system, the subtraction curve reaches its maximum at $t=0.8$ µs and slightly decays within a few microseconds (Figure 4.14a). This decay of the magnetic field effect is rationalized by recombination of radical pair via spin relaxation from the T_{+1} and T_{-1} states to the S-T_0 mixed spin states. Therefore, this type of slow decay observed in the subtracted curve should be observed in cases where the escaping rate constant of the radical pair, k_{esc}, is comparable to the rate constant of the electron spin relaxation, k_{rlx}. In contrast, such features have not been observed in the riboflavin-Trp system (Figure 4.14b). Thus, the magnetic field effect on the riboflavin-indole system was twice as large as that of riboflavin and Trp (Horiuchi et al., 2003). This result showed the difference in the dynamics of radicals in micelles (Horiuchi et al., 2003). The escape rate of the cation radicals generated from Trp was much faster than that generated from indole (Horiuchi et al., 2003). Horiuchi et al. (2003) tentatively concluded that the difference in the magnetic field effect is due to a large difference in the escape rate of cation radicals generated from indole derivatives.

As an example of another artificial system, the magnetic field effects of the system in the mixture of the hen egg white lysozyme (Trp-containing protein) and the flavin mononucleotide have been reported

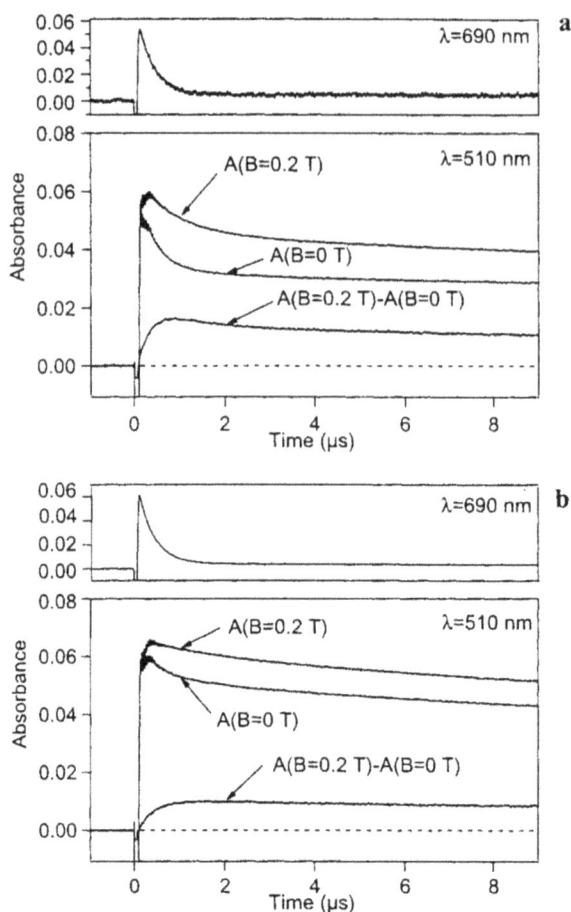

FIGURE 4.14   Time profiles of the transient absorption observed at 690 nm and those observed at 510 nm with and without magnetic field, $A(B=0.2$ T) and $A(B=0$ T), respectively (Horiuchi et al., 2003). The subtractions $A(B=0.2$ T)$-A(B=0$ T) are superimposed. (a) Time profiles were observed in the system of 0.18 mM riboflavin and 1 mM indole. (b) Time profiles observed in the system of 0.18 mM riboflavin and 1 mM tryptophan. (Reproduced with permission from Horiuchi et al., 2003, Copyright 2003, Springer Nature.)

(Miura et al., 2003). In this system, the magnetic field effect of 0.25 T in the mixture of the flavin mononucleotide and Trp as a free amino acid was about 2% of the yield of the free radical formation (Miura et al., 2003). In the riboflavin–hen egg white lysozyme system, the magnetic field effect increased to 5%–7%. These results are assumed to reflect the slower diffusion and higher collision probability in large protein molecules (Miura et al., 2003). In the flavin mononucleotide–hen egg white lysozyme system, the magnetic field effect was up to 13% and decreased rapidly upon the addition of NaCl (Miura et al., 2003). The only difference between flavin mononucleotide and riboflavin is a phosphoric acid group (Miura et al., 2003). Coulombic interaction between the protein surface and the ionic phosphoric acid group at the side chain of the flavin mononucleotide molecule plays an important role in the dynamics of the radical pair (Miura et al., 2003). These results suggest that the calculation of the radical pair dynamics containing molecular dynamics and intermolecular interactions is an interesting subject in the investigation of the chemical kinetics and magnetic spin effects in biological environments (Miura et al., 2003).

As for the methods for detecting the spin dynamics of radical pairs, the measurements of time-resolved EPR or electron spin resonance are used in addition to the transient absorption measurements.

This is a method of directly detecting the spin polarization state (non-Boltzmann distribution), and in the case of the spin dynamics of radical pair, both absorption (*A*) and emission (*E*) signals of opposite phases appear. Furthermore, the singlet-born radical pair from the excited singlet state of the flavin induced by a magnetic field showed the spectral shape of *E/A* spin polarization (emission [*E*] in the low magnetic field side and absorption [*A*] in the high magnetic field side) (Weber et al., 2010). In contrast, triplet-born radical pair from the excited triplet state of the flavin induced by a magnetic field indicated the spectral shape of *A/E* spin polarization (Weber et al., 2010).

In the in vitro biological systems, the magnetic field-induced E/A spin polarization has been detected in *Xenopus laevis* CRY-DASH (*Xl*CRY-DASH) (Biskup et al., 2009). In addition, the magnetic field-induced E/A spin polarization has also been detected in the natural form of CRY-DASH of the cyanobacterium *Synechocystis* sp. (*Scy*CRY-DASH), but in the case of the mutant in which Trp was replaced by phenylalanine (Phe), these signals were not detected for some sequences (Biskup et al., 2011).

The singlet-born radical pair from the excited singlet state of the flavin induced by a magnetic field was observed in natural flavoproteins such as FAD (Biskup et al., 2009, 2011), but not observed in artificial systems. The triplet-born radical pair from the excited triplet state of the flavin induced by a magnetic field was detected in artificial systems (Horiuchi et al., 2003). The distance between Trp and flavin is close enough to the state of the flavin, and the electron transfer reactions from Trp occur in the excited singlet state. In contrast, in the conventional artificial systems, the distance between Trp and flavin is so far, and therefore, the electron transfer reactions from Trp occur greatly in the excited triplet state with a long life.

In the case of the migratory bird CRY, its structure has not been clarified yet. In contrast, the molecular dynamics structures of *Arabidopsis thaliana* cryptochrome 1 (*At*CRY1) have been determined in detail (Brautigam et al., 2004; Huang et al., 2006; Kattnig et al., 2016, Figure 4.15).

Figure 4.15 gives an impression of the positions of the two radicals in *At*CRY1 in the average structure and in the two extreme conformations, $r=r_{min}$ and $r=r_{max}$ (Kattnig et al., 2016). The variation in the radical separation is mainly attributable to the greater mobility of the TrpH$_C^{\bullet+}$ radical compared to the more snugly bound FAD$^{\bullet-}$ radical located in the center of the protein (Kattnig et al., 2016).

Recently, it has been reported that the magnetic compass of migratory birds has a "quantum effect," which may be related to the detection of the magnetic direction required for migration. The quantum effect also could be related to the photosynthesis of green sulfur bacteria, in which multiple pathways responsible for photosynthesis simultaneously achieve different energy states (Lee et al., 2007; Engel et al., 2007; Ishizaki and Fleming, 2009, 2011; Ishizaki et al., 2010; Sarovar et al., 2010; Ishizaki and Fleming, 2011).

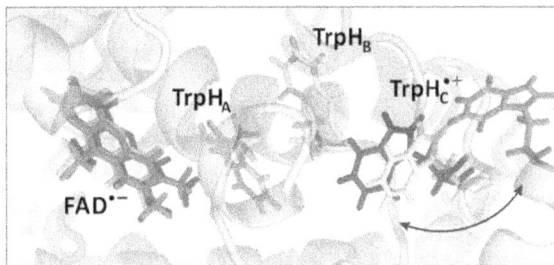

FIGURE 4.15 Molecular dynamics structures of *Arabidopsis thaliana* cryptochrome 1 (*At*CRY1) in the region of the FAD and Trp triad (Kattnig et al., 2016). The positions of the two components of the radical pair (FAD$^{\bullet-}$ and TrpH$_C^{\bullet+}$), the proximal and medial Trp (TrpH$_A$ and TrpH$_B$), and the backbone in the average structure (r_{av}=1.89 nm) are shown. The positions of the two radicals, and the backbone in the region of TrpH$_C^{\bullet+}$, are shown for the structures with, respectively, the largest (r_{max}=2.43 nm) and smallest (r_{min}=1.69 nm) radical separations. The curved arrow indicates the range of motion of TrpH$_C^{\bullet+}$. (Reproduced with permission from Kattnig et al., 2016, Copyright 2016, IOP Publishing.) It is licensed under the Creative Commons Attribution 3.0 International.

Ishizaki and Fleming (2009) have investigated theoretically the spatial and temporal dynamics of the electronic energy transfer through the Fenna–Matthews–Olson (FMO) pigment-protein complex of green sulfur bacteria *Chlorobaculum tepidum* in order to address the robustness and role of the quantum coherence under physiological conditions. They suggested that the excitation energy of bacterio-chlorophyll (BChl) in the FMO complex that works in the first step of photosynthesis is transferred while maintaining a "quantum superposition" that simultaneously realizes conflicting states (Ishizaki and Fleming, 2009). Until then, the photosynthetic electronic energy transfer was thought to classically diffuse between BChls according to the gradient of the energy level generated in the complex, but in reality, the multiple pathways were superposed to simultaneously realize different energy states as a whole (Ishizaki and Fleming, 2009). Consequently, it was clarified that the electronic energy transfer was observed from one BChl to another BChl like a vibrating wave (Ishizaki and Fleming, 2009). The transfer of this excitation energy towards the reaction center occurred with a near unity quantum yield (Ishizaki and Fleming, 2009). The electrons of the BChl were excited to change the energy state, which changed the state of the adjacent BChl and transfers energy (Ishizaki and Fleming, 2009). Thus, the experiments have confirmed that multiple pathways were entangled (Ishizaki and Fleming, 2009). They speculated that the quantum phenomenon might be related to the role of the "rectifier" that controls the flow of energy (Ishizaki and Fleming, 2009).

Ishizaki and Fleming (2009) presented and discussed the numerical results regarding the dynamics of the electronic energy transfer in the FMO complex of *C. tepidum* (Ishizaki and Fleming, 2009). The complex is a trimer made of identical subunits, each containing seven BChls. They clarified that the energy is transferred through two primary transfer pathways in the FMO complex: baseplate → BChls 1 → 2 → 3 → 4 and baseplate → BChls 6 → 5, 7, 4 → 3. Here, seven bacteriochlorophyll (BChl) molecules belonging to the monomeric subunit of the Fenna–Matthews–Olson (FMO) pigment-protein complex (Ishizaki and Fleming, 2009). The complex is oriented with BChl 1 and 6 towards the baseplate protein, whereas BChl 3 and 4 define the target region in contact with the reaction center complex. The spiral strands are α-helices that are part of protein environment. It is reasonable to think that these pathways are not explored each time by quantum computation but are determined by the structure of proteins and the arrangement of BChl molecules (Ishizaki and Fleming, 2009). The energy received by BChl 1 transfers to BChls 2 and 3, but in reality there is almost no energy level difference between BChls 1 and 3, and normally the energy of BChl 3 easily returns to BChl 1 due to thermal fluctuation (Ishizaki and Fleming, 2009). However, in reality, the energy level of BChl 2 is relatively high, and it can overcome local energetic traps, so no backflow occurs (Ishizaki and Fleming, 2009). It may seem that the electronic energy transfer from BChl 1 to BChl 2 does not occur for the same reason, but since BChls 1 and 2 are strongly entangled, such a classical pathway does not occur (Ishizaki and Fleming, 2009). The electronic energy can overcome local energetic traps and transfer to BChl 2 through quantum superposition or entanglement (Ishizaki and Fleming, 2009). On the other hand, the quantum entanglement between BChls 2 and 3 is relatively weak and the pathway is almost classical, so the energy cannot overcome local energetic traps (Ishizaki and Fleming, 2009). In order to elucidate the extremely high efficiency of the light capture system, it will be necessary to consider not only the electronic energy transfer speed but also the rectification mechanism and one-way pathway by the combination of quantum and classical physics (Ishizaki and Fleming, 2009). Thus, they suggested that quantum coherence allows the FMO complex to work as a rectifier for unidirectional energy flow from the chlorosome antenna to the reaction center complex (Ishizaki and Fleming, 2009). It is reasonable to suppose that quantum coherence can overcome local energetic traps to aid the subsequent trapping of electronic energy by the BChl molecules in contact with the reaction center complex (Ishizaki and Fleming, 2009).

These multiple states seem to play a role in the efficient use of light energy. Electrons and pigments in living organisms are always surrounded by macroscopic substances such as surrounding tissues and proteins. Until now, it has been the "common sense" of physics that microscopic substances such as electrons cannot maintain their state even if they are in multiple states once they come into contact with the macroscopic world. However, in reality, quantum multiple states can be realized even with macroscopic substances such as large molecules and crystals. There has been active discussion in the field of physics and biophysics, and special articles have been published (Ball, 2011; Vedral, 2011). This field has evolved into a promising research area called "Quantum Biology" (Ball, 2011; Al-Khalili and McFadden, 2014; Solov'yov et al., 2014).

The possibility has emerged that multiple states such as singlet and triplet states are actually realized in living organisms. It's like a "Schrodinger's bird" (Vedral, 2011). That is, in the eyes of migratory birds, there is a "quantum entanglement," in which the spin becomes zero as a whole, and it is thought that when light is absorbed, the quantum entanglement collapses, and consequently, the GMF can be sensed (Vedral, 2011). Vedral (2011) estimates that the quantum effect in the bird's eye lasts for 100 μs. The duration of the quantum effect is surprisingly long since in the case of the artificial electron system, the duration of the quantum effect is usually shorter than 50 μs (Vedral, 2011). The observation of long-lived electronic quantum coherence in a photosynthetic light-harvesting system (Engel et al., 2007) has led to much effort being devoted to the elucidation of the quantum mechanisms of the photosynthetic excitation energy transfer (Ishizaki and Fleming, 2009; 2011; Ishizaki et al., 2010; Sarovar et al., 2010; Ishizaki and Fleming, 2011). According to Vedral (2011), "We still don't know how natural biological systems sustain such a long-term quantum effect, but if we can elucidate the mechanism in detail, we will develop a method to prevent quantum computers from being destroyed by decoherence." In addition, together with the photosynthesis of green sulfur bacteria, Oka (2015) anticipates that the study of the quantum effect can be expected to provide insights that lead to the solution of energy problems.

What happens if an oscillating magnetic field is applied in addition to the GMF? As shown in Figure 4.6, in actual migratory birds, the magnetic compass in the body does not function normally when another oscillating magnetic field including anthropogenic electromagnetic noise is applied in addition to the GMF (Thalau et al., 2005; Kavokin, 2009; Ritz et al., 2009; Wiltschko et. al., 2011; Engels et al., 2014; Kavokin et al., 2014; Xu et al., 2014a; Hiscock et al., 2017). That is, the magnetic sensor exerts the function of migration in a normal state, but it is considered that the magnetic sensor causes an abnormality in the function in an environment accompanied by an electromagnetic abnormality.

Bojarinova et al. (2020) proposed that the RPM fails to explain the obtained results quantitatively; the observed sensitivity thresholds of the magnetic compass to oscillating magnetic fields in European robins and garden warblers are two orders of magnitude less than what the existing theory would give even with the most liberal choice of parameters (Thalau et al., 2005; Kavokin, 2009; Ritz et al., 2009; Wiltschko et. al., 2011; Engels et al., 2014; Kavokin et al., 2014; Xu et al., 2014a; Hiscock et al., 2017). Their experiments have demonstrated the insensitivity of the bird magnetic compass to oscillating magnetic fields applied locally to the eyes (Bojarinova et al., 2020). Their results do not necessarily deny the key role of radical pair reactions in magnetoreception. However, it definitely disproves the radical pair model, which in particular suggests that: (1) magnetic sensitivity of photochemical reactions in CRY molecules situated in the retina produces a visual image of the GMF, (2) the compass is disrupted by oscillating magnetic fields due to its effect on electron spins in radical pairs formed by the CRY (Bojarinova et al., 2020). If it were so, the magnetic-field-induced image would have been destroyed by the oscillating magnetic fields that we applied to the central part of the retina. Their findings, therefore, point out the existence of other, so far unknown, components of the avian magnetoreception system (Bojarinova et al., 2020).

The light-dependent magnetic compass sense of the night-migratory European robin is thought to rely on magnetically sensitive chemical reactions of radical pairs in cryptochromes (*Er*CRYs) located in the birds' eyes (Ritz et al., 2000; Zapka et al., 2009; Hore and Mouritsen, 2016; Xu et al., 2021). Ren et al. (2021) have investigated the angular precision of the light-dependent compass sense of the night-migratory European robin using the information theory approach devised by Hiscock et al. (2019). The information theory approach was developed that provides a strict lower bound on the precision with which a bird could estimate its head direction using only geomagnetic cues and a CRY-based radical pair sensor (Hiscock et al., 2019). The leading hypothesis is that the GMF can alter the course of photochemical reactions in the birds' eyes even though the energies involved are a million times smaller than the thermal energy, $k_B T$ (Hiscock et al., 2019). By means of this lower bound, Ren et al. (2021) showed how the performance of the compass sense could be optimized by adjusting the orientation of CRY molecules within photoreceptor cells, the distribution of cells around the retina, and the effects of the GMF on the photochemistry of the radical pair. Their results indicated how the precision of this compass could be optimized to make the best use of the relatively small number of photons available to these nocturnal migrants (Ren et al., 2021).

Xu et al. (2021) speculated that CRY4 molecules are not expected to show substantial responses to the GMF strength in vitro unless they are anchored, aligned, and associated with the appropriate signaling partners. None of the proteins required for these interactions is currently known (Wu et al., 2020). Furthermore, magnetic sensitivity could be enhanced by receptor alignment (Efimova and Hore, 2008), biochemical amplification (Wu et al., 2020), and neural processing (Zapka, 2009). These aspects can also be optimized by evolution and could substantially improve the magnetic sensitivity of the robin's magnetic sense (Xu et al., 2021).

Xu et al. (2021) showed that the photochemistry of CRY4 from the night-migratory European robin (*Erithacus rubecula*) (*ErCRY4*) is magnetically sensitive in vitro, and more so than CRY4 from two non-migratory bird species, chicken (*Gallus gallus*) and pigeon (*Columba livia*). Site-specific mutations of *ErCRY4*, which is a protein that is expressed in double-cone and long-wavelength single-cone photoreceptor cells in the eyes of night-migratory European robins (Günther et al., 2018), revealed the roles of four successive flavin-tryptophan (Trp) radical pairs in generating magnetic field effects and in stabilizing potential signaling states in a way that could enable sensing and signaling functions to be independently optimized in night-migratory birds (Xu et al., 2021). To determine whether CRY4 acts as a magnetoreceptor molecule in vivo, direct manipulations of this protein in the eyes of night-migratory songbirds would be required (Xu et al., 2021).

To explain the biological effects of weak magnetic fields, some molecular transduction mechanisms have been proposed (Binhi and Prato, 2018; Bialas et al., 2019). While for animal navigation/orientation, the main hypothesis is a specialized magnetic sense associated with pairs of radicals located in the retina of the eye, nonspecific effects could occur due to the interaction of magnetic fields with the magnetic moments of rotating molecules dispersed in the organism. Indeed, Binhi and Prato (2018) have shown that the precession of the magnetic moments of these rotating molecules can be slowed due to a mixing of the quantum levels of magnetic moments called the "Level Mixing Mechanism (LMM)" inducing a magnetic field dependence that is in good agreement with experiments in which biological effects arise in response to the reversal of the magnetic field vector.

The RPM-mediated magnetic field effects emerge from the anisotropic hyperfine interactions between the radicals' electron spins and associated nuclei (Schulten et al., 1978; Ritz et al., 2000). These interactions can govern the spin dynamics when inter-radical interactions, such as the dipolar and exchange interactions, are small (Hiscock et al., 2016b) or mutually balanced (Efimova and Hore, 2008). Radical-radical couplings generally inhibit magnetosensitivity at low fields by lifting the zero-field degeneracy of singlet and triplet states, thereby impeding field-dependent singlet–triplet conversion, and also by inducing spin relaxation (Timmel et al., 1998; Kattnig et al., 2016).

Furthermore, recent calculations showed that electron-electron dipolar (EED) interactions can abolish the "quantum needle," a sharp feature in the directional magnetic field effect that was predicted to boost the acuity of the compass (Hiscock et al., 2016b, 2017), and may nullify the Larmor resonance (Hiscock et al., 2016a, 2017), which is a phenomenon observed in some behavioral studies employing radiofrequency (RF) magnetic fields to test for the RPM (Ritz et al., 2009). Nonetheless, the majority of past theoretical works on CRY-mediated magnetoreception have omitted EED coupling to facilitate calculations on spin systems too large to be treated otherwise, instead of focusing on hyperfine-induced effects as the sine qua non of low field magnetic field effects (Hiscock et al., 2016b; Atkins et al., 2019).

Thus, prior studies have often neglected the EED coupling from this hypothesis. By simulating the radical pair models, Babcock and Kattnig (2020) showed that EED interactions suppress the anisotropic response to the GMF by the RPM in CRY, and that this attenuation is unlikely to be mitigated by the mutual cancellation of the EED and electronic exchange coupling, as previously suggested. Babcock and Kattnig (2020) then demonstrated that this limitation may be overcome by extending the conventional model to include a third, non-reacting radical. They predicted that hyperfine effects could work in concert with three-radical dipolar interactions to tailor a superior magnetic response, thereby providing a new principle for magnetosensitivity with applications for sensing, navigation, and the assessment of biological magnetic field effects (Babcock and Kattnig, 2020).

In recent years, it is known that electromagnetic phenomena can be observed with earthquakes and that electromagnetic noise is generated in a wide frequency band as one of the earthquake precursors (reviewed by Conti et al., 2021). It is not clear whether the magnetic sensor of the animal is involved in the anomalous phenomenon, but the possibility cannot be ruled out because abnormal animal behaviors and physiological conditions have been observed before the large earthquake (Ikeya et al., 1996; Yokoi et al., 2003; Garstang, 2009; Li et al., 2009; Yanai et al., 2012; Hayakawa, 2013; Whitehead and Ulusoy, 2013; Fidani et al., 2014; Yamauchi et al., 2014, 2017; Orihara et al., 2019). However, informative critical reviews on abnormal animal behaviors for the earthquake precursor have been published (Woith et al., 2018; Conti et al., 2021). Woith et al. (2018) reviewed 180 publications regarding abnormal animal behavior before earthquakes and analyze and discuss them with respect to (1) magnitude–distance relations, (2) foreshock activity, and (3) the quality and length of the published observations. More than 700 records of claimed animal precursors related to 160 earthquakes are reviewed with unusual behavior of more than 130 species. The precursor time ranges from months to seconds prior to the earthquakes, and the distances from a few to hundreds of kilometers. However, only 14-time series were published, whereas all other records are single observations. The time series are often short (the longest is 1 year), or only small excerpts of the full data set are shown. The probability density of foreshocks and the occurrence of animal precursors are strikingly similar, suggesting that at least parts of the reported animal precursors are in fact related to foreshocks. Another major difficulty for a systematic and statistical analysis is the high diversity of data, which are often only anecdotal and retrospective. Thus, the review of Woith et al. (2018) clearly demonstrated strong weaknesses or even deficits in many of the published reports on possible abnormal animal behavior. Conti et al. (2021) reported that animal magnetic sensitivity that could be even greater than the instrumental sensitivity of measurements made in various test campaigns to study earthquake precursors would cut across so many different species that they do not share other important sensory characteristics. Conti et al. (2021) proposed that systematic monitoring campaigns with continuous bio-logging of animal collectives, including movements and physiological parameters, could yield valuable insights on this topic.

If the model can be constructed by imitating the high-sensitivity magnetic sensor of migratory birds, it can be expected that new innovations can be made in predicting and measuring environmental changes accompanied by magnetic anomalies including earthquakes (Oka, 2015). Basic research related to magnetic sensing of migratory birds has not been clarified yet and has been the subject of debate in recent years, especially in the field of physics and biophysics, and it is considered to have high potential as innovative and applied research. In this way, it is expected that an environment-friendly lifestyle and system will be created by learning from the environment and life on Earth, as well as by utilizing the knowledge in various research fields.

References

Agliassa, C., Narayana, R., Bertea, C.M., Rodgers, C., and Maffei, M.E. 2018a. Reduction of the geomagnetic field delays *Arabidopsis thaliana* flowering time through downregulation of flowering-related genes. *Bioelectromagnetics* **39**(5):361–374.

Agliassa, C., Narayana, R., Christie, J.M., and Maffei, M.E. 2018b. Geomagnetic field impacts on cryptochrome and phytochrome signaling. *J Photochem Photobiol B Biol* **185**:32–40.

Ahmad, M., and Cashmore, A.R. 1993. HY4 gene of *A. thaliana* encodes a protein with characteristics of a blue-light photoreceptor. *Nature* **366**(6451):162–166.

Ahmad, M., Galland, P., Ritz, T., Wiltschko, R., and Wiltschko, W. 2007. Magnetic intensity affects cryptochrome-dependent responses in *Arabidopsis thaliana*. *Planta* **225**:615–624.

Al-Khalili, J., and McFadden, J. 2014. *Life on the Edge: The Coming of Age of Quantum Biology*. Bantam Press, London, UK.

Alexander, Y.R., Roman, V.C., Anna, A., Kirill, V.K., Nikita, C., Michael, L.F., and Luba, A.A. 2020. Searching for magnetic compass mechanism in pigeon retinal photoreceptors. *PLoS One* **15**(3):e0229142.

Ali, S.S., Maeda, K., Murai, H., and Azumi, T. 1997. Surprisingly large magnetic field effect in the electron-transfer reaction of 4,4'-bipyridine with triethylamine in acetonitrile. *Chem Phys Lett* **267**(5–6):520–524.

Antill, L.M., and Woodward, J.R. 2018. Flavin adenine dinucleotide photochemistry is magnetic field sensitive at physiological pH. *J Phys Chem Lett* **9**:2691–2696.

Atkins, C., Bajpai, K., Rumball, J., and Kattnig, D.R. 2019. On the optimal relative orientation of radicals in the cryptochrome magnetic compass. *J Chem Phys* **151**:065103.

Babcock, N.S., and Kattnig, D.R. 2020, Electron-electron dipolar interaction poses a challenge to the radical pair mechanism of magnetoreception. *J Phys Chem Lett* **11**(7):2414–2421.

Ball, P. 2011. Physics of life: the dawn of quantum biology. *Nature* **474**(7351):272–274.

Banerjee, R., Schleicher, E., Meier, S., Viana, R.M., Pokorny, R., Ahmad, M., Bittl, R., and Batschauer, A. 2007. The signaling state of *Arabidopsis* cryptochrome 2 contains flavin semiquinone. *J Biol Chem* **282**(20):14916–14922.

Barnes, F.S., and Greenebaum, B. 2015. The effects of weak magnetic fields on radical pairs. *Bioelectromagnetics* **36**(1):45–54.

Barnes, F.S., and Greenebaum, B. 2016. Some effects of weak magnetic fields on biological systems: RF fields can change radical concentrations and cancer cell growth rates. *IEEE Power Electron magaz* 60–68.

Batchelor, S.N., Kay, C.W.M., McLauchlan, K.A., and Shkrob, I.A. 1993. Time-resolved and modulation methods in the study of the effects of magnetic fields on the yields of free-radical reactions. *J Phys Chem* **97**:13250–13258.

Beason, R.C., and Semm, P. 1987. Magnetic responses of the trigeminal nerve system of the bobolink (*Dolichonyx oryzivorus*). *Neurosci Lett* **80**:229–234.

Berndt, A., Kottke, T., Breitkreuz, H., Dvorsky, R., Hennig, S., Alexander, M., and Wolf, E. 2007. A novel photoreaction mechanism for the circadian blue-light photoreceptor *Drosophila* cryptochrome. *J Biol Chem* **282**(17):13011–13021.

Bertea, C.M., Narayana, R., Agliassa, C., Rodgers, C.T., and Maffei, M.E. 2015. Geomagnetic field (Gmf) and plant evolution: investigating the effects of Gmf reversal on *Arabidopsis thaliana* development and gene expression. *J Vis Exp* **105**:53286.

Bialas, C., Barnard, D.T., Auman, D.B., McBride, R.A., Jarocha, L.E., Hore, P.J., Dutton, P.L., Stanley, R.J., and Moser, C.C. 2019. Ultrafast flavin/tryptophan radical pair kinetics in a magnetically sensitive artificial protein. *Phys Chem Chem Phys* **21**(25):13453–13461.

Binhi, V.N., and Prato, F.S. 2018. Rotations of macromolecules affect nonspecific biological responses to magnetic fields. *Sci Rep* **8**(1):13495.

Biskup, T., Hitomi, K., Getzoff, E.D., Krapf, S., Koslowski, T., Schleicher, E., and Weber, S. 2011. Unexpected electron transfer in cryptochrome identified by time-resolved EPR spectroscopy. *Angew Chem Int Ed Engl* **50**(52):12647–12651.

Biskup, T., Schleicher, E., Okafuji, A., Link, G., Hitomi, K., Getzoff, E.D., and Weber, S. 2009. Direct observation of a photoinduced radical pair in a cryptochrome blue-light photoreceptor. *Angew Chem Internat Ed* **48**(2):404–407.

Blankenship, R.E., Schaafsma, T.J., and Parson, W.W. 1977. Magnetic field effects on radical pair intermediates in bacterial photosynthesis. *Biochem Biophys Acta* **461**(2):297–305.

Bojarinova, J., Kavokin, K., Pakhomov, A., Cherbunin, R., Anashina, A., Erokhina, M., Ershova, M., and Chernetsov, N. 2020. Magnetic compass of garden warblers is not affected by oscillating magnetic fields applied to their eyes. *Sci Rep* **10**(1):3473.

Bolte, P., Bleibaum, F., Einwich, A., Günther, A., Liedvogel, M., Heyers, D., Depping, A., Wöhlbrand, L., Rabus, R., Janssen-Bienhold, U., and Mouritsen, H. 2016. Localisation of the putative magnetoreceptive protein Cryptochrome 1b in the retinae of migratory birds and homing pigeons. *PLoS One* **11**(3):e0147819.

Bolte, P., Einwich, A., Seth, P.K., Chetverikova, R., Heyers, D., Wojahn, I., Janssen-Bienhold, U., Federle, R., Hore, P., Dedek, K., and Mouritsen, H. 2021. Cryptochrome 1a localisation in light- and dark-adapted retinae od several migratory and non-migratory birds species: no signs of light-dependent activation. *Ethol Ecol Evol* **33**(3):248–272.

Bovey, F.A. 1988. *Nuclear Magnetic Resonance Spectroscopy*. 2nd Edition. Academic Press Inc, San Diego, CA.

Bradlaugh, A., Munro, A., Jones, A.R., and Baines, R. 2021. Exploiting the fruitfly, *Drosophila melanogaster* to identify the molecular basis of cryptochrome-dependent magnetosensitivity. *Quant Rep* **3**(1):127–136.

Brautigam, C.A., Smith, B.S., Ma, Z., Palnitkar, M., Tomchick, D.R., Machius, M., and Deisenhofer, J. 2004. Structure of the photolyase-like domain of cryptochrome 1 from *Arabidopsis thaliana*. *Proc Natl Acad Sci USA* **101**(33):12142–1247.

Brocklehurst, B. 1976. Magnetic field effect on the pulse shape of scintillations due to geminate recombination of ion pairs. *Chem Phys Lett* **44**:245–248.

Brocklehurst, B. 2002. Magnetic fields and radical reactions: recent developments and their role in nature. *Chem Soc Rev* **31**(5):301–311.

Brocklehurst, B., and McLauchlan, K.A. 1996. Free radical mechanism for the effects of environmental electromagnetic fields on biological systems. *Int J Radiat Biol* **69**(1):3–23.

Brudler, R., Hitomi, K., Daiyasu, H., Toh, H., Kucho, K., Ishiura, M., Kanehisa, M., Roberts, V.A., Todo, T., Tainer, J.A., and Getzoff, E.D. 2003. Identification of a new cryptochrome class: structure, function, and evolution. *Mol Cell* **11**(1):59–67.

Cashmore, A.R. 2003. Cryptochromes: enabling plants and animals to determine circadian time. *Cell* **114**(5):537–543.

Castello, P., Jimenez, P., and Martino, C.F. 2021. The role of pulsed electromagnetic fields on the radical pair mechanism. *Bioelectromagnetics* **42**(6):491–500.

Chen, J.R., and Ke, S.C. 2018. Magnetic field effects on coenzyme B_{12}- and B_6-dependent lysine 5,6-aminomutase: switching of the J-resonance through a kinetically competent radical-pair intermediate. *Phys Chem Chem Phys* **20**(18):13068–13074.

Cintolesi, F., Ritz, T., Kay, C.W.M., Timmel, C.R., and Hore, P.J. 2003. Anisotropic recombination of an immobilized photoinduced radical pair in a 50-μT magnetic field:A model avian photomagnetoreceptor. *Chem Phys* **294**:385–399.

Conti, L., Picozza, P., and Sotgiu, A. 2021. A critical review of ground based observations of earthquake precursors. *Front Earth Sci* **9**:676766.

Cranfield, J., Belford, R., Debrunner, P., and Schulten, K. 1994. A perturbation treatment of oscillating magnetic fields in the radical pair mechanism. *Chem Phys* **182**:1–18.

Damiani, M.J., Nostedt, J.J., and O'Neill, M.A. 2010. Impact of the N5-proximal Asn on the thermodynamic and kinetic stability of the semiquinone radical in photolyase. *J Biol Chem* **286**(6): 4382–4391.

Denzau, S., Nießner, C., Rogers, L.J., and Wiltschko, W. 2013a. Ontogenetic development of magnetic compass orientation in domestic chickens (*Gallus gallus*). *J Exp Biol* **216**:3143–3147.

Denzau, S., Nießner, C., Wiltschko, R., and Wiltschko, W. 2013b. Different responses of two strains of chickens to different training procedures for magnetic directions. *Anim Cogn* **16**(3):395–403.

Di Valentin, M., Bisol, A., Agostini, G., and Carbonera, D. 2005. Electronic coupling effects on photo-induced electron transfer in carotene-porphyrin-fullerene triads detected by time-resolved EPR. *J Chem Inf Model* **45**:1580–1588.

Dodson, C.A., Hore, P.J., and Wallace, M.I. 2013. A radical sense of direction: signalling and mechanism in cryptochrome magnetoreception. *Trends Biochem Sci* **38**:435–446.

Du, X.L., Wang, J., Pan, W. S., Liu, Q. J., Wang, X.J., and Wu, W.J. 2014. Observation of magnetic field effects on transient fluorescence spectra of cryptochrome 1 from homing pigeons. *Photochem Photobiol* **90**:989–996.

Dudley, K. H., Ehrenberg, A., Hemmerich, P., and Müller, F. 1964. Spektren und Strukturen der am Flavin-Redox-System beteiligten Partikeln. *Helv Chim Acta* **47**:1354–1383.

Efimova, O., and Hore, P.J. 2008. Role of exchange and dipolar interactions in the radical pair model of the avian magnetic compass. *Biophys J* **94**(5):1565–1574.

Emery, P., So, W.V., Kaneko, M., Hall, J.C., and Rosbash, M. 1998. CRY, a *Drosophila* clock and light-regulated cryptochrome, is a major contributor to circadian rhythm resetting and photosensitivity. *Cell* **95**(5):669–679.

Engel, G.S., Calhoun, T.R., Read, E.L., Ahn, T.K., Mancal, T., Cheng, Y.C., Blankenship, R.E., and Fleming, G.R. 2007. Evidence for wavelike energy transfer through quantum coherence in photosynthetic systems. *Nature* **446**(7137):782–786.

Engels, S., Schneider, N.L., Lefeldt, N., Hein, C.M., Zapka, M., Michalik, A., Elbers, D., Kittel, A., Hore, P.J., and Mouritsen, H. 2014. Anthropogenic electromagnetic noise disrupts magnetic compass orientation in a migratory bird. *Nature* **509**(7500):353–356.

Evans, E.W., Dodson, C.A., Maeda, K., Biskup, T., Wedge, C.J., and Timmel, C.R. 2013. Magnetic field effects in flavoproteins and related systems. *Interface Focus* **3**(5):20130037.

Evans, E.W., Kattnig, D.R., Henbest, K.B., Hore, P.J., Mackenzie, S.R., and Timmel, C.R. 2016. Sub-millitesla magnetic field effects on the recombination reaction of flavin and ascorbic acid radicals. *J Chem Phys* **145**(8):085101.

Falkenberg, G., Fleissner, G., Schuchardt, K., Kuehbacher, M., Thalau, P., Mouritsen, H., Heyers, D., Wellenreuther, G., and Fleissner, G. 2010. Avian magnetoreception: elaborate iron mineral containing dendrites in the upper beak seem to be a common feature of birds. *PLoS One* **5**(2):e9231.

Fidani, C., Freund, F., and Grant, R. 2014. Cows come down from the mountains before the (M_w=6.1) earthquake colfiorito in September 1997; a single case study. *Animals (Basel)* **4**(2):292–312.

Fleissner, G., Holtkamp-Rotzler, E., Hanzlik, M., Winklhofer, M., Fleissner, G., Petersen, N., and Wiltschko, W. 2003. Ultrastructural analysis of a putative magnetoreceptor in the beak of homing pigeons. *J Comp Neurol* **458**(4):350–360.

Fleissner, G., Stahl, B., Thalau, P., Falkenberg, G., and Fleissner, G. 2007. A novel concept of Fe-mineral-based magnetoreception: histological and physicochemical data from the upper beak of homing pigeons. *Naturwissenschaften* **94**(8):631–642.

Fogle, K.J., Baik, L.S., Houl, J.H., Tran, T.T., Roberts, L., Dahm, N.A., Cao, Y., Zhou, M., and Holmes, T.C. 2015. Cryptochrome-mediated phototransduction by modulation of the potassium ion channel beta-subunit redox sensor. *Proc Natl Acad Sci USA* **112**:2245–2250.

Freire, R., Munro, U., Rogers, L.J., Sagasser, S., Wiltschko, R., and Wiltschko, W. 2008. Different responses in two strains of chickens (*Gallus gallus*) in a magnetic orientation test. *Anim Cogn* **11**(3):547–552.

Freire, R., Munro, U.H., Rogers, L.J., Wiltschko, R., and Wiltschko, W. 2005. Chickens orient using a magnetic compass. *Curr Biol* **15**:R620–R621.

Garstang, M. 2009. Precursor tsunami signals detected by elephants. *Open Conservat Biol J* **3**:1–3.

Gegear, R.J., Casselman, A., Waddell, S., and Reppert, S. M. 2008. Cryptochrome mediates light-dependent magnetosensitivity in *Drosophila*. *Nature* **454**(7207):1014–1018.

Giachello, C.N.G., Scrutton, N.S., Jones, A.R., and Baines, R.A. 2016. Magnetic fields modulate blue-light-dependent regulation of neuronal firing by cryptochrome. *J Neurosci* **36**:10742–10749.

Gindt, Y.M., Vollenbroek, E., Westphal, K., Sackett, H., Sancar, A., and Babcock, G.T. 1999. Origin of the transient electron paramagnetic resonance signals in DNA photolyase. *Biochemistry* **38**(13): 3857–3866.

Giovani, B., Byrdin, M., Ahmad, M., and Brettel, K. 2003. Light-induced electron transfer in a cryptochrome blue-light photoreceptor. *Nat Struct Biol* **10**(6):489–490.

Goez, M., Henbest, K.B., Windham, E.G., Maeda, K., and Timmel, C.R. 2009. Quenching mechanisms and diffusional pathways in micellar systems unravelled by time-resolved magnetic-field effects. *Chem Eur J* **15**(24):6058–6064.

Grissom, C.B. 1995. Magnetic field effects in biology: a survey of possible mechanisms with emphasis on radical-pair recombination. *Chem Rev* **95**(1):3–24.

Günther, A., Einwich, A., Sjulstok, E., Feederle, R., Bolte, P., Koch, K.W., Solov'yov, I.A., and Mouritsen, H. 2018. Double-cone localisation and seasonal expression pattern suggest a role in magnetoreception for European robin cryptochrome 4. *Curr Biol* **28**(2):211–223.e4.

Hammad, M., Albaqami, M., Pooam, M., Kernevez, E., Witczak, J., Ritz, T., Martino, C., and Ahmad, M. 2020. Cryptochrome mediated magnetic sensitivity in *Arabidopsis* occurs independently of light-induced electron transfer to the flavin. *Photochem Photobiol Sci* **19**(3):341–352.

Harkins, T.T., and Grissom, C.B. 1994. Magnetic field effects on B_{12} ethanolamine ammonia lyase: evidence for a radical mechanism. *Science* **263**(5149):958–960.

Harkins, T.T., and Grissom, C.B. 1995. The magnetic field dependent step in B_{12} ethanolamine ammonia lyase is radical-pair recombination. *J Am Chem Soc* **117**(1):566–567.

Hata, N. 1976. The effect of external magnetic field on the photochemical reaction of isoquinoline N-oxide. *Chem Lett* 547–550.

Hata, N. 1978. The effect of external magnetic field on the photochemical reaction of isoquinoline N-oxide and 2-cyanoquinoline N-oxide. *Chem Lett* **7**(12):1359–1362.

Hata, N. 1985. The photochemical magnetic field effect on isoquinoline N-oxide. *Bull Chem Soc Japan (BCSJ)* **58**(4):1088–1093.

Hata, N. 1986. Photochemical magnetic-field effects of isoquinoline N-oxide in various alcohols. *Bull Chem Soc Japan (BCSJ)* **59**(9):2723–2728.

Hata, N., and Nishida, N. 1985. Photochemical magnetic field effects of 4-methyl-2-quinolinecarbonitrile. *Bull Chem Soc Japan (BCSJ)* **58**(12):3423–3430.

Hata, N., Ono, Y., and Nakagawa, F. 1979. The effect of external magnetic field on the photochemical reaction of isoquinoline N-oxide in various alcohols. *Chem Lett* **8**(5):603–606.

Hata, N., and Yagi, A. 1983. Radical ion pair mechanism of the photochemical isomerization of isoquinoline N-oxide in hydroxylic solvents, including the magnetic field effect. *Chem Lett* **12**(3):309–312.

Hayakawa, M. 2013. Possible electromagnetic effects on abnormal animal behavior before an earthquake. *Animals (Basel)* **3**(1):19–32.

Hayashi, H. 1982. The magnetic field effects on chmical reactions (written in Japanese). *J Phys Soc Jpn (JPSJ)* **37**(10):853–855.

Hayashi, H. 2004. *Introduction to Dynamic Spin Chemistry: Magnetic Field Effects on Chemical and Biochemical Reactions*. World Scientific Publishing Co., Singapore, p. 21.

Hayashi, H., and Nagakura, S. 1978. The theoretical study of external magnetic field effect on chemical reactions in solution. *Bull Chem Soc Jpn* **51**(10):2862–2866.

Hayashi, H., Itoh, K., and Nagakura, S. 1966. The determination of singlet-triplet separation from the anomalous hyperfine structure observed with a radical pair. *Bull Chem Soc Japan (BCSJ)* **39**:199.

Henbest, K.B., Kukura, P., Rodgers, C.T., Hore, P.J., and Timmel, C.R. 2004. Radio frequency magnetic field effects on a radical recombination reaction: a diagnostic test for the radical pair mechanism. *J Am Chem Soc* **126**:8102–8103.

Henbest, K.B., Maeda, K., Hore, P.J., Joshi, M., Bacher, A., Bittl, R., Weber, S., Timmel, C.R., and Schleicher, E. 2008. Magnetic-field effect on the photoactivation reaction of *Escherichia coli* DNA photolyase. *Proc Natl Acad Sci USA* **105**:14395–14399.

Heyers, D., Manns, M., Luksch, H., Güntürkün, O., and Mouritsen, H. 2007. A visual pathway links brain structures active during magnetic compass orientation in migratory birds. *PLoS One* **2**:e937.

Heyers, D., Zapka, M., Hoffmeister, M., Wild, J.M., and Mouritsen, H. 2010. Magnetic field changes activate the trigeminal brainstem complex in a migratory bird. *Proc Natl Acad Sci USA* **107**(20):9394–9399.

Hiscock, H.G., Hiscock, T.W., Kattnig, D.R., Scrivener, T., Lewis, A.M., Manolopoulos, D.E., and Hore, P.J. 2019. Navigating at night:fundamental limits on the sensitivity of radical pair magnetoreception under dim light. *Q Rev Biophys* **52**:e9.

Hiscock, H.G., Kattnig, D.R., Manolopoulos, D.E., and Hore, P.J. 2016a. Floquet theory of radical pairs in radiofrequency magnetic fields. *J Chem Phys* **145**:124117.

Hiscock, H.G., Mouritsen, H., Manolopoulos, D.E., and Hore, P.J. 2017. Disruption of magnetic compass orientation in migratory birds by radiofrequency electromagnetic fields. *Biophys J* **113**(7):1475–1484.

Hiscock, H.G., Worster, S., Kattnig, D.R., Steers, C., Jin Y., Manolopoulos, D.E., Mouritsen, H., and Hore, P.J. 2016b. The quantum needle of the avian magnetic compass. *Proc Natl Acad Sci USA* **113**(17):4634–4639.

Hochstoeger, T., Al Said, T., Maestre, D., Walter, F., Vilceanu, A., Pedron, M., Cushion, T.D., Snider, W., Nimpf, S., Nordmann, G.C., Landler, L., Edelman, N., Kruppa, L., Durnberger, G., Mechtler, K., Schuechner, S., and Ogris, E. 2020. The biophysical, molecular, and anatomical landscape of pigeon CRY4: a candidate light-based quantal magnetosensor. *Sci Adv* **6**(33):eabb9110.

Hoff, A.J., Rademaker, H., Van Grondelle, R., and Duysens, L.N.M. 1977. On the magnetic field dependence of the yield of the triplet state in reaction centers of photosynthetic bacteria. *Biochem Biophys Acta* **460**(3):547–554.

Hogben, H.J., Efimova, O., Wagner-Rundell, N., Timmel, C.R., and Hore, P.J. 2009. Possible involvement of superoxide and dioxygen with cryptochrome in avian magnetoreception: origin of Zeeman resonances observed by in vivo EPR spectroscopy. *Chem Phys Lett* **480**(1–3):118–122.

Hore, P.J. 2019. Upper bound on the biological effects of 50/60 Hz magnetic fields mediated by radical pairs. *eLife* **8**:e44179.

Hore, P.J., and Mouritsen, H. 2016. The radical-pair mechanism of magnetoreception. *Annu Rev Biophys* **45**:299–344.

Horiuchi, M., Maeda, K., and Arai, T. 2003. Magnetic field effect on electron transfer reactions of flavin derivatives associated with micelles. *Appl Magn Reson* **23**:309–318.

Huang, Y., Baxter, R., Smith, B.S., Partch, C.L., Colbert, C.L., and Deisenhofer, J. 2006. Crystal structure of cryptochrome 3 from *Arabidopsis thaliana* and its implications for photolyase activity. *Proc Natl Acad Sci USA* **103**(47):17701–17706.

Ikeya, M., Furuta, H., Kajiwara, N., and Anzai, H. 1996. Ground electric field effects on rats and sparrows: seismic anomalous animal behaviors (SAABs). *Jpn J Appl Phys* **35**(8):4587–4594.

Ishizaki, A., Calhoun, T.R., Schlau-Cohen, G.S., and Fleming, G.R. 2010. Quantum coherence and its interplay with protein environments in photosynthetic electronic energy transfer. *Phys Chem Chem Phys* **12**(27):7319–7337.

Ishizaki, A., and Fleming, G.R. 2009. Theoretical examination of quantum coherence in a photosynthetic system at physiological temperature. *Proc Natl Acad Sci USA* **106**(41):17255–17260.

Ishizaki, A., and Fleming, G.R. 2011. On the interpretation of quantum coherent beats observed in two-dimensional electronic spectra of photosynthetic light harvesting complexes. *J Phys Chem B* **115**(19):6227–6233.

Itoh, K., Hayashi, H., and Nagakura, S. 1969. Determination of singlet-triplet separation of a weakly interacting radical pair from ESR spectrum. *Mol Phys* **17**:561–577.

Iwata, T., Zhang, Y., Hitomi, K., Getzoff, E.D., and Kandori, H. 2010. Key dynamics of conserved asparagine in a cryptochrome/photolyase family protein by fourier transform infrared spectroscopy. *Biochemistry* **49**(41):8882–8891.

Kavokin, K. 2009. The puzzle of magnetic resonance effect on the magnetic compass of migratory birds. *Bioelectromagnetics* **30**(5):402–410,

Kavokin, K., Chernetsov, N., Pakhomov, A., Bojarinova, J., Kobylkov, D., and Namozov, B. 2014. Magnetic orientation of garden warblers (*Sylvia borin*) under 1.4 MHz radiofrequency magnetic field. *J R Soc Interface* **11**(97):20140451.

Kobayashi, Y., Ishikawa, T., Hirayama, J., Daiyasu, H., Kanai, S., Toh, H., Fukuda, I., Tsujimura, T., Terada, N., Kamei, Y., Yuba, S., Iwai, S., and Todo, T. 2000. Molecular analysis of zebrafish photolyase/cryptochrome family: two types of cryptochromes present in zebrafish. *Genes Cells* **5**:725–738.

Kubo, Y., Akiyama, M., Fukada, Y., and Okano, T. 2006. Molecular cloning, mRNA expression, and immunocytochemical localization of a putative blue-light photoreceptor CRY4 in the chicken pineal gland. *J Neurochem* **97**:1155–1165.

Kaptein, R. 1972, Chemically induced dynamic nuclear polarization. VIII. Spin dynamics and diffusion of radical pairs. *J Am Chem Soc* **94**(18):6251–6262.

Kattnig, D.R., and Hore, P.J. 2017. The sensitivity of a radical pair compass magnetoreceptor can be significantly amplified by radical scavengers. *Sci Rep* **7**(1):11640.

Kattnig, D.R., Sowa, J.K., Solov'Yov, I.A., and Hore, P.J. 2016. Electron spin relaxation can enhance the performance of a cryptochrome-based magnetic compass sensor. *New J Phys* **18**:063007.

Kerpal, C., Richert, S., Storey, J.G., Pillai, S., Liddell, P.A., Gust, D., Mackenzie, S.R., Hore, P.J., and Timmel, C.R. 2019. Chemical compass behaviour at microtesla magnetic fields strengthens the radical pair hypothesis of avian magnetoreception. *Nat Commun* **10**(1):3707.

Kiontke, S., Göbel, T., Brych, A., and Batschauer, A. 2020. DASH-type cryptochromes-solved and open questions. *Biol Chem* **401**(12):1487–1493.

Klar, T., Pokorny, R., Moldt, J., Batschauer, A., and Essen, L.O. 2007. Cryptochrome 3 from *Arabidopsis thaliana*: structural and functional analysis of its complex with a folate light antenna. *J Mol Biol* **366**(3):954–964.

Kleine, T., Lockhart, P., and Batschauer, A. 2003. An *Arabidopsis* protein closely related to *Synechocystis* cryptochrome is targeted to organelles. *Plant J* **35**: 93–103.

Klevanik, A. 1996. Magnetic-field effects on primary reactions in Photosystem I. *Biochimica et Biophysica Acta* **1275**:237–243.

Kodis, G., Liddell, P.A., Moore, A.L., Moore, T.A., and Gust, D. 2004. Synthesis and photochemistry of a carotene-porphyrin-fullerene model photosynthetic reaction center. *J Phys Org Chem* **17**:724–734.

Langmesser, S., Tallone, T., Bordon, A., Rusconi, S., and Albrecht, U. 2008. Interaction of circadian clock proteins PER2 and CRY with BMAL1 and CLOCK. *BMC Mol Biol* **9**:41.

Lau, J.C., Rodgers, C.T., and Hore, P.J. 2012. Compass magnetoreception in birds arising from photo-induced radical pairs in rotationally disordered cryptochromes. *J R Soc Interface* **9**(77):3329–3337.

Lee, A.A., Lau, J.C., Hogben, H.J., Biskup, T., Kattnig, D.R., and Hore, P.J. 2013. Alternative radical pairs for cryptochrome-based magnetoreception. *J R Soc Interface* **11**(95):20131063.

Lee, H, Cheng, Y.C., and Fleming, G.R. 2007. Coherence dynamics in photosynthesis: protein protection of excitonic coherence. *Science* **316**(5380):1462–1465.

Levy, C., Zoltowski, B.D., Jones, A.R., Vaidya, A.T., Top, D., Widom, J., Young, M.W., Scrutton, N.S., Crane, B.R., and Leys, D. 2013. Updated structure of *Drosophila* cryptochrome. *Nature* **495**(7441):E3–E4.

Lewis, A.M., Fay, T.P., Manolopoulos, D.E., Kerpal, C., Richert, S., and Timmel, C.R. 2018. On the low magnetic field effect in radical pair reactions. *J Chem Phys* **149**(3):034103.

Li Y, Liu Y, Jiang Z, Guan J, Yi G, Cheng S, Yang B, Fu T, and Wang Z. 2009. Behavioral change related to Wenchuan devastating earthquake in mice. *Bioelectromagnetics* **30**(8):613–620.

Liedvogel, M., Maeda, K., Henbest, K., Schleicher, E., Simon, T., Timmel, C.R., Hore, P.J., and Mouritsen, H. 2007. Chemical magnetoreception: bird cryptochrome 1a is excited by blue light and forms long-lived radical-pairs. *PLoS One* **2**(10):e1106.

Light, P., Salmon, M., and Lohmann, K.J. 1993. Geomagnetic orientation in loggerhead sea turtles: evidence for a inclination compass. *J Exp Biol* **182**:1–10.

Lin, C., and Todo, T. 2005. The cryptochromes. *Genome Biol* **6**(5):220.1–220.9.

Lin, C., Robertson, D.E., Ahmad, M., Raibekas, A.A., Jorns, M.S., Dutton, P.L., and Cashmore, A.R. 1995. Association of flavin adenine dinucleotide with the *Arabidopsis* blue light receptor CRY1. *Science* **269**(5226):968–970.

Liu, B., Liu, H., Zhong, D., and Lin, C. 2010. Searching for a photocycle of the cryptochrome photoreceptors. *Curr Opin Plant Biol* **13**(5):578–586.

Liu, H., Liu, B., Zhao, C., Pepper, M., and Lin, C. 2011. The action mechanisms of plant cryptochromes. *Trends Plant Sci* **16**(12):684–691.

Lohmann, K.J., and Lohmann, C.M.F. 1993. A light-independent magentic compass in the leatherback sea turtle. *Biol Bull* **185**:149–151.

Maeda, K., Henbest, K.B., Cintolesi, F., Kuprov, I., Rodgers, C.T., Liddell, P.A., Gust, D., Timmel, C.R., and Hore, P.J. 2008. Chemical compass model of avian magnetoreception. *Nature* **453**(7193): 387–390.

Maeda, K., Robinson, A.J., Henbest, K.B., Hogben, H.J., Biskup, T., Ahmad, M., Schleicher, E., Weber, S., Timmel, C.R., and Hore, P.J. 2012. Magnetically sensitive light-induced reactions in cryptochrome are consistent with its proposed role as a magnetoreceptor. *Proc Natl Acad Sci USA* **109**(13):4774–4779.

Maeda, K., Storey, J.G., Liddell, P.A., Gust, D., Hore, P.J., Wedge, C.J., and Timmel, C.R. 2015. Probing a chemical compass: novel variants of low-frequency reaction yield detected magnetic resonance. *Phys Chem Chem Phys* **17**(5):3550–3559.

Maeda, K., Wedge, C.J., Storey, J.G., Henbest, K.B., Liddell, P.A., Kodis, G., Gust, D., Hore, P.J., and Timmel, C.R. 2011. Spin-selective recombination kinetics of a model chemical magnetoreceptor. *Chem Commun* **47**(23):6563–6565.

Maier, E.J. 1992. Spectral sensitivities including the ultraviolet of the passeriform bird *Leiothrix lutea*. *J Comp Physiol A* **170**:709–714.

Malhotra, K., Kim, S.T., Batschauer, A., Dawut, L., and Sancar, A. 1995. Putative blue-light photoreceptors from *Arabidopsis thaliana* and *Sinapis alba* with a high degree of sequence homology to DNA photolyase contain the two photolyase cofactors but lack DNA repair activity. *Biochemistry* **34**(20):6892–6899.

Mazzotta, G., Rossi, A., Leonardi, E., Mason, M., Bertolucci, C., Caccin, L., Spolaore, B., Martin, A. J., Schlichting, M., Grebler, R., Helfrich-Forster, C., Mammi, S., Costa, R., and Tosatto, S. C. 2013. Fly cryptochrome and the visual system. *Proc Natl Acad Sci USA* **110**(15):6163–6168.

Mitsui H., Maeda T., Yamaguchi C., Tsuji Y., Watari R., Kubo Y., Okano K. and Okano T. 2015. Overexpression in yeast, photocycle, and in vitro structural change of an avian putative magnetoreceptor cryptochrome4. *Biochemistry* **54**(10):1908–1917.

Miura, T., Maeda, K., and Arai, T. 2003. Effect of coulomb interaction on the dynamics of the radical pair in the system of flavin mononucleotide and hen egg-white lysozyme (HEWL) studied by a magnetic field effect. *J Phys Chem B* **107**:6474–6478.

Möller, A., Gesson, M., Noll, C., Phillips, J., Wiltschko, R., and Wiltschko, W. 2001. Light-dependent magnetoreception in migratory birds previous exposure to red light alters the response to red light. In: *Orientation and Navigation—Birds, Humans and Other Animals*. Royal Institute of Navigation, Oxford, pp. 6-1–6-6.

Möller, A., Sagasser, S., Wiltschko, W., and Schierwater, B. 2004. Retinal cryptochrome in a migratory passerine bird: a possible transducer for the avian magnetic compass. *Naturwissenschaften* **91**(12):585–588.

Mora, C.V., Acerbi, M.L., and Bingman, V.P. 2014. Conditioned discrimination of magnetic inclination in a spatial-orientation arena task by homing pigeons (*Columba livia*). *J Exp Biol* **217**(23):4123–4131.

Mora, C.V., Davison, M., Wild, J.M., and Walker, M.M. 2004. Magnetoreception and its trigeminal mediation in the homing pigeon. *Nature* **432**(7016):508–511.

Mouritsen, H. 2018. Long-distance navigation and magnetoreception in migratory animals. *Nature* **558**(7708):50–59.

Mouritsen, H., and Hore, P.J. 2012. The magnetic retina: light-dependent and trigeminal magnetorecep-tion in migratory birds. *Curr Opin Neurobiol* **22**(2):343–352.

Mouritsen, H., and Ritz, T. 2005. Magnetoreception and its use in bird navigation. *Curr Opin Neurobiol* **15**(4):406–414.

Mouritsen, H., Feenders, G., Liedvogel, M., Wada, K., and Jarvis, E.D. 2005. Night-vision brain area in migratory songbirds. *Proc Natl Acad Sci USA* **102**:(23)8339–8344.

Mouritsen, H., Janssen-Bienhold, U., Liedvogel, M., Feenders, G., Stalleicken, J., Dirks, P., and Weiler, R. 2004. Cryptochromes and neuronal-activity markers colocalize in the retina of migratory birds during magnetic orientation. *Proc Natl Acad Sci USA* **101**(39):14294–14299.

Muheim, R., Bäckman, J., and Akesson, S. 2002. Magnetic compass orientation in European robins is dependent on both wavelength and intensity of light. *J Exp Biol* **205**:3845–3856.

Muheim, R., and Liedvogel, M. 2015. The light-dependent magnetic compass. In: *Photobiology, The Science of Light and Life, 323.* Björn, L.O. (Ed.), Springer Science+Business Media, New York, pp. 323–334.

Müller, P., and Ahmad, M. 2011. Light-activated cryptochrome reacts with molecular oxygen to form a flavin-superoxide radical pair consistent with magnetoreception. *J Biol Chem* **286**(24):21033–21040.

Müller, F., Brüstlein, M., Hemmerich, P., Massey, V., and Walker, W.H. 1972. Light-absorption studies on neutral flavin radicals. *Eur J Biochem* **25**(3):573–580.

Murakami, M., Maeda, K., and Arai, T. 2002. Structure and kinetics of the intermediate biradicals gen-erated from intramolecular electron transfer reaction of FAD studied by an action spectrum of the magnetic field effect. *Chem Phys Lett* **362**:123–129.

Murakami, M., Maeda, K., and Arai, T. 2005. Dynamics of intramolecular electron transfer reac-tion of FAD studied by magnetic field effects on transient absorption spectra. *J Phys Chem A* **109**(26):5793–5800.

Muus, L., Atkins, P., McLauchlan, K., and Pedersen, J. 1977. *Chemically Induced Magnetic Polarization.* D. Reidel Publishing, Dordrecht.

Neil, S.R., Li, J., Sheppard, D.M., Storey, J., Maeda, K., Henbest, K.B., Hore, P.J., Timmel, C.R., and Mackenzie, S.R. 2014. Broadband cavity-enhanced detection of magnetic field effects in chemical models of a cryptochrome magnetoreceptor. *J Phys Chem B* **118**(15):4177–4184.

Nielsen, C., Kattnig, D.R., Sjulstok, E., Hore, P.J., and Solov'yov, I.A. 2017. Ascorbic acid may not be involved in cryptochrome-based magnetoreception. *J R Soc Interface* **14**(137):20170657.

Nießner, C., Denzau, S., Gross, J.C., Peichl, L., Bischof, H.J., Fleissner, G., Wiltschko, W., and Wiltschko, R. 2011. Avian ultraviolet/violet cones identified as probable magnetoreceptors. *PLoS One* **6**(5):e20091.

Nießner, C., Denzau, S., Stapput, K., Ahmad, M., Peichl, L., Wiltschko, W., and Wiltschko, R. 2013. Magnetoreception: activated cryptochrome 1a concurs with magnetic orientation in birds. *J R Soc Interf* **10**(88):20130638.

Nießner, C., Gross, J.C., Denzau, S., Peichl, L., Fleissner, G., Wiltschko, W., and Wiltschko, R. 2016. Seasonally changing cryptochrome 1b expression in the retinal ganglion cells of a migrating pas-serine bird. *PLoS One* **11**(3):e0150377.

Nohr, D., Franz, S., Rodriguez, R., Paulus, B., Essen, L.O., Weber, S., and Schleicher, E. 2016. Extended electron-transfer in animal cryptochromes mediated by a tetrad of aromatic amino acids. *Biophys J* **111**(2):301–311.

Oka, Y. 2015. Spin control related to chemical compass of migratory birds (in Japanese with abstract in English). *Magnet Japan* **10**(3):140–145.

Orihara, Y., Kamogawa, M., Noda, Y., and Nagao, T. 2019. Is Japanese folklore concerning deep-sea fish appearance a real precursor of earthquakes? *Bull Seismol Soc Am* **109**(4):1556–1562.

Ozturk, N., Selby, C.P., Song, S.H., Ye, R., Tan, C., Kao, Y.T., Zhong, D., and Sancar, A. 2009. Comparative photochemistry of animal type 1 and type 4 cryptochromes. *Biochemistry* **48**:8585–8593.

Pedersen, J.B., Nielsen, C., and Solov'yov, I.A. 2016. Multiscale description of avian migration: from chemical compass to behaviour modeling. *Sci Rep* **6**:36709.

Phillips, J.B. 1986a. Two magnetoreception pathways in a migratory salamander. *Science* **233**(4765): 765–767.

Phillips, J.B. 1986b. Magnetic compass orientation in the eastern red-spotted newt (*Notophthalmus viridescens*). *J Comp Physiol A* **158**(1):103–109.

Pinzon-Rodriguez, A., Bensch, S., and Muheim, R. 2018. Expression patterns of cryptochrome genes in avian retina suggest involvement of Cry4 in light-dependent magnetoreception. *J R Soc Interface.* 15:20180058.

Player, T.C., and Hore, P.J. 2019. Viability of superoxide-containing radical pairs as magnetoreceptors. *J Chem Phys* **151**(22):225101.

Prato, F.S., Desjardins-Holmes, D., Keenliside, L.D., DeMoor, J.M., Robertson, J.A., and Thomas, A.W. 2013. Magnetoreception in laboratory mice: sensitivity to extremely low-frequency fields exceeds 33 nT at 30 Hz. *J Roy Soc Interface* **10**(81):20121046.

Ramsey, N. 1956. *Molecular beams.* Clarendon Press, Oxford, pp. 237.

Rappl, R., Wiltschko, R., Weindler, P., Berthold, P., and Wiltschko, W. 2000. Orientation behavior of garden warblers, *Sylvia borin*, under monochromatic light of different wavelengths. *Auk* **117**:256–260.

Ren, Y., Hiscock, H.G., and Hore, P.J. 2021. Angular precision of radical pair compass magnetoreceptors. *Biophys J* **120**(3):547–555.

Ritz, T., Adem, S., and Schulten, K. 2000. A model for photoreceptor-based magnetoreception in birds. *Biophys J* **78**(2):707–718.

Ritz, T., Thalau, P., Phillips, J.B., Wiltschko, R., and Wiltschko, W. 2004. Resonance effects indicate a radical-pair mechanism for avian magnetic compass. *Nature* **429**(6988):177–180.

Ritz, T., Wiltschko, R., Hore, P.J., Rodgers, C.T., Stapput, K., Thalau, P., Timmel, C.R., and Wiltschko, W. 2009. Magnetic compass of birds is based on a molecule with optimal directional sensitivity. *Biophys J* **96**(8):3451–3457.

Roberts, R.G. 2016. Living life on a magnet. *Plos Biol.* **14**(8):e2000613.

Rodgers, C.T. 2007. *D Phil thesis.* University of Oxford, Oxford.

Rodgers, C.T. 2009. Magnetic field effects in chemical systems. *Pure Appl Chem* **81**(1):19–43.

Rodgers, C.T., and Hore, P.J. 2009. Chemical magnetoreception in birds: the radical pair mechanism. *Proc Natl Acad Sci USA* **106**(2):353–360.

Rodgers, C.T., Norman, S.A., Henbest, K.B., Timmel, C.R., and Hore, P.J. 2007. Determination of radical re-encounter probability distributions from magnetic field effects on reaction yields. *J Amer Chem Soc* **129**(21):6746–6755.

Rotov, A.Y., Cherbunin, R.V., Anashina, A., Kavokin, K.V., Chernetsov, N., Firsov, M.L., and Astakhova, L.A. 2020. Searching for magnetic compass mechanism in pigeon retinal photoreceptors. *PLoS One* **15**(3):e0229142.

Runcorn, S.K. 1969. The paleomagnetic vector field. In: *The Earth's Crust and Upper Mantle, Vol. 13.* Hart, P.J. (Ed.), The American Geophysical Union, Washington DC, pp. 447–457.

Sadiek, G., Huang, Z., Aldossary, O., and Kais, S. 2008. Nuclear-induced time evolution of entanglement of two-electron spins in anisotropically coupled quantum dot. *Mol Phys* **106**:1777–1786.

Sagdeev, R.Z., Molin, Yu.N., Salikhov, K.M., Leshina, T.V., Kamha, M.A., and Shein, S.M. 1973. Effects of magnetic field on chemical reactions. *Org Magn Reson* **5**(12):603–605.

Sakaguchi, Y., and Hayashi, H. 1982. Laser-photolysis study of the photochemical processes of carbonyl compounds in micelles under high magnetic fields. *Chem Phys Lett* **87**(6):539–543.

Sakaguchi, Y., Hayashi, H., and Nagakura, S. 1980a. Classification of the external magnetic field effects on the photodecomposition reaction of dibenzoyl peroxide. *Bull Chem Soc Jpn* **53**(1):39–42.

Sakaguchi, Y., Nagakura, S., and Hayashi, H. 1980b. External magnetic field effect on the decay rate of benzophenone ketyl radical in a micelle. *Chem Phys Lett* **72**(3):420–423.

Sakaguchi, Y., Nagakura, S., Minoh, A., and Hayashi, H. 1981. Magnetic isotope effect upon the decay rate of the benzophenone ketyl radical in a micelle. *Chem Phys Lett* **82**(2):213–216.

Salikhov, K. 1996. *Magnetic Isotope Effect in Radical Reactions.* Springer-Verlag, Vienna, Austria.

Salikhov, K.M., Molin, Yu.N., Sagdeev, R.Z., and Buchachenko, A.L. 1984. *Spin Polarization and Magnetic Effects in Radical Reactions.* Elsevier, Amsterdam.

Sancar, A. 2003. Structure and function of DNA photolyase and cryptochrome blue-light photorceptors. *Chem Rev* **103**:2203–2237.

Sarovar, M., Ishizaki, A., Fleming, G.R., and Whaley, K.B. 2010. Quantum entanglement in photosynthetic light harvesting complexes. *Nat Phys* **6**:462–467.

Schulten, K., Swenberg, C.E., and Weller, A. 1978. A biomagnetic sensory mechanism based on magnetic field modulated coherent electron spin motion. *Z Phys Chem* **111**:1–5.

Semm, P., and Beason, R.C. 1990. Responses to small magnetic variations by the trigeminal system of the Bobolink. *Brain Res Bull* **25**:735–740.

Sheppard, D.M., Li, J., Henbest, K.B., Neil, S.R., Maeda, K., Storey, J., Schleicher, E., Biskup, T., Rodriguez, R., Weber, S., Hore, P.J., Timmel, C.R., and Mackenzie, S.R. 2017. Millitesla magnetic field effects on the photocycle of an animal cryptochrome. *Sci Rep* **7**:42228.

Solov'yov, I.A., Chandler, D.E., and Schulten, K. 2007. Magnetic field effects in *Arabidopsis thaliana* cryptochrome-1. *Biophys J* **92**(8):2711–2726.

Solov'yov, I.A., Ritz, T., Schulten, K., and Hore, P.J. 2014. A chemical compass for bird navigation. In: *Quantum Effects in Biology.* Mohseni, M., Omar, Y., Engel, G.S., and Plenio, M.B. (Eds.), Cambridge University Press, Cambridge, pp. 218–236.

Song, S.H., Dick, B., Penzkofer, A., Pokorny, R., Batschauer, A., and Essen, L.O. 2006. Absorption and fluorescence spectroscopic characterization of cryptochrome 3 from *Arabidopsis thaliana. J Photochem Photobiol B* **85**(1):1–16.

Spexard, M., Thoing, C., Beel, B., Mittag, M., and Kottke, T. 2014. Response of the sensory animal-like cryptochrome aCRY to blue and red light as revealed by infrared difference spectroscopy. *Biochemistry* **53**:1041–1050.

Steiner, U.E., and Ulrich, T. 1989. Magnetic field effects in chemical kinetics and related phenomena. *Chem Rev* **89**:51–147.

Takeuchi, T., Kubo, Y., Okano, K., and Okano, T. 2014. Identification and characterization of cryptochrome4 in the ovary of western clawed frog Xenopus tropicalis. *Zool Sci* **31**:152–159.

Tanimoto, Y., Hayashi, H., Nagakura, S., Sakuragi, H., and Tokumaru, K. 1976. The external magnetic field effect on the singlet sensitized photolysis of dibenzoyl peroxide. *Chem Phys Lett* **41**(2):267–269.

Thalau, P., Ritz, T., Stapput, K., Wiltschko, R., and Wiltschko, W. 2005. Magnetic compass orientation of migratory birds in the presence of a 1.315 MHz oscillating field. *Naturwissenschaften* **92**(2):86–90.

Timmel, C., and Henbest, K. 2004. A study of spin chemistry in weak magnetic fields. *Philos Trans A Math Phys Eng Sci* **362**(1825):2573–2589.

Timmel, C.R., and Hore, P.J. 1996. Oscillating magnetic field effects on the yields of radical pair reactions. *Chem Phys Lett* **257**:401–408.

Timmel, C.R., Till, U., Brocklehurst, B., Mclauchlan, K.A., and Hore, P.J. 1998. Effects of weak magnetic fields on free radical recombination reactions. *Mol Phys* **95**:71–89.

Tu, D.C., Batten, M.L., Palczewski, K., and Van Gelder, R.N. 2004. Nonvisual photoreception in the chick iris. *Science* **306**(5693):129–131.

Turro, N.J., and Chow, M.F. 1979. Magnetic field effects on the thermolysis of endoperoxides of aromatic compounds. Correlations with singlet oxygen yield and activation entropies. *J Am Chem Soc* **101**(13):3701–3703.

van Dijk, B., Carpenter, J.K.H., Hoff, A.J., and Hore, P.J. 1998. Magnetic field effects on the recombination kinetics of radical pairs. *J Phys Chem B* **102**:464–472.

Vedral, V. 2011. Living in a quantum world. *Sci Am* **304**(6):38–43.

Wang, K.F., and Ritz, T. 2006. Zeeman resonances for radical-pair reactions in weak static magnetic fields. *Mol Phys* **104**(10–11):1649–1658.

Watari, R., Yamaguchi, C., Zemba, W., Kubo, Y., Okano, K., and Okano, T. 2012. Light-dependent structural change of chicken retinal Cryptochrome4. *J Biol Chem* **287**:42634–42641.

Weber, S., Biskup, T., Okafuji, A., Marino, A.R., Berthold, T., Link, G., Hitomi, K., Getzoff, E.D., Schleicher, E., and Norris, J.R. Jr. 2010. Origin of light-induced spin-correlated radical pairs in cryptochrome. *J Phys Chem B* **114**(45):14745–14754.

Weber, S., Kay, C.W., Mögling, H., Möbius, K., Hitomi, K., and Todo, T. 2002. Photoactivation of the flavin cofactor in *Xenopus laevis* (6–4) photolyase: observation of a transient tyrosyl radical by time-resolved electron paramagnetic resonance. *Proc Natl Acad Sci USA* **99**(3):1319–1322.

Weller, A., Nolting, F., and Staerk, H. 1983. A quantitative interpretation of the magnetic field effect on hyperfine-coupling-induced triplet fromation from radical ion pairs. *Chem Phys Lett* **96**(1):24–27.

Werner, H.J., Staerk, H., and Weller, A. 1978. Solvent, isotope, and magnetic field effects in the geminate recombination of radical ion pairs. *J Chem Phys* **68**(5):2419–2426.

Whitehead, N.E., and Ulusoy, Ü. 2013. Macroscopic anomalies before the September 2010 $M=7.1$ earthquake in Christchurch, New Zealand. *Nat Hazards Earth Syst Sci* **13**:167–176.

Wiltschko, R., and Wiltschko, W. 1995. *Magnetic Orientation in Animals*. Springer Verlag, Berlin Heidelberg, New York.

Wiltschko, R., and Wiltschko, W. 1998. Pigeon homing: effect of various wavelengths of light during displacement. *Naturwissenschaften* **85**:164–167.

Wiltschko, R., and Wiltschko, W. 2006. Magnetoreception. *BioEssays* **28**:157–168.

Wiltschko, R., and Wiltschko, W. 2014. Sensing magnetic directions in birds: radical pair processes involving cryptochrome. *Biosensors (Basel)* **4**(3):221–242.

Wiltschko, R., and Wiltschko, W. 2019. Magnetoreception in birds. *J R Soc Interf* **16**:20190295.

Wiltschko, R., and Wiltschko, W. 2021. The discovery of the use of magnetic navigational information. *J Comp Physiol A* doi: 10.1007/s00359-021-01507-0.

Wiltschko, R., Nießner, C., and Wiltschko, W. 2021. The magnetic compass of birds: the role of cryptochrome. *Front Physiol* **12**:667000.

Wiltschko, R., Nohr, D., and Wiltschko, W. 1981. Pigeons with a deficient sun compass use the magnetic compass. *Science* **214**(4518):343–345.

Wiltschko, R., Stapput, K., Thalau, P., and Wiltschko, W. 2010. Directional orientation of birds by the magnetic field under different light conditions. *J R Soc Interf* **7**(Suppl 2):S163–S177.

Wiltschko, W., Freire, R., Munro, U., Ritz, T., Rogers, L., Thalau, P., and Wiltschko, R. 2007. The magnetic compass of domestic chickens, *Gallus gallus*. *J Exp Biol* **210**(13):2300–2310.

Wiltschko, W., Gesson, M., Stapput, K., and Wiltschko, R. 2004. Light dependent magnetoreception in birds: interaction of at least two different receptors. *Naturwissenschaften* **91**(3):130–134.

Wiltschko, W., Gesson, M., and Wiltschko, R. 2001. Magnetic compass orientation of European robins under 565 nm green light. *Naturwissenschaften* **88**(9):387–390.

Wiltschko, W., Munro, U., Ford, H., and Wiltschko, R. 1993. Red light disrupts magnetic orientation of migratory birds. *Nature* **364**(6437):525–527.

Wiltschko, W., Munro, U., Ford, H., and Wiltschko, R. 2006b. Bird navigation: what type of information does the magnetite-based receptor provide? *Proc Biol Sci* **273**(1603):2815–2820.

Wiltschko, W., Stapput, K., Thalau, P., and Wiltschko, R. 2006a. Avian magnetic compass: fast adjustment to intensities outside the normal functional window. *Naturwissenschaften* **93**(6):300–304.

Wiltschko, W., Traudt, J., Güntürkün, O., Prior, H., and Wiltschko, R. 2002. Lateralization of magnetic compass orientation in a migratory bird. *Nature* **419**(6906):467–470.

Wiltschko, W., and Wiltschko, R. 1972. Magnetic compass of European robins. *Science* **176**(4030):62–64.

Wiltschko, W., and Wiltschko, R. 1999. The effect of yellow and blue light on magnetic compass orientation in European robins, *Erithacus rubecula*. *J Comp Physiol A* **184**:295–299.

Wiltschko, W., and Wiltschko, R. 2001. Light-dependent magnetoreception: the behavior of European robins, *Erithacus rubecula*, under monochromatic light of various wavelengths and intensities. *J Exp Biol* **204**(19):3295–3302.

Wiltschko, W., and Wiltschko, R. 2002. Magnetic compass orientation in birds and its physiological basis. *Naturwissenschaften* **89**(10):445–445.

Wiltschko, W., and Wiltschko, R. 2005. Magnetic orientation and magnetoreception in birds and other animals. *J Comp Physiol A* **191**(8):675–693.

Wiltschko, W., and Wiltschko, R. 2007. Magnetoreception in birds: two receptors for two different tasks. *J Ornithol* **148**(Suppl 1):S61–S76.

Wiltschko, W., Wiltschko, R., and Munro, U. 2000a. Light-dependent magnetoreception in birds: does directional information change with light intensity? *Naturwissenschaften* **87**(1):36–40.

Wiltschko, W., Wiltschko, R., and Munro, U. 2000b. Light-dependent magnetoreception in birds: the effect of intensity of 565 nm green light. *Naturwissenschaften* **87**(8):366–369.

Wiltschko, W., Wiltschko, R., and Ritz, T. 2011. The mechanism of the avian magnetic compass. *Procedia Chem* **3**:276–284.

Wilzeck, C., Wiltschko, W., Güntürkün, O., Buschmann, J.O., Wiltschko, R., and Prior, H. 2010. Learning of magnetic compass directions in pigeons. *Anim Cogn* **13**:443–451.

Woith, H., Petersen, G. M., Hainzl, S., and Dahm, T. 2018. Review: can animals predict earthquakes? *Bull Seismol Soc Am* **108**(3A):1031–1045.

Woodward, J.R., Foster, T.J., Jones, A.R., Salaoru, A.T., and Scrutton, N.S. 2009. Timeresolved studies of radical pairs. *Biochem Soc Trans* **37**:358–362.

Woodward, J.R., Timmel, C.R., McLauchlan, K.A., and Hore, P.J. 2001. "Radio frequency magnetic field effects on electron-hole recombination. *Phys Rev Lett* **87**(7):077602.

Wu, H., Scholten, A., Einwich, A., Mouritsen, H., and Koch, K.W. 2020. Protein-protein interaction of the putative magnetoreceptor cryptochrome 4 expressed in the avian retina. *Sci Rep* **10**(1):7364.

Wu, L.Q., and Dickman, D. 2011. Magnetoreception in an avian brain in part mediated by inner ear lagena. *Curr Biol* **21**:418–423.

Wu, L.Q., and Dickman, D.J. 2012. Neural correlates of a magnetic sense. *Science* **336**(6084):1054–1057.

Xu, B.M., Zou, J., Li, H., Li, J.G., and Shao, B. 2014a. Effect of radio frequency fields on the radical pair magnetoreception model. *Phys Rev E Stat Nonlin Soft Matter Phys* **90**(4):042711.

Xu, C., Lv, Y., Chen, C., Zhang, Y., and Wei, S. 2014b. Blue light-dependent phosphorylations of cryptochromes are affected by magnetic fields in *Arabidopsis*. *Adv Space Res* **53**(7):1118–1124.

Xu, C.X., Wei, S.F., Lu, Y., Zhang, Y.X., Chen, C.F., and Song, T. 2013. Removal of the local geomagnetic field affects reproductive growth in *Arabidopsis*. *Bioelectromagnetics* **34**(6):437–442.

Xu, C.X., Yin, X., Lv, Y., Wu, C.Z., Zhang, Y.X., and Song, T. 2012. A near-null magnetic field affects cryptochrome-related hypocotyl growth and flowering in *Arabidopsis*. *Adv Space Res* **49**(5):834–840.

Xu, J, Jarocha, L.E., Zollitsch, T., Konowalczyk, M., Henbest, K.B., Richert, S., Golesworthy, M.J., Schmidt, J., Déjean, V., Sowood, D.J.C., Bassetto, M., Luo, J., Walton, J.R., Fleming, J., Wei, Y., Pitcher, T.L., Moise G, Herrmann, M., Yin, H., Wu, H., Bartölke, R., Käsehagen, S.J., Horst, S., Dautaj, G., Murton, P.D.F., Gehrckens, A.S., Chelliah Y., Takahashi, J.S., Koch, K.W., Weber, S., Solov'yov, I.A., Xie, C., Mackenzie, S.R., Timmel, C.R., Mouritsen, H., and Hore, P.J. 2021. Magnetic sensitivity of cryptochrome 4 from a migratory songbird. *Nature* **594**(7864):535–540.

Yamauchi, H., Hayakawa, M., Asano, T., Ohtani, N., and Ohta, M. 2017. Statistical evaluations of variations in dairy cows' milk yields as a precursor of earthquakes. *Animals (Basel)* **7**(3):19.

Yamauchi, H., Uchiyama, H., Ohtani, N., and Ohta, M. 2014. Unusual animal behavior preceding the 2011 earthquake off the Pacific coast of Tohoku, Japan: a way to predict the approach of large earthquakes. *Animals (Basel)* **4**(2):131–145.

Yanai, S., Semba, Y., and Endo, S. 2012. Remarkable changes in behavior and physiology of laboratory mice after the massive 2011 Tohoku earthquake in Japan. *PLoS One* **7**(9):e44475.

Yang, H.Q., Wu, Y.J., Tang, R.H., Liu, D., Liu, Y., and Cashmore, A.R. 2000. The C termini of *Arabidopsis* cryptochromes mediate a constitutive light response. *Cell* **103**(5):815–827.

Yokoi, S., Ikeya, M., Yagi, T., and Nagai, K. 2003. Mouse circadian rhythm before the Kobe earthquake in 1995. *Bioelectromagnetics* **24**(4):289–291.

Yuan, Q., Metterville, D., Briscoe, A.D., and Reppert, S.M. 2007. Insect cryptochromes: gene duplication and loss define diverse ways to construct insect circadian clocks. *Mol Biol Evol* **24**:948–955.

Zapka, M., Heyers, D., Hein, C.M., Engels, S., Schneider, N.L., Hans, J., Weiler, S., Dreyer, D., Kishkinev, D., Wild, J.M., and Mouritsen, H. 2009. Visual but not trigeminal mediation of magnetic compass information in a migratory bird. *Nature* **461**(7268):1274–1277.

Zaporozhan, V., and Ponomarenko, A. 2010. Mechanisms of geomagnetic field influence on gene expression using influenza as a model system: basics of physical epidemiology. *Int J Environ Res Public Health* 7:938–965.

Zeugner, A., Byrdin, M., Bouly, J.P., Bakrim, N., Giovani, B., Brettel, K., and Ahmad, M. 2005. Light-induced electron transfer in *Arabidopsis* cryptochrome-1 correlates with in vivo function. *J Biol Chem* **280**(20):19437–19440.

Zhang, Y.T., Berman, G.P., and Kais, S. 2015. The radical pair mechanism and the avian chemical compass: quantum coherence and entanglement. *Int J Quantum Chem* **115**:1327–1341.

Zoltowski, B.D., Chelliah, Y., Wickramaratne, A., Jarocha, L., Karki, N., Xu, W., Mouritsen, H., Hore, P.J., Hibbs, R.E., Green, C.B., and Takahashi, J.S. 2019. Chemical and structural analysis of a photoactive vertebrate cryptochrome from pigeon. *Proc Natl Acad Sci USA* **116**(39):19449–19457.

Zoltowski, B.D., Vaidya, A.T., Top, D., Widom, J., Young, M.W., and Crane, B.R. 2011. Structure of full-length *Drosophila* cryptochrome. *Nature* **480**(7377):396–399.

5

Magnetoreception in Plants

Massimo Emilio
Maffei

5.1 Introduction

As sessile organisms, plants have evolved both constitutive and inducible responses to the changing environment. Several environmental factors affect plant growth and development. Among them, the Earth magnetic field or geomagnetic field (GMF) is an environmental component of our planet. It is fairly homogeneous and relatively weak. The strength of the GMF at the surface of the Earth ranges from <30 µT in an area that includes most of South America and South Africa (the so-called South Atlantic anomaly) to over 60 µT around the magnetic poles in northern Canada, the south of Australia, and in parts of Siberia (Occhipinti et al., 2014). Plants that are known to sense different wavelengths of light, respond to gravity, and react to touch and electrical signals cannot avoid the presence of the GMF (Maffei, 2014). While plant phototropism, gravitropism, hydrotropism, and autostraightening have been thoroughly documented (Harmer and Brooks, 2018), possible effects of the GMF on plant growth and development are still a matter of discussion. Nevertheless, a growing body of evidence indicates that plants do react to varying magnetic field (MF) fluxes at values both below and above the GMF (Teixeira da Silva and Dobranszki, 2016; Radhakrishnan, 2019).

Reduction of the GMF to Near Null Magnetic Field (NNMF; about 30 nT) has been shown to influence many plant biological processes (Xu et al., 2013; Maffei, 2014). In previous studies aimed at evaluating the effect of GMF reversal on plants, we found differential root/shoot responses in plant morphology and in the expression of some genes (e.g., *Cruciferin 3*, *Copper Transport Protein1*, and *Redox Responsive Transcription Factor1*) (Bertea et al., 2015). This finding is in agreement with the current view that the magnetic reaction could change the ratio of redox states in the cryptochrome photocycle to alter the biological activity of cryptochrome (Pooam et al., 2019; Hammad et al., 2020). We also found that the GMF impacted the flowering time by differentially regulating leaf and floral meristem genes (Agliassa et al., 2018a) and by altering the signaling of cryptochrome and phytochrome. In particular, blue light exposure led to a partial association between the GMF-induced changes in gene expression and an alteration in cryptochrome activation (Agliassa et al., 2018b). The GMF also affected plant mineral nutrition by influencing both root ion uptake and ion channel activity (Narayana et al., 2018; Islam et al., 2020a). Similar results have been obtained in Arabidopsis and other plant species (Rakosy-Tican et al., 2005; Xu et al., 2015, 2017; Azizi et al., 2019; Jin et al., 2019; Pooam et al., 2019; Radhakrishnan, 2019). More recently, we carried out a time-course microarray experiment to identify genes that are differentially regulated by the GMF in shoot and roots. We found that the GMF regulates genes in both shoot and roots, suggesting that both organs can sense the GMF. However, 49% of the genes were regulated in a reverse direction in these organs, meaning that the resident signaling networks define the up- or down-regulation of specific genes. The set of GMF-regulated genes strongly overlapped with various stress-responsive genes, implicating the involvement of one or more common signals, such as reactive oxygen species, in these responses. The biphasic dose response of GMF-responsive genes indicates a hormetic response of plants to the GMF (Paponov et al., 2021). Therefore, plants can sense and respond to the GMF using the signaling networks involved in stress responses.

In this chapter, I will highlight some of the basic mechanisms proposed to be involved in plant magnetoreception and summarize the plant responses to varying MF intensities both dependent and independent from the presence of light.

5.2 Mechanism of Magnetoreception in Plants

Three different mechanisms of magnetoreception have been described: a mechanism involving radical pairs (i.e., magnetically sensitive chemical intermediates that are formed by photoexcitation of cryptochrome (Guo et al., 2018)), which has been demonstrated both in animals (Hore and Mouritsen, 2016) and in plants (Pooam et al., 2019); the presence of MF sensory receptors present in cells containing ferromagnetic particles, as has been shown in magnetotactic bacteria (Kornig et al., 2014); and the detection of minute electric fields by electroreceptors in the ampullae of Lorenzini in elasmobranch animals (Kempster et al., 2012).

Of the three possible mechanisms of magnetoreception, only the radical pair mechanism of chemical magnetosensing adequately explains the alterations in the MF by the rates of redox reactions and subsequently altered concentrations of free radicals and ROS observed in plants, animals, and humans (Bertea et al., 2015; Pooam et al., 2019, 2020b; Albaqami et al., 2020). The theory underlying the radical pair mechanism predicts that MFs similar in strength to the GMF are too weak to trigger cellular biochemical reactions; however, these MFs are able to interact with short-lived reaction intermediates that affect the reaction rates of biochemical reactions. Examples include photoreceptors (e.g., cryptochromes) and redox reactions that can be initiated by metabolic factors. This modulation of cryptochrome signaling and/or redox reactions can alter ROS synthesis in the cells (Pooam et al., 2020a).

5.2.1 The Radical-Pair Mechanism

Spin interactions have profound effects on chemical reactions despite the energies involved are orders of magnitude smaller than the thermal energy, k_BT (Hayashi, 2004). It is known that applied MFs and magnetic isotope substitution can alter the rates and product yields of free-radical reactions with the formation of transient paramagnetic intermediates in non-equilibrium electron spin states. The most common sources of spin-chemical effects are organic radical pairs (RPs). Typically formed in a singlet (S) or a triplet (T) state by a reaction that conserves electron spin, RPs interconvert coherently between their S and T states as a result of the Zeeman, hyperfine, exchange, and dipolar interactions of the electrons and the nuclear spins to which they are coupled (Hore et al., 2020). Applied MFs alter the extent and timing of the $S \leftrightarrow T$ interchange and hence the yields of products formed spin-selectively from the S and T states (Jones, 2016).

A typical situation considered in the RP mechanism (RPM) is the production of a spin-correlated RP, let us assume from an electronically excited triplet state, yielding an RP with initially parallel electron spins. The spin motion is visible in the vector representations of the RP spin states shown in Figure 5.1.

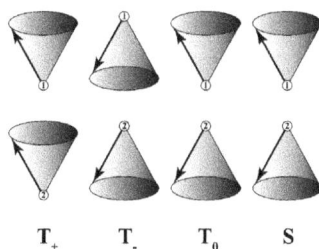

FIGURE 5.1 Vector representation of radical pair spin triplet states T_+, T_-, T_0, and singlet state S. (Adapted from Steiner and Ulrich (1989).)

The individual electron spins are confined to cones oriented along the axis of quantization, either upward (α-spin) or downward (β-spin). The resultant of the two electron spins is either 1, and oriented parallel (T_+), perpendicular (T_0), or antiparallel (T_-) to the axis of quantization, or 0 (S). The corresponding spin eigenfunctions are $\alpha_1\alpha_2$ for T_+, $\beta_1\beta_2$ for T_-, $\frac{1}{\sqrt{2}}(\alpha_1\beta_2 + \beta_1\alpha_2)$ for T_0, and $\frac{1}{\sqrt{2}}(\alpha_1\beta_2 - \beta_1\alpha_2)$ for S. Figure 5.1 suggests that there is some phase relation between the spins of different electrons. However, the phase relation distinguishing S and T_0 operates between the spin function products $\alpha_1\beta_2$ and $\beta_2\alpha_1$, and not between the spins of single electrons (Steiner and Ulrich, 1989).

The main driving force for electronic spin motion is isotropic hyperfine coupling, and the quantum-mechanical determination of accurate isotropic hyperfine coupling constants essentially relies on the precise calculation of the electron density at each nucleus with a nonzero magnetic moment (Fermi contact first-order interaction (Chipman and Rassolov, 1997)). The active magnetic field B is made up as a vector sum of the external magnetic field B_0 and an effective magnetic field resulting from the sum of the hyperfine couplings of the various nuclear spins in the corresponding radical.

$$B = B_0 + B_{hfc}$$

In the absence of an MF, the total spin angular momentum of electrons and nuclei must be conserved, so that a change in electron spin must be compensated by a change of nuclear spin. In the condition of zero external field, any transition between the spin substates of the RP is possible. As the external field strength increases, the resultant field B is more and more determined by B_0 so that the directions of the precession axes of the two spins coincide, precluding transitions between T_+, T_- and S, T_0 (so-called spin-flip transitions). However, the precession frequency difference due to the B_{hfc} component parallel to B_0 is retained, and S-T_0 transitions (so-called rephasing transitions) are not suppressed by the external MFs (Steiner and Ulrich, 1989).

Considering the MF dependence (MFD) of chemical reaction yields caused by the RPM we can consider three cases or combinations of them, as suggested by Sakaguchi et al. (1980). Figure 5.2 depicts some phenomenological cases of MFD of reaction yields. In the first case, the suppression of the hyperfine coupling-induced $S \leftrightarrow T_{+,-}$ transitions by the MF shows a clear saturation behavior of the product yield (R_S). Here, the parameters to be specified are $B_{1/2}$, the field where half of the saturation effect is obtained, and B_s, the region of beginning saturation, although the latter is not very exactly defined. The yield increases suddenly at a low field and remains almost constant above B_s. In the second case, a rather monotonously rising MFD curve requires very high MFs for obtaining a saturation. The yield decreases with increasing MF by the S-T_0 conversion through the electronic Zeeman term. In the combination of case 1 and case 2 (case 1,2), the yield first increases with the increasing MF. Such curves may be characterized by the field values B_m of the maximum and B_c of the zero-line crossing. In the last case (case 3), RPs are subjected to a moderate but constant exchange interaction and may be characterized by the field values B_m, corresponding to the maximum, and ΔB corresponding to the width of the resonance.

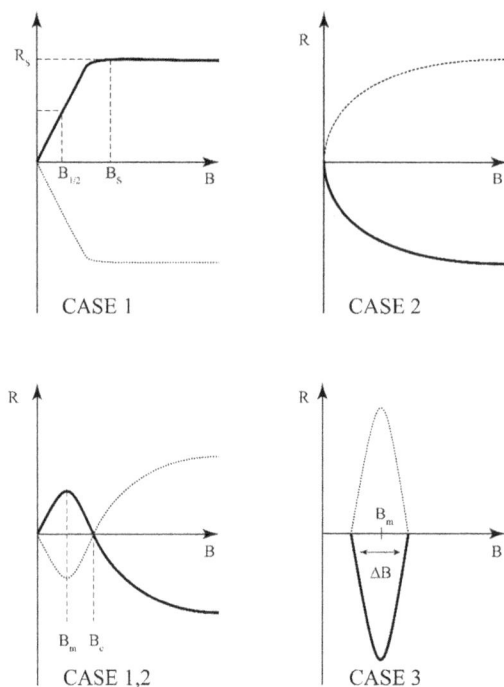

FIGURE 5.2 Phenomenological cases of MFD of reaction yields. (Adapted from Steiner and Ulrich (1989).) Dotted lines indicate sign inversion of effects when changing the precursor multiplicity in the RPM.

According to the RPM, the singlet-triplet conversion rate is slowed down at low fields (hyperfine mechanism) and increased in high fields (Δg mechanism). When an MF exceeds that of the hyperfine interaction, the triplet states $T_{+,-}$ are separated from the remaining states energy levels due to the Zeeman interaction:

$$E_z = \mu_B g_j B_{ext} m_j$$

(where μ_B is the Bohr magneton, g_j is the Landé g-factor, B_{ext} is the applied MF, and m_j is the magnetic moment). Therefore, the spin-dependent kinetics becomes MF-dependent.

The Zeeman interaction with the external MF modifies the energy levels of the pair, changing the extent and frequency of $S \leftrightarrow T$ interconversion, and so leads to magnetic field effects (MFEs) on the lifetime of the radical pair and on the yields of the reaction products formed from it (Zollitsch et al., 2018). The MFE develops during the lifetime of the spin-correlated radical pair (RP1) and is inherited by downstream products. As such, the time- and wavelength-dependence of the MFE can provide information on $F^{\cdot-}$, even when its signature in the spectral deconvolution of the ΔA data is difficult to observe (Zollitsch et al., 2018).

One of the most studied MFE is the flavin-based RP system. In these proteins, photoexcitation of the flavin adenine dinucleotide (FAD) cofactor in its fully oxidized redox state, FAD^+, produces a singlet excited state, $^1FAD^\cdot$, which is then reduced on a picosecond timescale by consecutive electron transfers along a chain of three tryptophan or guanosine residues, generating the singlet radical pair, $^1[FAD^{\cdot-} Trp^{\cdot+}]$ (Evans et al., 2015). In radical pairs formed by photo-induced electron transfer reactions in sensor proteins, the magnetic sensitivity arises from a combination of the coherent quantum spin dynamics and the spin-selective reactivity of a pair of spin-correlated radicals, which cause the yield of a signaling state to depend on the intensity and direction of the external MF. An RP-based sensor may also be

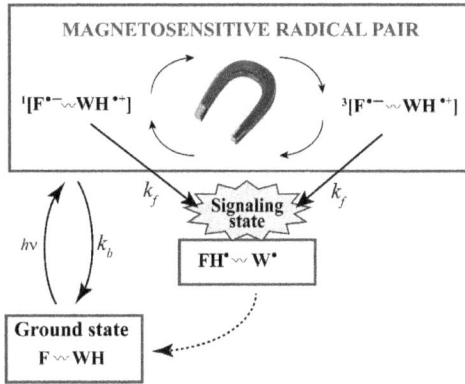

FIGURE 5.3 The classical RPM entails three essential steps. In the case of cryptochrome, F is the flavin and W is tryptophan. The spin-selective reaction channel is charge recombination (rate constant k_b) within the singlet state of the RP which regenerates the ground, or resting state of the protein. The magnetosensitive RP state also undergoes proton transfer reactions, which occur with equal rate constants (k_f) for singlet and triplet pairs, which is assumed to be, or to lead to, the biochemical signaling state of the protein. (Adapted from Kattnig and Hore (2017).)

enhanced by interactions with its fluctuating environment, a property apparently found in many areas of "quantum biology" (Kattnig and Hore, 2017).

In the current model, the spin-selective reaction channel is charge recombination (rate constant k_b) within the singlet state of the RP which regenerates the ground state of the protein. The RP state also undergoes proton transfer reactions, which occur with equal rate constants (k_f) for singlet and triplet pairs, to produce a secondary, long-lived radical pair state, S, which is assumed to be, or to lead to, the biochemical signaling state of the protein. The interaction of the electron spins with the MF can induce a significant change in the yield of S if k_b^{-1} (the characteristic time of singlet recombination) is comparable to or shorter than k_f^{-1} (the time required for the formation of S), small compared to the electron spin relaxation time (~1 µs or possibly longer) and longer than the coherent singlet-triplet interconversion time. These conditions mean that the radicals must not be too far apart, otherwise charge recombination will be too slow. Figure 5.3 shows the classical RPM.

Recent comprehensive calculations for the FAD·−/W·+ RP revealed that the key quantity, the differences in the yield of signaling state at different field directions, might be perplexingly small (Kattnig et al., 2016b). According to Hore et al., the fact that the compass performance in animals surpasses these predictions (e.g., with respect to the acuity of the sensor, its function under very low light intensities, or its sensitivity to weak radio frequency MFs) suggests the presence of a powerful, yet unknown, amplification process and remarkable resilience to decoherence (Hore and Mouritsen, 2016; Kattnig et al., 2016a). The same group recently proposed an extended reaction scheme (Figure 5.4), that is predicted to greatly (by a factor of 10 and more) enhance the compass sensitivity via the so-called "chemical Zeno effect" (Letuta and Berdinskii, 2015), the effect of spin-dependent recombination on the singlet–triplet RP evolution rate and frequency. The chemical Zeno effect changes the S–T recombination rate and, thus, has an effect on the yield of recombination products, quantum beat frequency, chemically induced nuclear polarization, and magnetic isotope effect (Letuta and Berdinskii, 2015). The model of Hore and co-workers relies on a spin-selective reaction of one of the two radicals of the primary pair and a spin-bearing, external scavenger. This scavenger is initially uncorrelated with respect to the RP, a situation that resembles f-pairs, but eventually acquires correlation as a result of its spin selective reactivity, i.e. the chemical Zeno effect. It has been shown that this additional reaction induces singlet–triplet conversion in the original RP and serves as a spin-selective reaction channel. As a consequence, and contrary to previous theories, spin-selective recombination of the primary RP (see Figure 5.3) is no longer essential, and the radicals could thus be farther apart than is necessary for efficient charge recombination.

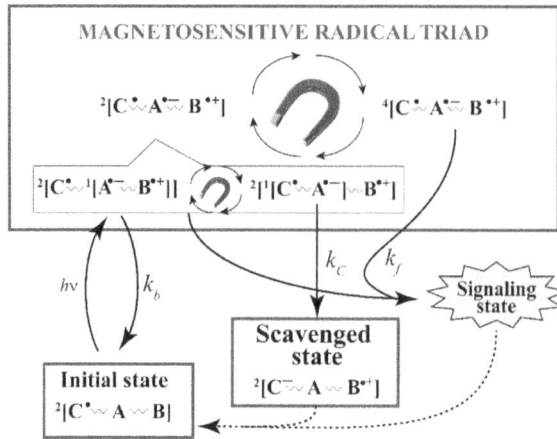

FIGURE 5.4 Scheme of magnetoreception extended to comprise a radical scavenging reaction of the primary RP. A$^{\bullet-}$ is generally thought to derive from a cryptochrome-bound FAD, implying the semiquinoid radical FAD$^{\bullet-}$ or its protonated form, FADH$^{\bullet}$. B$^{\bullet+}$ appears to be more elusive; different schemes and radicals are discussed in the literature. C$^{\bullet}$ is the scavenger radical. (Adapted from Kattnig (2017).)

This means that tryptophan tetrads (rather than triads) or systems involving freely diffusing radicals can also give rise to sizable MF effects and that the detrimental effects of inter-radical exchange and dipolar interactions can be minimized. The spin dynamics in these three-radical systems are characterized by doublet-quartet conversions (instead of the conventional singlet–triplet interconversions characteristic of the classical RP mechanism). The doublet state can be formulated as singlet in the A/B- or A/C-manifold in combination with a doublet radical, as shown in Figure 5.4. These singlet substates are, however, not mutually exclusive, i.e., orthogonal.

The remarkable resilience to fast relaxation in the second radical, B$^{\bullet+}$, which, for $k_b > 0$ even affords relaxation-enhancement, widens the scope of the RPM to include constituent radicals with spin relaxation rates that are too fast to elicit a significant MFE via the conventional RPM. In the context of plant magnetoreception, this prospect appears particularly relevant to reaction schemes involving the superoxide radical, which has been advocated by several studies but suffers from fast spin relaxation. An increase in the intensity of the ambient MF from 33–44 to 500 μT enhanced growth inhibition in *Arabidopsis thaliana* under blue light, when cryptochromes are the mediating photoreceptor, but not under red light when the mediating receptors are phytochromes showing that higher plants are sensitive to the MF in responses that might be linked to cryptochrome-dependent signaling pathways (Ahmad et al., 2007). A number of recent theoretical works suggest that cryptochromes might also be the receptors responsible for the sensing of the GMF (e.g., in insects, migratory birds, or migratory fish).

5.3 Light-Dependent Magnetoreception

Plants show both light-dependent (Xu et al., 2015; Vanderstraeten et al., 2018; Pooam et al., 2019) and light-independent (Dhiman and Galland, 2018; Agliassa and Maffei, 2019) magnetoreception, which may reflect a differential ability of plant organs to interact with light and the GMF. For instance, leaves are constantly exposed to both light and GMF, whereas roots perceive the GMF but only a low light fluence when close to the soil surface and no light when they grow deep in the ground.

Cryptochromes, which are blue-light photoreceptors, are involved in magnetoreception and the effect of MFs was reported in cryptochrome phosphorylation (Xu et al., 2014; Agliassa et al., 2018b; Pooam et al., 2019; Hammad et al., 2020). Phytochromes, the red-light photoreceptors, are known to interact with cryptochrome at many levels and enhance and maintain plant downstream responses to blue light (Ahmad and Cashmore, 1997). Phytochrome *phyAphyB* deficient mutants showed a visibly enhanced response to applied MFs (Pooam et al., 2019). Therefore, plants can use different cues to perceive variations in the MF and trigger signal transduction events eventually leading to biochemical and developmental changes.

The photocycle underlying magnetoreception has not yet been unequivocally identified. MFE studies on isolated cryptochromes and closely related photolyases have ascribed the magnetosensitivity to the RP [FAD$^{\bullet-}$/W$^{\bullet+}$], produced through a sequence of swift electron transfer steps involving the tryptophan triad (Maeda et al., 2012). In animals, the transferability of this finding to in vivo conditions has been questioned by several behavioral and histochemical studies suggesting that the key step involves the reoxidation of the fully reduced FADH$^-$, likely by molecular oxygen (Muller and Ahmad, 2011). Support for this hypothesis is predominantly drawn from the light dependence of the compass sense. For instance, in birds, pre-exposure to white light could generate orientation under green light (~560 nm) even though oxidized FAD is not excitable below ~500 nm (Wiltschko et al., 2010). This finding has been attributed to the secondary photoreduction of the semiquinone radical FADH$^{\bullet}$ to FADH$^-$, which can indeed be facilitated by green light. Along the same lines, histochemical studies suggested that Cry1a in the retina of chickens could be photoactivated by green light, and that structural changes connected to the C-terminal region of the cryptochrome, which could be relevant in signaling, are triggered in the fully reduced form (Nießner et al., 2014). It was also argued that the efficient charge separation in animal cryptochromes and closely related animal photolyases containing a tryptophan tetrad would preclude magnetosensitivity in the photoinduced flavin-tryptophan RPs because the rate of spin-selective recombination was too low (Cailliez et al., 2016). The only currently hypothesized RP that may potentially explain these results is FADH$^{\bullet}$ / O$_2^{\bullet-}$, generated in the reoxidation of the fully (photo)reduced FADH$^-$.

Most of the knowledge on MFE in plants comes from the work of Hore, Ahmad, and Solov'yov and their co-workers.

The activity of cryptochrome-1 in *A. thaliana* is enhanced by the presence of a weak external MF, confirming the ability of cryptochrome to mediate MF responses. As noticed, cryptochrome's signaling is tied to the photoreduction of FAD. The spin chemistry of this photoreduction process, which involves electron transfer from a chain of three tryptophans, can be modulated by the presence of an MF with the RPM. In *Arabidopsis*, the RPM in cryptochrome can produce an increase in the protein's signaling activity of about 10% for MF on the order of 5 G (500 µT), which is consistent with experimental results (Solov'yov et al., 2007).

Despite a variety of supporting evidence, it is still not clear whether cryptochromes have the properties required to respond to magnetic interactions orders of magnitude weaker than the thermal energy, $k_B T$. It has been shown that the kinetics and quantum yields of photo-induced flavin-tryptophan RPs in cryptochrome are indeed magnetically sensitive. The mechanistic origin of the MFE has been suggested, its dependence on the strength of the MF measured, and the rates of relevant spin-dependent, spin-independent, and spin-decoherence processes determined. Therefore, cryptochrome appears to fit for purpose as a chemical magnetoreceptor (Maeda et al., 2012).

One of the most stimulating observations in plant evolution is a correlation between the occurrence of GMF reversals (or excursions) and the moment of the radiation of angiosperms (Figure 5.5) (Occhipinti et al., 2014). This led to the hypothesis that alterations in GMF polarity may play a role in plant evolution. *A. thaliana* exposed to artificially reversed GMF conditions in the presence of light showed significant effects on plant growth and gene expression, supporting the hypothesis that the GMF reversal contributes to inducing changes in plant development that might justify a higher selective pressure, eventually leading to plant evolution (Bertea et al., 2015).

In *A. thaliana* seedlings grown under NNMF in the presence of light, flowering time was found to be delayed compared with seedlings grown in normal GMF (Xu et al., 2013, 2015, 2017, 2018). Moreover, the transcription level of a few flowering-related genes also changed (Xu et al., 2012). Furthermore, the biomass accumulation of plants in NNMF was significantly suppressed at the time when plants were switching from vegetative growth to reproductive growth compared to that of plants grown in normal

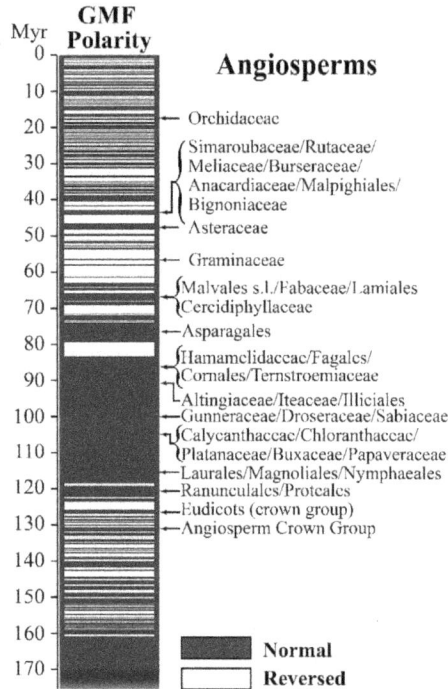

FIGURE 5.5 Geomagnetic field reversals and angiosperm evolution. In the direct comparison of GMF polarity and diversion of angiosperms it is interesting to note that most of the diversion occurred during periods of normal magnetic polarity. (From Maffei (2014).)

GMF. This was caused by a delay in flowering of plants in NNMF, which resulted in a significant reduction in the harvest index of plants in NNMF compared with that of control plants. Therefore, the removal of the local GMF negatively affects the reproductive growth of *A. thaliana*, which thus affects the yield and harvest index (Xu et al., 2013). Since timing of flowering is crucial to the life cycle of plants, it is not surprising that plants constantly monitor environmental signals to adjust the timing of the floral transition (Capovilla et al., 2015), but it is amazing that this is exquisitely sensitive to the GMF. Plant flowering time is controlled by several genes, including circadian clock-associated genes (Hara et al., 2014), genes involved both in the transition from the vegetative to the reproductive phase (Gu et al., 2013) and in the precise control of flowering (Song et al., 2014), and microRNA regulation (Spanudakis and Jackson, 2014; Hong and Jackson, 2015). Current models provide us with a basis on which to address a number of fundamental issues for a better understanding of the molecular mechanisms by which plants respond to environmental stimuli to control flowering time (Fornara et al., 2010).

A. *thaliana* plants grown under NNMF show a significant and consistent downregulation of gene expression in early induction times for *CCA1, CO, FD, FKF1, FRI, FT, GA20ox1, GA20ox2, LFY, LHY, TOC1, TSF*, and *WUS. AP1, GI*, and *STM* were downregulated at later times, whereas a significant upregulation was found for *FLC* during early floral induction, but the gene was downregulated during later stages of floral development. In the floral meristem, despite its repressing activity on flowering, *FLC* was significantly downregulated and during early times of flowering, *LFY, SVP, SDG26*, and, particularly, *FD* showed a significant upregulation, whereas *AGL24* was significantly downregulated in early times and upregulated during flowering. *SOC1* regulation occurred only during late flowering, whereas *LFY* downregulation occurred later. *GA2ox1* and *GA20ox1* were mildly downregulated in the early phase of floral development, whereas a strong downregulation was observed for GA20ox2 during early flowering.

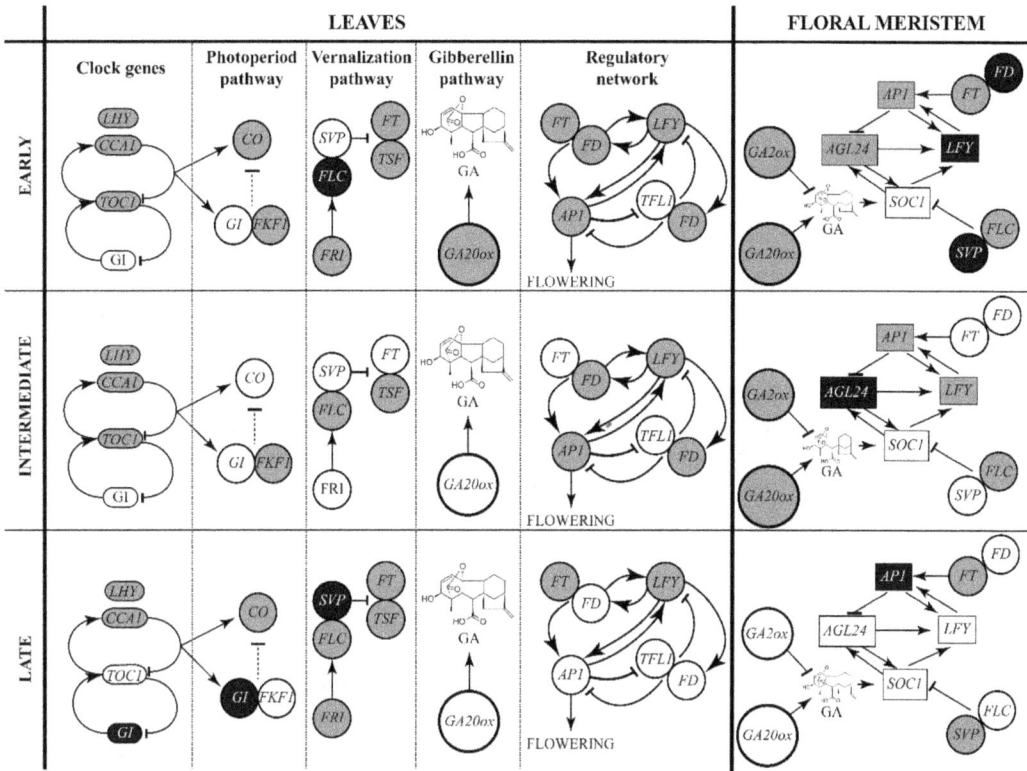

FIGURE 5.6 Schematic representation of gene expression patterns in *Arabidopsis thaliana* leaves and floral meristem under near-null magnetic field (NNMF). Leaf gene regulation of clock, photoperiod pathway, vernalization pathway, gibberellin pathway, and regulatory network are depicted during early, intermediate, and late stages of flowering. An early downregulation of clock, photoperiod, gibberellin, and vernalization pathways is accompanied by a downregulation of *AP1* and *GA20ox*. In the floral meristem, NNMF determines an early downregulation of the gibberellin pathway, *AGL24* and *AP1*, with a significant upregulation of *LFY*, *FD*, and *SVP*. In intermediate times, *AGL24* is upregulated, whereas at late times *AP1* is upregulated. The gibberellin pathway is downregulated in early and intermediate times, whereas no regulation is found in late times. In both leaves and floral meristem data, upregulation is shown in black areas, downregulation in gray areas, and no regulation in white. According to Fornara et al. (2010), Jaeger et al. (2013), and Valentim et al. (2015). (Modified from Agliassa et al. (2018a).)

NAC050 and NAC052 were also significantly downregulated. Therefore, exposure of *A. thaliana* to NNMF causes a delay in the transition to flowering due to a combined regulation of leaves and floral meristem genes (Figure 5.6).

An early downregulation of clock, photoperiod, gibberellin, and vernalization pathways is accompanied by a downregulation of *AP1* and *GA20ox*. *FLC* is upregulated by NNMF in early flowering induction. In the floral meristem, the strong downregulation of *FT* and *FLC* in early phases of floral development is accompanied by the downregulation of the gibberellin pathway and the upregulation of *FD*, *SVP*, and the transcription factor *LFY*. The common downregulation of *AP1* in both floral meristem and leaves is associated with the delay in flowering. In the floral meristem and leaves, the progressive upregulation of *AGL24*, *AP1*, *GI*, and *SVP* from early to late phase of plant development is correlated to the delay of flowering. These events are followed by the progressive reduction of gibberellin pathway downregulation. These results indicate that NNMF does not prevent flowering, and that variations of the MF are sufficient to modulate specific genes in the early stages of flower induction that are associated with the observed delay. However, the gene expression regulation might not reflect the post-translational

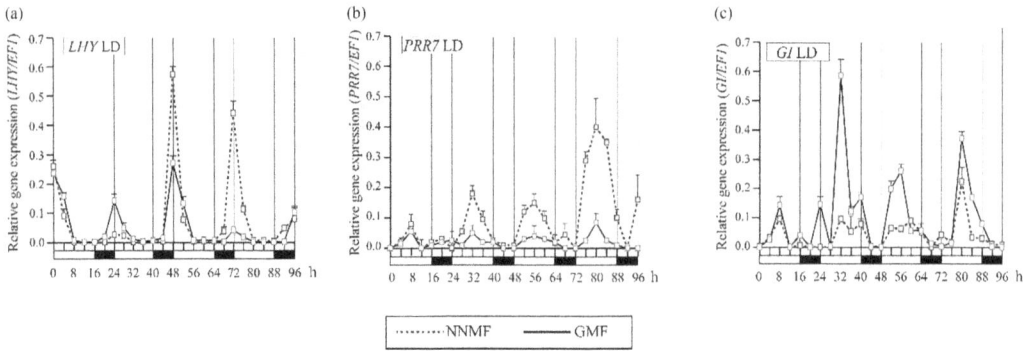

FIGURE 5.7 Time course of LHY, PRR7, and GI relative expression in *Arabidopsis thaliana* grown under GMF and NNMF in long day conditions (LD). *LHY* (A) and *PRR7* (B) under LD conditions always show increased gene expressions when exposed to NNMF, with respect to GMF. *GI* (C) always shows a reduced gene expression under NNMF, when compared to GMF. In all plots, white boxes indicate the light phase, whereas black boxes indicate the dark phase. (Modified from Agliassa and Maffei (2019).)

modifications that lead to the production of proteins involved in flowering control, and proteomic studies are necessary to better asses the role of NNMF on flowering control.

Plant endogenous clock consists of self-sustained interlocked transcriptional/translational feedback loops whose oscillation regulates many circadian processes, including gene expression. Its free-running rhythm can be entrained by external cues, which can influence all clock parameters. The quantitative expression (qRT-PCR) of three clock genes (*LHY, GI*, and *PRR7*) in time-course experiments under long day conditions in *A. thaliana* seedlings exposed to GMF and NNMF conditions reveals that reduction of GMF to NNMF prompted a significant increase of the gene expression of *LHY* and *PRR7*, whereas an opposite trend was found for *GI* gene expression. Exposure of *Arabidopsis* to NNMF altered clock gene amplitude, regardless of the presence of light, by reinforcing the morning loop (Figure 5.7). Therefore, these results are consistent with the existence of a plant magnetoreceptor that affects the *Arabidopsis* endogenous clock (Agliassa and Maffei, 2019).

To comprehensively investigate the influence of the GMF on *A. thaliana* photoreceptor signaling, wild-type (WT) *Arabidopsis* seedlings and photoreceptor-deficient mutants (*cry1cry2, phot1, phyA*, and *phyAphyB*) were exposed to NNMF and GMF under different light wavelengths. For the first time, the influence of the GMF on photoreceptor signaling both under red and blue light was shown (Agliassa et al., 2018b). Overall, despite the absence of a GMF-induced changes in *Arabidopsis* seedling photomorphogenesis, a significant GMF-dependent differential shoot/root regulation of genes expressed following photoreceptor activation after 72 hours exposure to GMF with respect to NNMF conditions was found. In particular, under blue light, the GMF regulation of gene expression appears to be partially dependent on cryptochrome activation, which is enhanced in terms of increased cry1 phosphorylation and cry2 degradation. Under red light, the GMF-dependent regulation of light-induced genes is partially mediated by phyA and phyB, whose activation is altered by cry1, cry2, and phot1 in their inactive form (Figure 5.8). If we consider that the red light response to GMF is not limited to phyA and phyB (Jeong et al., 2016), the contribution of other phytochromes to this response cannot be excluded. Therefore, despite the involvement of cryptochrome, and the possibility of a cryptochrome-based RPM, magnetoreception in *Arabidopsis* appears to be different from the mechanism thought to be responsible for the ability of migratory songbirds to detect the direction of the GMF. These results suggest also that other processes besides photoreceptor activation could be probably involved in GMF perception (Agliassa et al., 2018b).

The reduction of the GMF to NNMF affects the accumulation of metals in plant tissues, mainly iron (Fe) and zinc (Zn) content, while the content of other metals such as copper (Cu) and manganese (Mn)

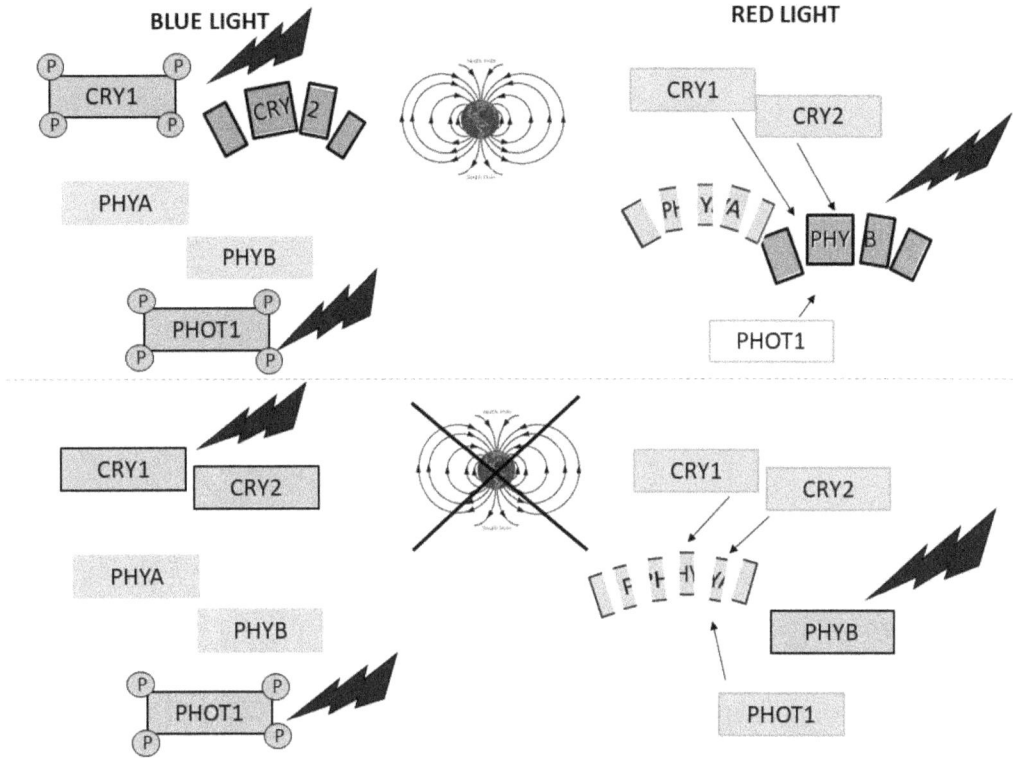

FIGURE 5.8 The GMF influences the photoreceptor activation and signaling in *Arabidopsis*. Under blue light, the GMF regulation of gene expression is mainly dependent on cryptochromes, whose activation is enhanced in terms of increased cry1 phosphorylation and cry2 degradation. By contrast, phot1 phosphorylation is not affected by the GMF. Under red light, cry1 and phot1 in their inactive form contribute to the GMF-dependent increase in phyB activation and the GMF-dependent decrease in phyA: phyB degradation is indeed enhanced by the GMF, whereas that of phyA is enhanced under NNMF conditions. (From Agliassa et al. (2018b).)

is not affected. Accordingly, Fe uptake genes were induced in the roots of NNMF-exposed plants and the root Fe reductase activity was affected by transferring GMF-exposed plant to NNMF condition. Under Fe deficiency, NNMF-exposed plants displayed a limitation in the activation of Fe-deficiency induced genes. Such an effect was associated with the strong accumulation of Zn and Cu observed under NNMF conditions (Islam et al., 2020a). Overall, these results provide evidence on the important role of the GMF on the iron uptake efficiency of plants and the importance of such findings is also associated to the variability of the GMF with latitude as well as global GMF inversions during life evolution (Maffei, 2014). The importance of the GMF on metals (e.g., Fe) uptake might be also linked to the evolution of Strategy I (Fe reductive-mechanism) and II (non-reductive mechanism) mechanisms in plants. Considering the importance of MF variations in different environments and in future space exploration where plants might experience MF conditions different from the GMF, our results stress the importance to deepen the investigation on the effect of MF on plant nutrients homeostasis in order to understand how magnetoreception occurs in plants and in turn how nutrients availability changes depend on GMF fluctuating values.

In the presence of light, the lipid content (fatty acids and surface alkanes, SA) and mineral nutrients of *Arabidopsis* were also affected by changes in MF. A progressive increase of SA with carbon numbers between 21 and 28 was found in plants exposed to NNMF from bolting to flowering developmental stages, whereas the content of some fatty acids significantly increased in rosette, bolting,

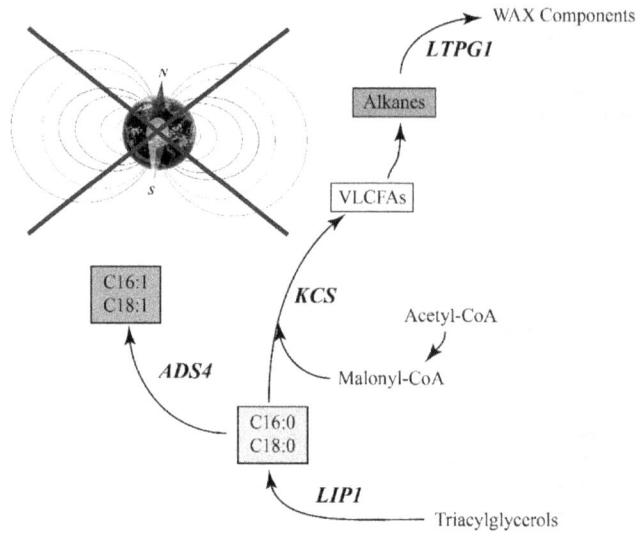

FIGURE 5.9 Plant lipid responses to reduced magnetic field.

and seed-set developmental stages. Variations in SA composition were correlated to the differential expression of several *Arabidopsis* 3-ketoacyl-CoAsynthase (KCS) genes, including *KCS1*, *KCS5*, *KCS6*, *KCS8*, and *KCS12*, a lipid transfer protein (*LTPG1*), and a lipase (*LIP1*). Ionomic analysis showed also a significant variation in some micronutrients (Fe, Co, Mn, and Ni) and macronutrients (Mg, K, and Ca) during the plant development of plants exposed to NNMF (Islam et al., 2020b). Figure 5.9 summarizes the effects of GMF reduction on lipid metabolism. These results and the results shown above indicate that *A. thaliana* responds to variations of the GMF which are perceived as is typical of abiotic stress responses.

The GMF regulates genes in both shoot and roots, suggesting that both organs can sense the GMF. However, 49% of the genes are regulated in a reverse direction in these organs, meaning that the resident signaling networks define the up- or downregulation of specific genes. The set of GMF-regulated genes strongly overlaps with various stress-responsive genes, implicating the involvement of one or more common signals, such as reactive oxygen species, in these responses. The biphasic dose response of GMF-responsive genes indicates a hormetic response of plants to the GMF that can sense and respond to the GMF using the signaling networks involved in stress responses (Paponov et al., 2021). It is interesting to note that (1) roots and shoot have different gene expression responses to the GMF, as almost 50% of the regulated genes were triggered in a reverse manner; (2) the effects of the GMF are related to activation of stress-responsive genes; and (3) the majority of identified GMF-responsive genes show biphasic dose-dependent expression, indicating a hormetic response of plants to MFs (Figure 5.10). The differential regulation of genes in roots and shoots might also be related to the different roles of plastids in the roots and shoots. Indeed, GO analysis has shown that genes related to chloroplast functions were overrepresented among the genes regulated by the GMF, indicating that different functions of plastids in roots and shoots can contribute to the differential responses of roots and shoots to the GMF. These observations support recent findings that chloroplasts are one of the main targets of MF effects in *Arabidopsis* (Jin et al., 2019). One important consideration is that the function of chloroplasts is related to the cellular redox status (Baier and Dietz, 2005), which can induce ROS imbalances and modulate the expression of genes induced by different stresses. Indeed, GO analysis has shown that variations in MFs affect the regulation of stress-responsive genes. Moreover, early gene modulations by MFs are associated with redox responses, implying that rapid rates of redox reactions, triggered by an MF, alter the metabolism of free radicals and ROS.

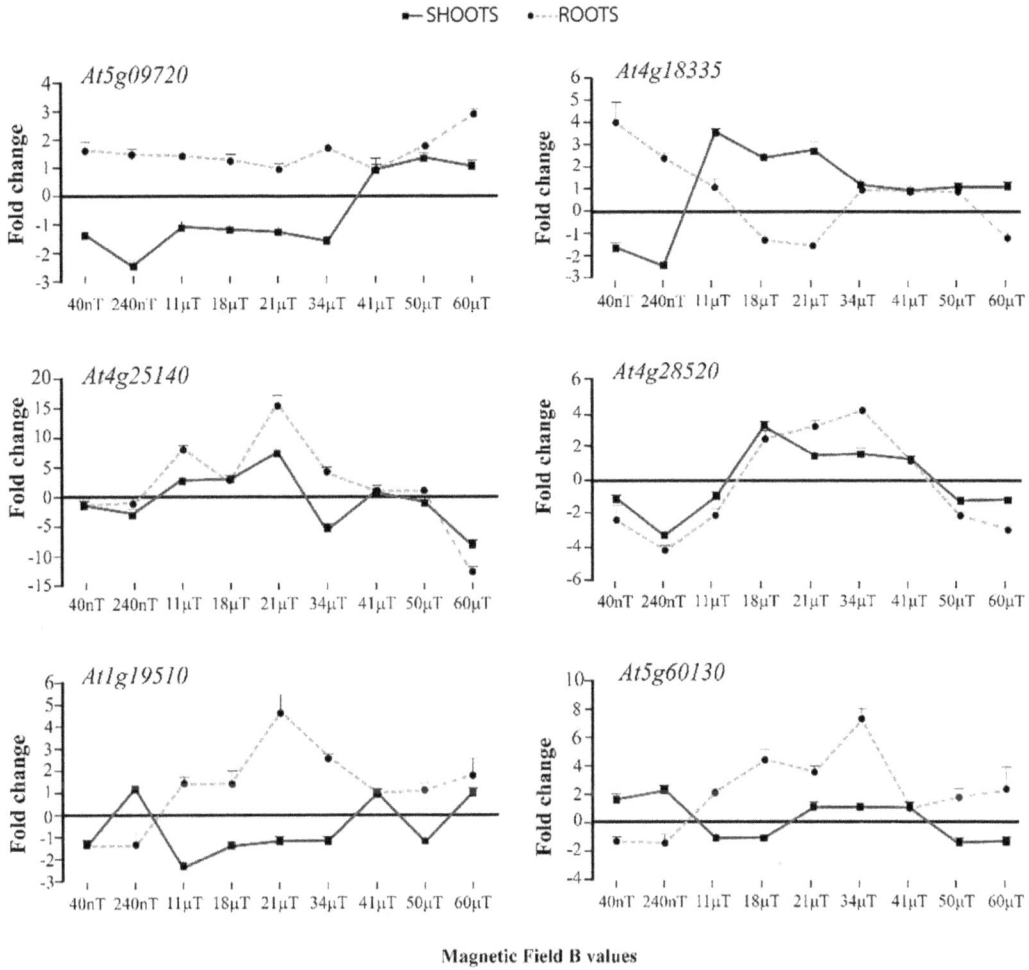

FIGURE 5.10 Differential gene expression of roots and shoots selected genes in response to varying MF intensity. The data are expressed as fold change in relation to controls (measured at 41 µT). In order to emphasize the visualization of data, fold change values below zero were plotted as −1/value, in order to obtain negative fold change values (indicating downregulation). Metric bars indicate standard deviation. (From Paponov et al. (2021).)

Cryptochrome modulates ROS in response to weak MFs through an alteration of the rate of redox reactions in the presence of an MF (Pooam et al., 2020a). This effect changes the cellular ROS production (in the nucleus and cytosol, and possibly also in other organelles) and is proposed to be similar in both plants and animals. Thus, the primary effect of an MF with respect to cryptochrome function has been postulated to be the modulation of ROS (Pooam et al., 2020a). This mechanism perfectly predicts an effect on cellular ROS signaling pathways; therefore, this hypothesis explains the ROS-related modulation of gene expression in response to MFs observed in recent studies (Paponov et al., 2021).

5.4 Light-Independent Magnetoreception

As alternative (but less efficient) pathways of photoreduction exist and are known to be active in vivo, the dark reoxidation is an interesting hypothesis to rationalize these findings in terms of a single, well-defined RP (Engelhard et al., 2014). The FADH$^{•}$ / O$_2^{-}$-model is principally endowed with a favorable property: the anisotropy of the singlet yield is markedly (by an order of magnitude) larger in RPs combining FAD with a partner radical with no significant hyperfine interactions, such as $O_2^{\cdot-}$, or better a hypothetic variety of it not subject to fast spin relaxation, conventionally denoted $Z^{\cdot}$ (Lee et al., 2014).

Cryptochromes undergo forward light-induced reactions involving electron transfer to excited state flavin to generate radical intermediates, which correlate with biological activity. A mechanism for the reverse reaction, namely dark reoxidation of protein-bound flavin in *Arabidopsis* cryptochrome (AtCRY1) by molecular oxygen, involves the formation of a spin-correlated FADH-superoxide RP (Muller and Ahmad, 2011). Under conditions of illumination, the cryptochrome photoreceptors are constantly cycling between inactive (oxidized) and activated (reduced) redox states, such that the net biological activity results from the sum of the light-induced (activating) and reverse (de-activating) redox reactions at any given timepoint. A model of the cryptochrome photocycle incorporating these elements and an estimation of the quantum efficiency of redox state interconversions both in vitro and in vivo has been recently derived (Procopio et al., 2016). However, when light and dark intervals are given intermittently, the plant MFE is observed even when the MF is given exclusively during the dark intervals between light exposures. This indicates that the magnetically sensitive reaction step in the cryptochrome photocycle must occur during flavin reoxidation, and likely involves the formation of reactive oxygen species (Pooam et al., 2019). A recent model of MFE on the cryptochrome photocycle involves activation of cryptochrome by flavin reduction which triggers conformational change leading to unfolding and subsequent phosphorylation of the C-terminal domain. The flavin is subsequently reoxidized by reaction with molecular oxygen that occurs independently of light (Ahmad, 2016). The effect of an applied MF on the cryptochrome photocycle occurs during the period of flavin reoxidation. The most likely effect is to alter the rate constant of reoxidation of the reduced flavin intermediates, and thereby alter the lifetime of the activated state (Figure 5.11). As discussed above, theoretical considerations have argued against a flavin/superoxide radical pair, which is formed in the course of flavin reoxidation as the magnetosensing intermediate in cryptochromes (Hore and Mouritsen, 2016); however, cryptochrome localized within living cells is in contact with many cellular metabolites, which, moreover, can move into the flavin pocket

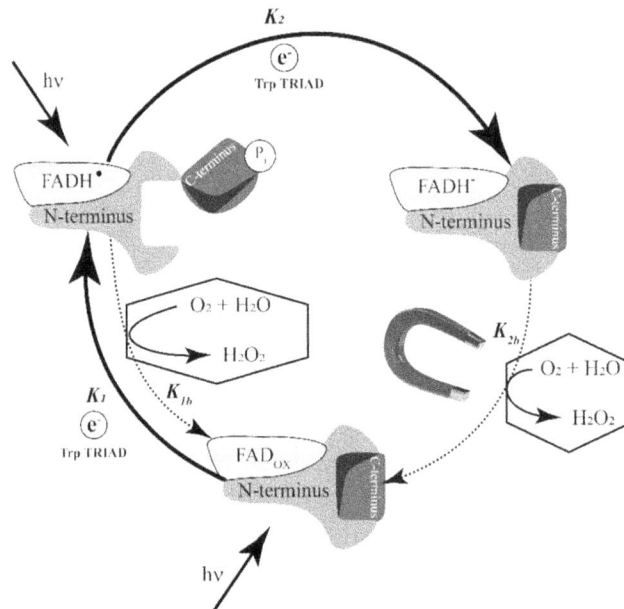

FIGURE 5.11   Model of MFE on the cryptochrome photocycle, see text for explanation. (Modified from Pooam et al. (2019).)

in close association with the flavin cofactor. Therefore, the possibility of the third-party cellular factors participating in the formation of RPs during the process of cryptochrome flavin reoxidation cannot be excluded (Pooam et al., 2019). In sum, there are at least two reaction steps in the course of cryptochrome photocycle that could in principle be altered by the magnetic fields: the step of flavin photoreduction and that of flavin reoxidation. In either case, the effect of the magnetic field would be to alter cryptochrome biological activity by changing the rate of formation of the active state (forward reaction) or the rate of disappearance of the active state (reoxidation reaction) (Pooam et al., 2020a).

The GMF was also found to impact photomorphogenic-promoting gene expression in etiolated seedlings, indicating the existence of a light-independent magnetoreception mechanism. In *Arabidopsis* in the absence of light, the most highly regulated gene in response to MF changes is *NDPK2* (Agliassa et al., 2018b), which is involved in the oxidative stress signaling (Kim et al., 2011). This result clearly implies the presence of a light-independent root magnetoreception mechanism that involves an oxidative response. These results are in agreement with previous studies on GMF reversal (Bertea et al., 2015). Root light-independent responses to MF variations have been demonstrated in plants under a continuous high gradient MF application, with a magnetophoretic plastid displacement and a consequent induction of root curvature (Kuznetsov et al., 1999). Therefore, these results indicate the possibility of a light-independent magnetoreception mechanism and further studies are necessary to better understand how roots are involved in magnetoreception.

The rhythmic expression of *Arabidopsis* clock genes was different under GMF with respect to NNMF under free rhythm running conditions in continuous darkness. The switching to free-running conditions under continuous darkness caused the internal clock oscillator resetting to its natural period length. Therefore, the GMF seems not to give any temporal signal to *Arabidopsis* clock under continuous darkness, thus excluding the GMF influence on the internal clock period as reported in animals (Bliss and Heppner, 1976). On the other hand, NNMF treatments showed that the internal clock gene amplitude was significantly ($p < 0.05$) different with respect to plants exposed to local GMF conditions, regardless of the light presence. Even though exposure to continuous darkness is known to reduce the amplitude of the clock rhythm (Salome et al., 2008), this was not observed for LHY (Figure 5.12b) and PRR7 (Figure 5.12d) under NNMF (Figure 5.12). Results exclude a possible role of the GMF as a ZT to *Arabidopsis* clock under continuous darkness and highlight the impact of NNMF on *Arabidopsis* clock gene amplitude, regardless of the presence of light. Dhiman and Galland (2018) have recently demonstrated that MF intensities from GMF to NNMF modulate *Arabidopsis* seedlings gene expression under both light and dark conditions. Furthermore, NNMF intensities can produce nonspecific biological effects on gene expression by affecting RNA polymerase rotation (Binhi and Prato, 2018).

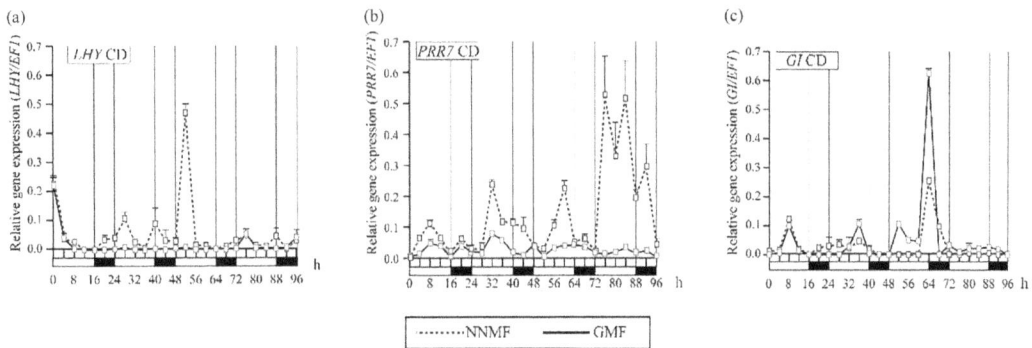

FIGURE 5.12 Time course of LHY, PRR7, and GI relative expression in *Arabidopsis thaliana* grown under GMF and NNMF in continuous darkness (CD). *LHY* (A) and *PRR7* (B) under CD conditions show increased gene expressions when exposed to NNMF, with respect to GMF. *GI* (C) shows a reduced gene expression under NNMF when compared to GMF. In all plots, white boxes indicate the light phase, whereas black boxes indicate the dark phase. (Modified from Agliassa and Maffei (2019).)

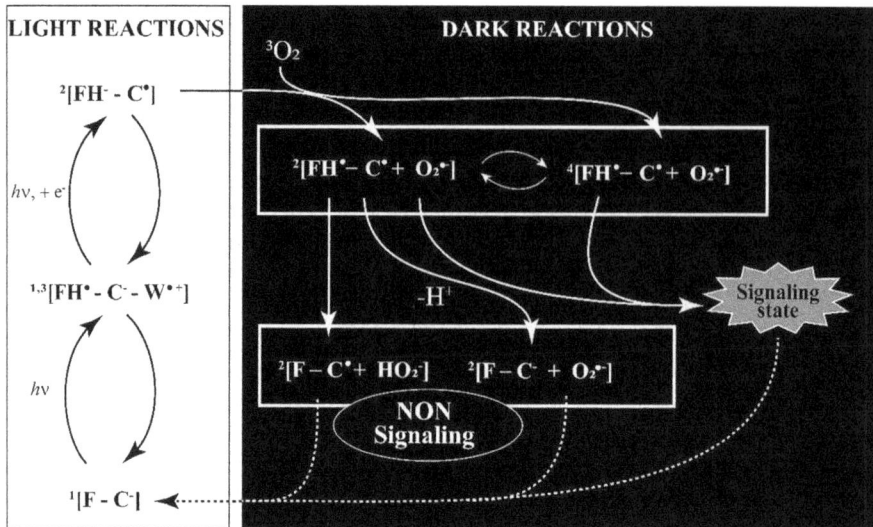

FIGURE 5.13 A tentative reaction scheme of an RP-based magnetic compass involving dark-state reoxidation of the fully reduced $FADH^-$, the superoxide radical anion ($O_2^{\cdot-}$)n and radical scavenging to avoid the detriment of fast spin relaxation associated with the latter. Details are discussed in the main text. (Adapted from Kattnig (2017).)

Recently, a tentative scheme has been proposed to show the dark-state reoxidation of the fully reduced $FADH^-$, the superoxide radical anion ($O_2^{\cdot-}$) and radical scavenging to avoid the detriment of fast spin relaxation associated with the latter. The spin-correlated $FADH^{\cdot}$ / $O_2^{\cdot-}$-RP is formed in a dark-state reaction from the fully reduced flavin, $FADH^-$, and molecular oxygen. Due to the $^3\Sigma$ ground state of O_2, this RP is generated exclusively in a (local) triplet state, which corresponds to both doublet and quartet states in the combined system including the initially uncorrelated radical scavenger. The doublet and quartet states interconvert via hyperfine interactions in $FADH^{\cdot}$ and $C^{\cdot}$ and the Zeeman interactions of all radicals. The MFE arises from the competition between the regeneration of the fully oxidized FAD and the spin-independent formation of the signaling state. It is assumed that the singlet recombination of the $FADH^{\cdot}$ / $O_2^{\cdot-}$-RP produces hydrogen peroxide (H_2O_2), possibly via the C4a-hydroperoxy-flavin, and that $C^{\cdot}$ reoxidizes $FADH^{\cdot}$ to FAD (Massey, 1994). Both assumptions are not critical as long as the products of these spin-selective reactions are disparate from the signaling state. Note that the flavin/tryptophan RP intermediate produced in the light-activation step is likely to be magnetosensitive as well. It is also assumed that the scavenger radical $C^{\cdot}$ is produced in the course of the photoreduction. It could be an oxidized electroactive residue within the protein (Kattnig, 2017) (Figure 5.13).

5.5 Magnetic Fields with a Higher Intensity with Respect to Geomagnetic Field

MFE depends on the strength of the MF, which can be classified as weak (<1 mT), moderate (1 mT to 1 T), strong (1–5 T), and ultrastrong (>5 T). Weak MF, as the GMF, can be perceived by animals and plants as described above. Strong and ultrastrong fields are of sufficient intensity to alter the preferred orientation of a variety of diamagnetic anisotropic organic molecules and their effects have been attributed to this mechanism. Moderate intensity static MFs (SMFs) influence those biological systems where function depends on the properties of excitable membranes. Inasmuch as these fields are insufficient to influence individual molecules, there must be something unique about biological membranes that make them susceptible to these fields in a way that would alter cell function (Rosen, 2003).

Diamagnetic anisotropy is the physical property of the membranes molecular structure that has the potential to be influenced by MFs. Many inorganic and virtually all organic compounds have some degree of diamagnetism. The relationship between the induced magnetic moment and the applied field characterizes the magnetic properties of a substance and is referred to as magnetic susceptibility, χ_m

$$\chi_m = \frac{M}{H}$$

where M is magnetization or magnetic moment per unit volume, and H is field strength. χ_m is dimensionless but with a negative sign for diamagnetic substances. The energetics for orientation in an MF are favorable for structures made up of a large number of parallel molecules and explains the effects of these fields on retinal rods, chloroplasts, lecithin vesicles, and synthetic phospholipid bilayers. Most of the diamagnetic anisotropy of lipids is contributed by their acyl chains and biological membranes, with their highly ordered phospholipid bilayer structure, would be expected to exhibit substantial diamagnetic properties (Rosen, 2003). The increased permeability of a phospholipid bilayer to low-molecular-weight solutes, during SMF exposure occurs with SMFs >10 mT and was evident within 1 min of exposure onset. This phenomenon has been attributed to deformational changes within the membrane because of its diamagnetic properties. In the presence of a moderate SMF, the enhanced diamagnetic anisotropy of biological membranes is sufficient to result in a slow reorientation of their phospholipid molecules (Rosen, 2003).

When a magnetic field is present in cell systems, three types of magnetic forces can act on subcellular components and ions:

1. the Lorentz force,

$$F_L = [vB]q$$

where B is the magnetic induction, q is the ion electric charge, and v is its velocity
2. the magnetic gradient force (Zablotskii et al., 2014)

$$F_{\nabla B} \propto \nabla B^2$$

where ∇ is the differential operator nabla
3. the concentration-gradient magnetic force (Svendsen and Waskaas, 2020)

$$F_{\nabla n} \propto B^2 \nabla n$$

(where ∇_n is the gradient of the concentration of the diamagnetic and paramagnetic species). There are hundreds of diamagnetic and paramagnetic species inside living cells, many of which may have very large concentration gradients (Sear, 2019) and the presence of a high MF (HMF), and a relatively large concentration gradient magnetic force is operative.

Cell behavior depends on the membrane potential and bioelectric signals control. For example, undifferentiated cells have low membrane potential that allows them to be depolarized and highly plastic. In contrast, cells that are mature, terminally differentiated, and quiescent tend to be hyperpolarized. Therefore, by changing the membrane voltage, one can also control cell rigidity and functions. Driving the cell membrane potential with an MF could represent a new tool for cell modulation. Unlike the electric field, the MF is not attenuated by living tissue and penetrates through the whole body. By considering the Nernst potential, the free-energy change for the diffusion of an electrolyte into the cell is

$$\Delta G = RT \ln\left(\frac{n_i}{n_o}\right) + zFV_m$$

where z is the ion valence ($z=+1$ for a positive, univalent ion), F is the Faraday constant, R is the gas constant, T is the absolute temperature, V_m is the potential difference between the two sides of the membrane, and n_o and n_i are the ion concentrations outside and inside a cell, respectively. By setting ΔG to zero, which is the case when the movement of ions is at equilibrium, one can derive the Nernst equation:

$$V_m = \frac{RT}{zF}\ln\left(\frac{n_o}{n_i}\right)$$

When a highly uniform MF ($\nabla B=0$) is applied to a cell, the magnetic concentration-gradient force acts on diamagnetic and paramagnetic ions and can either assist or oppose ion movements through the membrane. The volume density of the concentration-gradient magnetic force is given by Hinds et al. (2001)

$$\vec{f} = \frac{\chi B^2}{2\mu_0}\vec{\nabla}n$$

where n is the molar concentration of ions with the molar magnetic susceptibility χ and μ_0 is the magnetic permeability. When the ions diffuse in the presence of an MF, the free-energy (per mole) change is

$$\Delta G = RT\ln\left(\frac{n_i}{n_o}\right) + zFV_m \mp \int_{\chi_0}^{\chi_i} vf(\chi)\,dx$$

where $v=1/n$ is the molar volume of a diffusing substance and x_0 and x_i are the coordinates defined by the conditions $n(\chi_0)=n_o$ and $n(x_i)=n_i$. The last term of this equation represents the work of the magnetic concentration-gradient forces when a mole of either diamagnetic or paramagnetic ions diffuses across a membrane. The "plus" and "minus" signs correspond to the two limiting cases; the magnetic force either assists or opposes the electric force exerted on ions moving across the membrane. Thus, changing the ion flux balance due to the magnetic concentration-gradient forces leads to changes in the cell membrane potential, as obtained from the equation

$$V_m(B) = \frac{1}{zF}\left(RT \mp \frac{\chi B^2}{2\mu_0}\right)\ln\left(\frac{n_o}{n_i}\right)$$

Estimations made from this equation for K^+ ions with the magnetic susceptibility $\chi=-188.5\times10^{-12}\,m^3/mol$, $n_i=140\,mM$, $n_o=5\,mM$, and $B=100$ T give a magnetic increase to the membrane potential as small as approximately $\pm0.026\,mV$, which is significantly smaller than the membrane potential created by K^+ ions without a magnetic field, $-89.1206\,mV$. Similar estimations for Ca^{2+} ions with the magnetic susceptibility $\chi=-150.8\times10^{-12}\,m^3/mol$, the concentrations $n_i=0.0001\,mM$ and $n_o=3\,mM$, and $B=100$ T give a magnetic increase of approximately $\pm0.0321\,mV$. Thus, for a strong MF with $B=100$ T, the estimated magnetic contribution to the membrane potential would be in the order of $10^{-2}\,mV$, while the minimum membrane potential change required to modify many cell functions is in the order of $1–10\,mV$. Therefore, this estimated change of the membrane potential seems to be too small to change anything in the cell. Relative changes in the V_m are obtained only at very high MF intensities (Figure 5.14). Therefore, the variations in channel activities found after exposure to GMF or NNMF intensities cannot be ascribed to the direct effect of the MF, being the consequence of the cascade of events after MF perception.

Several opinions have been suggested and calculations reflect that if the magnetic force on mobile ions is considered, magnetic flux density can be effective as long as it is >20 T. However, much smaller values have been observed to be effective in practice.

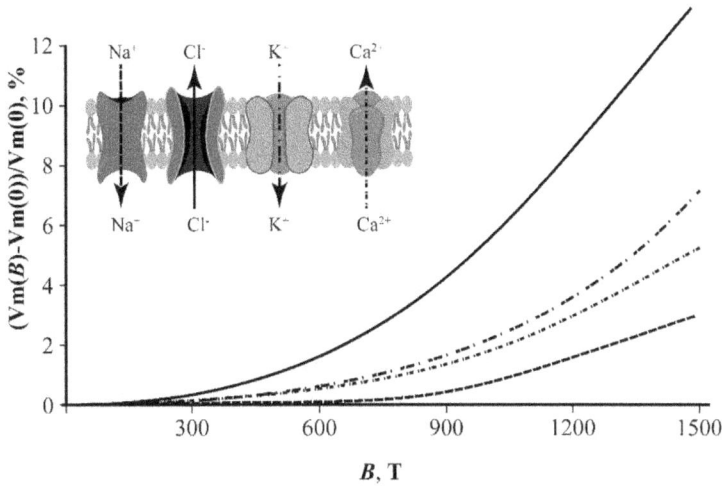

FIGURE 5.14 Relative changes of the membrane potential $[(V_m(B) - V_m(0))/V_m(0)] \times 100\%$, versus B for individual contributions of K^+, Na^+, Ca^{2+}, and Cl^- ions as calculated from the last equation. (Adapted from Zablotskii et al. (2021).)

The possibility that strong, SMFs might have an influence on biological processes has been discussed, and reports implicate high MFs in alterations of the cleavage plane during cell division (Denegre et al., 1998) and other cellular disorders (Valiron et al., 2005). Nevertheless, the common viewpoint is that presently achievable SMFs do not have a lasting effect on biological systems (Paul et al., 2006). In *Arabidopsis*, high MFs may compromise some aspects of the transcriptional machinery, and effectively arrest the process. This field dependence may suggest that magnetic orientation or *magnetophoresis* plays a role in the seemingly dual nature of the response. The biomacromolecules involved in signal transduction and gene regulation may experience forces and/or torques that are induced by the presence of the MF. In addition, the macromolecules may experience magnetophoresis due to forces generated by inhomogeneities in the applied MF.

According to Paul et al. (2006), MFE is sufficient to perturb the delicate conformational dynamics involved in aspects of gene regulation, thereby resulting in the differential expression of a variety of genes in the plant. The magnetic orientation effects are estimated to be 10–100 times larger than the magnetophoretic forces, and thus it is likely that magnetophoresis plays a minor role in the induction or repression of gene expression. Nominally 15 T is the threshold of field strength required to initiate such stress response, indicating that macromolecular orientation plays a role in field strengths >15 T. Therefore, exposure to MFs above 15 T induces the perturbation of metabolic processes in the presence of strong MFs and may be useful for guiding future research designed to calibrate safe exposure standards for living organisms (Paul et al., 2006).

Another interesting aspect is that nonuniform MFs may exert a ponderomotive force on amyloplasts which can result in intracellular magnetophoresis. Thus, plants can perceive an MF of sufficient intensity and gradient and respond to the resulting amyloplast displacement as they do to gravity. In fact, by using high gradient MFs it is possible to induce curvature in roots, an effect defined magnetotropism, although it seems that the cause of the growth response is a ponderomotive force and not the MF (Penuelas et al., 2004). Magnetic gradients are therefore an important tool to study localized mass interactions independent of gravity effects on the entire organism. The forces generated by a magnetic system are sufficient to provide a directional stimulus (i.e., induce curvature) even under weightlessness conditions and that depends on the distance, the magnetic gradient affects starch particles similar to gravity (Hasenstein et al., 2013).

5.6 Conclusions

Variations in the MF intensity influence many plant biological processes. The mechanism of plant magnetoreception/perception is still unknown; however, the RPM of chemical magnetosensing appears to adequately explain the MFEs observed in plants. The flavin-based RP system is one of the most studied MFE and an RP-based sensor may also be enhanced by interactions with its fluctuating environment, a property apparently found in many areas of "quantum biology." The only currently hypothesized RP that may potentially explain plant MFE is FADH$^{\bullet}$ / O$_2^-$, generated in the reoxidation of the fully (photo)reduced FADH$^-$. Unlike the electric field, the MF is not attenuated by living tissue and penetrates through the whole plant body. Relative changes in the V_m are obtained only at very high MF intensities. Therefore, the variations in channel activities found after exposure to GMF or NNMF intensities cannot be ascribed to the direct effect of the MF, being the consequence of the cascade of events after MF perception. Exposure to MFs above 15 T induces the perturbation of metabolic processes in the presence of strong MFs and may be useful for guiding future research designed to calibrate safe exposure standards for living organisms. Plants show both light-dependent and light-independent magneto perception, and recent data suggest that different organs may perceive MFE in a differential way, with a typical hormetic behavior. Further studies are required to dissect the plant signaling cascade of events triggered by MF variations, which remain still far from being deciphered.

References

Agliassa, C., and Maffei, M.E. (2019). Reduction of geomagnetic field (GMF) to near null magnetic field (NNMF) affects some *Arabidopsis thaliana* clock genes amplitude in a light independent manner. *J Plant Physiol* 232, 23–26. doi: 10.1016/j.jplph.2018.11.008.

Agliassa, C., Narayana, R., Bertea, C.M., Rodgers, C.T., and Maffei, M.E. (2018a). Reduction of the geomagnetic field delays *Arabidopsis thaliana* flowering time through downregulation of flowering-related genes. *Bioelectromagnetics* 39(5), 361–374. doi: 10.1002/bem.22123.

Agliassa, C., Narayana, R., Christie, J.M., and Maffei, M.E. (2018b). Geomagnetic field impacts on cryptochrome and phytochrome signaling. *J Photochem Photobiol B* 185, 32–40. doi: 10.1016/j.jphotobiol.2018.05.027.

Ahmad, M. (2016). Photocycle and signaling mechanisms of plant cryptochromes. *Curr Opin Plant Biol* 33, 108–115. doi: 10.1016/j.pbi.2016.06.013.

Ahmad, M., and Cashmore, A.R. (1997). The blue-light receptor cryptochrome 1 shows functional dependence on phytochrome A or phytochrome B in *Arabidopsis thaliana*. *Plant J* 11(3), 421–427. doi: 10.1046/j.1365-313X.1997.11030421.x.

Ahmad, M., Galland, P., Ritz, T., Wiltschko, R., and Wiltschko, W. (2007). Magnetic intensity affects cryptochrome-dependent responses in *Arabidopsis thaliana*. *Planta* 225(3), 615–624. doi: 10.1007/s00425-006-0383-0.

Albaqami, M., Hammad, M., Pooam, M., Procopio, M., Sameti, M., Ritz, T., et al. (2020). *Arabidopsis* cryptochrome is responsive to Radiofrequency (RF) electromagnetic fields. *Sci Rep* 10(1). doi: 10.1038/s41598-020-67165-5.

Azizi, S.M.Y., Sarghein, S.H., Majd, A., and Peyvandi, M. (2019). The effects of the electromagnetic fields on the biochemical components, enzymatic and non-enzymatic antioxidant systems of tea *Camellia sinensis* L. *Physiol Mol Biol Plants* 25(6), 1445–1456. doi: 10.1007/s12298-019-00702-3.

Baier, M., and Dietz, K.J. (2005). Chloroplasts as source and target of cellular redox regulation: A discussion on chloroplast redox signals in the context of plant physiology. *J Exp Botany* 56(416), 1449–1462.

Bertea, C.M., Narayana, R., Agliassa, C., Rodgers, C.T., and Maffei, M.E. (2015). Geomagnetic field (GMF) and plant evolution: Investigating the effects of GMF reversal on *Arabidospis thaliana* development and gene expression. *J Visual Exp* 105, e53286.

Binhi, V.N., and Prato, F.S. (2018). Rotations of macromolecules affect nonspecific biological responses to magnetic fields. *Sci Rep* 8. doi: 10.1038/s41598-018-31847-y.

Bliss, V.L., and Heppner, F.H. (1976). Circadian activity rhythm influenced by near zero magnetic field. *Nature* 261, 411. doi: 10.1038/261411a0.

Cailliez, F., Müller, P., Firmino, T., Pernot, P., and de la Lande, A. (2016). Energetics of photoinduced charge migration within the tryptophan tetrad of an animal (6–4) photolyase. *J Am Chem Soc* 138(6), 1904–1915. doi: 10.1021/jacs.5b10938.

Capovilla, G., Schmid, M., and Pose, D. (2015). Control of flowering by ambient temperature. *J Exper Bot* 66(1), 59–69. doi: 10.1093/jxb/eru416.

Chipman, D.M., and Rassolov, V.A. (1997). New operators for calculation of indirect nuclear spin-spin coupling constants. *J Chem Phys* 107(14), 5488–5495. doi: 10.1063/1.474253.

Denegre, J.M., Valles, J.M., Jr., Lin, K., Jordan, W.B., and Mowry, K.L. (1998). Cleavage planes in frog eggs are altered by strong magnetic fields. *Proc Nat Acad Sci U S A* 95(25), 14729–14732. doi: 10.1073/pnas.95.25.14729.

Dhiman, S.K., and Galland, P. (2018). Effects of weak static magnetic fields on the gene expression of seedlings of *Arabidopsis thaliana*. *J Plant Physiol* 231, 9–18. doi: doi.org/10.1016/j.jplph.2018.08.016.

Engelhard, C., Wang, X., Robles, D., Moldt, J., Essen, L.O., Batschauer, A., et al. (2014). Cellular metabolites enhance the light sensitivity of *Arabidopsis* cryptochrome through alternate electron transfer pathways. *Plant Cell* 26(11), 4519–4531. doi: 10.1105/tpc.114.129809.

Evans, E.W., Li, J., Storey, J.G., Maeda, K., Henbest, K.B., Dodson, C.A., et al. (2015). Sensitive fluorescence-based detection of magnetic field effects in photoreactions of flavins. *Phys Chem Chem Phys* 17(28), 18456–18463. doi: 10.1039/c5cp00723b.

Fornara, F., de Montaigu, A., and Coupland, G. (2010). SnapShot: Control of flowering in *Arabidopsis*. *Cell* 141(3), 550–552.

Gu, X., Wang, Y., and He, Y. (2013). Photoperiodic regulation of flowering time through periodic histone deacetylation of the florigen gene FT. *PLoS Biol*. 11(9). doi: 10.1371/journal.pbio.1001649.

Guo, J.P., Wan, H.Y., Matysik, J., and Wang, X.J. (2018). Recent advances in magnetosensing cryptochrome model systems. *Acta Chim Sinica* 76(8), 597–604. doi: 10.6023/a18040173.

Hammad, M., Albaqami, M., Pooam, M., Kernevez, E., Witczak, J., Ritz, T., et al. (2020). Cryptochrome mediated magnetic sensitivity in *Arabidopsis* occurs independently of light-induced electron transfer to the flavin. *Photochem Photobiol Sci* 19(3), 341–352. doi: 10.1039/c9pp00469f.

Hara, M., Kamada, H., and Mizoguchi, T. (2014). Co-Expressed with clock genes LHY and CCA1 1 (CEC1) is regulated by LHY and CCA1 and plays a key role in phase setting of GI in *Arabidopsis thaliana*. *Plant Biotechnol* 31(1), 35. doi: 10.5511/plantbiotechnology.13.1108a.

Harmer, S.L., and Brooks, C.J. (2018). Growth-mediated plant movements: Hidden in plain sight. *Curr Opin Plant Biol* 41, 89–94. doi: 10.1016/j.pbi.2017.10.003.

Hasenstein, K.H., John, S., Scherp, P., Povinelli, D., and Mopper, S. (2013). Analysis of magnetic gradients to study gravitropism. *Am J Bot* 100(1), 249–255.

Hayashi, H. (2004). *Introduction to Dynamic Spin Chemistry* Singapore: World Scientific Publisher.

Hinds, G., Coey, J.M.D., and Lyons, M.E.G. (2001). Influence of magnetic forces on electrochemical mass transport. *Electrochem Commun* 3(5), 215–218. doi: doi: 10.1016/S1388-2481(01)00136-9.

Hong, Y., and Jackson, S. (2015). Floral induction and flower formation-the role and potential applications of miRNAs. *Plant Biotechnol J* 13(3), 282–292. doi: 10.1111/pbi.12340.

Hore, P.J., and Mouritsen, H. (2016). The radical-pair mechanism of magnetoreception. *Annu Rev Biophys* 45, 299–344.

Hore, P.J., Ivanov, K.L., and Wasielewski, M.R. (2020). Spin chemistry. *J Chem Phys* 152(12), 120401. doi: 10.1063/5.0006547.

Islam, M., Maffei, M.E., and Vigani, G. (2020a). The geomagnetic field is a contributing factor for an efficient iron uptake in *Arabidopsis thaliana*. *Front Plant Sci* 11, 325. doi: 10.3389/fpls.2020.00325.

Islam, M., Vigani, G., and Maffei, M.E. (2020b). The geomagnetic field (GMF) modulates nutrient status and lipid metabolism during *Arabidopsis thaliana* plant development. *Plants (Basel)* 9(12), 1729. doi: 10.3390/plants9121729.

Jaeger, K.E., Pullen, N., Lamzin, S., Morris, R.J., and Wigge, P.A. (2013). Interlocking feedback loops govern the dynamic behavior of the floral transition in *Arabidopsis*. *Plant Cell* 25(3), 820–833.

Jeong, A.R., Lee, S.S., Han, Y.J., Shin, A.Y., Baek, A., Ahn, T., et al. (2016). New constitutively active phytochromes exhibit light-independent signaling activity. *Plant Physiol* 171(4), 2826–2840. doi: 10.1104/pp.16.00342.

Jin, Y., Guo, W., Hu, X.P., Liu, M.M., Xu, X., Hu, F.H., et al. (2019). Static magnetic field regulates *Arabidopsis* root growth via auxin signaling. *Sci Rep* 9. doi: 10.1038/s41598-019-50970-y.

Jones, A.R. (2016). Magnetic field effects in proteins. *Mol Phy* 114(11), 1691–1702. doi: 10.1080/00268976. 2016.1149631.

Kattnig, D.R. (2017). Radical-pair-based magnetoreception amplified by radical scavenging: Resilience to spin relaxation. *J Phys Chem B* 121(44), 10215–10227. doi: 10.1021/acs.jpcb.7b07672.

Kattnig, D.R., and Hore, P.J. (2017). The sensitivity of a radical pair compass magnetoreceptor can be significantly amplified by radical scavengers. *Sci Rep* 7. doi: 10.1038/s41598-017-09914-7.

Kattnig, D.R., Evans, E.W., Dejean, V., Dodson, C.A., Wallace, M.I., Mackenzie, S.R., et al. (2016a). Chemical amplification of magnetic field effects relevant to avian magnetoreception. *Nat Chem* 8(4), 384–391. doi: 10.1038/nchem.2447.

Kattnig, D.R., Solov'yov, I.A., and Hore, P.J. (2016b). Electron spin relaxation in cryptochrome-based magnetoreception. *Phys Chem Chem Phys* 18(18), 12443–12456. doi: 10.1039/c5cp06731f.

Kempster, R.M., McCarthy, I.D., and Collin, S.P. (2012). Phylogenetic and ecological factors influencing the number and distribution of electroreceptors in elasmobranchs. *J Fish Biol* 80(5), 2055–2088. doi: 10.1111/j.1095-8649.2011.03214.x.

Kim, Y.H., Kim, M.D., Choi, Y.I., Park, S.C., Yun, D.J., Noh, E.W., et al. (2011). Transgenic poplar expressing Arabidopsis NDPK2 enhances growth as well as oxidative stress tolerance. *Plant Biotechnol J* 9(3), 334–347.

Kornig, A., Winklhofer, M., Baumgartner, J., Gonzalez, T.P., Fratzl, P., and Faivre, D. (2014). Magnetite crystal orientation in magnetosome chains. *Adv Funct Mater* 24(25), 3926–3932. doi: 10.1002/adfm.201303737.

Kuznetsov, O.A., Schwuchow, J., Sack, F.D., and Hasenstein, K.H. (1999). Curvature induced by amyloplast magnetophoresis in protonemata of the moss *Ceratodon purpureus*. *Plant Physiol* 119(2), 645–650.

Lee, A.A., Lau, J.C.S., Hogben, H.J., Biskup, T., Kattnig, D.R., and Hore, P.J. (2014). Alternative radical pairs for cryptochrome-based magnetoreception. *J R Soc Interface* 11(95), 20131063. doi: doi:10.1098/rsif.2013.1063.

Letuta, A.S., and Berdinskii, V.L. (2015). Chemical Zeno effect-A new mechanism of spin catalysis in radical triads. *Dokl Phys Chem* 463, 179–181. doi: 10.1134/s0012501615080059.

Maeda, K., Robinson, A.J., Henbest, K.B., Hogben, H.J., Biskup, T., Ahmad, M., et al. (2012). Magnetically sensitive light-induced reactions in cryptochrome are consistent with its proposed role as a magnetoreceptor. *Proc Natl Acad Sci U S A* 109(13), 4774–4779.

Maffei, M.E. (2014). Magnetic field effects on plant growth, development, and evolution. *Front Plant Sci* 5, 445. doi: 10.3389/fpls.2014.00445.

Massey, V. (1994). Activation of molecular oxygen by flavins and flavoproteins. *J Biol Chem* 269(36), 22459–22462.

Muller, P., and Ahmad, M. (2011). Light-activated cryptochrome reacts with molecular oxygen to form a flavin-superoxide radical pair consistent with magnetoreception. *J Biol Chem* 286(24), 21033–21040. doi: 10.1074/jbc.M111.228940.

Narayana, R., Fliegmann, J., Paponov, I., and Maffei, M.E. (2018). Reduction of geomagnetic field (GMF) to near null magnetic field (NNMF) affects *Arabidopsis thaliana* root mineral nutrition. *Life Sci Space Res* 19, 43–50. doi: 10.1016/j.lssr.2018.08.005.

Nießner, C., Denzau, S., Peichl, L., Wiltschko, W., and Wiltschko, R. (2014). Magnetoreception in birds: I. Immunohistochemical studies concerning the cryptochrome cycle. *J Exp Biol* 217(Pt 23), 4221–4224. doi: 10.1242/jeb.110965.

Occhipinti, A., De Santis, A., and Maffei, M.E. (2014). Magnetoreception: An unavoidable step for plant evolution? *Trends Plant Sci* 19(1), 1–4. doi: 10.1016/j.tplants.2013.10.007.

Paponov, I.A., Fliegmann, J., Narayana, R., and Maffei, M.E. (2021). Differential root and shoot magnetoresponses in *Arabidopsis thaliana*. *Sci Rep* 11(1), 9195. doi: 10.1038/s41598-021-88695-6.

Paul, A.-L., Ferl, R.J., and Meisel, M.W. (2006). High magnetic field induced changes of gene expression in Arabidopsis. *BioMagn Res Technol* 4(1), 7. doi: 10.1186/1477-044X-4-7.

Penuelas, J., Llusia, J., Martinez, B., and Fontcuberta, J. (2004). Diamagnetic susceptibility and root growth responses to magnetic fields in *Lens culinaris*, *Glycine soja*, and *Triticum aestivum*. *Electromagn Biol Med* 23(2), 97–112.

Pooam, M., Arthaut, L.D., Burdick, D., Link, J., Martino, C.F., and Ahmad, M. (2019). Magnetic sensitivity mediated by the *Arabidopsis* blue-light receptor cryptochrome occurs during flavin reoxidation in the dark. *Planta* 249(2), 319–332. doi: 10.1007/s00425-018-3002-y.

Pooam, M., El-Esawi, M., Aguida, B., and Ahmad, M. (2020a). Arabidopsis cryptochrome and Quantum biology: New insights for plant science and crop improvement. *J Plant Biochem Biotechnol* 29(4), 636–651. doi: 10.1007/s13562-020-00620-6.

Pooam, M., Jourdan, N., El Esawi, M., Sherrard, R.M., and Ahmad, M. (2020b). HEK293 cell response to static magnetic fields via the radical pair mechanism may explain therapeutic effects of pulsed electromagnetic fields. *PLoS One* 15(12), e0243038-e0243038. doi: 10.1371/journal.pone.0243038.

Procopio, M., Link, J., Engle, D., Witczak, J., Ritz, T., and Ahmad, M. (2016). Kinetic modeling of the Arabidopsis Cryptochrome Photocycle: FADH accumulation correlates with biological activity. *Front Plant Sci* 7. doi: 10.3389/fpls.2016.00888.

Radhakrishnan, R. (2019). Magnetic field regulates plant functions, growth and enhances tolerance against environmental stresses. *Physiol Mol Biol Plants* 25(5), 1107–1119. doi: 10.1007/s12298-019-00699-9.

Rakosy-Tican, L., Aurori, C.M., and Morariu, V.V. (2005). Influence of near null magnetic field on in vitro growth of potato and wild Solanum species. *Bioelectromagnetics* 26(7), 548–557.

Rosen, A.D. (2003). Mechanism of action of moderate-intensity static magnetic fields on biological systems. *Cell Biochem Biophys* 39(2), 163–173. doi: 10.1385/cbb:39:2:163.

Sakaguchi, Y., Hayashi, H., and Nagakura, S. (1980). Classification of the external magnetic field effects on the photodecomposition reaction of dibenzoyl peroxide. *Bull Chem Soc Jpn* 53(1), 39–42. doi: 10.1246/bcsj.53.39.

Sear, R.P. (2019). Diffusiophoresis in cells: A general nonequilibrium, nonmotor mechanism for the metabolism-dependent transport of particles in cells. *Phys Rev Lett* 122(12), 128101. doi: 10.1103/PhysRevLett.122.128101.

Solov'yov, I.A., Chandler, D.E., and Schulten, K. (2007). Magnetic field effects in *Arabidopsis thaliana* cryptochrome-1. *Biophys J* 92(8), 2711–2726.

Song, Y.H., Estrada, D.A., Johnson, R.S., Kim, S.K., Lee, S.Y., MacCoss, M.J., et al. (2014). Distinct roles of FKF1, GIGANTEA, and ZEITLUPE proteins in the regulation of CONSTANS stability in *Arabidopsis* photoperiodic flowering. *Proc Natl Acad Sci U S A* 111(49), 17672–17677. doi: 10.1073/pnas.1415375111.

Spanudakis, E., and Jackson, S. (2014). The role of microRNAs in the control of flowering time. *J Exper Bot* 65(2), 365–380. doi: 10.1093/jxb/ert453.

Steiner, U.E., and Ulrich, T. (1989). Magnetic field effects in chemical kinetics and related phenomena. *Chem Rev* 89(1), 51–147. doi: 10.1021/cr00091a003.

Svendsen, J.A., and Waskaas, M. (2020). Mathematical modelling of mass transfer of paramagnetic ions through an inert membrane by the transient magnetic concentration gradient force. *Phys Fluids* 32(1), 013606. doi: 10.1063/1.5130946.

Teixeira da Silva, J.A., and Dobranszki, J. (2016). Magnetic fields: How is plant growth and development impacted? *Protoplasma* 253(2), 231–248. doi: 10.1007/s00709-015-0820-7.

Valentim, F.L., van Mourik, S., Pose, D., Kim, M.C., Schmid, M., van Ham, R.C.H.J., et al. (2015). A quantitative and dynamic model of the arabidopsis flowering time gene regulatory network. *PLos One* 10(2). doi: 10.1371/journal.pone.0116973.

Valiron, O., Peris, L., Rikken, G., Schweitzer, A., Saoudi, Y., Remy, C., et al. (2005). Cellular disorders induced by high magnetic fields. *J Magn Reson Imaging* 22(3), 334–340. doi: doi: 10.1002/jmri.20398.

Wiltschko, R., Stapput, K., Thalau, P., and Wiltschko, W. (2010). Directional orientation of birds by the magnetic field under different light conditions. *J R Soc Interface* 7(Suppl 2), S163–S177. doi: 10.1098/rsif.2009.0367.focus.

Xu, C.X., Yin, X., Lv, Y., Wu, C.Z., Zhang, Y.X., and Song, T. (2012). A near-null magnetic field affects cryptochrome-related hypocotyl growth and flowering in *Arabidopsis. Adv Space Res* 49(5), 834–840.

Xu, C.X., Wei, S.F., Lu, Y., Zhang, Y.X., Chen, C.F., and Song, T. (2013). Removal of the local geomagnetic field affects reproductive growth in *Arabidopsis. Bioelectromagnetics* 34(6), 437–442.

Xu, C., Lv, Y., Chen, C., Zhang, Y., and Wei, S. (2014). Blue light-dependent phosphorylations of cryptochromes are affected by magnetic fields in *Arabidopsis. Adv Space Res* 53(7), 1118–1124.

Xu, C., Li, Y., Yu, Y., Zhang, Y., and Wei, S. (2015). Suppression of *Arabidopsis* flowering by near-null magnetic field is affected by light. *Bioelectromagnetics* 36(6), 476–479. doi: 10.1002/bem.21927.

Xu, C., Yu, Y., Zhang, Y., Li, Y., and Wei, S. (2017). Gibberellins are involved in effect of near-null magnetic field on *Arabidopsis* flowering. *Bioelectromagnetics* 38(1), 1–10. doi: 10.1002/bem.22004.

Xu, C.X., Zhang, Y.X., Yu, Y., Li, Y., and Wei, S.F. (2018). Suppression of *Arabidopsis* flowering by near-null magnetic field is mediated by auxin. *Bioelectromagnetics* 39(1), 15–24. doi: 10.1002/bem.22086.

Zablotskii, V., Lunov, O., Novotná, B., Churpita, O., Trošan, P., Holáň, V., et al. (2014). Down-regulation of adipogenesis of mesenchymal stem cells by oscillating high-gradient magnetic fields and mechanical vibration. *Appl Phys Lett* 105(10), 103702. doi: 10.1063/1.4895459.

Zablotskii, V., Polyakova, T., and Dejneka, A. (2021). Modulation of the cell membrane potential and intracellular protein transport by high magnetic fields. *Bioelectromagnetics* 42(1), 27–36. doi: 10.1002/bem.22309.

Zollitsch, T.M., Jarocha, L.E., Bialas, C., Henbest, K.B., Kodali, G., Dutton, P.L., et al. (2018). Magnetically sensitive radical photochemistry of non-natural flavoproteins. *J Am Chem Soc* 140(28), 8705–8713. doi: 10.1021/jacs.8b03104.

6

Geomagnetic Field Effects on Living Systems

Hideyuki Okano

Shoogo Ueno

6.1 Introduction

The existence of the Earth's magnetic field (geomagnetic field, GMF) is indispensable for the origin and evolution of life. This is because when a constant stream of potentially harmful charged particles (mainly protons and electrons), which have reached the Earth blown by the solar wind, and galactic cosmic rays (GCRs), including ultraviolet (UV)-, X-, and γ-rays, is shielded and protected by the barrier of the Earth's magnetosphere (geomagnetosphere) (see ESA, 2016b), living organisms can only survive. In an environment where damaging charged particles and GCRs could reach, their genes were damaged and they could not survive and proliferate. It is inferred that primitive organisms such as cyanobacteria could survive even in an environment near the sea surface where light reaches. So when and how was the GMF to protect living organisms against harmful radiations generated?

The GMF intensity was about half of the current one 4.2 billion years ago (Ga = giga-annum = 10^9 years), but it became about the current intensity 4.1–4.0 Ga, and again about half the current one 3.9–3.3 Ga, and since about 2.7–2.1 Ga, it remains as large as the present intensity (Tarduno et al., 2007, 2010). The magnetosphere is the region above the ionosphere that is defined by the extent of the GMF in space. It extends several tens of thousands of kilometers into space, protecting the Earth from the charged particles of the solar wind and GCRs that would otherwise strip away the upper atmosphere, including the ozone layer that protects the Earth from the potentially harmful UV-B (280–315 nm) radiation. The ionosphere is the ionized part of Earth's upper atmosphere, from about 48 to 965 km altitude (Zell, 2020). The ionosphere is ionized by solar radiation. It plays an important role in atmospheric electricity and

DOI: 10.1201/9781003181354-6

forms the inner edge of the magnetosphere. It has practical importance because, among other functions, it influences radio propagation to distant places on the Earth (Rawer, 1993). The coupling currents or field-aligned currents flow along magnetic field (MF) lines between the magnetosphere and ionosphere. One of the manifestations of the coupling currents is the auroral oval (e.g., Nagata and Kokubun, 1962).

We can see the schematic of the Earth's inner core and outer core motion and the GMF generation from the following source image: USGS FAQs (2013). This diagram shows the relationship between the motion of conducting fluid organized into rolls by the Coriolis force, and the MF that the motion generates (see USGS FAQs, 2013). The current understanding of the origin of the GMF is that in the outer core of the Earth, a liquid metal containing iron and nickel as the main component generates electric currents (eddy currents) by heat convection while receiving the effect of rotation, and these electric currents generated the GMF (Hale, 1987; Selkin et al., 2000; Smirnov et al., 2003).

The inner core is primarily a solid ball with a radius of about 1,220 km, which is about 20% of Earth's radius (Monnereau et al., 2010). The temperature of the inner core can be estimated from the melting temperature of impure iron (Fe) at the pressure which Fe is under at the boundary of the inner core about 330 GPa (Giga-Pascals). Anzellini et al. (2013) obtained experimentally a substantially higher temperature for the melting point of Fe, $6{,}230 \pm 500$ K. Fe can be solid at such high temperatures only because its melting temperature increases dramatically at pressures of that magnitude (Aitta, 2006, 2008). Static compression experiments showed that the hexagonal close-packed structure of Fe is stable up to 377 GPa and 5,700 K, corresponding to inner core conditions (Tateno et al., 2010). The observed weak temperature dependence of the c/a axial ratio suggests that hcp-Fe is elastically anisotropic at core temperatures (Tateno et al., 2010). The solid inner core has a faster rotation rate toward the east relative to the mantle (Alboussière et al., 2010; Dumberry and Mound, 2010). This should have generated a stronger and possibly more stable "dipolar" MF counteracting a decreasing solar ionizing flux (Doglioni et al., 2016). The heavy elements started to sink, and the inner core initiated to solidify at about 1 (1.5–0.5) Ga due to the Earth's cooling. Since the solid inner core rotates faster than the external core, this possibly would have generated a stronger dipolar MF and a thicker atmospheric shield. While the intensity of the magnetic dipole may have slightly increased, the high-energy solar flux hitting the Earth was decreasing. X- and UV-rays were 100–1,000 times higher in the early stages with respect to the present solar radiation.

Doglioni et al. (2016) speculated that the development of life on Earth was significantly affected by the growth of the solid inner core and the natural evolution of the Earth. The fast diversification of basal eukaryotes may have been triggered by a more stable atmosphere as a consequence of the stronger magnetic shield exerted by the newly developed rotating solid inner core and weaker X-, γ-, and UV-rays (Doglioni et al., 2016). The GMF strength should increase in the core itself, possibly being related also to the tidally sheared liquid outer core (Buffett, 2010).

It was announced that a jet-stream of rapidly moving liquid iron is moving at around 50 km per year (Amos, 2016; Livermore et al., 2017). The liquid outer core must be convective in order to maintain the MF against "Ohmic dissipation" of an electric current (Buffett, 2010; Jackson and Livermore, 2009; Jackson et al., 2011). Buffett (2010) determined that the average MF in the liquid outer core is about 2.5 mT, which is about 40 times the maximum strength at the surface. He started from the known fact that the Moon and Sun cause tides in the liquid outer core, just as they do on the oceans on the surface (Buffett, 2010). He observed that motion of the liquid through the local MF creates electric currents, which dissipate energy as heat according to Ohm's law (Buffett, 2010). This dissipation, in turn, damps the tidal motions and explains previously detected anomalies in Earth's nutation, which is, in this case, the variation over time of the orientation of the axis of rotation of the Earth. From the magnitude of the latter effect, he could calculate the MF (Buffett, 2010). The field inside the inner core presumably has a similar strength.

The MF generated by the flow in the direction of rotation is called the "toroidal MF," and the important cause of the convection is in the outer core. As convection, both "thermal convection" and "compositional convection" are estimated to occur (Jones, 2015). The mantle determines how much heat is released from the core. A spirally rotating fluid flow is also essential in order to generate the GMF. The fluid flow is induced in the liquid outer core due to the "Coriolis force" effect caused by the rotation of the Earth (Kageyama and Sato, 1997). That is, due to the Coriolis force, fluid motion and convection currents are organized in columns along the rotation axis (Kageyama and Sato, 1997). Thus, the spiral flow of conducting fluid is created by the combination of convection currents and the Coriolis force.

The GMF varies depending on the region. This is because the GMF increases in strength due to the increased density of MF lines at the north (N) and south (S) magnetic poles, and weakens in the sparse equatorial region of MF lines. This indicates that the shape of the GMF is a dipole MF. The main GMF is a dipole generated in the Earth's outer core (Elsasser, 1950). The dipole field constitutes ~90% of the main GMF. However, since the GMF has partially non-dipole MF components, spatial variations are magnified. These variations in GMF strength are particularly remarkable in areas where the GMF in the South Atlantic Ocean is extremely weak. These areas are the "South Atlantic Anomaly" shown as the patch (ESA, 2012).

The variation of ionospheric plasma is strongly correlated with solar radiation and geomagnetic activity (Datta and Das, 2020). However, some other reports have pointed out that extreme tropospheric events such as thunderstorms and heavy lightning can also change the ionospheric total electron content distribution (Lay et al., 2013). Lightning which is the discharge between separated positively charged ice particles and negatively charged small other particles generates huge energy and couples through a quasi-electrostatic or electric field (EF) and electromagnetic (EM) pulsed field (Datta and Das, 2020). This generated energy then reaches to lower ionosphere and produces reionization by heating up this medium (Datta and Das, 2020).

Xiong et al. (2016) addressed the close relationship between ionospheric plasma irregularities, i.e., "ionospheric thunderstorms" and Global Positioning System (GPS) signal total interruption of Swarm satellites mainly at the satellite altitude of ~500 km. Thus, GPS navigation systems on low-orbiting satellites sometimes black out when they fly over the equator between Africa and South America (ESA, 2016a). According to an ESA announcement (ESA, 2016a), the GPS link was broken 166 times during the first 2 years of Swarm project since 2013, of which 161 coincided with ionospheric thunderstorms. The high-resolution observations from the satellite helped to link these outages to ionospheric thunderstorms 300–600 km in the Earth's atmosphere (DTU Space, 2016). Surprisingly, the area where satellite failures such as GPS losses are concentrated is in perfect coincidence with South Atlantic Anomaly (ESA, 2012).

According to Tarduno et al. (2010), MF and stronger solar wind strengths suggest important modifications during the first billion years of Earth evolution. Studies of the evolution of solar-type stars suggest that the Sun could emit high-energy flux (X-rays to UV-rays) up to 1,000 times stronger than the present flux during the early stages, gradually decreasing to 6 times 3.5 Ga, and to 2.5 times 2.5 Ga (Ribas et al., 2005, 2010). Therefore, besides the shielding effect of the Earth's magnetosphere, the Earth was hit by a much stronger ionizing flux in the past, which could have inhibited the evolution of "surface life" during the earlier stages of the Earth's history (Doglioni et al., 2016).

The evolution of life is affected by fluctuations and variations of the GMF intensity together with atmospheric oxygen (O_2) level and UV radiation (Meert et al., 2016). More recently, it has been suggested that under the condition of the GMF intensity minima, UV radiation reaching to Earth's surface could influence mammalian evolution with the loci of extinction controlled by the geometry of stratospheric ozone (O_3) depletion (Channell and Vigliotti, 2019). It has been reported that when the GMF reverses, the GMF is estimated to weaken (Brown et al., 2007; Valet and Plenier, 2008; Ferk and Leonhardt, 2009) and the changes in the last 200 years have been potentially diminishing declines. The magnetic poles are also moving, and it is estimated that the N pole has moved 1,100 km over 170 years, but it is not fully understood how much harmful charged particles and GCRs will radiate to the Earth. Therefore, it is unclear whether the GMF reversals will cause the mass extinction of living organisms on Earth.

Many living organisms can perceive and respond to the GMF. Several animals use the GMF as a magnetic compass (reviewed by R. Wiltschko and Wiltschko, 2012, 2021) and plants use the GMF as a physiological modulator (reviewed by Maffei, 2014). Furthermore, it has been reported that human health effects of environmental MFs including extremely low-frequency (ELF)-MFs ranging from 1 to 300 Hz have been discussed on the basis of the "radical pair mechanism" (RPM; reviewed by Juutilainen et al., 2018).

In this chapter, we review the current understanding of the effects of fluctuations and variations of the GMF on living organisms. We discuss the further possibility that the effects of the GMF on living organisms. We propose a hypothesis for explaining the link between the GMF reversals and the extinction and evolution of life on Earth and make proposals for future research.

6.2 Magnetic Sense

6.2.1 Primitive Magnetic Sense

Apart from gravity, the GMF is the only ubiquitous and relatively permanent element of the environment, thus being a great source of information for organisms (Erdmann et al., 2021). For certain kinds of bacteria, the only way to survive is to sense the GMF and move to a place suitable for survival. The magnetosensitivity of aquatic bacteria have been found in 1958 by an Italian medical doctor, Salvatore Bellini who was working in the Institute of Microbiology at the University of Pavia, Italy (Bellini, 1963a,b). He serendipitously discovered magnetosensitive bacteria at that time while examining water samples from sources around Pavia for pathogens (Frankel, 2009). Whereas Bellini was the first discoverer of magnetosensitive bacteria, his discovery was lost until an American scientist, Blakemore rediscovered the phenomenon in 1975 (Blakemore, 1975). These bacteria can sense the GMF and swim north or south. Due to their characteristics, these bacteria are called "magnetotactic bacteria." Magnetosensitivity and magnetotaxis of bacteria are the same behavior (Frankel, 2009). Blakemore (1982) had the advantage of electron microscopy by which he discovered the "magnetosomes." This instrument was apparently unavailable to Bellini at that time (Frankel, 2009). Nevertheless, it is clear that Bellini's manuscripts document a valid scientific discovery, and support his claim to the discovery of magnetosensitive/magnetotactic bacteria (Frankel, 2009).

Subsequent research by Frankel et al. (1979) shows that magnetotactic bacteria contain 10–20 microscopic ferromagnetic substances in their bodies called "magnetite" (~50 nm) and they are thus aligned passively with the GMF lines. Endogenous magnetite has a structure in which fatty acid covers magnetite (Fe_3O_4) or iron sulfide (Fe_3S_4), and each is covered with an organic thin film, forming a long chain-shaped or beaded structure of a single magnetic domain (~0.05–1.2 μm) called "magnetosomes" as a whole (Schüler, 2002). It is the "magnetic compass needle" that exists inside the body just like nano biomagnets (Uebe and Schüler, 2016). Magnetosomes detect the deviation from the direction of the MF as torque. Torque is the rotational force, which is the minimum energy when bacteria swim in the direction of the MF lines. Therefore, an external MF acting on magnetic moments can arrange and rotate these small bacteria according to the MF lines. This happens in both horizontal and vertical MFs (Kalmijn and Blakemore, 1978). The general properties of magnetite depend on the size and shape of the particles (first described by Kirschvink and Gould, 1981). Spin interactions cause the spins of adjacent atoms to align, thus forming domains with all spins parallel. Even smaller particles are superparamagnetic: at room temperature, their magnetic moment fluctuates as a result of thermal agitation, but it can easily be aligned by an external MF (Kirschvink and Walker, 1985).

Key functions of magnetosome biogenesis are encoded by about 30 genes that are clustered in a genomic "magnetosome island," although additional auxiliary functions are contributed by general cellular metabolic and regulatory pathways, including aerobic and anaerobic respiration (Uebe and Schüler, 2016). Surprisingly, a non-magnetotactic bacterium has recently been "magnetized" through the heterologous expression of genes that encode the magnetosome biogenesis pathway, which is a proof-of-principle demonstration that non-magnetotactic bacteria that are more facile for laboratory investigation could be "magnetized" to provide new models for genetic dissection and synthetic biology (Uebe and Schüler, 2016).

Magnetotactic bacteria do not require O_2 for respiration (anaerobic). Because it is dangerous for them to have a lot of harmful light and O_2 near the water surface, they must dive in the less oxygenated mud of lakes and seabeds. Therefore, it has been reported that anaerobic magnetotactic bacteria respond to high O_2 levels by swimming downward into areas with low or no O_2 toward geomagnetic north in the Northern Hemisphere (Blakemore, 1975, 1982; Moench and Konetzka, 1978) and geomagnetic south in the Southern Hemisphere (Blakemore et al., 1980; Kirschvink, 1980). Magnetotactic bacteria will dive downward, but this does not mean that they can sense gravity and move in the direction indicating "compass information" directly below. It has been found that magnetotactic bacteria aiming downwards can swim and orient themselves along the direction of the MF lines that are inclined from the horizontal direction. Magnetotactic bacteria can sense the "inclination" or "dip angle" with a higher degree of accuracy, which is the angle between the horizontal plane (H) and the total field vector (F) (World Data Center for Geomagnetism, Kyoto).

In contrast, the opposite cases of the overwhelming majority of other Northern or Southern Hemisphere magnetotactic bacteria have been found. As is the case with the bacteria in the Northern Hemisphere that swim south along the direction of the MF lines (Simmons et al., 2006), the bacteria in the Southern Hemisphere swim north along the direction of the MF lines (Leão et al., 2016). These findings support the idea that magnetotaxis is more complex than previously thought, and may be modulated by factors other than O_2 concentration and redox gradients in sediments and water columns (Leão et al., 2016).

The discovery of magnetite had been done after the discovery of magnetotactic bacteria. Magnetite was the constituent element of the chiton teeth, which was discovered in 1962 by an American scientist, Lowenstam (Lowenstam, 1962). One tooth of the chiton is one of the largest magnetites that living organisms possess, and its size is 1 mm (1 million times that of in vivo magnetite). It is reported that chitons may be responsible for natural magnetizations on the order of 10^{-10} T in marine sediments, whereas mud magnetotactic bacteria could produce remanence near 10^{-12} T in both marine and fresh water sediments (Kirschvink and Lowenstam, 1979). However, so far, there are no reports that chitons use their teeth to sense the GMF, and it is not yet clear why their teeth are magnets. In the subsequent research, these small biomagnets were used not only for chitons and bacteria but also for nematodes, *Caenorhabditis elegans* (Cranfield et al., 2004) that have made important contributions to molecular genetics and developmental biology. Nematodes, *C. elegans*, are model organisms, which reflect their evolutionary relationships with familiar animals such as pigeons, migratory birds, salmon and dolphins, and furthermore, humans. Moreover, in *C. elegans*, the physiological recordings of the magnetosensory amphid finger (AFD) bilateral neuron pair showed that neuronal responses saturate at larger than GMF intensity, ~65 μT (Wu and Dickman, 2012; Vidal-Gadea et al., 2015). *C. elegans* uses the left-right pair of AFD sensory neurons that project sensory structures to the tip of the worm's head (Clites and Pierce, 2017). If the origin of magnetite lies in magnetotactic bacteria, it is exactly the realization of the "endosymbiotic theory" (anaerobic eukaryotes coexisted by swallowing aerobic bacteria and evolved into the present eukaryotic cells), which was proposed in 1967 by an American biologist "Lynn Margulis" (Sagan, 1967).

6.2.2 Magnetic Sense of Animals

Suppose you have a geomagnetic sensor, such as a magnetic compass, inside your body. It seems likely that you know the direction indicating "compass information," and more specifically, the location indicating "map information" of the latitude and longitude coordinates where you are. So where should we go to reach our destination? To get this answer, you need something like a map that tells you the positional relationship between your current location and your destination. By detecting the GMF, it shows its position exactly on the "magnetic map" (first described by Wallraff, 1999), and it is used as the magnetic map like a car navigation system or GPS as several mobile or migratory animals that travel long distances for their living.

Surprisingly, the ability of long-distance traveling animals, such as homing pigeons, migratory birds, salmon, tuna, sea turtles, and whales, "not to get lost" seems to be related to this magnetic sensor. Moreover, it has been experimentally reported that various animal species also use this "magnetic sense" or "magnetoreception" (a sense which provides magnetic compass capabilities) to move only a short distance (daily <~3 km). They can figure out their "whereabouts" and "destination" with the accuracy that they might have a GPS built in their bodies. Recently, a series of experiments have tested how animals would respond when they were "virtually displaced" by exposing them to the MF of a distant site. Some kinds of animals (see below) showed headings that compensated this "virtual magnetic displacement," thus indicating a large-scale magnetic map: eastern red-spotted newts, *Notophthalmus viridescens* (Fischer et al., 2001), spiny lobsters, *Panulirus argus* (Boles and Lohmann, 2003), green sea turtles, *Chelonia mydas* (K.J. Lohmann et al., 2004; Luschi et al., 2007), lesser whitethroats, *Sylvia curruca* (Henshaw et al., 2010), Australian silvereyes, *Zosterops lateralis* (Deutschlander et al., 2012), Eurasian reeds warblers, *Acrocephalus scirpaceus* (Chernetsov et al., 2008, 2017; Kishkinev et al., 2013, 2015), and bonnethead sharks, *Sphyrna tiburo* (Keller et al., 2021). In contrast, adult and juvenile European robins (*Erithacus rubecula*) and adult garden warblers (*Sylvia borin*) under the same experimental conditions did not respond to this virtual magnetic displacement, suggesting significant variation in how navigational maps are organized in different songbird migrants (Chernetsov et al., 2020).

In addition, more recently, the following domestic or laboratory animal species that don't migrate for long distances can perceive and respond to the GMF: domestic dogs (Hart et al., 2013; Martini et al., 2018; Benediktová et al., 2020; Yosef et al., 2020), cows and deer (Begall et al., 2008; Burda et al., 2009), rodents (Mather and Baker, 1981; Malkemper et al., 2015, Norimoto and Ikegaya, 2015), domestic chickens (Freire et al., 2005, 2008; Wiltschko et al., 2007; Denzau et al., 2013a,b), zebra finch (Voss et al., 2007; Pinzon-Rodriguez and Muheim, 2017), zebrafish (Skauli et al., 2000; Sherbakov et al., 2005; Takebe et al., 2012; Osipova et al., 2016), carp (Hart et al., 2012), and domestic insects such as cockroaches (Vácha, 2006; Vácha et al., 2008, 2009; Bazalova et al., 2016; Slaby et al., 2018), fruit flies (Dommer et al., 2008; Gegear et al., 2008, 2010; Yoshii et al., 2009; Fedele et al., 2014a,b; Marley et al., 2014; Bae et al., 2016; Qin et al., 2016; Lee et al., 2018; Oh et al., 2020; Bradlaugh et al., 2021), and nematodes, *Caenorhabditis elegans* (Wu and Dickman, 2012; Vidal-Gadea et al., 2015; Clites and Pierce, 2017).

Moreover, it has been demonstrated that mole rats also have magnetoreception, and they use the MF azimuth to determine compass heading (first described by Burda et al., 1990). Surprisingly, it is suggested that mole rats perceive MFs with their minute eyes, probably relying on magnetite-based receptors in the cornea (Caspar et al., 2020). Thus, in addition to migratory animals, recently it has been reported that many non-migratory animals possess magnetoreception, and therefore, it is conceivable that magnetoreception is one of the senses that many animal species have.

Magnetic alignment, the preference to align the body axis in a certain angle relative to the GMF lines, is expressed by a variety of vertebrates during diverse behaviors, often during grazing and resting, and is regarded as a clear indicator of magnetoreceptive abilities (Begall et al., 2013). In the case of domestic dogs' sensitivity to MFs, Hart et al. (2013) found that dogs during defecation and urination preferred to excrete with the body being aligned along the N-S axis under calm GMF conditions. This directional behavior was abolished under unstable GMF (Hart et al., 2013). The best predictor of the behavioral switch was the rate of change in declination, i.e., the polar orientation of the GMF (Hart et al., 2013). This declination compass was also observed in American cockroaches, *Periplaneta americana* (Bazalova et al., 2016), fruit flies, *Drosophila melanogaster* (Lee et al., 2018), and Eurasian reed warblers, *Acrocephalus scirpaceus* (Chernetsov et al., 2017).

Oberbauer et al. (2021) performed a similar study using methodology analogous to that in the original paper by Hart et al. (2013). However, Oberbauer et al. (2021) did not detect any preference for body alignment during defecation and urination, although they utilized a greater number of dogs, collected the data within a brief time window making the data very comparable, and had a more balanced representation of individual dogs when compared to the previous study by Hart et al. (2013). Thus, their study did not demonstrate preferential alignment to any geomagnetic orientation which emphasized to the researchers the need for scientific replication (Oberbauer et al., 2021). Domestic dogs may live in an artificial living environment, in which case they are likely to be affected by anthropogenic and artificial EMF fields, which may interfere with their ability to sense the natural GMF (Burda et al., 2009).

Benediktova et al. (2020) equipped 27 hunting dogs with GPS collars and action cams, let them freely roam in forested areas, and analyzed components of homing in over 600 trials. As a result, the "compass run" was significantly oriented along the N-S geomagnetic axis, suggesting that its orientation was independent of the direction of the dog owner (Benediktova et al., 2020). Noteworthy, scouting dogs in unfamiliar locations cannot use visual landmarks to recalibrate a path integration system (Benediktova et al., 2020). Therefore, in the absence of familiar landmarks, the compass run may serve to recalibrate a path integration system relative to the GMF (Benediktova et al., 2020). Performing such a compass run significantly increased homing efficiency. Benediktova et al. (2020) proposed that this run is instrumental for bringing the mental map into the register with the magnetic compass and establishing the heading of the animal.

Interestingly, in the case of magnetic alignment of wild red foxes (*Vulpes vulpes*), which belong to the family of Canidae, Červený et al. (2011) found that magnetic alignment to the north enhances the precision of hunting attacks in high vegetation and under snow cover. It will also be interesting to examine the relationship between geomagnetic sensing and certain kinds of behaviors in wild canines, e.g., Australian wild dog Dingoes (*Canis lupus dingo*), African wild dogs (*Lycaon pictus*), black-backed jackals (*Canis mesomelas*), North American wolves (*Canis lupus*), coyotes (*Canis latrans*), and Japanese raccoon dogs (*Nyctereutes procyonoides viverrinus*).

More recently, in the case of the Suidae, it was reported that the wild boars (*Sus scrofa*) in the herds had a highly significant axial preference to align themselves approximately along the magnetic N-S axis, with a slight shift toward the east (Červený et al., 2017). A similar and equally strong axial N-S preference was revealed for the orientation of wild boar beds (Červený et al., 2017). In warthogs (*Phacochoerus africanus*), the same axial N-S preference became apparent (Červený et al., 2017). Surprisingly, their pairs showed antiparallel body orientation and the antiparallel orientation of pairs can be interpreted as an antipredator strategy (Červený et al., 2017).

In the case of the avian magnetic compass, it is conceivable that an efficient magnetic compass was an important precondition for some species to adapt to a migratory lifestyle (Wiltschko and Wiltschko, 2021). The most important function of the avian magnetic compass is to provide a "directional reference system": it acts as a reference for recording the direction of the outward journey to obtain the home direction in inexperienced young birds (Wiltschko and Wiltschko, 1978) and also provides the reference for the innate migratory direction in first-time migrants (e.g., Wiltschko and Gwinner, 1974; Beck and Wiltschko, 1988).

A recent advanced study on head-mountable microstimulators coupled with a digital geomagnetic compass demonstrated that blind rats were able to find food in a maze using food-associated geomagnetic information from a head-mounted magnetic compass (Norimoto and Ikegaya, 2015). Furthermore, a recent interesting study suggested that using the fruit fly, *Drosophila melanogaster* as a model organism, which has a geomagnetic declination compass for horizontal orientation (Lee et al., 2018), prenatal exposure to a specific geographic MF during development affected adult responses to the matching field gradient through downward movements associated with foraging (Oh et al., 2020). This same behavior occurred spontaneously in the progeny of the next generation (Oh et al., 2020). These findings implicated that imprinting on the MF of a natal area assists magnetoreceptive organisms and their offspring in recognizing locations suitable for foraging and reproduction (Oh et al., 2020).

Inexperienced sea turtle migrants cannot have a detailed map of their migration route but could have inherited simple cue values for the goal and/or a few "signposts" and associated these with adaptive behaviors, such as the responses of hatchling sea turtles to magnetic parameters (Lohmann et al., 2001, 2007, 2013). Inexperienced bird migrants usually follow experienced companions or rely on a simple clock-and-compass strategy (vector navigation) using only an innate circannual clock and compass orientation programs, but no map (Mouritsen, 2018). For example, in the case of juvenile whitethroats, *Sylvia communis*, it is reported that magnetic information is not necessary for establishing the appropriate migratory direction when natural celestial cues are available in the pre-migratory period (Rabøl and Thorup, 2006). Thus, it remains unclear exactly which combination of sensory parameters including celestial cues triggers the start and stop of the first natural migration (Mouritsen, 2018).

R. Wiltschko and Wiltschko (2021) have also noticed and reviewed that specific magnetic conditions or changes in conditions may act as "signposts," triggering specific responses. These are spontaneous, obviously innate responses, in contrast to the navigational processes based on the learned "map" (Wiltschko and Wiltschko, 2021). Although not many examples have been discovered yet, these examples show that the GMF is involved in a variety of phenomena where it is advantageous that certain things happen at specific locations (Wiltschko and Wiltschko, 2021).

Appropriate vertical movement is critical for the survival of flying animals. Although negative geotaxis (moving away from the ground surface) driven by gravity has been extensively studied, much less is understood concerning a static regulatory mechanism for inducing positive geotaxis (moving toward the ground surface). Using the *Drosophila melanogaster* as an above-mentioned model organism, Bae et al. (2016) showed that the GMF induces positive geotaxis and antagonizes negative gravitaxis. Remarkably, the GMF acts as a sensory cue for an appetite-driven associative learning behavior through the GMF-induced positive geotaxis. This GMF-induced positive geotaxis requires the three geotaxis genes, such as *cry*, *pyx*, and *pdf*, and the corresponding neurons residing in Johnston's organ of the fly's antennae (Bae et al., 2016).

It has been shown that in principle it is possible to obtain MF effects for MFs as weak as the GMF of ~50 μT (Maeda et al., 2008; Henbest et al., 2008). It is gradually becoming known that a wide variety of animal species perceive and respond to MFs, but the mechanisms of their magnetic sense are not clarified yet in detail. One plausible hypothesis about the magnetic sense of animals is that a very small magnetic material called "magnetite" in the animal's body works like a compass needle (first described by Kirschvink and Gould, 1981). As typical examples, magnetites, which are present in the nose of rainbow trout (Walker et al., 1997) and the upper beak of pigeons (Fleissner et al., 2003, 2007; Falkenberg et al., 2010), have been considered to be responsible for magnetic reception. Moreover, in addition to magnetites, another type of iron minerals, i.e., "maghemites" were also found in subcellular compartments within sensory dendrites of the upper beak of several bird species (Solov'yov and Greiner, 2007, 2009). Thus, the iron minerals in the beak were found in the form of crystalline maghemite (γ-Fe_2O_3) platelets arranged in chains inside the dendrite, and assemblies of magnetite (Fe_3O_4) nanoparticles attached to the cell membrane (Solov'yov and Greiner, 2007, 2009). Here, maghemite can be considered as an Fe(II)-deficient magnetite (Cornell and Schwertmann, 2003). As one of the mechanisms for iron mineral-based magnetoreception, it was suggested that in an external MF, the maghemite platelets become magnetized and enhance the local MF in the cell by orders of magnitude (Fleissner et al., 2007). Thus the magnetite clusters will experience an attractive (repulsive) force inducing their displacement, what might induce primary receptor potential via strain-sensitive membrane channels leading to a certain orientation effect (Solov'yov and Greiner, 2007, 2009).

However, it has been found that magnetite-containing cells are a type of "macrophage" (macrophage is known to have a vital role in host defense and iron homeostasis) (Wang and Pantopoulos, 2011), and although they are rich in iron ions, they do not project onto nerve cells (Treiber et al., 2012). However, it is unknown whether the magnetite-containing macrophages are involved in magnetic sensing.

With regard to the transduction link between nanoparticles of iron oxides (most frequently magnetite, Fe_3O_4) and nerve cells, a number of distinguished researchers proposed the following transduction mechanism (Cadiou and McNaughton, 2010; Winklhofer and Kirschvink, 2010; Eder et al., 2012; Vácha, 2020). The strength of the Fe Oxide Particles model lies in its simplicity which is underlined by its simplicity with other known cellular mechanisms sensitive to mechanical forces and tensions. If microscopic particles with either a permanent or induced magnetic dipole are exposed to the surrounding GMF, mechanical forces act on them.

Cell membranes can be extremely sensitive to a force, and there are a number of models as to how the movement of magnetite particles could affect the function of ion channels, and thereby result in the conversion of the magnetic energy into a change to the membrane potential (Vácha, 2017, based on Shaw et al., 2015). Here, two hypothetical representations of how Fe oxide particles may transform magnetic energy into physical force via opening the cation channel in the cell membrane, hence enabling neural signalization (Vácha, 2017, based on Shaw et al., 2015). A cluster of superparamagnetic particles attached by a cytoskeletal filament to the gating domain of force-gated channel moves along the GMF force. The chain of single-domain grains having stable moment rotates accordingly to the external field and the filament transfers the torque to the channel.

This model has several characteristic features that are well met in certain animals (Vácha, 2017): Most important, (1) such a magnetoreceptor does not require light to function; (2) it is a receptor sensitive on principle to the polarity of the MF, that is, distinguishing magnetic north from south; (3) its magnetic properties may also be impacted by a strong and short pulse of the external field; and (4) another supporting argument is the discovery of Fe oxide particles in animal tissues. Their occurrence and properties have been documented relatively extensively in invertebrates and in social insects in popular (Wajnberg et al., 2010).

Magnetite was also found in specialized cells in honeybees (Gould et al., 1978) and pigeons (Wallcott et al., 1979). In pigeons, the detector is localized in innervated tissue in the ethmoid sinus near the nose (the upper beak) (Wallcott et al., 1979). Moreover, it has been reported that night-migratory songbirds have a magnetic compass in their eyes and a second magnetic sense in the ophthalmic branch of the trigeminal nerve (V1) (Kishkinev et al., 2013). The second magnetic sense is assumed to be magnetite iron oxide nanoparticles-based and located in the upper beak (first described by Beason and Semm, 1987). In this case, Solov'yov and Greiner (2007) suggested that the pull or push to the magnetite assemblies, which are connected to the cell membrane, may reach a value of ~0.2 pN, and this value would be sufficient to excite specific mechanoreceptive membrane channels in the nerve cells.

Electrophysiological recordings from V1 and the trigeminal ganglion of the bobolink (*Dolichonyx oryzivorus*) by Beason and Semm claimed that V1 transmits magnetic information (Beason and Semm, 1987; Semm and Beason, 1990), but the relevance of these studies is questionable, since, despite several serious attempts, nobody has managed to replicate them (Mouritsen and Hore, 2012). In two recent independent studies, migratory European robins (Heyers et al., 2010) and non-migratory homing pigeons (Wu and Dickman, 2011) were exposed to changing magnetic stimuli, and neuronal activity in the V1-recipient regions in the hindbrain was quantified using ZENK (Heyers et al., 2010) or c-fos (Wu and Dickman, 2011) expression. The results suggest that neurons in the trigeminal nuclei are most probably activated by magnetic stimuli. Thus, V1 seems to transmit magnetic information into the brain in different avian species, but the biological function of this information remains unknown (Heyers et al., 2010). It has often been assumed that this information is perceived by a putative magnetic map organ associated with the trigeminal nerve (Fleissner et al., 2003, 2007; Mora et al., 2004; Falkenberg et al., 2010; Heyers et al., 2010), even though the receptors remain to be identified with certainty (Mouritsen, 2012; Treiber et al., 2012). Kishkinev et al. (2013) found that the ability of Eurasian reed warblers (*Acrocephalus scirpaceus*) which are typical long-distance night-migratory songbirds, to correct for an eastward displacement, requires intact V1. Thus, Kishkinev et al. (2013) suggested that some kind of map information is transmitted via V1 into the brain. In the case of the non-migratory homing pigeons, Lefeldt et al. (2014) suggested that the trigeminal system is involved in processing MF information and that V1 transmits this information from V1-associated magnetosensors to the brain.

Thus, magnetite organs have since been found in a strikingly long list of vertebrates, including fish, amphibians, reptiles, birds, and mammals (Wiltschko and Wiltschko, 2005). Theoretical calculations (Kirschvink and Gould, 1981) suggest that these organs have more than enough magnetite crystals to measure map location, at least in most places on the globe. Thus, many magnetites have been found in the body of a variety of animal species including humans (Kirschvink et al., 1992a,b; Gould, 2008a,b), but how do they convey information such as the direction of the MF to the brain? How has the simple magnetite sensor evolved as the magnetic-based sophisticated GPS? That is unclear and enigmatic.

Another strong candidate hypothesis is that the mechanism is being clarified in migratory bird research. It is said that a blue light-sensing protein called "cryptochrome (CRY)" in the retina of the eye plays an important role as a magnetic sensor or magnetoreceptor (first described by Möller et al., 2004). CRY proteins are also components of the central circadian clockwork (Yuan et al., 2007), and are closely related to the light-dependent DNA repair enzymes, the photolyases (Cashmore, 2003). It is reported that the CRY is fingered as the smoking gun in the exquisite magnetic reception of birds (Roberts, 2016). As a putative magnetoreceptor, four different isoform CRYs (CRY1a, CRY1b, CRY2, CRY4a, and CRY4b) have been identified in the retinae of several bird species. CRY1a (Liedvogel et al., 2007; Nießner et al., 2011) is found from garden warblers (*Sylvia borin*) and European robins (*Erithacus rubecula*), CRY1b (Bolte et al., 2016; Nießner et al., 2016) is from European robins, migratory northern wheatears (*Oenanthe oenanthe*), and homing pigeons (*Columba livia*), CRY2 (Mouritsen et al., 2004) is from migratory garden warblers, and CRY4 (Günther et al., 2018; Pinzon-Rodriguez et al., 2018; Hochstoeger et al., 2020; Wu et al., 2020) is from European robins. In particular, more recently, R. Wiltschko et al. (2021) reviewed and speculated that CRY1a (termed as *gw*CRY1a) appears to be the most likely receptor molecule for magnetic compass information due to its location in the outer segments of the UV cones with their clear oil droplets. In the case of cockroaches, however, CRY2 mediates sensitivity to the magnetic declination in American cockroaches, *Periplaneta americana* (Bazalova et al., 2016).

These CRYs could generate free radical pairs (RPs) for quantum-assisted magnetic sensing, which play a key role as a kind of "quantum compass" (Hiscock et al., 2016), and are deeply involved in a magnetic sense. That is, CRY-dependent magnetoreception is currently proposed to be a result of light-initiated electron transfer chemistry in the protein, which is magnetically sensitive by virtue of the RPM (Rodgers and Hore, 2009; Dodson et al., 2013). "Spin-correlated RPs" can undergo coherent mixing between singlet and triplet spin states, which have different reactive fates, and this mixing process can be modulated by MFs (Woodward et al., 2009). In the principle of magnetoreception mechanism according to the model of RPM, it is not possible to distinguish whether the directions of electron spins are opposite or the same, so in principle, it is possible to detect the inclination of the MF lines, but which is N or S polarity. It is impossible to obtain information on the direction itself. Therefore, it is called an "inclination compass" or "axial compass" in the magnetic sense that only the information on the dip angle (World Data Center for Geomagnetism, Kyoto), which is the orientation component of the magnetic vector, can be obtained. Thus, the avian magnetic compass does not distinguish between magnetic "N" and "S" as indicated by polarity, but between "poleward" where the MF lines point to the ground, and "equatorward," where they point upward (Wiltschko and Wiltschko, 2005).

The geomagnetic variables of the intensity of the field and the inclination of its lines, as well as "magnetic anomalies," are used by animals to ascertain their position. Here, the magnetic anomaly is an area where the GMF is spatially distorted and presents opportunities to assess how the homing or migration behavior of animals is affected by spatial variation of geomagnetic parameters (Dennis et al., 2007). The magnetic anomalies are mainly produced by magnetized rocks that change the field value on the Earth's surface (Moskowitz et al., 2015). It can be thus assumed that animals have a biological analog of a GPS; the difference is that it is not based on satellite signals but on the GMF (Lohmann, 2010). If local MF intensity does play a role in their navigational map, then behavior undertaken by animals to determine their locations relative to goals may be influenced by the orientation of the local MF intensity. To test this idea, Dennis et al. (2007) conducted the following experiment: pigeons were released at unfamiliar sites and flight trajectories recorded by GPS-based tracking devices. Release sites were located in or around the Auckland Junction Magnetic Anomaly in New Zealand (Dennis et al., 2007). In total, 92 complete flight trajectories of pigeons were obtained and 59 out of the 92 (64.1%) trajectories exhibited significant alignment: 29 trajectories (31.5%) were aligned parallel to the field, 33 (35.7%) were aligned perpendicular to the field, and 42 (45.7%) were aligned both parallel and perpendicular to the field (some individuals exhibited more than one type of response) (Dennis et al., 2007). Thus, many pigeons initially flew, in some cases up to several kilometers, in directions parallel and/or perpendicular to the bearing of the local MF intensity (Dennis et al., 2007). This behavior occurred irrespective of the homeward direction and significantly more often than what was expected by random chance (Dennis et al., 2007). It is suggested that pigeons when homing detect and respond to spatial variation in the GMF (Dennis et al., 2007).

A potential complication for all strategies of magnetic map navigation similar to GPS is that the GMF is not static but instead changes gradually over time. This change in field elements, known as "secular variation" (Skiles, 1985), means that the MF existing at a given location will not necessarily remain exactly the same over the life span of a long-lived animal. Similarly, the pattern of isolines throughout a given geographic region gradually changes. Although the GMF changes over time, strong selective pressure presumably acts to ensure a continuous match between the responses of animals and the fields that mark critical locations in migratory routes at any point in time (Lohmann and Lohmann, 1998; Lohmann et al., 1999, 2001).

W. Wiltschko and Wiltschko (1972) discovered the inclination compass (as described above) in European robin and speculated with great insight that on the whole, this magnetic compass represents a highly flexible direction-finding system. W. Wiltschko and Wiltschko (1972) estimated that its ability to adjust to a varying intensity range makes it independent of any secular variation in total intensity, and the fact that it does not use the polarity of the MFs, so-called "polarity compass," means that it is not affected by the GMF reversals that have taken place several times since the phylogenetic origin of birds (Runcorn, 1969).

Behavioral studies in European robins (Zapka et al., 2009) strongly suggest that a forebrain region named "Cluster N" (Mouritsen et al., 2005), which receives input from the eyes via the thalamofugal visual pathway (Heyers et al., 2007), is involved in processing magnetic compass information. Several studies on the migratory birds' brains showed that bilateral lesions of Cluster N disable magnetic orientation, and therefore, Cluster N is assumed to be a light-processing forebrain region (Möller et al., 2004; Mouritsen et al., 2004; Zapka et al., 2009). Moreover, it is presumed that it is the cryptochrome (CRY) of the retina that meets the conditions under which the electron transfer reaction occurs at the photoreceptors on the retina sphere (Ritz et al., 2000).

The vast majority of mobile animal species have magnetoreception in the form of inclination compass. Insects including butterflies and flies may use a CRY-based chemical magnetic sensor in their antennae. Birds may sense MFs using magnetosensory cells in the inner ear (first described by Harada et al., 2001) and beak (first described by Beason and Semm, 1987) with an iron-based mechanism, and in eyes with a CRY-based RPM (Ritz et al., 2009). In particular, in birds, it is assumed that the inclination compass is based primarily on RPM (Wiltschko et al., 2005). However, the exact identity of the magnetically sensitive RP in CRY or "CRY-based RPM" is currently unknown. Presumably, the influence of the MF in some way affects the concentration of a CRY signaling state that, in turn, results in a neurophysiological response. However, there exists very little evidence of the signal transduction mechanism that might link magnetically sensitive chemistry in CRY to an organism response. In this context, Marley et al. (2014) examined the MF effect on seizure response in *Drosophila* larvae. Embryos were exposed to light for 100 ms every second between 11 and 19 hours after egg laying, but exposed to the MF throughout embryogenesis. After hatching, larvae were transferred to vials and maintained in complete darkness and in the absence of any applied MF until ~3 days later when wall climbing third instar larvae were tested for seizure-like behavior (Marley et al., 2014). They revealed that 100 mT static MF (SMF) significantly increased the effect of blue light (470 nm) on seizure severity in larvae compared to light pulses alone (Marley et al., 2014). The MF effect on seizure duration was shown to be *Drosophila melanogaster* CRY (*Dm*CRY)-dependent: being abolished in a cry03 null (cry-/-) background and rescued by transgenic expression of UAS-cry in a cry null (Marley et al., 2014). Prolongation of seizure duration was also prevented by prior ingestion of typical antiepileptic drugs (e.g., phenytoin and gabapentin), consistent with an effect on neuronal activity (Marley et al., 2014). Exposing the embryos to a 100 mT MF in the first instance has a range of benefits over the mT field exposures immediately relevant to animal magnetoreception (Marley et al., 2014). They inferred that this study paves the way for assessing the influence of the amplitude and orientation of the GMF intensity in the μT range on seizure duration in *Drosophila* larvae (Marley et al., 2014).

Thus, both the "magnetite theory" and the "CRY theory" (CRY-based RPM theory) are powerful theories, and there is evidence that some organisms may use a combination of these two mechanisms (reviewed by Roberts, 2016; Wiltschko and Wiltschko, 2021), but there are many unclear points. Recently, protein complex, so-called "MagR/CRY complex," which is present in retinal cells such as pigeons, has been found to behave like a magnetic compass (Qin et al., 2016). Briefly, the complex composed of the above-mentioned blue-light photoreceptor protein CRY and the protein-coding by the CG8198 gene, called the magnetic receptive protein (MagR), has an extremely rare property of responding and orienting to weak MFs similar to that of the GMF (Qin et al., 2016). Due to this property, the MagR/CRY complex has been noted as a potential causative agent of animal magnetic sensitivity. Qin et al. (2016) showed through biochemical and biophysical methods that the MagR/CRY complex is stable in the retina of pigeons and can also form in butterflies, rats, whales, and human retinal cells. They note that it is unclear how the MagR/CRY complex can sense MFs and whether the MagR/CRY complex is involved in animal magnetic sensing (Qin et al., 2016). However, the discovery of protein complexes for magnetic compasses has the potential to create a wide range of new approaches to MF-induced macromolecular manipulation, as well as cell behavior.

Pang et al. (2017) investigated that MagR expression alone could achieve cellular activation by SMFs up to 1.2 mT. Despite systematically testing different ways of measuring intracellular calcium and different SMF exposure protocols, it was not possible to detect any cellular or neuronal responses to SMF exposure in MagR-expressing HEK cells or primary neurons from the dorsal root ganglion and the hippocampus (Pang et al., 2017). By contrast, in neurons, co-expressing MagR and channel rhodopsin, artificial visible light but not SMF exposure increased calcium influx ($[Ca^{2+}]_i$) in hippocampal neurons (Pang et al., 2017). The discovery that MagR/CRY is a putative magneto-responsive protein complex does not directly imply that MagR itself may induce a neuronal response in transfected cells (Pang et al., 2017). While the possibility exists that MagR, when associated with other proteins such as CRY or linked to other channels such as TRV4 may be used for magnetogenetics, these results suggested that more factors seem necessary, in addition to the expression of MagR alone, for MagR to be used as a tool for neuronal modulation via MFs (Pang et al., 2017). Moreover, regarding other prospects, the search for the molecular principle of the signaling mechanism from the receptor molecule or structure to the activated neural cell remains the cardinal question of current magnetoreception research (Vácha, 2017).

Nordmann et al. (2017) assumed that magnetoreception relies on a single receptor, or indeed a single mechanism. At conferences, CRY often faces magnetite on the scientific battlefield, but this conflict is illusionary (Nordmann et al., 2017). Selective pressure in diverse environments, from the oceans to the air, may have facilitated the evolution of a multiplicity of magnetoreceptors (Nordmann et al., 2017). This mystery might therefore have more than one solution (Nordmann et al., 2017).

In another aspect, in the case of elasmobranch magnetoreception in sharks, skates, and rays, "EM induction theory" has been proposed as a feasible mechanism by which elasmobranchs could perceive the GMF through the electroreceptor system, so-called "ampullae of Lorenzini" (first described by Murray, 1960). Elasmobranchs, which are evolutionarily much older than teleosts, have the ampullae of Lorenzini, and perceive the electric current induced by the fish itself and the electric current induced by the water movement when the fish or the water moves in a constant MF or SMF (Brown and Ilyinsky, 1978). However, it is not completely understood if the electroreceptor system receives only electric information, or if it can receive magnetic information as well (Formicki et al., 2019).

According to the principles of the active mode of EM induction (Kalmijn, 1981, 1982; Paulin, 1995; Molteno and Kennedy, 2009), it is the horizontal polarity component of the GMF that induces vertical electric fields (EFs) that could convey information regarding MF directionality. "The EM induction-based magnetoreceptor system" is used to gain a compass heading regarding the direction of travel. More recently, Keller et al. (2021) conducted magnetic displacement experiments on wild-caught bonnethead sharks (*Sphyrna tiburo*) and show that magnetic map cues can elicit homeward orientation.

Keller et al. (2021) further showed that the use of a magnetic map to derive positional information may help explain aspects of the genetic structure of bonnethead populations in the northwest Atlantic and this ability may contribute to population-level processes (Escatel-Luna et al., 2015; Fields et al., 2016; Gonzalez et al., 2019; Díaz-Jaimes et al., 2020). There is other information useful for movements, such as ocean currents and tides, but Keller et al. (2021) insisted that the MF is more stable than other information, so it is likely to be more useful for navigation. These findings complement recent research that has shown elasmobranchs likely have a polarity compass (Newton and Kajiura, 2020a). The combination of magnetic map and compass senses would likely be highly adaptive and allow the evolution of complex movement patterns that are a hallmark of elasmobranch life histories (Keller et al., 2021). Their results are significant because for 50 years researchers have highlighted the importance of determining whether sharks and rays use the GMF to aid in orientation and navigation. Multiple species of elasmobranchs have been shown capable of detecting various components of the MF (Kalmijn, 1981, 1982; Molteno and Kennedy, 2009; Anderson et al., 2017; Newton and Kajiura, 2017, 2020a), and this research provides ecologic context for how these abilities may be used (Keller et al., 2021).

Recently, in the case of the birds, Nimpf et al. (2019) suggested that a putative mechanism of magnetoreception by EM induction in the pigeon inner ear. Nimpf et al. (2019) reported the presence of a splice isoform of a voltage-gated calcium channel ($Ca_V1.3$) in the pigeon inner ear that has been shown to mediate electroreception in skates and sharks (Bellono et al., 2018). Nimpf et al. (2019) proposed that pigeons detect MFs by EM induction within the semicircular canals that are dependent on the presence of apically located voltage-gated cation channels in a population of electrosensory hair cells.

It is suggested that the ampullae of Lorenzini in shark and highly sensitive electrosensory system to detect MF-induced EFs may not be the sole sensory receptor structures used to perceive MF stimuli, and that an EM induction-based magnetoreceptor structure capable of perceiving changes in MF intensity may be located in the nasal olfactory capsules of sharks as putative magnetoreceptor structures (Anderson et al., 2017). These elasmobranchs are one of the more electrosensitive species, and generally, they are primarily responsive to both DC and AC low-intensity EFs between 0.02 and 100 µV/cm and frequencies of 0–15 Hz (Sisneros and Tricas, 2002; Bedore and Kajiura, 2013). Anderson et al. (2017) behaviorally conditioned sandbar sharks (*Carcharhinus plumbeus*) to respond to weak magnetic stimuli (> 0.03 µT) that generated electrical artifacts of 73 nV/cm. Moreover, sharks and stingrays are well known for their sensitivity to EM fields (EMFs) themselves (Kalmijn, 1981, 1982; Newton and Kajiura 2017, 2020a,b; Anderson et al., 2017). Therefore, as a future study, Newton and Kajiura (2020b) proposed that if chondrichthyans do use their electroreceptors to detect, encode, and perceive magnetic stimuli, the next step is to uncover how these fishes might distinguish between natural geomagnetic and bioelectric cues.

As a research experiment conducted to examine the effect on the MF generated by the high-voltage transmission line, the conclusion that the effect could not be detected has been reported. One example is a preliminary study of a project of hydrokinetic (HK) technologies that use the rapids of the Mississippi River to generate 8,000 MW of electricity, led by the US Federal Energy Regulatory Commission (Cada et al., 2011, 2012). Laboratory experiments conducted in the fiscal year 2010 found no evidence that three common freshwater taxa, i.e., the snail (*Elimia clavaeformis*), the clam (*Corbicula fluminea*), and the fathead minnow (*Pimephales promelas*) were either attracted to or repelled by an SMF (~36 mT at the strongest point) (Cada et al., 2011). Similarly, further experiments in the fiscal year 2011 with juvenile sunfish (*Lepomis* spp.), channel catfish (*Ictalurus punctatus*), and striped bass (*Morone saxatilis*) did not detect a significant change in position relative to controls (Cada et al., 2012). These results suggested that the predicted 60 Hz EMF (~166 mT at the strongest point) that may be created by a single submerged DC transmission cable from a hydrokinetic project would not seriously affect the behavior of common freshwater species (Cada et al., 2012). The variable EMF associated with AC currents caused little or no behavioral effects in American paddlefish (*Polyodon spathula*) that is known to be highly sensitive to EFs (Cada et al., 2012). However, another fish of known EMF sensitivity, lake sturgeons (*Acipenser fulvescens*) displayed temporarily altered swimming behavior when exposed to variable MFs (Cada et al., 2012). Other than the brief reactions by sturgeon to the variable fields reported here, no long-term changes in behavior or mortalities were observed (Cada et al., 2012). It is possible that the frequencies and intensities of the induced electrical signals created by the strong, 60 Hz EMF in the experiments were beyond the range that is readily detected by paddlefish (Cada et al., 2012).

The electricity produced by offshore wind turbines is transmitted by cables over long distances (European Wind Energy Association, 2009). The electric current generated produces MFs. Studies of possible effects of artificial SMFs have been carried out on various species under various experimental conditions (European Wind Energy Association, 2009). Anthropogenic and artificial EMFs could interact with marine organisms to produce detectable changes (European Wind Energy Association, 2009). Usually, however, only very slight differences in control groups have been recorded (European Wind Energy Association, 2009).

As the stock of sharks and rays is declining worldwide greatly, preserving efforts have been made to reduce their catch. When a magnet is attached to the fishing gear, the bycatch of sharks and rays is significantly reduced by attaching magnets to basket fishing gear and hooks. Since the MF intensity of the attached magnets is quite strong (e.g., 150 mT) relative to the GMF or localized MFs, it is thought that sharks and rays with a sensitive MF sensation dislike strong MFs (Stoner and Kaimmer, 2008; Richards et al., 2018).

The MF may affect mollusks, crustaceans, fish, and marine mammals that use the GMF for orientation during navigation. But it is still unknown whether the MFs associated with wind turbines influence marine organisms (Gill, 2005). Electrosensitive species could be attracted or repelled by the EFs generated by submarine cables. Special attention must be paid in areas of breeding, feeding, or nursing because of the congregation or dispersion of sensitive individuals in the benthic community (Gill, 2005).

It is reported that the survival rate of several benthic organisms exposed to SMF of 3.7 mT for several weeks as well as the reproduction rate of mussels living under these SMF conditions for 3 months did not present significant differences with the control group (Bochert and Zettler, 2004). From these results, conclusions are that SMFs of power cable transmissions don't seem to influence the orientation, movement, or physiology of the tested benthic organisms (Köller et al, 2006; Meißner, 2006).

Trański et al. (2005) reported that the catch rate of spinycheek crayfishes (*Orconectes limosus*) was improved by attaching a magnet to the fishing gear, so the influence of the SMF on crayfish sheltering behavior was particularly apparent. The average catch increased 1.6 times when a magnet was attached to the entrance of a tubular trap for crayfishes (Trański et al., 2005). The MFs at 1 cm and 10 cm from the entrance of the tubular trap were 410 and 190 μT, respectively (Trański et al., 2005), and therefore, it seems likely that the magnetic sensory threshold can be estimated.

Fish in the Gulf of Mexico position themselves over buried oil pipelines off the shore of Texas, orienting themselves directly above the buried pipeline at a height of 1–3 m above the seabed and perpendicular to the axis of the pipeline (Arnason et al., 2002). Presumably, they are responding to some EMF stimuli, such as remnant magnetism in pipeline sections, voltage gradient induced by corrosion protection devices, or transient signals induced into the pipeline by remote lightning or solar wind-induced magnetic storms (Arnason et al., 2002).

The EMFs of both types of cable (bipolar and concentric) used in marine wind farms are small or zero. It has been reported that the EMFs of submarine cables have no significant impacts on the marine environment (Köller et al, 2006; Meißner, 2006). Studies with a long-term perspective are necessary to confirm the negligible impact of EMFs of wind energy on marine ecosystems (Köller et al, 2006; Meißner, 2006). Based on the findings obtained on the MF sensing mechanism of living organisms, further multifaceted and detailed research will be required on the basis of the effects of EMF on marine organisms.

6.2.3 Magnetic Sense of Plants

Plants known to sense light of various wavelengths, respond to gravity, and respond to contact and electrical signals are subject to the effects of the GMF (Maffei, 2014). Therefore, it has been reported that plants use the GMF as a physiological modulator (Maffei, 2014). As a pioneer study on the effects of MFs on plant growth, it was reported that very low MFs less than the GMF were capable of delaying both organ formation and development of barley (*Hordeum vulgare*) seedlings (Lebedev et al., 1977). By using ferromagnetic shields, the influence of weak, alternating MF, which was adjusted to the "cyclotron frequency" of Ca^{2+} and K^+, was studied on the fusion of tobacco (*Nicotiana tabacum*) and soybean (*Glycine max*) protoplasts (Nedukha et al., 2007). It was observed that in these conditions protoplasts fusion increased its frequency two to three times with the participation of Ca^{2+} in the induction of protoplast fusion (Nedukha et al., 2007). Artificial shielding of the GMF caused a significant decrease in the cell number with enhanced DNA content in root and shoot of onion (*Allium cepa*) meristems (Nanushyan and Murashov, 2001).

Putative magnetoreceptor, CRY was first found in a flowering plant *Arabidopsis thaliana* in 1993 (Ahmad and Cashmore, 1993), termed as *At*CRY1, which played an important role in photomorphogenic responses (Lin and Todo, 2005). Recently, the role for *At*CRY1 has been investigated as a candidate for magnetoreceptor. For example, the growth-inhibiting influence of blue light on the *Arabidopsis* is moderated by MFs in a way that may use the radical pair mechanism (Ahmad et al., 2007). Consistent with the theoretical framework of the radical-pair mechanism, *At*CRY1 has been shown to form magnetically sensitive radical pairs after photoexcitation of a flavin adenine dinucleotide (FAD) cofactor (Maeda et al., 2012). Moreover, the removal of the local GMF negatively affects the reproductive growth of *Arabidopsis*, which thus affects the yield and harvest index (Xu et al., 2012, 2013), and delays the flowering time through downregulation of flower-related genes (Xu et al., 2012; Agliassa et al., 2018a). The expression changes of three *At*CRY1-signaling related genes, PHYB, CO, and FT, suggest that the effects of a near null MF are CRY-related, which may be revealed by a modification of the active state of CRY and the subsequent signaling cascade plant CRY has been suggested to act as a magnetoreceptor (Xu et al., 2012). Artificial reversal of the GMF has confirmed that *Arabidopsis* can respond not only to MF intensity but also to MF direction and polarity (Bertea et al., 2015). Moreover, the GMF was found to impact photomorphogenic-promoting gene expression in etiolated seedlings of *Arabidopsis*, indicating the existence of a light-independent root magnetoreception mechanism (Agliassa et al., 2018b).

With regard to exposure of plants to MFs higher than the GMF, the MFs ranging ~1–30 mT have been reported to produce changes in quantum yield of flavin semiquinone radicals in *At*CRY1 (Maeda et al., 2012). However, most of the attention has been focused on seed germination of important crops like wheat, rice, and legumes, and many other physiological effects on plants of high MFs described plant responses in terms of growth, development, photosynthesis, and redox status (Maffei, 2014). The examination of the effects of 7 mT SMF combined with 20 kV/m EF on redox status was performed on shallot (*Allium ascalonicum*) leaves and the combined exposures increased lipid peroxidation and H_2O_2 levels (Cakmak et al., 2012). Hence, the combined exposures have distinct impacts on the antioxidant system leaves (Cakmak et al., 2012). Photosynthesis, stomatal conductance and chlorophyll content increased in corn plants (*Zea mays*) exposed to SMFs of 100 and 200 mT, compared to control under irrigated and mild stress conditions (Anand et al., 2012). In tomato (*Solanum lycopersicum*), a significant delay in the appearance of first symptoms of geminivirus, and early blight and a reduced infection rate of early blight were observed in the plants from exposed seeds to increased SMFs from 100 to 170 mT (De Souza et al., 2006).

Mass extinction events profoundly reshaped Earth's biota during the early and late Mesozoic and terrestrial plants were among the most severely affected groups (Maffei, 2014). Several plant families were wiped out, while some new families emerged, and eventually, became dominant (Maffei, 2014). The behavior of the GMF during the Mesozoic and Late Paleozoic, or more precisely between 86 and 276.5 Ma, is of particular interest. Its "virtual dipole moment (VDM)," originating from a geocentric dipole, seems to have been significantly reduced ($\approx 4 \times 1{,}022$ Am²), compared to the present values (Shcherbakov et al., 2002).

One of the working hypotheses to explain these mass extinctions is the cease of the GMF when the geomagnetic pole was reversed (Maffei, 2014). Because the GMF strength is strongly reduced during polarity transitions when compared to stable normal or reversed polarities (Maffei, 2014). It has been proposed that these variations might be correlated to plant evolution (Occhipinti et al., 2014). In the case of plant evolution, the greater incidence of high-energy particles and direct effects of the GMF on the biological system during the GMF reversal and excursion periods might contribute to alteration that eventually led to mass extinction (Maffei, 2014). Because plants, in general, do not change their orientation and habitat once germinated, there might be distinctive action of the terrestrial magnetism on the growth and physiology of plants (Yamashita et al., 2004). Magnetoreception might be a driving force contributing to plant evolution, but in order to prove such a hypothesis, further studies should be needed to confirm that some plant genes are affected by GMF reversals (Maffei, 2014).

By partially simulating GMF reversals with reduced MF, Narayana et al. (2018) suggested that GMF reduction to near null MF has a significant effect on ion content and ion transport gene expression in *Arabidopsis*. A few minutes after exposure to near null MF, plants respond with modulated root content and gene expression of all nutrient ions under study, indicating the presence of a plant magnetoreceptor that responds immediately to MF variations by modulating channels, transporters, and genes involved in mineral nutrition (Narayana et al., 2018). With time, the content of the nutrient ions decreases and is followed by the typical physiological responses of plants exposed to near null MF, including delay of flowering time (Xu et al., 2012; Agliassa et al., 2018a), photoreceptor signaling (Xu et al., 2014; Agliassa et al., 2018b; Vanderstraeten et al., 2018), and seed germination (Soltani et al., 2006). It is interesting to note that the response to near null MF is very rapid, which suggests that some ion channel and transport activity might be dependent on magnetoreception systems not necessarily related to gene expression (Narayana et al., 2018). Ongoing studies are evaluating the role of ferromagnetic, paramagnetic, and diamagnetic metals on plant magnetoreception (Narayana et al., 2018).

6.2.4 Magnetic Sense of Humans

First of all, historically, how did we get to know the GMF? It was the ancient Chinese people who realized that there was the GMF on the Earth, which exerted a mysterious force to make magnetic materials point to the "S" pole in the Northern Hemisphere (William, 2007). The magnetic compass was not, at first, used for navigation, but for geomancy and fortune-telling by the Chinese. In China during the Han Dynasty between the second century BC and first century AD, primitive compasses were used as designators of direction that the Chinese primarily used to order and harmonize their environments and lives (Merrill and McElhinny, 1983). Today we are familiar with this kind of use of direction as part of "Feng Shui," an ancient Chinese practice that has evolved into a decorating trend. Although the ancient Chinese people did not understand why this phenomenon happened, they can be called "the discoverers of the GMF" in the fact that they used the south pointer compass as a tool to know the direction of the GMF. Magnetic compasses were later adapted for navigation during the Song Dynasty in the eleventh century (William, 2007).

An English physician, physicist, and natural philosopher, William Gilbert first revealed that the main origin of the GMF is inside the Earth (Gilbert, 1600). He didn't know the mechanisms, but he was regarded as the first discoverer of the GMF in the fact that he noticed that if the Earth were magnets, he could explain the world distribution of the inclination (dip angle) (Gilbert, 1893). In a bar magnet, the N and S poles are arranged in a straight line, but even if the magnet is a sphere, an MF similar to that in the case of a bar magnet appears. By measuring the direction of the MF around a spherical magnet, he found that the already known distribution of the inclination of the GMF at the time was coincidental with that of the MF around a spherical magnet.

In addition, a German mathematician and physicist, Carl Friedrich Gauss accurately measured the GMF strength (Bühler, 1987). He contributed greatly to the elucidation of the mystery of the GMF. He also knew the method used to analyze gravity in celestial mechanics, so he applied it to the description of the GMF. In the 1830s, he undertook research on the origin of the GMF. He focused on a "dipole" in which the GMF becomes weaker at $1/r^3$ with respect to the distance r. Moreover, he found that the GMF could be a superposition of infinite fields such as "quadrupole" whose strength becomes weaker at $1/r^4$ and "octuple" whose strength becomes weaker at $1/r^5$. In the gravitational field, there is a single "monopole" field whose strength becomes weaker at $1/r^2$, but in the GMF, such a monopole has not yet been discovered and at least two poles always appear in pairs. He used this method to analyze the GMF data measured around the world and proved that the main components of the GMF originated inside the Earth, not from space. Furthermore, it was clarified that about 80% of the components of the GMF can be explained by the dipole MF originating from the inside of the Earth. Generally speaking, he is considered to be one of the discoverers of the GMF.

Gauss has also developed a device that accurately measures the GMF. This device quickly spread all over the world, and continuous measurements were made all over the world in the 1840s. In Japan, measurements of the GMF began in Tokyo in 1883, but it became difficult to measure the GMF because the noise of the EMFs became too strong due to the expansion of electric transmission lines and trams. Therefore, since 1913, the Geomagnetic Observatory, which was relocated to Kakioka, Ishioka City, Ibaraki Prefecture, has been measuring the GMF for more than 100 years (Suganuma, 2020).

It has been confirmed that various animals possess magnetoreception that perceives the direction, strength, and location of the GMF. In many migratory birds, photosensitive chemical reactions involving the cryptochrome (CRY) flavoprotein were thought to play an important role in the ability to sense the GMF. In the case of *Drosophila*, CRY1 found in these flies mediates light-dependent magnetoreception in a wavelength-dependent manner (Gegear et al., 2008, 2010; Fedele et al., 2014a,b). Surprisingly, this finding was later extended to CRY2 by showing that monarch butterfly and human CRY2 overexpressed in CRY-deficient flies could restore magnetosensitivity and its light-dependency (Gegear et al., 2010; Foley et al., 2011). In the case of monarch butterflies, although monarch butterfly CRY2 has been shown to mediate light-dependent magnetosensitivity in the *Drosophila* cellular environment, similar to human CRY2 (Gegear et al., 2010; Foley et al., 2011,2014a), monarch butterflies respond to a reversal of the inclination of the GMF in a UV-A/blue light and CRY1, but not CRY2, dependent manner, and both antennae and eyes, which express CRY1, are magnetosensory organs (Wan et al., 2021).

Humans are widely assumed not to have a magnetic sense for the GMF and also higher intensity SMFs (Phillips et al., 2010). A research team of the University of Massachusetts Medical School examined the light-dependent magnetosensing potential of human CRY2 (hCRY2) (Foley et al., 2011). To test whether the hCRY2 protein has similar magnetic sensory abilities, researchers have developed a transgenic *Drosophila* model that lacks the native CRY but instead expresses hCRY2 (Foley et al., 2011). Thus, they used hCRY2, which is abundantly expressed in the human retina, as a transgenic approach (Foley et al., 2011). Using a previously developed behavioral system (Gegear et al., 2008, 2010), when the CRY2 protein expressed in the human retina is transplanted into *Drosophila*, they examined whether these transgenic flies could sense and respond to the MF generated by the electric coil (Foley et al., 2011).

The results showed that the hCRY-rescued transgenic flies can sense the MF (Foley et al., 2011). However, it remains unclear whether this function is translated into a biological response downstream of the human retina (Foley et al., 2011). These findings demonstrated that hCRY2 has the molecular capability to function in a magnetic sensing system and may pave the way for further investigation into human magnetoreception (Foley et al., 2011). The light sensitivity of the human visual system has been shown to be remarkably sensitive to the modulated direction of the GMF (Thoss et al., 2000, 2002). Alternating MFs significantly altered electroencephalogram (EEG) signals in several brain regions, indicating the presence of "magnetosensory evoked potentials" (Carrubba et al., 2007; Mulligan and Persinger, 2012).

Do we humans really have the ability to sense the GMF? However, it has been estimated that humans lack at the very least an active perception of the GMF (Nordmann et al., 2017). Nordmann et al. (2017) emphasized that this complicates the design of experiments that aim to unravel specific aspects of the sensory modality, and compromises an experimenter's ability to detect obvious artifacts. The application of magnetic stimuli risks EM induction in a recording electrode, which makes the interpretation of electrophysiological data challenging (Nordmann et al., 2017). Anything from nT strength RF interference to the chosen perfume of experimenters can render an assay useless (Engels et al., 2014; Pinzon-Rodriguez and Muheim, 2017).

Recent experimental results suggested that humans may also possess magnetoreception (Chae et al., 2019; Wang et al., 2019). How accurately can humans perceive the GMF without using any tools such as magnetic compasses? If so, what is the physiological mechanism? Recently, South Korean researchers have demonstrated in a self-rotatory chair experiment that humans sense the GMF and turn toward food (Chae et al., 2019) within a Faraday cage, which is a conductive enclosure that shields its contents from EFs (Engels et al., 2014). The results demonstrated that starved men were largely oriented toward the ambient/modulated magnetic north or east, rather than women (Chae et al., 2019). This is a direction that was previously related to food, without other useful clues such as sight or sound. The orientation was reproduced under blue light but was abolished under blindfold or longer wavelength light (>500 nm), indicating that blue light is required for magnetic orientation (Chae et al., 2019). Importantly, the reversal of the vertical component of the GMF seemed to orient in the S magnetic direction, allowing food-induced blood glucose levels to act as a motive for sensing the direction of the MF (Chae et al., 2019). The results show that male humans rely on blue light to sense the GMF, suggesting that the geomagnetic orientations are mediated by an inclination compass (Chae et al., 2019). This study suggests that blue light-dependent human magnetoreception occurs in the eyes in a manner that appears to involve the brain and glucose, whereas the magnetoreceptor and magnetoreception mechanism (radical pair [RP] or otherwise) remains to be established (Chae et al., 2019). The male-specific magnetic orientation might have originated from prehistoric male ancestors who were dominantly responsible for gathering or hunting for food, and the varying level of individual orientation could be a diverging trait from the evolutionary process to date (Chae et al., 2019).

A joint research team of the California Institute of Technology (Caltech), and the University of Tokyo provided human subjects with "artificial MFs" that have the same strength as the GMF with changes only in the direction and inclination of the MF (Wang et al., 2019). The directional changes of the MF were perceived "unconsciously" by the subjects because the brainwaves showed specific reactions (Wang et al., 2019). The experiment was conducted at Caltech in Pasadena, California, USA. The research team made an MF exposure device that changes only in the direction and inclination of the GMF, and the recruited participants living in the local Pasadena (34 individuals; their genders, races, and ages ranging 18–68 years are variable) (Wang et al., 2019). EEG signals were measured at 64 points on the head surface in the electromagnetically shielded room within a Faraday cage in a dark condition while simulating the MF with the same intensity as the GMF and changing only the direction and inclination of the MF (Wang et al., 2019).

As a result, none of the participants answered that they noticed the change in the MF (Wang et al., 2019). However, only when they were stimulating with MF in the same direction and inclination as they were exposed to in Pasadena, where they usually live, the α-wave component of the EEG (8–13 Hz) decreased, leading to the response termed as α-wave event-related desynchronization (α-ERD) (Wang et al., 2019). The EEG data revealed that certain MF rotations could trigger strong and reproducible brain responses (Wang et al., 2019). One EEG pattern known from existing research, called α-ERD, typically shows up when a person suddenly detects and processes a sensory stimulus (Peng et al., 2012). The brains were "concerned" with the unexpected change in the MF direction, and this triggered the α-wave reduction (Wang et al., 2019). The researchers insisted that such α-ERD patterns in response to simple magnetic rotations are powerful evidence for human magnetoreception (Wang et al., 2019). The participants' brains only responded when the vertical component of the field was pointing downwards at ~60° (while horizontally rotating), as it does naturally in Pasadena (60° inclination in the Northern Hemisphere), California (Wang et al., 2019). They did not respond to unnatural directions and inclinations of the MF, such as when it pointed upwards (Wang et al., 2019). The researchers suggest the response is tuned to natural stimuli, reflecting a biological mechanism that has been shaped by natural selection (Wang et al., 2019). In this study, only 4 out of 34 participants showed an α-wave reduction in response to magnetic rotations. The research team told that it was expected that only 4 out of 34 participants possess magnetoreception (Wang et al., 2019). Just as not everyone excels in art and mathematics, the researchers think it's not strange that there are individual differences in the ability to sense MFs (Wang et al., 2019).

It has been reported that human health effects of environmental MFs including extremely low-frequency (ELF) MFs (1–300 Hz) have been discussed on the basis of the RPM (reviewed by Juutilainen et al., 2018). The RPM appears to be involved in sensing and responding to the static GMF (~50 µT) by animals. Evidence from numerous studies suggests that cancer-related biological processes can be affected by MFs≥100 µT, 50–60 Hz. It is unreasonable to assume that the plausible mechanism by which these effects occur is the modification by RP reaction on a specific target molecule (such as CRY) involved in the biological regulatory mechanism. Therefore, the results in MFs≥100 µT do not directly explain the epidemiological relationship between childhood leukemia and MFs≥0.4 µT ELF (Juutilainen et al., 2018). It remains unclear how it could explain human health effects of ELF-MFs<1 µT (Juutilainen et al., 2018).

6.3 Change of the Geomagnetic Field

6.3.1 Pole Shift

In 1905, a French physicist, Bernard Brunhes found some rocks, in an ancient lava flow at Pontfarin in the commune of Cézens (part of the Cantal département) in France, are magnetized reversely to the present GMF (Brunhes, 1905a,b, 1906). His observations made it clear that the GMF, providing MF values (total intensity, inclination and declination), is dynamic with frequent and aperiodic reversals of N and S poles (Howell, 1990; Dunlop, 1997). About 20 years later after the report by Brunhes, a Japanese geophysicist, Motonori Matuyama (a professor at Kyoto Imperial University at that time) in 1926 measured the paleomagnetic field of basalt samples from "Genbudo cave" (see also https://www.facebook.com/genbuguide/photos/a.427376960700781/3883154781789631/) in Hyogo Prefecture and Yakuno in Kyoto Prefecture in Japan, together with the Korean Peninsula and Manchuria, and he found that the orientation of the paleomagnetic field was divided into two groups (Matuyama, 1929). That is, the first group, including basalt samples from Yakuno, close to the current geomagnetic direction (declination: northward, inclination: downward), and in contrast, the second group, including basalt samples from Genbudo cave, indicating the opposite direction (Matuyama, 1929).

There is also evidence that the GMF was in two states. It has been reported that the estimated age is 1.65 Ma from Genbudo cave samples, and 0.3–0.4 Ma from Yakuno samples using the K-Ar method (Furuyama et al., 1993). Since Genbudo cave samples are estimated to be in the early Quaternary, and Yakuno samples are much closer to modern times, Matuyama reported that the GMF reversal from the second group to the first group was in the relatively short period of the Quaternary (Matuyama, 1929). Thus, he found that the GMF takes these two states of "reverse" polarity and "normal" polarity, and the period required for the "GMF reversal transition" is relatively short. His discovery had a great impact on our understanding of the mechanisms of GMF reversals. At that time, radiometric dating did not exist and it would not have been easy to date rocks. It seems plausible that great insight was needed to obtain these results. In the 1950s, the phenomenon of the self-reversal of "thermoremanent magnetization" of the rock itself was discovered (Nagata et al., 1952, 1953), so there was a time when the discussion was continued to deny the GMF reversal. The GMF reversal was only established in the first half of 1960 by showing that rocks of the same era had magnetization in the same direction globally.

In 1964, an American geophysicist, Allan Verne Cox and co-workers published a groundbreaking paper entitled "Reversals of the geomagnetic field" (Cox et al., 1964). They measured the paleomagnetism of rocks collected from all over the world and at the same time dated these rocks as evidence of controversial topics of GMF reversals. As a result, rock data collected from all over the world showed that rocks of the same period have the same normal or reverse paleomagnetic polarity regardless of their locations. These results suggest that the GMF reversals are not largely dependent on the self-reversal of thermoremanent magnetization of the rock itself, and that the dipole polarity of the GMF reversed many times in the past. Since then, advances in measurement technology have increased the accuracy of restoring the past GMF intensity using volcanic rocks of various ages, and more reliable methods have been developed for measuring the past long-term GMF intensity (Kono and Nagata, 1967; Kono, 1971).

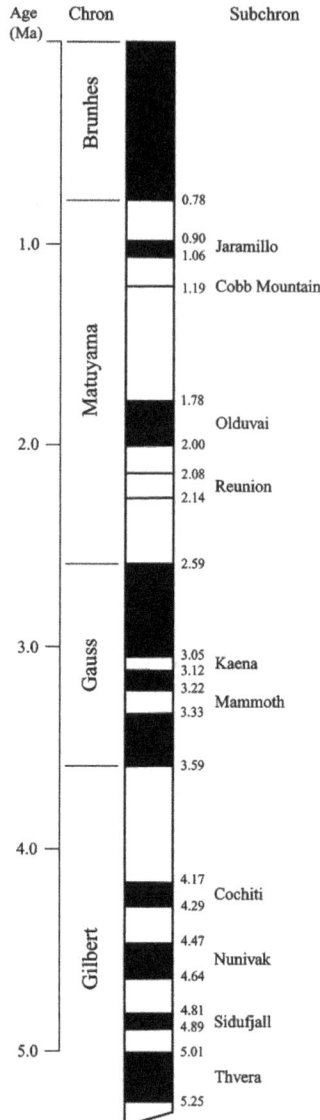

FIGURE 6.1 Reversals of the GMF during the past 5 million years (5 Ma) (Mankinen and Wentworth, 2003). (Reproduced with permission from Mankinen and Wentworth (2003), Copyright 2003, USGS; Mankinen and Dalrymple (1979), Copyright 1979, John Wiley & Sons.)

Polarity chrons or epochs are long periods during which the GMF remained oriented in one direction for most of the duration of the chron (Mankinen and Wentworth, 2003). As shown in Figure 6.1, the reversals of the GMF during the past 5 million years (5 Ma) are illustrated by Mankinen and Wentworth (2003), modified from Mankinen and Dalrymple (1979).

The periods during which the geomagnetic polarity was "normal" (that is the same as at present) are indicated in black, whereas periods of reversed polarity are indicated in white (Mankinen and Wentworth, 2003) (Figure 6.1). Most chrons are interrupted by shorter periods, called "subchrons," during which the field flips to the opposite of the dominant orientation during the longer parent chron (Mankinen and Wentworth, 2003).

The current chron, called the "Brunhes normal polarity chron," began about 774,000 years ago (774 ka) (ka is an abbreviation for kilo annum, 1 ka=10^3 years) (Valet et al., 2019; Simon et al., 2019). The immediate prior chron from 2.581 to 0.774 million years (Myr) ago (2.581–0.774 Ma) (Ma is an abbreviation for Mega annum, 1 Ma=10^6 years) is the "Matuyama reversed polarity chron." It was named after Motonori Matuyama, but at the time when he proposed "the theory of geomagnetic reversal" in 1929 (Matuyama, 1929), like the theory of ground motion and continental movement, it was not accepted at first. Subsequently, the chron from 3.58 to 2.581 Ma is the "Gauss normal" (Clague et al., 2006), and the chron from 5.894 to 3.58 Ma is the "Gilbert reverse" (Hill et al., 2006).

The typical periods of GMF reversals are as follows: the Gilbert–Gauss reversal, 3.58 Ma (Hill et al., 2006); the Gauss–Matuyama reversal, 2.581 Ma (Clague et al., 2006); the Matuyama–Brunhes (M–B) reversal, 774 ka (Valet et al., 2019; Simon et al., 2019). As described above, the rate of reversals in the GMF has varied widely over time. It is not yet known when and how the GMF reversal will occur, but the mechanism by which the Earth becomes a magnet has been elucidated by the "dynamo theory (geodynamo theory)" (Elsasser, 1950; Bullard and Gellman, 1954). This dynamo theory was first proposed by a German-born American physicist, Walter M. Elsasser in 1946 (Elsasser, 1950), and is the first mathematical model to show that the GMF is generated by the current (~3×10^9 A) induced by the convection in the Earth's outer core (Elsasser, 1950). The dynamo theory was proposed not only by Elsasser but also by a British geophysicist, Edward Bullard during the mid-1900s (Bullard and Gellman, 1954). Bullard showed that the movement of fluid in the outer core can generate the GMF (Bullard and Gellman, 1954; Massey, 1980).

There is a dynamo (generator) in the Earth's outer core (Elsasser, 1950), from which an MF is generated, creating dipoles that cover the entire Earth. In the 1990s, research on the Earth's dynamo (geodynamo), which had been developed only in theory for a long time, became possible on a large scale due to advances in computers, and research began to clarify more specific mechanisms of geomagnetic generation. The key area was located on the core-mantle boundary (CMB) (Glatzmaier and Olson, 2005). Moreover, observations by artificial satellites have revealed that there are magnetic flux patches on this CMB that are opposite to the normal orientation. These are called "reverse magnetic flux patches" (Glatzmaier and Olson, 2005). The largest reverse magnetic flux patches extend from the southern tip of Africa to below the southern tip of South America. As a result of comparison with past observations, it was found that reverse magnetic flux patches were formed one after another on the CMB (Glatzmaier and Olson, 2005). It seems plausible that the reverse magnetic flux patches are formed when the MF lines are affected by the Coriolis force of the Earth's rotation, and/or by the MF in the east-west direction (Glatzmaier and Olson, 2005).

In 1995, the team of Satoshi Kageyama and Tetsuya Sato of National Institute for Fusion Science, Japan (Kageyama et al., 1995), and the team of Gary A. Glatzmaier of the Los Alamos National Laboratory and Paul H. Roberts of the University of California, Los Angeles (Glatzmaier and Roberts, 1995), announced that both teams succeeded in simulating the geodynamo independently. Each team has developed a program that can simultaneously calculate the temperature, pressure, density, fluid movement, and even the generated MF in the center of the Earth, and using state-of-the-art supercomputers, they have succeeded in simulating the geodynamo, which has a dominant dipole MF closer to the real Earth (Suganuma, 2020). These were the first simulations of the geodynamo, albeit with some simple calculations (Suganuma, 2020). The three-dimensional MF structure simulated by supercomputers was completely different from the GMF structure that had been imagined from the time of Gilbert to the present day (Suganuma, 2020). As a result of calculations to reproduce hundreds of thousands of years, a dipole MF was generated, reverse magnetic flux patches were formed on the CMB, and then the MF was reversed (Glatzmaier and Roberts, 1995; Glatzmaier and Olson, 2005). However, no one can be sure that these simulations are occurring in the core of the Earth, because all calculations use approximate values (Glatzmaier and Olson, 2005). The heat convection in the core should be complex and there should be many small turbulences (Glatzmaier and Olson, 2005). It is impossible to handle this turbulence three-dimensionally with current supercomputers (Glatzmaier and Olson, 2005). While research on

simulation methods incorporating the effects of turbulence is progressing, the development of experimental equipment that imitates the geodynamo is also progressing (Glatzmaier and Olson, 2005). This is
an experiment to clarify the dynamo mechanism by rotating a huge container containing liquid sodium
(Glatzmaier and Olson, 2005). Geodynamo models are a powerful tool for testing various other hypotheses for the Earth's core. However, it is a matter of concern that many geodynamo models might be using
unphysical basic state buoyancy profiles, prescribing either uniform heat flux throughout the core, or
even worse, heat flux that increases from the inner core boundary to the CMB (Sreenivasan, 2010).

More recent studies on the geodynamo simulations have revealed in detail that convection of the
liquid iron (ferrofluid) in the outer core (~3,000 km below the surface), and electrical properties of iron
are responsible for the geodynamo that generates the GMF (Yong et al., 2019). Molten iron moves at
a speed of ~1 mm/s, and when it cuts the GMF lines, it produces a voltage that reinforces the original
MF (Gubbins, 2008). Fluid motion is driven by buoyancy resulting from a density gradient caused by
the slow cooling of the whole Earth (Gubbins, 2008). The core solidifies from the center to the outside,
and the light elements of the liquid separate and rise, and the heat flow promotes convection (Gubbins,
2008).

What is interesting is that even if the direction of the GMF is reversed, there is almost no change in
the direction of the molten iron flow in the outer core. Even if the GMF is reversed, the overall molten
iron flow does not reverse. GMF reversal can occur irregularly without giving an external trigger. It
seems likely that the GMF reversal may occur due to small things such as the slight disturbance of the
rotation speed of the Earth.

6.3.2 Chibanian

A geological layer indicating the latest GMF reversal of 774 ka, during the Matuyama–Brunhes (M–B)
reversal near the terminal Matuyama reverse, was found in a cliff wall along the Yoro River in Tabuchi,
Ichihara City, Chiba Prefecture in Japan (Kazaoka et al., 2015). This cliff wall in Chiba composite section of the Boso Peninsula was with an exposed layer of marine deposits and mineral debris (Kazaoka
et al., 2015). This layer contains the volcanic ash of Mt. Kiso Ontake erupted 774 ka, located in central
Japan, and is accordingly named Ontake-Byakubi tephra bed (Takeshita et al., 2016). The Byakubi tephra
zone (Byk A–E) is located within thick and massive siltstones in the Tabuchi section, and represents a
set of five individual tephra beds (Kazaoka et al., 2015). The most remarkable is the Byk-E bed, which
varies from 1 to 3 cm in thickness and consists of white, glassy, fine-grained ash (Kazaoka et al., 2015).
The traces of the last GMF reversal 774 thousand years ago (ka) in Chibanian stratum have been shown
by Kazaoka et al. (2015) and Suganuma et al. (2018) (see also https://www.facebook.com/town.otaki/
photos/pcb.1093824400728659/1093822697395496). A layer contains volcanic ash from Mt. Kiso Ontake
(Ontake-Byakubi Tephra Bed) erupted 774 ka. The layer was found in a cliff wall along the Yoro River in
Tabuchi, Ichihara City, Chiba Prefecture in Japan. The Byakubi (Byk) zones observed in this layer during and after the during the M–B reversal are classified as follows: Byk-B, Fine sand grain scoria with
normal polarity; Byk-C, Medium sand grain scoria with unstable polarity; Byk-E, Tephra bed, White silt
grain volcanic ash with reversal polarity. The M–B boundary, the primary marker of the GSSP, is located
~0.8 m above the Byk-E tephra bed. The thick siltstones are interpreted to have been deposited in warm
oceanic conditions based on the pelagic gastropod assemblages (Ujihara, 1986). The M–B boundary is
located ~0.8 m above the Byk-E tephra bed (Kazaoka et al., 2015; Suganuma et al., 2018). As shown in
Figure 6.2, lithofacies across the Lower–Middle Pleistocene boundary in the Tabuchi section are photo-
graphed by Kazaoka et al. (2015).

Geologically the Chiba composite section is located in the middle part of the Kokumoto Formation
as a member formation of the Kazusa Group. The geological epoch, the Middle Pleistocene from ~774
to 129 ka was named "Chibanian" (Chiba era) after the Japanese Prefecture Chiba, home to the city of
Ichihara (The Geological Society of Japan, 2010; Suganuma et al., 2018; Suganuma, 2020). The quaternary part of the international chronostratigraphic chart is shown by the Geological Society of Japan

FIGURE 6.2 Lithofacies across the Lower–Middle Pleistocene boundary in the Tabuchi section (Kazaoka et al., 2015). The white ash stratum of Byakubi-E (Byk-E) tephra bed is ~2 cm. The Byakubi tephra zone (Byk A–E) is located within thick and massive siltstones in the Tabuchi section and represents a set of five individual tephra beds. The most remarkable is the Byk-E bed (1–3 cm in thickness) and consists of white, glassy, fine-grained ash. (Reproduced with permission from Kazaoka et al. (2015) Copyright 2015, Elsevier.)

(2010). The name of Chibanian was approved by the International Union of Geological Sciences, and was certified as a Global Boundary Stratotype Section and Point (GSSP) on January 17, 2020, (Suganuma, 2020). Other than this Chiba composite section, it is known that the formations that prove the latest GMF reversal of 774 ka exist in Montalbano Jonico section (Bertini et al., 2015; Marino et al., 2015; Maiorano et al., 2016; Simon et al., 2017), and Valle di Manche section (Capraro et al., 2015, 2017; Macri et al., 2018) in southern Italy.

More recently, the M–B boundary is estimated to be dated to 772.9 ka (Suganuma et al., 2015, 2018; Okada et al., 2017; Simon et al., 2019; Haneda et al., 2020), and the polarity switch is completed within 1.1 ± 0.4 kyr (1σ) (Suganuma et al., 2018). The most detailed sedimentary record revealed that the average stratigraphic position of the M–B boundary between the Chiba, Yoro-Tabuchi, and Yanagawa sections, and the TB-2 sediment core drilled 190 m northeast of the Chiba composite section (Hyodo et al., 2016), lies 1.1 ± 0.3 m (1σ) above the Byk-E tephra bed, and the average age of the M–B boundary in the Chiba composite section is estimated as 772.9 ± 5.4 ka (1σ) (Haneda et al., 2020). The GSSP defining the base of the Chibanian Stage and Middle Pleistocene Subseries is placed at the base of the regionally wide-spread and significant Byk-E tephra bed in the Chiba section, with an astronomical age of 774.1 ka (The Geological Society of Japan, 2010; Suganuma et al., 2018; Suganuma, 2020). The GSSP lies 1.1 m below the directional midpoint of the M–B boundary (772.9 ka, duration 1.7 kyr) which therefore serves as its primary guide (Suganuma et al., 2021).

Overall, there is no bias in the polarities of the GMF found so far, such as either normal or reverse being long or short. According to the dynamo theory, the GMF reversal is thought to be due to "spontaneous convective instability of the liquid outer core." In addition, since the convective velocity of the outer core is ~20 km per year, it is estimated that a period of several hundred years is required for GMF reversal (Anderson, 1989). However, the frequency of long-term GMF reversals has a tendency to occur as follows (Biggin et al., 2012). In the last 800 ka, the GMF reversal has occurred only once in the M–B reversal (Biggin et al., 2012). However, in the past 2.5 Ma (million years ago), GMF reversals have occurred more than 11 times, i.e., five times in ~1 Ma (Biggin et al., 2012).

Could GMF reversals be caused by meteorite or comet impacts? Extraterrestrial impacts such as meteorite impacts may also trigger the GMF reversals. This "meteorite impact theory on the origin of geomagnetic reversals" is a theme that has been discussed for a long time, but no clear evidence has been presented so far, and it has been denied for a long time. Gilder et al. (2018) indicated that the impact events producing large impact craters, e.g., the Manicouagan crater (Quebec, Canada), Nördlinger Ries crater (Germany), Rochechouart crater (France), and Mistastin crater (Labrador, Canada), had no observable effect on the geodynamo (Koch et al., 2012; Eitel et al., 2014, 2016; Hervé et al., 2015). Hence, geodynamo

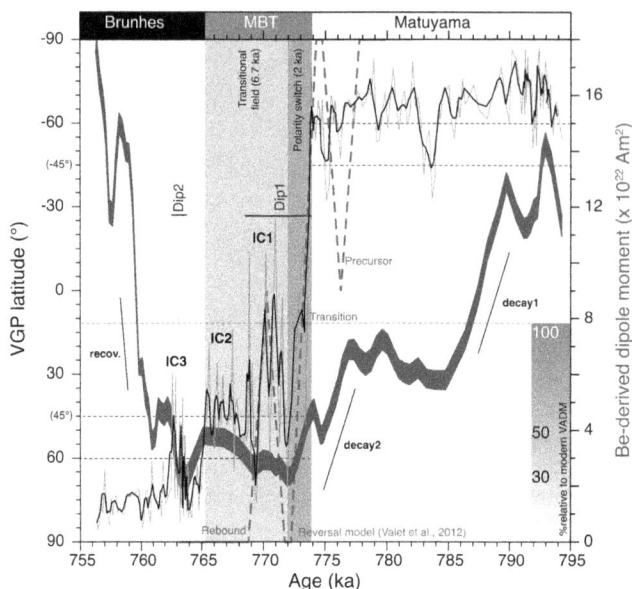

FIGURE 6.3 The GMF fluctuations in the Chiba composite section (Simon et al., 2019). They were indicated using [10]Be-derived dipole moment and virtual geomagnetic pole (VGP) latitude over the M–B transition. The black line corresponds to a three-point running mean of the VGPs. The "polarity switches" that explain most of the angular deviation span 2 ka (773.9–771.9 ka), but the "transitional field" prevails until 765.2 ka (two instability clusters, IC1–IC2). The largest VGP deviations correspond to the first low dipole intensity interval (Dip1) interval (773.7–767.4 ka). The loading bar shows dipole moment percent relative to present value. The results are compared with a schematic reversal path illustrating the three-phase succession of reversal (precursor, transit, and rebound) described by Valet et al. (2012) from reversal volcanic sequences. MBT indicates Mid-Brunhes Transition. (Reproduced with permission from Simon et al. (2019), Copyright 2019, Elsevier. It is licensed under the Creative Commons Attribution 4.0 International.)

disturbances from impacts should be considered improbable during the Phanerozoic when impact frequency and meteorite sizes were lower than in the Precambrian (French, 1998). Recently, a meteorite impact event was revealed just before the M–B transition. In this meteorite impact, no crater was found for a long time. However, in 2020, a joint research team of from Singapore's Nanyang Technological University, Thailand's Ministry of Natural Resources and Environment, and the University of Madison-Wisconsin, etc., discovered a crater in Southern Laos, Indochina, finally confirming the existence of this meteorite impact event that happened ~790 ka (Sieh et al., 2019). Comparing the timing of this meteorite impact with the GMF reversal record of the Chiba composite session reveals an interesting fact (Simon et al., 2019).

As shown in Figure 6.3, the GMF fluctuations in the Chiba composite section are presented by Simon et al. (2019).

That is, the virtual geomagnetic pole (VGP) latitude and dipole moment variations, both of which are an indicator of GMF fluctuations in the Chiba composite section, were compared through time using the chronology of Suganuma et al. (2018). It is estimated that this meteorite impact around 790 ka, "immediately before the decline in the GMF intensity," which is the process of the beginning of GMF reversal, before the timing of the reversal. During a meteorite impact, small glass particles called "microtektites" (Glass, 1990) are scattered over a wide area. If this microtektite is also contained in the ancient strata of the Chiba composite section, it is considered that the timing of this meteorite impact event can be determined directly from the strata of the Chiba composite section. Currently, the search for microtektites has begun in the strata of the Chiba composite section (Suganuma et al., 2018). In the future, if microtektites are found, the relationship between meteorite impact event and GMF reversal will be clear (Suganuma et al., 2018). This research approach could provide a major breakthrough in solving the mystery of GMF reversal (Suganuma et al., 2018).

6.3.3 Superchron

Around 15 Ma, GMF reversals occurred very frequently but randomly, occurring on average about once every 100 ka (0.1 Ma). Further back in the Cretaceous, there was a period of 42 Myr (~83–125 Ma), in which GMF reversals have never occurred and the stable state persisted (Biggin et al., 2012). The period, in which GMF reversals have not occurred and one polarity has been maintained for tens of millions of years, is termed "superchron" (Eide and Torsvik, 1996). As other superchrons, the Ordovician (Moyero) Reversed Superchron (ORS; ~460–490 Ma), and the Permian–Carboniferous (Kiaman) Reversed Superchron (PCRS; ~267–313 Ma) are known even before the Cretaceous Normal Superchron (CNS; ~83–125 Ma) (Biggin et al., 2012). Examining the trends of GMF reversals over the past 600 Ma suggests that the frequency of GMF reversals gradually increases after the superchron and tends to peak thousands of years before the next superchron (Biggin et al., 2012). In this way, the frequency of GMF reversals fluctuates slowly on timescales of thousands to hundreds of millions of years, and it is assumed that superchrons occur at a rate of <200 Myr in their process (Biggin et al., 2012). The available data show that the GMF reversal frequency has varied significantly over time, from the zero value during superchrons to epochs, when seven to ten reversals could occur for 1 Myr (Opdyke and Channell, 1996; Biggin et al., 2012; Pavlov and Gallet, 2005, 2010). As shown in Figure 6.4, records of geomagnetic polarity reversal frequency and dipole moment since the Cambrian are presented by Biggin et al. (2012).

Including superchron, the frequency of GMF reversals is fluctuating greatly and several hypotheses about the mechanisms have been proposed. One of the most plausible hypotheses is "mantle convection theory" (Biggin et al., 2012). The Earth has been cold since its birth. In the process, the mantle, which occupies 80% of the Earth's volume, is gently convected, transporting heat from the outer core to the Earth's surface. Since the mantle is a huge heat transporter for the Earth, when the mantle becomes active, the heat flow to the Earth's surface increases. As shown in Figure 6.5, the average reversal frequency and eruption ages of large igneous provinces (LIPs) (offset by +50 Myr) that have not yet been subducted are presented by Biggin et al. (2012).

Since the Cretaceous, mantle convection has gradually become inactive, no extreme volcanic eruptions have occurred, and it is presumed that the Earth has gradually cooled. Intermittent upstream from the lower part of the mantle, called "mantle plume," is thought to play a major role in mantle convection changes over a long timescale. Mantle convection, including upstream of mantle plume, also has periodic fluctuations, which are estimated to be 200 Myr. In particular, huge one is called "superplume" (Larson, 1991), and it is thought that when the superplume rises from the bottom of the mantle, large-scale magma and volcanic activity occur on the Earth's surface.

Mantle plume heads leaving the CMB may reflect enhanced heat flow out of the core potentially increasing reversal frequency tens of Myr before the resulting eruption of the "large igneous provinces (LIPs)." As shown by Figure 6.6, the easterly distribution of the Middle-Late Permian LIPs in Pangea is presented by Isozaki (2009), which is modified from Isozaki (2007a).

LIPs are often accompanied by large flood basalt eruptions. Allowing for an average rise-time of 50 Myr produces a broad correlation that would associate GMF reversal hyperactivity in the mid-Jurassic with widespread LIP emplacement in the mid-Cretaceous. In the period 0–50 Myr, mantle plume heads that had left the CMB would not yet have reached the surface.

When the supercontinent Pangea split in the Jurassic, multiple plumes rose side by side in the N-S direction, and the crevices are connected to create a new ocean, the Atlantic Ocean. Apart from the LIPs on the Atlantic coast, multiple plumes also hit the eastern part of Pangea in the Permian, forming each LIP. Typical examples of flood basalt are Siberia and Mt. Emei Scenic Area, Emeishan in China.

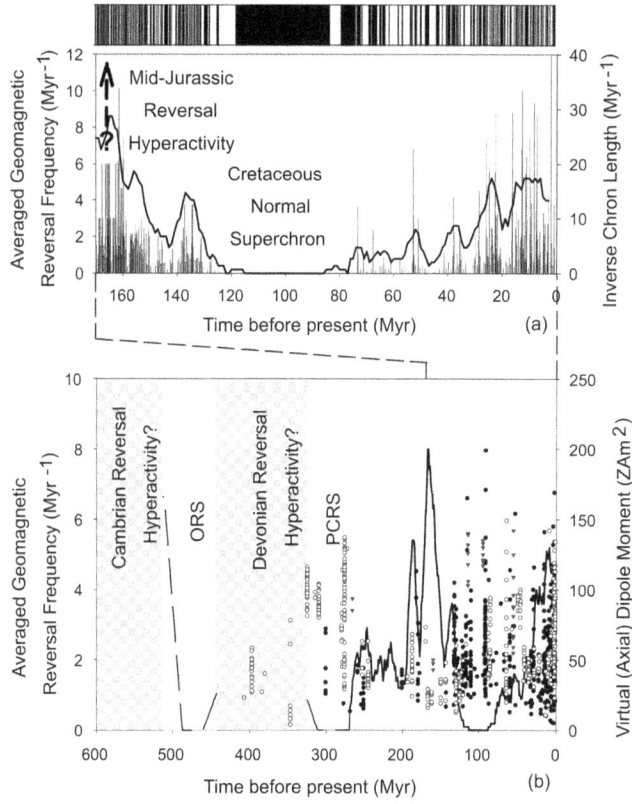

FIGURE 6.4 Records of geomagnetic polarity reversal frequency and dipole moment since the Cambrian (Biggin et al., 2012). (a) The marine magnetic anomaly record (MMA), and plots of inverse chron length (black bars), and reversal frequency. The black and white stripes on the top represent the year of geomagnetic polarity. Black is the same normal polarity as at present, and white is the reverse polarity. Cretaceous Normal Superchron (~83–125 Ma). (b) Reversal frequency from the MMA and magnetostratigraphic studies (chequered area indicates insufficient data) alongside virtual (axial) dipole moment (spatially normalized field intensity) measurements. ORS, Ordovician (Moyero) Reversed Superchron (~460–490 Ma); PCRS, Permian–Carboniferous (Kiaman) Reversed Superchron (~267–313 Ma). (Reproduced with permission from Biggin et al. (2012), Copyright 2012, Springer Nature.)

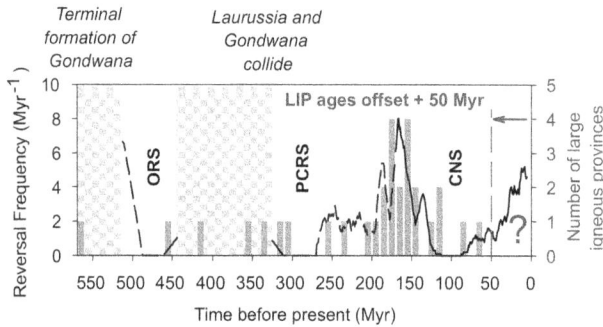

FIGURE 6.5 The average reversal frequency and eruption ages of large igneous provinces (LIPs) (offset by +50 Myr) that have not yet been subducted (Biggin et al., 2012). ORS, Ordovician (Moyero) Reversed Superchron (~460–490 Ma); PCRS, Permian–Carboniferous (Kiaman) Reversed Superchron (~267–313 Ma); CNS, Cretaceous Normal Superchron (~83–125 Ma). (Reproduced with permission from Biggin et al. (2012), Copyright 2012, Springer Nature.)

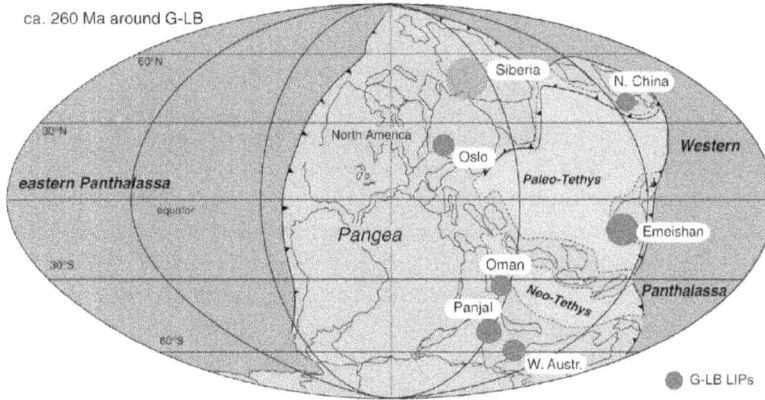

FIGURE 6.6 The easterly distribution of the Middle-Late Permian large igneous provinces (LIPs) in Pangea (Isozaki, 2009). Paleogeographic base map is after Maruyama et al. (1989). W. Austr., Western Australia; N. China, North China. (Reproduced with permission from Isozaki (2009), Copyright 2009, Elsevier.)

The estimated changes in CO_2 concentrations during the Phanerozoic were shown by Rohde (2006). The CO_2 concentration in the atmosphere started rapidly decreasing from the Late Ordovician and subsequently never reached the Early Paleozoic (Cambrian) values during the Phanerozoic (Rohde, 2006). Three estimates are based on geochemical modeling: GEOCARB III (Berner and Kothavala, 2001), COPSE (Carbon-Oxygen-Phosphorus-Sulfur-Evolution), (Bergmann et al., 2004), and another geochemical model (Rothman, 2002). These are compared to the carbon dioxide (CO_2) measurement database of Royer et al. (2004), and a 30 Myr filtered average of those data (see Rohde, 2006; Cm, Cambrian; O, Ordovician; S, Silurian; D, Devonian; C, Carboniferous; P, Permian; Tr, Triassic; J, Jurassic; K, Cretaceous; Pg, Paleogene; N, Neogene). Direct determination of past CO_2 levels relies primarily on the interpretation of carbon isotopic ratios in fossilized soils (paleosols) or the shells of phytoplankton and through interpretation of stomatal density in fossil plants (Rohde, 2006). The Ordovician (O) greenhouse climate was replaced by a glacial one with several glacial periods (Saltzman and Young, 2005). Huge volcanoes occurred frequently in the Cretaceous (K) (Hofmann et al., 2000; Nordt et al., 2003). As a result, volcanic activity released a large amount of CO_2 into the atmosphere, causing the Earth's surface to warm greatly.

Thus, it has been explained that past cooling and warming conditions of the Earth's surface are caused by changes in the concentration of CO_2 in the atmosphere. In addition, recent artificial satellite data suggest that the amount of clouds affected temperature changes of the Earth's surface. It has been found that the formation of the cloud requires the formation of a large number of cloud particle nuclei by ionization of atmospheric molecules by GCRs (first described by Svensmark and Friis-Christensen, 1997). When a large amount of GCRs enter into the atmosphere, clouds are generated over a wide area and the Earth's surface becomes cold, whereas when the inflow is small, the Earth's surface continues to be sunny and warms. The influx of GCRs into the atmosphere is determined by two factors: the strength and distance of sources such as supernovae and the GMF strength. The probability of encountering a supernova explosion changes depending on the position of the solar system in the galaxy. On the other hand, the GMF strength plays the role in the influx of GCRs into the atmosphere as a magnetic shield, and therefore, when the GMF strength is sufficiently high, GCRs cannot easily flow into the atmosphere and vice versa (Ueno et al., 2019).

Kirkby (2007) speculated that GCRs forcing of the climate acts simultaneously and with the same sign throughout the entire globe and operates on all time scales from days to hundreds of millions of years. For this reason, even a relatively small forcing can lead to a large climatic response over time (Kirkby, 2007). However, the hypothesis about this mechanism has not been proved (see discussion).

The geodynamo simulations suggested that active heat transport from the outer core to the mantle may result in more active convection in the outer core, resulting in instability in the geodynamo. Hence, it is speculated that the activation of mantle convection could cause GMF reversals together with the activation of volcanic activity.

Additional reports now understand that the solid inner core is growing in the Earth's core, which is the source of convective energy in the liquid outer core (Lister and Buffett, 1995; Jones, 2015; Lythgoe et al., 2015). The Earth has been a heat engine that has been cooling since its birth, but the inner core is growing as it cools. When the solid inner core grows, the light elements contained in the liquid outer core are released, and convection occurs when they rise toward the mantle (Zhang et al., 2016). The compositional convection (the driving force is the density difference due to the difference in composition) and the thermal convection (the driving force is the density difference due to heat) are considered to be main geodynamo energy sources (Lister and Buffett, 1995; Lythgoe et al., 2015). Moreover, latent heat released when the inner core solidifies can also cause thermal convection. Just changing the thermal conditions at the CMB can alone account for the full spectrum of temporal variations in the GMF reversal rate (Driscol and Olson, 2011; Olson et al., 2013). However, it is speculated that the driven effect of the compositional convection on the geodynamo might be larger than that of the thermal convection (Lythgoe et al., 2015).

According to a recent study published in the journal Nature Geoscience, the inner core is growing lopsidedly (Frost et al., 2021). One half of the sphere, the eastern half under Indonesia's Banda Sea, accrues 60% more iron crystals than its western counterpart, which is located under Brazil (Frost et al., 2021). Frost's team (Daniel Frost, a leading seismologist at the University of California, Berkeley) created geomagnetic growth models and conducted mineral physics calculations that tracked the inner core's growth over the last billion years (Frost et al., 2021). According to Science News (Woodward, 2021), briefly, they found that its lopsided nature likely began as soon the core formed.

"Every layer in the Earth is controlled by what's above it, and influences what's below it," Frost told Live Science from Specktor (2021). If iron is crystallizing more quickly on one side of the inner core than the other, it means that the outer core is cooling faster on that side (Woodward, 2021). So the mantle on that side, in turn, must be cooling the outer core faster than the mantle on the other side (Woodward, 2021). The genesis of that cooling chain, Frost said, could be Earth's tectonic plates (Woodward, 2021). When one plate pushes up against another, one subducts, or sinks, below the other (Woodward, 2021). The subducting plate cools the mantle in that area of the planet (Woodward, 2021). Earth's core plays a key role in protecting the planet from dangerous solar wind and GCRs (Woodward, 2021). Swirling iron in the outer core generates an MF that stretches all the way from there to the space surrounding our planet (Woodward, 2021). That swirl happens, in part, because of a process in which hotter, lighter material from the outer core rises into the mantle above (Woodward, 2021). There, it swaps places with cooler, denser mantle material, which sinks into the core below (Woodward, 2021). This convection also happens between the inner and outer core, so if various parts of the outer and inner core are cooling at different rates, that could affect how much heat gets exchanged at the boundary, which might have an impact on the swirling engine powering Earth's protective sheath (Woodward, 2021). "The question is, does this change the strength of the MF?" Frost told Live Science (Specktor, 2021). Questions this big are beyond the scope of the team's new paper, but Frost said he has begun work on new research with a team of geomagnetists to investigate some possibilities (Specktor, 2021).

Other geodynamo energy sources are precession and tidal interactions (Olson, 2016). Tides distort the CMB, and precession derives their energy from the Earth's rotation. Precession is caused by the torques on the Earth's equatorial bulge and the Earth's axis of rotation precesses once every 26 kyr. In another aspect, the GMF reversal frequency would have been very sensitive to variations in the vigor of the convection in the Earth's liquid outer core. Yadav et al. (2016) expected that flow in the outer core might be primarily governed by (1) the Coriolis force due to the Earth's rotation, (2) the buoyancy force driving the convection, and (3) the Lorentz force due to the GMF. That is, these expectations suggested that the outer core flow might be governed by a balance between Lorentz force, rotational force, and buoyancy, which is called "MAC balance for Magnetic, Archimedean, and Coriolis," with only minute roles for viscous and inertial force (Yadav et al., 2016). A relevant geodynamo model has to be in such a force balance (Yadav et al., 2016). Contemporaneous simulations invoke high viscosities to suppress flow turbulence to keep the computational costs manageable. The unrealistically large viscosity in these simulations is a major concern (Yadav et al., 2016). They showed that the state-of-the-art simulations with a viscosity that is lower than in most simulations, but still much larger than in the Earth's core, can approach a realistic force balance (Yadav et al., 2016). Their simulations produce many properties that have been theoretically predicted in the past (Yadav et al., 2016).

6.3.4 Decrease of Geomagnetic Field

The GMF has changed on different time scales in geological history, and the present normal polarity started around 774 thousand years ago (774 ka) during Chibanian (~774–129 ka). The changes in the last 200 years have been potentially diminishing declines. An eminent GMF reversal would not be so unexpected (De Santis et al., 2004). The magnetic poles are also moving, and it is estimated that the N pole has moved 1,100 km over 170 years. According to analysis of ESA (European Space Agency) on the distribution of GMF up to June 2014, it seems plausible that global GMF intensity is not weakened evenly, and there is a great difference depending on the place (De Michelis et al., 2017). Thus, changes in the GMF reflect changes in the deep Earth, and there are two cases of extreme changes: reversals and excursions. During the former event, the GMF strength decreases, the poles rapidly reverse polarity, and the polarity reverses as mentioned above (Brown et al., 2007; Valet and Plenier, 2008; Ferk and Leonhardt, 2009).

Meanwhile, in the case of excursions, the magnetic pole moves but eventually returns to the original polarity (flip-flopping poles). Excursion generally refers to an event in which the position of the geomagnetic pole is 45° or more away from the N pole or S pole. Excursion occurs in a very short geological period, so it is difficult to find it from fragmentary paleomagnetic records, and it is still a mysterious phenomenon (Suganuma, 2020). Many of the excursion records have been found in seafloor and lake bottom sediments where estimates of continuous GMF fluctuations are possible, and fortunately, they have also been found in paleomagnetic lava (Suganuma, 2020). So far, there is no consensus on the frequency of excursions, but it is estimated that it may have occurred 23 times in the ~800 kyr since the M–B boundary (Oda, 2005).

Oda (2005) showed the excursions and relative MF intensity, which are compiled from Channell et al. (1998), Channell (1999), Guyodo and Valet (1999), and Channell and Kleiven (2000). The data of GMF intensity variations were obtained for the last 800 kyr. The data of relative GMF intensity variations were obtained from Ocean Drilling Program (ODP) Site 983 in the northern North Atlantic Ocean. The relative GMF intensity was calculated by standardizing the natural residual magnetization with the isothermal residual magnetization. Oda (2005) found that many excursions tend to occur during the period of low GMF strength. This suggests that when the magnetic dipole component, which is the main driving force of the GMF, weakens, the non-dipole component becomes relatively dominant, causing excursion.

The GMF intensity over the past few centuries has been declining strongly, triggering a GMF reversal. It can be inferred that by analyzing the past excursion events, the present GMF is not in the early stages of reversals or excursions and will recover without an extreme event such as an excursion or reversal (Brown et al., 2018). Brown et al. (2018) inferred that for excursions to occur, a weakening of the field across much of the globe spreading from multiple sources is required, and not just localized weakening expanding from a South Atlantic Anomaly-like feature. With special reference to the geomagnetic excursions, however, the N magnetic pole is heading from Canada into Siberia, and recently crossed the International Date Line (Livermore et al., 2020). That is, since the first in situ measurements in 1831 of its location in the Canadian arctic, the pole has drifted inexorably toward Siberia, accelerating between 1990 and 2005 from its historic speed of 0–15 km/year to its present speed of 50–60 km/year. In late October 2017, the N magnetic pole crossed the international date line, passing within 390 km of the geographic pole, and is now moving southwards (Livermore et al., 2020). Its rapid motion, plus other shifts in the GMF, has forced scientists to revise magnetic models that guide navigation. The scientists are working to understand why the GMF is changing so dramatically. Geomagnetic pulses, like the one that happened in 2016, might be traced back to "hydromagnetic" waves arising from deep in the core (Aubert, 2018). And the fast motion of the N magnetic pole could be linked to a high-speed jet of liquid iron beneath Canada (Livermore et al., 2017). Many geomagnetic excursions tend to occur during the period of low GMF intensity. Most of the known geomagnetic excursions contain GMF reversals, and the GMF strength is greatly reduced as much as the GMF reversals (Ferk and Leonhardt, 2009). The differences between excursions and reversals are that the duration of the excursions is as short as 1–2 kyr or less, and the observed GMF fluctuations of the excursions are regional. In the case of the GMF reversals, global intensity variations appear more coherent (Brown et al., 2007). Not only GMF reversals but also known excursions always occur when the magnetic dipole component (the main key player in the GMF generation) reduced to <50% of its strength (Guyodo and Valet, 1999). This suggests that reversals and excursions are not sudden abnormal phenomena, but stochastic phenomena that occur in the rhythm of the geodynamo itself. Using the observed GMF model over the past 7 kyr, it has been confirmed that when the magnetic dipole intensity is reduced to 50% or less, the non-dipole intensity increases relative to the dipole intensity, and eventually, geomagnetic excursion occurs (Brown et al., 2007; Valet and Plenier, 2008).

The most representative geomagnetic excursion known today is the "Laschamp excursion" about 41–42 kyr ago (41–42 ka, range: 41,4±2 ka) during the end of the Last Glacial Period, and was first recognized in the late 1960s as a GMF reversal (Bonhommet and Babkine, 1967). The name of Laschamp comes from the place Laschamp lavaflows in the Clermont-Ferrand district of France (Chaîne des Puys, France), where Brunhes lived. The Laschamp excursion is said to have occurred at the time of the weakest GMF in the last 100 ka, and it is reported that the GMF strength at that time dropped to nearly 10% of the current strength level.

Geodynamo simulations have suggested "pessimistic" and "optimistic" predictions for the life on Earth regarding the consequences of decrease in the GMF. Generally speaking, as mentioned above, the GMF protects the life on Earth from harmful rays from the universe. Therefore, if the GMF decreases too much, it will be harmful to many lives on Earth, including humans. It is predicted that the magnetic pole may be exchanged, and the GMF reversal transition can occur unexpectedly in a short time scale; one surprisingly abrupt "centennial reversal transition" occurred in 144±58 years (2σ) (Chou et al., 2018). By investigating the structure of rocks on ancient Earth, Chinese geologists reported that the S and N poles had already swapped places, and the reversing process has recently begun again (Chou et al., 2018). The first sign of magnetic pole reversal is attributed to the weakening of the GMF intensity (Chou et al., 2018). It is found that the GMF has weakened by 15% in the last 200 years and is becoming more and more unstable (Chou et al., 2018). However, it is impossible to predict the specific time of reversal (Chou et al., 2018). This is because it is very difficult to know whether the weakening of the GMF is sudden or gradual (Chou et al., 2018). It has been warned that the Earth will lose its barriers from the solar wind and GCRs during a magnetic pole change (Chou et al., 2018). Consequently, it may cause a rapid change in the flora and fauna on Earth and may have a very harmful effect on humans (Chou et al., 2018). In addition, all digital technology will be destroyed and artificial satellites and power lines will fail (Chou et al., 2018). The Earth will be exposed to the powerful solar wind, lose water, be bombarded by "γ-ray burst," and experience the same environmental conditions as Mars today, with a significant loss of the GMF due to pole shift (Brown et al., 2007; Valet and Plenier, 2008; Ferk and Leonhardt, 2009).

6.3.5 The Extinction of Neanderthals

The following hypothesis concerning the Laschamp excursion has recently been proposed for the relationship between the geomagnetic excursion and the extinction of living organisms. It has been hypothesized that the Laschamp excursion was associated with the extinction of Neanderthals, *Homo neanderthalensis* in Europe as shown in Figure 6.7 (Channell and Vigliotti, 2019).

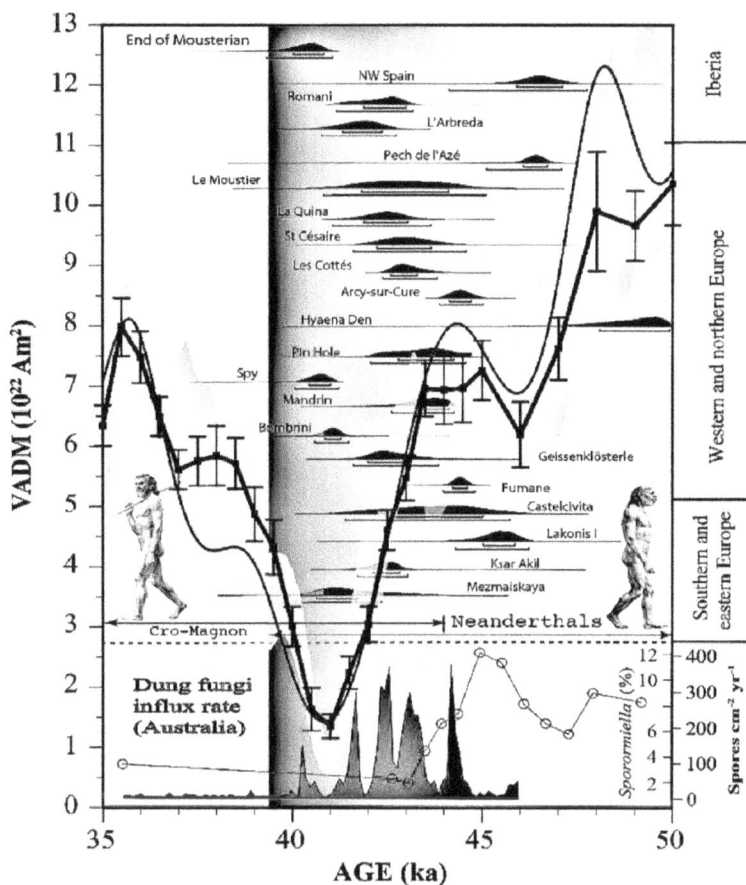

FIGURE 6.7 Timing of the Laschamp excursion and the extinction of Neanderthals (Channell and Vigliotti, 2019). VADM, Virtual Axial Dipole Moment. (Reproduced with permission from Channell and Vigliotti (2019), Copyright 2019, John Wiley & Sons.)

According to this report, the extinction of Neanderthals is estimated to be about 41–39 ka (Channell and Vigliotti, 2019). The GMF intensity at that time was estimated from the mean value of the "Virtual Axial Dipole Moment (VADM)" originating from a geocentric axial dipole (Korte and Constable, 2005a,b). VADM reflects the absolute paleomagnetic field strength derived from volcanic rocks (Korte and Constable, 2005a,b). In more detail, VADM is intensity of an imaginary axial (along the Earth's rotation axis) geocentric (located in the center of the Earth) dipole that would produce the estimated archaeo-/palaeointensity at the sampling site (Korte and Constable, 2005a,b). It is calculated from the archaeo-/palaeointensity of a sample as estimated by measurements in the laboratory and the magnetic co-latitude of the sampling site (Korte and Constable, 2005a,b). Here, the timing of this extinction is almost the same as the peak period of the Laschamp excursion (Channell and Vigliotti, 2019).

It is believed that the Laschamp excursion caused an extreme decrease in the GMF intensity, which destroyed the stratosphere ozone (O_3) layer, resulting in a sharp increase in UV-B (280–315 nm) radiation, which made the Neanderthals extinct (Channell and Vigliotti, 2019). By the time the Laschamp excursion occurred about 41 ka, the Cro-Magnon Man, a modern human *Homo sapiens*, already existed and coexisted with the Neanderthals (Channell and Vigliotti, 2019). It is supposed that modern humans are less susceptible to UV radiation than Neanderthals, and therefore have escaped extinction due to differences in the aryl hydrocarbon receptor, a transcription factor involved in the regulation of UV sensitivity (Channell and Vigliotti, 2019). Moreover, in this period, mammalian fossils in Australia and Eurasia record an important die-off of other large mammals probably due to the same reason (Channell and Vigliotti, 2019). In the case of Australia, fossil occurrences and dung-fungal proxies indicate that episodes of the Late Quaternary extinction of mammalian megafauna occurred close to the Laschamp and Blake excursions (Channell and Vigliotti, 2019). In addition, in the Americas and Europe, a large mammalian die-off appears to have occurred ~13 ka (Channell and Vigliotti, 2019). This is because fossil and dung-fungal evidence for the age of the Late Quaternary extinction coincides with a prominent decline in the decrease in the GMF intensity (Channell and Vigliotti, 2019). Both die-offs can be linked to minima in the GMF strength implying that variations of UV radiation flux to the Earth's surface influenced mammalian evolution (Channell and Vigliotti, 2019). For the last ~200 ka, estimates of the timing of branching episodes in the human evolutionary tree, from modern and fossil mitochondrial DNA and Y-chromosomes, can be linked to minima in field strength which implies a long-term role for UV radiation in human evolution (Channell and Vigliotti, 2019).

According to Cooper et al. (2021), the Laschamp excursion in combination with the Grand Solar Minima initiated substantial changes in the concentration and circulation of the atmospheric ozone and increased atmospheric ionization and UV radiation levels, leading to global climate shifts that caused major environmental changes. It was also suggested that those environmental changes could have sparked a chain of events leading to the extinction of large mammals in Australia and Europe, and possibly to the extinction of *Homo neanderthalensis* and subsequent success of *Homo sapiens* (Channell and Vigliotti, 2019; Cooper et al., 2021). Similarly, large mammals' extinctions in North America and Europe 13,000 years ago could be linked to geomagnetic excursions identified in sediments of Brunhes Chron (Erdmann et al., 2021).

More recently, Cooper et al. (2021) created a precisely dated radiocarbon record around the time of the Laschamp excursion about 41 ka from the annual rings of ancient New Zealand giant swamp kauri trees (*Agathis australis*, 42,000-year-old subfossil). They precisely characterized the geomagnetic transition and performed global chemistry-climate modeling and detailed radiocarbon dating of paleoenvironmental records to investigate impacts (Cooper et al., 2021). This geological archives record revealed a substantial increase in the cosmogenic isotopes ^{14}C content of the atmosphere culminating during the period of weakening MF strength preceding the polarity switch (Cooper et al., 2021). The radiocarbon was supplied to the Earth by continuous GCRs, and the radiocarbon left in the annual rings indicated that a large amount of the radiocarbon was brought to the Earth during this period (Cooper et al., 2021). Combined with an unusually quiet Sun, known as the "Grand Solar Minima" that is believed to have occurred during the Laschamp excursion, a large amount of GCRs could have caused a notable drop in stratospheric ozone, shifting wind flows and climate patterns (Cooper et al., 2021).

The GMF at that time has been estimated to be weakened to <28% of the current strength. According to this record, however, the GMF became the weakest about 42.2 ka before the Laschamp excursion and its intensity was only 0%–6% of the present (Laj et al., 2014; Brown et al., 2018; Liu et al., 2020). This GMF reduction resulted in more GCRs reaching the Earth, causing greater production of the cosmogenic isotopes ^{14}C (Cooper et al., 2021). Thus, due to the almost disappearance of the GMF and the loss of the "protective shield" of the Earth, charged GCRs fell unobstructed and ionized fine particles in the Earth's atmosphere, and the ionized atmosphere destroyed the ozone layer and synchronous global climate change, such as glacial maxima and aridification, was caused (Cooper et al., 2021). Comparing this record with that of the Laschamp excursion already measured in ice cores, etc., it coincided with the period when ice beds and glaciers expanded in North America and the generating mechanism of wind zones and tropical cyclones changed significantly (Cooper et al., 2021). During this period, the large fauna (Megafauna) became extinct at the same time in mainland Australia and Tasmania. While the Neanderthals became extinct (recalibrated at 40.9–40.5 ka), murals in caves suddenly appeared around the world (Cooper et al., 2021). Cooper et al. (2021) think that it is very premature to connect this event to milestones in human evolution that occurred around that time, such as the demise of the Neanderthals or increase in cave painting though the link is worth investigating (Voosen, 2021).

6.3.6 Mass Extinctions of Life on Earth

Raup and Sepkoski (1982) identified five mass extinctions of life on Earth. Sepkoski's evolutionary faunas of marine animals are shown by Sepkoski (1984) as "Big Five":

1. Ordovician–Silurian (O–S) extinction events: ~450–440 Ma at the O–S boundary (Sepkoski, 1984; Baez, 2006). This is the second-largest of the five major extinctions. Two independent studies simultaneously suggested the cause of the mass extinction was due to global warming, related to volcanism, and anoxia, and not due, as considered earlier, to cooling and glaciation (Hall, 2020; Bond and Grasby, 2020).
2. Late Devonian (Late D) extinction: ~375–360 Ma near the Devonian–Carboniferous boundary (Sepkoski, 1984; Baez, 2006).
3. Permian–Triassic (P–Tr) extinction event: ~252 Ma at the P–Tr boundary (Sepkoski, 1984; Baez, 2006; St. Fleur, 2017). This is the Earth's largest extinction. The highly successful marine arthropod, the trilobite, became extinct. The "Great Dying" had enormous evolutionary significance (Erwin, 2006).
4. Triassic–Jurassic (Tr–J) extinction event: ~201.3 Ma at the Tr–J boundary (Sepkoski, 1984; Baez, 2006). Most non-dinosaurian archosaurs, most therapsids, and most of the large amphibians were eliminated, leaving dinosaurs with little terrestrial competition. After that, the Mesozoic era, characterized by ammonites and dinosaurs, began.
5. Cretaceous–Paleogene (K–Pg) extinction event: ~66 Ma at the K–Pg boundary (Sepkoski, 1984; Baez, 2006). This mass extinction of organisms at the K–Pg boundary was triggered by the collision of a giant meteorite with a diameter of about 10 km on the Yucatan Peninsula in Mexico (Schulte et al., 2010; Renne et al., 2013). All non-avian dinosaurs became extinct during that time (Fastovsky and Sheehan, 2005). Mammals and birds, the latter descended from theropod dinosaurs, emerged as dominant large land animals.

In addition to the five major mass extinctions, there are numerous minor ones as well. Regarding the sixth extinction, for example, it has been suggested that the deep environmental transformations that took place during the Pliocene–Quaternary boundary not only altered patterns of species diversity and composition but also might have brought a major shift in the functioning of marine ecosystems (Garreaud et al., 2010). Moreover, the ongoing mass extinction caused by human activity is sometimes called the sixth extinction (Mason, 2015).

In contrast to the P–T (or P–Tr, Permian–Triassic) boundary issue, not much attention has been paid to the G–L boundary event; however, the significance of the G–L boundary event was re-emphasized from a different aspect relevant to the superocean Panthalassa (Isozaki et al., 2007a). The timing of the end-Guadalupian extinction apparently coincides with the onset of the superanoxia in Panthalassa, i.e., another global scale geologic phenomenon across the P–T boundary (Isozaki, 1997, 2007b). In addition to the faunal turnover in mid-oceanic plankton (radiolarians) detected in deep-sea chert, shallow marine sessile benthos (fusulines) also sharply declined in diversity across the G–L boundary in mid-Panthalassan paleo-atoll complex (Isozaki and Ota, 2001; Ota and Isozaki, 2006). These positively suggest the global nature of the G–L boundary extinction and causal environmental change (Isozaki et al., 2007a).

Regarding the causes of mass extinctions, there are some plausible large-scale events such as astronomical impacts with a giant meteorite at the K–Pg boundary (Schulte et al., 2010; Renne et al., 2013), and various environmental changes due to volcanic activity during the formation and division of supercontinents in the P–T boundary, where "plume tectonics" is regarded as promising (Maruyama, 1994).

As shown in Figure 6.8, the pattern of geomagnetic polarity change in the Permian demonstrates a clear contrast between the early Permian (Cisuralian) to the middle Guadalupian and the late Guadalupian to Lopingian (Isozaki, 2009, modified from Gradstein et al., 2004).

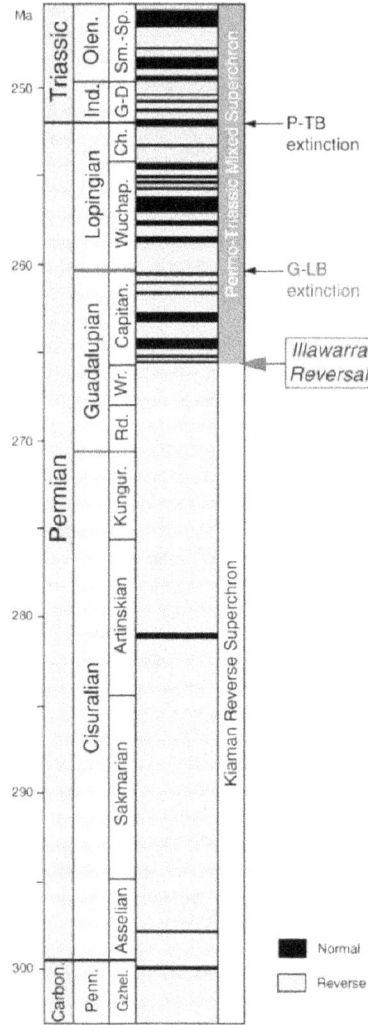

FIGURE 6.8 The Permian magnetostratigraphy and the Illawarra Reversal (Isozaki, 2009, modified from Gradstein et al., 2004). G–LB, Guadalupian–Lopingian boundary; P–TB, Permian–Triassic boundary. (Reproduced with permission from Isozaki (2009), Copyright 2009, Elsevier.)

The Kiaman Reverse Superchron ended ~50 million years (Myr) before the G–L boundary, and since then it has suddenly changed to an era in which the polarity changes frequently. This change in the GMF reversal pattern is called the "Illawarra Reversal" placed at 265 Ma around the Wordian–Capitanian boundary in the Guadalupian (Isozaki, 2009).

The Illawarra Reversal and the G–L boundary event record the significant transition processes from the Paleozoic to Mesozoic–Modern world (Isozaki, 2009). One of the major global environmental changes in the Phanerozoic occurred almost simultaneously in the latest Guadalupian, as recorded in (1) mass extinction, (2) ocean redox change, (3) sharp isotopic excursions (C and Sr), (4) sea-level drop, and (5) plume-related volcanism (Isozaki, 2009). The Illawarra Reversal event records that a major modulation has occurred in the process of the geodynamo mechanism. The Illawarra Reversal represents the most prominent marker in magnetostratigraphic correlation of the Late Paleozoic. Isozaki (2009) estimated that this major change in the stability in the GMF reflects a mode change in geodynamo in the

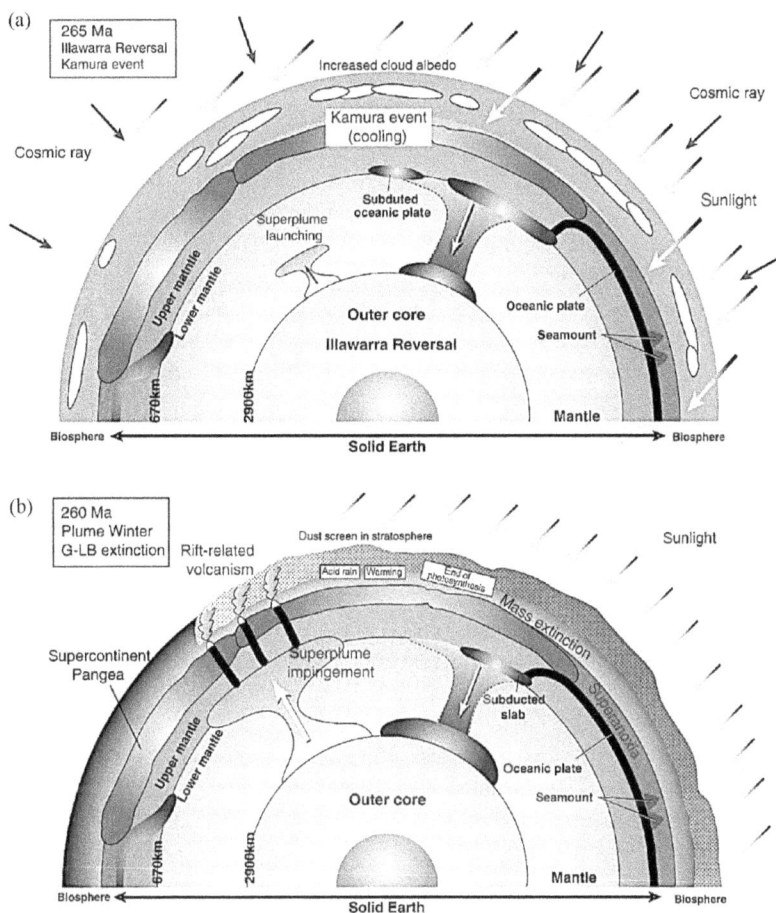

FIGURE 6.9 Plume Winter scenario for the late Guadalupian event from the core to the surface (Isozaki, 2009, revised from Isozaki, 2007a). Mantle superplume represents the largest flow of material and of energy within the Earth. Up-rising hot plumes paired with a down-going cold subducted slabs (megalith) form a large-scale whole-mantle convection cell. The launching of a superplume from core-mantle boundary (CMB) was the most likely ultimate cause of the mass extinction at the Guadalupian–Lopingian boundary (G–LB). (a) In the beginning of the Capitanian (~265 Ma), the launching of a mantle superplume caused the Illawarra Reversal in the outer core's geodynamo and the Kamura cooling event on the surface. (b) In the end the Capitanian (~260 Ma), the impingement of the plume head at the bottom of Pangea, continental rifting occurred and large igneous provinces (LIPs) formed to start "Plume Winter." Through various volcanic hazards, the G–LB mass extinction occurred, and subsequently long-term oceanic anoxia (superanoxia) started afterward. (Reproduced with permission from Isozaki (2009), Copyright 2009, Elsevier.)

outer core inside the Earth that can be caused solely by the appearance of a thermal instability on the CMB, such as launching a superplume.

As shown in Figure 6.9, Plume Winter scenario for the late Guadalupian event from the core to the surface is presented by Isozaki (2009), revised by Isozaki (2007a).

Isozaki (2009) pointed out that the Illawarra Reversal occurred nearly 5 Myr earlier than the G–L boundary turmoil on the surface including the mass extinction. The 5 Myr time lag may represent a travel time of a superplume from the CMB to the surface (Isozaki, 2009). When the convection is stable, the GMF strength is high, but when the convection is disturbed, the GMF strength decreases. When the polarity of the GMF is reversed, the GMF strength generally decreases when the reversal is repeated frequently (Brown et al., 2007; Valet and Plenier, 2008; Ferk and Leonhardt, 2009).

It is important to investigate the carbon isotope composition of limestone in seawater in order to estimate the history of biological production. Across the P–T boundary and the G–L boundary, the mid-oceanic paleo-atoll carbonates recorded secular change in stable carbon isotope composition. Musashi et al. (2001) and Isozaki et al. (2007b) documented the secular change in carbonate carbon isotopic ratio ($\delta^{13}C_{carb}$) of mid-Panthalassa across the P–T boundary and the G–L boundary, respectively. Besides the boundary negative shifts both at P–T boundary and G–L boundary properly predicted from previous studies (e.g., Holser et al., 1989; Wang et al., 2004), a unique high productivity interval in the Capitanian (late Guadalupian) was newly detected on the basis of the appreciable length of high positive $\delta^{13}C_{carb}$ (between +5‰ and +6‰) interval (Isozaki et al., 2007a).

As shown in Figure 6.10, the schematic diagram indicating the late Guadalupian Kamura event documented by high positive $\delta^{13}C_{carb}$ values at Kamura in Japan is presented by Isozaki et al. (2007a), which is modified from Isozaki et al. (2007b).

B: Two possible paths (broken lines) for the Guadalupian secular change of $\delta^{13}C_{carb}$ values were shown by Korte et al. (2005); the lower for the Tethyan domain, the upper for the Delaware basin in Texas. The Capitanian Kamura event recorded much higher positive $\delta^{13}C_{carb}$ values between +5.0‰ and +7.0‰ in Kamura, suggesting the positive excursion of the global context in the late Guadalupian. G–LB, Guadalupian–Lopingian boundary; P–TB, Permian–Triassic boundary; Road, Roadian; Wor, Wordian. (Reproduced with permission from Isozaki et al. (2007a), Copyright 2007, Elsevier.)

Such high positive values over +5.0‰ are quite rare in the Phanerozoic record except for several unique events in the Paleozoic (e.g., Veizer et al., 1999; Saltzman, 2005). It was found in Japan that the first major isotopic change occurred in the late Guadalupian, and Isozaki et al. (2007a, b) named this Capitanian episode "Kamura event" after the discovery site, Kamura section, Takachiho-cho, Nishiusuki-gun, Miyazaki Prefecture, in Kyushu. The Kamura event emphasized its significance of global cooling and relevant extinction of large fusulines and gigantic bivalves in low-latitude Panthalassa (Isozaki et al., 2007a). In the fusuline-tuned section, the waning history of the Kamura cooling event was clearly documented in high resolution, whereas the earlier history including the onset timing was not yet revealed, owing to the absence of continuous exposure in the previously studied section. This left a big chasm in the understanding of the major environmental change in the late Guadalupian, in particular, the cause and processes of the Kamura cooling event.

The main extinction occurred not at the G–L boundary per se but in a much lower horizon in the midst of the positive $\delta^{13}C_{carb}$ excursion interval. Thus an appreciable time has elapsed between the end-Guadalupian extinction and the following radiation of the Lopingian fauna in shallow mid-Panthalassa as shown by Isozaki et al. (2007a). It is noteworthy that a strange condition has appeared in the middle of the superocean around the Wordian–Capitanian boundary because the Kamura event may mark the first episode of large isotopic excursion in the Permian as shown by Isozaki et al. (2007a).

Wei et al. (2014) hypothesized that GMF reversals cause O_2 level drops and subsequent mass extinctions. As shown in Figure 6.11, temporal evolution of reversal rate, O_2 level, and marine diversity over the Phanerozoic are presented by Wei et al. (2014).

Because of several data gaps in this database, the relative reversal rate from an older database (dashed line, McElhinny, 1971) is also plotted to show the trend of reversal rate for a reference. The reversal rate (Ogg et al., 2008), atmospheric O_2 level (Berner, 2009), and marine diversity (Alroy, 2010) show a strong correlation in support of their hypothesis. During the second, third, and fourth mass extinctions identified by the diversity drops, the reversal rates increased and the O_2 level decreased. In contrast, when the reversal rate remained at zero or very low, namely during the superchrons (Merrill and Mcfadden, 1999), the diversity increased, and the O_2 level also increased for three out of four superchrons. However, during first and fifth mass extinctions, the reversal rate also increased despite that the O_2 level just remained at low level or had no discernable change. These features suggest that some mass extinctions might be explained by their hypothesis: increasing GMF reversals continually

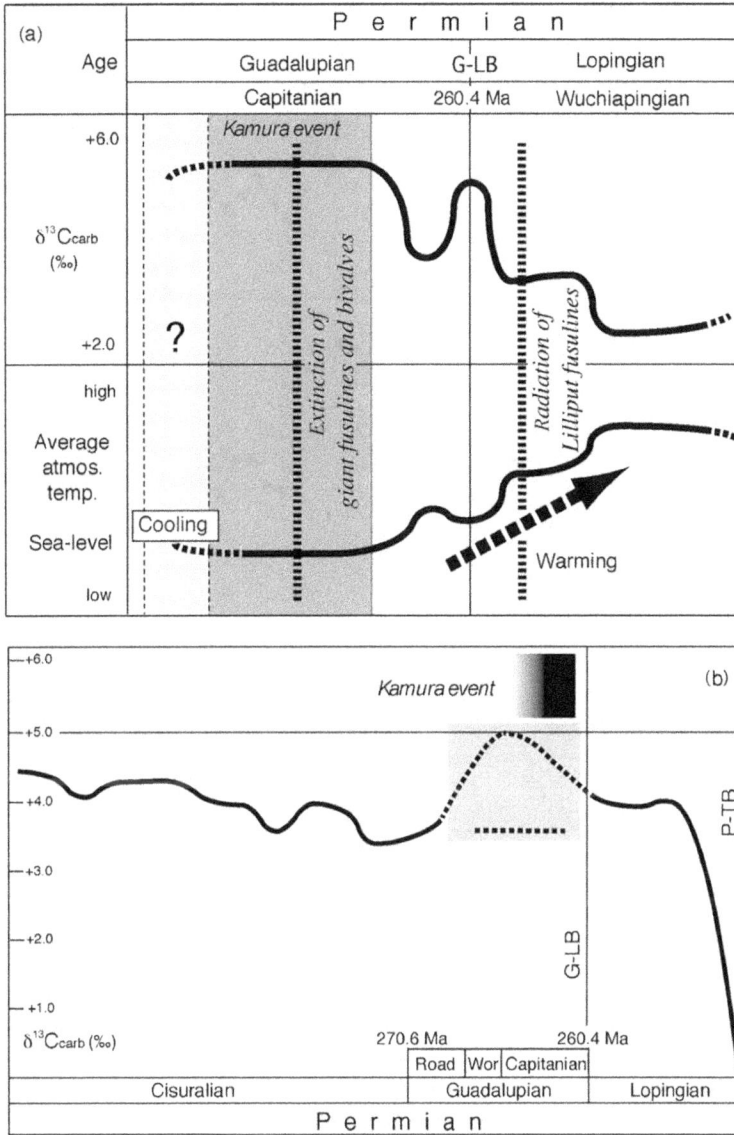

FIGURE 6.10 Schematic diagram indicating the late Guadalupian Kamura event documented by high positive $\delta^{13}C_{carb}$ values at Kamura in Japan. (Isozaki et al., 2007a, modified from Isozaki et al., 2007b) (a), and the composite Permian secular curve of $\delta^{13}C_{carb}$ values (modified from Korte et al., 2005) (b). (a) Note that the Guadalupian large fusuline and bivalve fauna became extinct in the middle of the Kamura cooling event, whereas the post-extinction radiation of the Lopingian small fusulines started during the subsequent warming period. In contrast to the waning history of the Kamura event, its onset timing and processes were unknown previously.

enhance oxygen escape, and this cumulative effect could cause a significant drop in O_2 level over a few millions years (Wei et al., 2014). The reversal rate was increasing and the diversity was decreasing over the last 40 Ma. Wei et al. (2014) speculated that the present Earth might be experiencing a new gradual pattern of mass extinction. However, the slight change of O_2 level cannot be regarded as a strong link between them. Wei et al. (2014) proposed that more data and work are required to understand these variations.

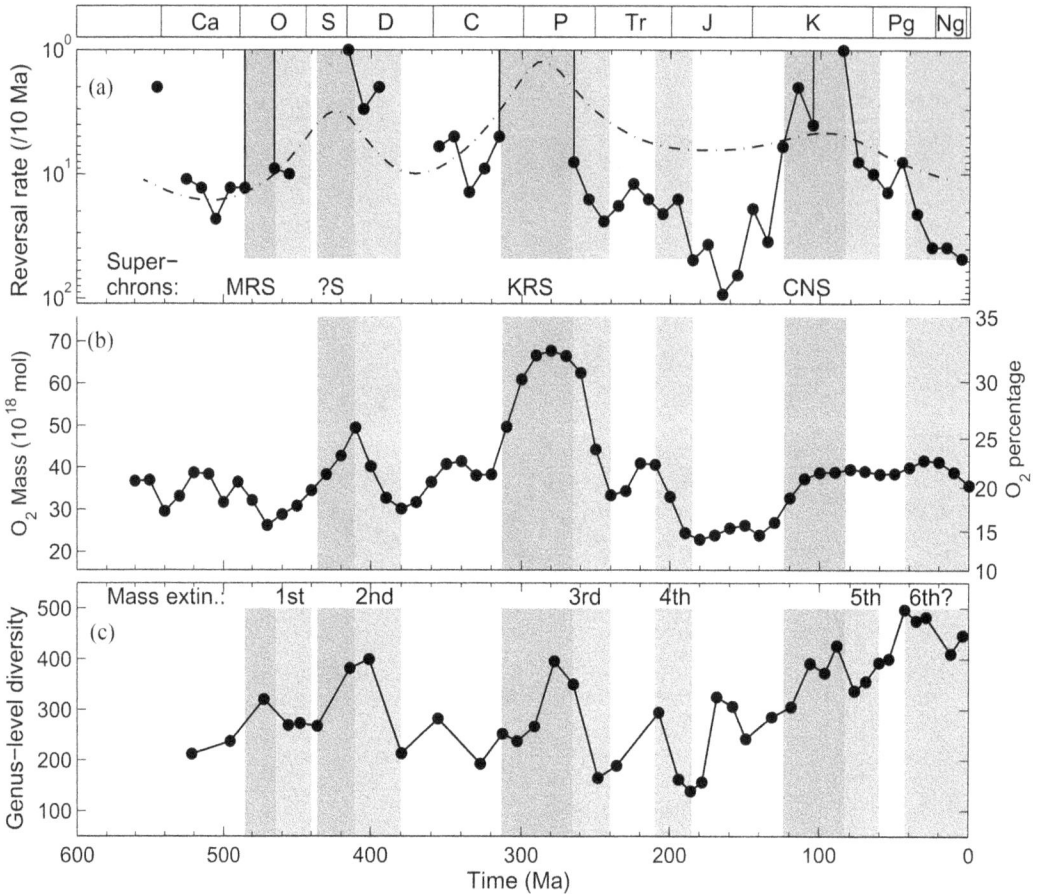

FIGURE 6.11 Temporal evolution of reversal rate, O_2 level, and marine diversity over the Phanerozoic (Wei et al., 2014). Ca, Cambrian; O, Ordovician; S, Silurian; D, Devonian; C, Carboniferous; P, Permian; Tr, Triassic; J, Jurassic; K, Cretaceous; Pg, Paleogene; Ng, Neogene. (a) GMF reversal rate. The solid line is total reversals within a 10 Myr bin (time period) from the new database (Ogg et al., 2008). The dashed line is the relative reversal rate to represent the trend of reversal rate from an older database (McElhinny, 1971). The blocks show the superchrons (Merrill and Mcfadden, 1999). KRS, Kiaman Reversed Superchron (~267–313 Ma); CNS, Cretaceous Normal Superchron (~83–125 Ma); MRS, Mayero Reversed Superchron (~463–481 Ma). (b) Modeled percentage and amount of atmospheric O_2 over time (Berner, 2009). (c) Number of marine genera (Alroy, 2010). The blocks show the gradual pattern of five well-known mass extinctions (the Ordovician–Silurian [O–S], the Late Devonian [Late D], the Permian–Triassic [P–Tr], the Triassic–Jurassic [Tr–J], and the Cretaceous–Paleogene [K–Pg] extinctions), and the sixth mass extinction (the Neogene [Ng] extinction) has not been confirmed. (Reproduced with permission from Wei et al. (2014), Copyright 2014, Elsevier. It is licensed under the Creative Commons Attribution 3.0 International.)

Moreover, Wei et al. (2014) simulated the oxygen escape rate for the Triassic-Jurassic (Tr-J) extinction event (the fourth mass extinction) using a modified Martian ion escape model with an input of quiet solar wind inferred from Sun-like stars. The simulation predicted that GMF reversals could enhance the oxygen escape rate by three to four orders only if the MF was extremely weak, even without consideration of space weather effects (Wei et al., 2014). Consequently, it is estimated that the global hypoxia might gradually kill numerous species (Wei et al., 2014).

6.3.7 The Cambrian Explosion of Life on Earth

In contrast to the mass extinction, Doglioni et al. (2016) proposed the hypothesis that "the Cambrian explosion of life on Earth" could have been enabled by the increase of the GMF intensity mainly leading to shield UV radiation. Marine prokaryotic organisms dominated the Earth for nearly 3 billion years since they were protected by water from UV radiation (Doglioni et al., 2016). Some microbes are the organisms that are most resistant to UV radiation (Doglioni et al., 2016).

The Cambrian explosion was a great flowering of life forms that occurred between roughly 542 and 530 Ma; during this period practically all major animal phyla (body plan), and many now-extinct ones were established. Highly differentiated multicellular animals such as various corals, shellfish, brachiopods, and trilobites are found in the Cambrian strata, but few animal fossils are found in the strata before the Cambrian period. In 1998, evolutionary biologist and paleontologist Andrew R. Parker of the Australian Museum in Sydney at that time proposed the "light switch theory" as the cause of the Cambrian explosion, which increases the selection pressure due to the emergence of eye-catching organisms (Parker, 1998). For the first time in the history of life on Earth, trilobites, organisms with eyes, were born, and by actively preying on others, it became advantageous for organisms without eyes (Parker, 1998). The theory is that organisms having acquired eyes and hard tissues are more able to adapt for defense against their predation. As a result, fossil records appear to have exploded in a short period of time (Parker, 1998). He estimates that the Cambrian explosion is a phenomenon in which many phyla acquired hard tissue all at once at the same time (Parker, 1998). An important criticism remained in the timing of the relationship between the rise of predation and the period when vision developed. In Parker's review of many of the debates about the transformation of life forms around the Precambrian–Cambrian boundary, Martin (2009) argues that predation played a critical role during the Cambrian explosion, but that Parker's emphasis on vision is misplaced because eyes did not develop until late in the spread of such predation.

It is conceivable that the cause of the Cambrian explosion could be related to the termination of the "snowball Earth," which is global glaciation that the whole Earth was frozen during the most severe ice age of Precambrian time. Here, the term "snowball Earth" was coined by a geobiologist, Joseph Kirschvink of the California Institute of Technology (Caltech), in a short paper published in 1992 within a lengthy volume concerning the biology of the Proterozoic Eon (Kirschvink, 1992). The major contributions from his insight were: (1) the recognition that the presence of banded iron formations is consistent with such a global glacial episode, and (2) the introduction of a mechanism by which to escape from a completely ice-covered Earth, specifically, the accumulation of CO_2 from volcanic outgassing leading to an ultra-greenhouse effect (Kirschvink, 1992). Moreover, Hoffman et al. (1998) suggested that the snowball Earth ended abruptly when subaerial volcanic outgassing raised atmospheric CO_2 to about 350 times the modern level. The rapid termination would have resulted in a warming of the snowball Earth to extreme greenhouse conditions (Hoffman et al., 1998). The CO_2 transfer to the ocean would result in the rapid precipitation of calcium carbonate in warm surface waters, producing the cap carbonate rocks observed globally (Hoffman et al., 1998). However, it was supposed that it would be an indirect relationship, if any, because the Cambrian explosion started at least 32 Myr after the end of the snowball Earth (Marshall, 2006). Moreover, the cold periods of the snowball Earth may even have delayed the evolution of large size organisms (Bengtson, 2002).

The Ediacaran gave way to the Cambrian explosion in 542 Ma after the Kotlinian Crisis (~550–542 Ma). New species with complex body plans, hard parts for defense, and sophisticated eyes have emerged rapidly. Burrowing also became more common and diverse, destroying the once widespread bacterial mats and sending O_2 to the seafloor, creating a new livable space. There are many factors that may explain why the Cambrian explosion occurred, but the researchers' idea of "flight from light" adds a novel possibility to the debate. David Harper, a paleontologist at Durham University in the United Kingdom who was not involved in the study, says that the researchers have opened up yet another exciting and imaginative area of research within which to frame and test new hypotheses for the origin and early evolution of animal-based communities (Randall, 2016).

6.4 Health Effects of the Geomagnetic Field (GMF)

6.4.1 Magnetic Field Deficiency Syndrome

It has been believed that some human energy is being absorbed from the GMF, so what if an MF reduction occurs on the Earth? A Japanese medical doctor, Kyoichi Nakagawa (Director of Isuzu Hospital) has spent more than 20 years investigating the effects of MFs on humans. He pointed out that there is an "MF deficiency syndrome" likely caused by a decline in the GMF intensity in humans and the syndrome can be corrected with the application of external MFs (Nakagawa, 1976). He noticed that the symptoms are stiffness of the shoulders, back and neck, lumbago, chest pains, headache, dizziness, insomnia, habitual constipation, general malaise, and so forth (Nakagawa, 1976). Nakagawa referred to scientific authorities who have proven that over the last two centuries the geomagnetic dipole strength has been decreasing at a rate of ~6.3% per century (Merrill et al., 1996).

Here, the absolute magnitude of the axial dipole component of the GMF from 1600 to 2020 (Merrill et al., 1996), according to three models—CALS3k.4 (Korte and Constable, 2011), gufm1 (Jackson et al., 2000), and IGRF-12 (Thébault et al., 2015)—is shown in Figure 6.12. In this figure, geomagnetic axial dipole strength increased linearly from ~33 μT in 1,600 to ~35 μT by about 1,700, and then decreased linearly and is now around 29.5 μT.

Moreover, together with the decrease of the GMF itself, Nakagawa pointed out that in modern industrialized life, metal buildings, cars, trains, etc. could shield the GMF, causing interference and loss of the GMF strength (Nakagawa, 1976). Therefore, living or working in steel frame or steel structure buildings could decrease the effects of the GMF on the human body (Nakagawa, 1976). Nakagawa (1976) hypothesized that if the GMF is artificially shut out, there would be adverse changes in the central nervous system and diurnal rhythm. Therefore, Nakagawa (1976) speculated that supplying the human body with MFs by various methods would improve and alleviate physical complaints. He investigated the therapeutic effects of stronger MFs (including SMFs and time-varying MFs) than the GMF, and found that MFs improved some symptoms when MFs were externally applied, and advocated that the ameliorated disease is actually MF deficiency syndrome (Nakagawa, 1976).

There is an average of 50 μT GMF, ranging from ~25 μT near the magnetic equator to ~65 μT at the magnetic poles throughout the Earth, but not only is outer space weightless due to the absence of gravity but also there is "almost no MF" outside the geomagnetosphere (Prölss, 2004). It is a well-established fact that bone mass, bone mineral density (BMD), and bone mineral content (BMC) decrease in the case of zero gravity. Jia et al. (2014) examined the physiological effects of "hypomagnetic field (HMF)" (<300 nT) on osteoporosis using animal models. Osteoporosis was induced in the femur by

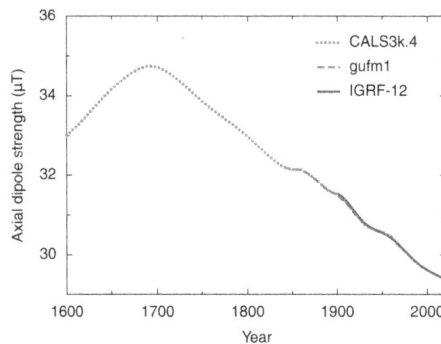

FIGURE 6.12 Geomagnetic axial dipole strength from 1600 to 2020, according to three models: CALS3k.4, gufm1, and IGRF-12. (Obtained from Wikimedia Commons, the free media repository. It is licensed under the Creative Commons Attribution 4.0 International.)

experimentally applying no exercise load to the hind limbs of the rat (Jia et al., 2014). Statistical comparisons of hindlimb unloading rats were performed between the "hindlimb unloading + HMF" group, in which the GMF was shielded, and the "hindlimb unloading" group, in which the GMF, was not shielded (Jia et al., 2014).

As a result, the BMD and BMC of hindlimb unloading rats were further significantly reduced by HMF, and the symptoms of osteoporosis were further exacerbated (Jia et al., 2014). The results suggest that HMF may stimulate osteoblasts to secret a receptor activator of nuclear factor-ΚB ligand (RANKL), and promote the maturation and activation of osteoclasts, eventually causing bone resorption (Jia et al., 2014). Thus, it is conceivable that this is because the osteoclasts were out of balance and bone resorption was promoted by HMF (Jia et al., 2014).

In healthy subjects without cardiovascular pathologies, Gurfinkel et al. (2016) examined the effects of HMF exposure ($\pm$10 nT) on the cardiovascular system and microcirculation and found a significant increase of the capillary blood velocity, and after turning off the HMF the capillary blood velocity returned to the baseline level. HMF exposure significantly reduced the heart rate (HR) to the end of 60-minute exposure to HMF (Gurfinkel et al., 2016). Hypomagnetic conditions also significantly decreased the diastolic blood pressure (DBP) (Gurfinkel et al., 2016). The capillary blood velocity of all tested subjects was enlarged by 17%, and the average duration of cardio intervals was increased by 88.7% during HMF exposure in comparison with sham conditions (Gurfinkel et al., 2016). These results demonstrated the effect only for brief exposure to HMF on the cardiovascular system and microcirculation (Gurfinkel et al., 2016). Gurfinkel et al. (2016) anticipated that during interplanetary missions, long-term stay in hypomagnetic conditions may have a significant impact on the health and working ability of deep space explorers.

Exposure to an HMF has been found to cause a decrease in amylolytic activity of the intestinal enzymes and in the activity of proteinases and glycosidases in Crucian carp *Carassius carassius* (Kuz'mina et al., 2015). When intracellular Ca^{2+} dependent proteinase (calpains) activity was analyzed following exposure to an HMF a significant decrease was observed in calpain activity and Ca^{2+} dependent proteases were inactivated in *Carassius carassius*, *Rutilus rutilus*, and *Carassius carpio* (Kantserova et al., 2017).

Zhang et al. (2004) found that wild-type Berlin flies raised in an HMF environment continuously for 10 successive generations were gradually significantly impaired in visual learning and memory (L/M), and finally the tenth generation of HMF flies became completely amnesiac. The reverse shift from HMF to natural GMF fully reversed the GMF-free induced amnesia after more than six consecutive generations. The control experiments show that the impairment of L/M could not be ascribed to apparent sensorimotor problems in HMF flies. This study suggests that *Drosophila* has the potential to develop into a new model organism for the study of the neurobiology of magnetism for multiple generations (Zhang et al., 2004).

Wan et al. (2014) investigated the effects of HMF on the growth, development, and reproduction of migratory insects compared with the normal GMF. Two species of nocturnal migratory planthopper, the small brown planthopper, *Laodelphax striatellus*, and the brown planthopper, *Nilaparvata lugens*, were investigated (Wan et al., 2014). HMF of ~500 nT was generated using the HMF Space System with MFs of two planthopper species. The results suggested that exposure of both species to HMF delayed egg and nymphal developmental durations and decreased adult weight and female fecundity (Wan et al., 2014). Compared with the GMF, the vitellogenin transcript levels of newly molted female adults and the number of eggs per female were significantly reduced in both species, indicating a negative effect on fertility under HMF (Wan et al., 2014). Overall, *N. lugens* seemed more sensitive to HMF than *L. striatellus* (Wan et al., 2014). It is estimated that HMF may also lead to the titer changes of juvenile hormone and ecdysone via CRY, i.e., decreased juvenile hormone may downregulate vestigial (Vg) gene expression, and decreased ecdysone may prolong the nymphal duration (Wan et al., 2014). These findings provided experimental evidence that HMF negatively affected the growth and development of both species, with particularly strong effects on reproduction (Wan et al., 2014).

In addition, more recently, Wan et al. (2020a) suggested that the expression of crucial migration-related traits in *N. lugens* can respond to changes in the GMF intensity encountered by migratory populations. Moreover, their results raise the possibility that *N. lugens* phenotypes vary in response to changes in the GMF intensity in a manner consistent with predicted trade-offs between reproduction in more permanent overwintering habitats in the south and migration to ephemeral northern habitats that are re-colonized annually as part of the seasonal migration cycle.

Furthermore, in *N. lugens*, Wan et al. (2020b) found that appetite-related regulatory pathways (neuropeptide signaling) and phenotypic outcomes (glucose levels and phloem ingestion) can be influenced by changes in the GMF intensity, which is suggested to finally result in decreases in adult body weight. The physiological and behavioral responses of *N. lugens* to strong changes in the GMF intensity highlight the GMF's role in the complicated system of maintaining animal energy homeostasis via appetite regulation (Wan et al., 2020b). Energy homeostasis is also critical for migratory animals that can undergo large-scale spatial displacements in a short time (Chapman et al., 2015). Previous works on fueling behavior of several migratory birds (Fransson et al., 2001; Kullberg et al., 2007; Henshaw et al., 2008) all suggested that a difference in the GMF intensity between the emigration and immigration areas could be important environmental cues for the regulation of biological processes during migration. Thus, the weight loss suggests an adaptive response of migratory *N. lugens* to the changes in the GMF intensity during S-N migration or N-S remigration (Wan et al., 2020b).

It is indicated that endogenous magnetic materials are present in the abdomen of *N. lugens* (Xie et al., 2011), as well as that of *L. striatellus* (Wan et al., 2014). According to the magnetite theory on a magnetite-based system, long time exposure to HMF may alter the status of clusters of superparamagnetic crystals anchored to neuronal membranes by cytoskeletal filaments. This would be predicted to deform the membrane, opening or closing ion channels (Johnsen and Lohmann, 2005, 2008; Lohmann, 2010) with potential impacts on the regulation of development and reproduction. Additionally, it has been reported that HMF can alter circadian rhythm probably through CRY (Zamoshchina et al., 2012). Therefore, like the "dual receptor magnetoreception hypothesis" (Lohmann, 2010), it is inferred that the two types of "magnetoreception side effects" respectively related to magnetite and CRY may operate together to affect the development and reproduction of the insect in the presence of MFs (Wan et al., 2014). However, it was reported that the absence of the GMF does not influence the development of the chick embryo (Janoutova et al., 2000), and therefore, it is supposed that the different results likely reflect variation among taxa in sensitivity to MFs (Wan et al., 2014).

Baek et al. (2014) reported that exposure to a specific ELF-MF (50 Hz, 1 mT) induced epigenetic changes during cell reprogramming by mediating histone modification, whereas an HMF (<5,000 nT) environment was detrimental to cell fate changes in epigenetic reprogramming (Baek et al., 2014). Here, three-axis Helmholtz coils were used to generate a zero-field environment by canceling the GMF (Baek et al., 2014). These results suggested that an HMF environment is critical for epigenetic changes during cell fate conversion (Baek et al., 2014). Furthermore, more recently, Baek et al. (2019) investigated whether HMFs affect cell fate determination during direct differentiation. They found that HMF conditions delayed cell fate conversion during the differentiation of mouse embryonic stem cells (mESCs), whereas there were no effects on mESCs in a pluripotent state (Baek et al., 2019). In addition, HMF conditions reduced the expression of markers corresponding to three germ layers (Pax6, Brachyury, and Gata4) during mESC differentiation and reduced embryoid body formation (Baek et al., 2019). They suggested that HMF conditions caused abnormal DNA methylation through the dysregulation of DNA methyltransferase3b (Dnmt3b) expression, eventually resulting in incomplete DNA methylation during differentiation, and that a specific ELF-MF is critical for the normal differentiation of mESCs (Baek et al., 2019). Therefore, they concluded that ELF-MFs play a role in establishing cellular identity, and it will be highly interesting to apply ELF-MFs in specific therapeutic paradigms for which epigenetic reorganization is a critical outcome (Baek et al., 2019).

Gurhan et al. (2021) investigated the effects of weak SMFs including HMF (0.5 µT) on HT-1080 human fibrosarcoma cells. Exposures to artificial SMFs for four consecutive days were varied from 0.5 to 600 µT for treated units, while exposures to control units were held at 45 µT of the GMF. Cell growth was significantly accelerated between 200 and 400 µT, and in contrast, was significantly decreased at 0.5

and 600 µT along with variations in the membrane potentials, mitochondrial calcium, reactive oxygen species (ROS), superoxide (O_2^-), nitric oxide (NO), H_2O_2, intercellular pH, and oxidative stress (Gurhan et al., 2021). Thus, they suggested that weak SMFs can accelerate and inhibit cell growth rates and induce alterations in ROS, which indicates that changes in ROS and oxidative stress are important for various cell functions (Gurhan et al., 2021). They speculated that $[Ca^{2+}]_i$ into mitochondria was one of the initial steps into the corresponding changes (Gurhan et al., 2021).

HMF effects have also been observed to influence animal behavior (Choleris et al., 2002; Prato et al., 2005). The behavioral effects were observed only when the ambient fields were shielded by µ-metal and not by canceling the fields with tri-axial coils. This is because the µ-metal shielding attenuates the ambient SMF and ELF-MFs, while the tri-axial coils cancel only the static component (Martino et al., 2010).

Binhi and Prato (2017) critically reviewed the biological effects of the HMF by referring to 137 publications, which reflect an unbiased reporting of results in this area. After analyzing experiments and theories, they concluded that despite the apparent non-specific biological effects of HMF, no regular relationship was found between the magnitude of this effect and the physical conditions of the experiment (Binhi and Prato, 2017). These analytical results suggest that there are no common biophysical MF "targets" that are identical in different organisms (Binhi and Prato, 2017). Common MF targets are estimated to exist only at major physical levels, in the form of "magnetic moments" of unknown nature (Binhi and Prato, 2017).

More recently, Zhang et al. (2021) summarized the literature related to the biological effects of HMF and calculated the magnitude of the effect. They concluded that HMF impairs multiple animal systems, especially in the central nervous system, and furthermore, HMF is a stress factor in plant growth and reproduction.

As mentioned above (Zaporozhan and Ponomarenko, 2010; Zamoshchina et al., 2012; Xue et al., 2021), CRY has been shown to participate in a molecular complex feedback loop to generate the circadian oscillation in response to MFs. In this context, multiple experimental studies have focused on the effects of HMF on circadian rhythms and found that HMF disrupts circadian rhythms and contributes significantly to health problems such as sleep disorders, metabolic changes, and neurological disorders (Xue et al., 2021). Recently, Xue et al. (2021) reviewed and emphasized that ground-based studies on interplanetary-like HMF have revealed changes in structure and function of the central nervous system, embryonic development, hormone release synchronization, and circadian rhythm in animals and cells. Inconsistency and seemingly contradictory observations are shown in the literature that may arise from differences in parameters of MF and characteristics of the study subject (Xue et al., 2021). Xue et al. (2021) proposed the need for additional research on a higher number of subjects to better understand health issues during travel beyond the Earth's orbit and into deep space.

6.4.2 Magnetic Storm and Its Related Diseases

Literally, the disturbance of the GMF is called a "geomagnetic storm." A geomagnetic storm is called a "magnetic storm" on the Earth's ionosphere (ESA, 2016b), and is also referred to as a "solar storm." Actually, it needs a large size for forming the magnetic storm. Can we know the generation of the magnetic storms by observing the aurora? The aurora borealis is the light generated when the electrons responsible for this transmitted electric current collide with atoms and molecules in the atmosphere (Aurora photo from National Geographic Science News, 2019). When the solar wind interacts with the GMF, it shows a wonderful show of light. The bright light reminds us of the importance of the GMF, which protects life on Earth from solar radiation. As above-mentioned, the coupling currents or field-aligned currents flow along MF lines between the magnetosphere and ionosphere. The auroral oval is one of the manifestations of the coupling currents (e.g., Nagata, and Kokubun, 1962). Since the MF generated by this aurora current is applied to the GMF that normally exists, the GMF changes to some extent depending on the intensity of the aurora.

No matter how magnificent the aurora is and the MFs around the Earth change, it is not called a magnetic storm by itself. Whether or not a magnetic storm is becoming larger in size is determined by changes in the MF at mid and low latitudes, not by the aurora. A westward current called "ring current" flows around the equatorial plane of the Earth. This current generates an MF in the opposite direction to the normal GMF, and thus by the interference effects, leads to weakening the MF at mid-low latitudes. If such a state lasts for 2–3 days, this is the magnetic storm. Of course, at that time, active aurora occurs continuously at high latitudes. The solar wind has the greatest effect on these magnetic storms.

The solar wind from the Sun reaches the Earth ~150 million kilometers away. The solar wind that reaches the Earth's orbit collides with the geomagnetosphere (Borovsky and Valdivia, 2018). The geomagnetosphere is also called a "magnetic comet" (Gonzales, 1969) because its shape with a long magnetic tail blown by the solar wind is just like the tail of a comet. The range of the geomagnetosphere extends tens of thousands of kilometers on the Sun side and like a tail on the other side, reaching more than hundreds of thousands of kilometers.

As shown in Figure 6.13, the solar wind and the geomagnetosphere are presented by Borovsky and Valdivia (2018).

Bow shock indicates the shock front along which the solar wind encounters the GMF (Borovsky and Valdivia, 2018) (Figure 6.13). A magnetic storm is a temporary disturbance of the geomagnetosphere caused by a solar wind shock wave and/or cloud of the MF that interacts with the GMF. The reason why the magnetosphere becomes asymmetrical in this way is that the GMF is affected by the solar wind and is blown away to the opposite side of the Sun. On the other hand, most of the solar wind that collides with the magnetosphere is bounced off and flows into outer space, but the large current generated at that time is transmitted to the sky near the N and S poles of the Earth. Thus, the GMF acts as a shield to deflect potentially harmful charged particles mainly coming from the solar wind.

The most well-known solar activity is the activity of sunspots, which repeats in a cycle of ~11 years. The dark spots are because the temperature is ~4,000°C, which is lower than the normal photosphere temperature of ~6,000°C on the Sun's surface. The activity of sunspots is considered to be caused by the periodic changes of the MF of the Sun. The greater the number of sunspots, the more active the solar activity, and such a state is called "the maximum period of solar activity." In contrast, the state where the number of sunspots is small is "the minimum period of solar activity." However, during the 70 years from 1645 to 1715, there was a period when the number of sunspots decreased drastically. This period is called the "Maunder Minimum"(1645–1715) (Nymmik, 2011; Riley, 2012). This name comes from Edward Walter Maunder, a solar astronomy researcher who studied the past records of the disappearance of the sunspots. During the 30 years during the Maunder Minimum, only about 50 sunspots were observed. Normally, except for during the Maunder Minimum, about 40,000–50,000 sunspots are observable.

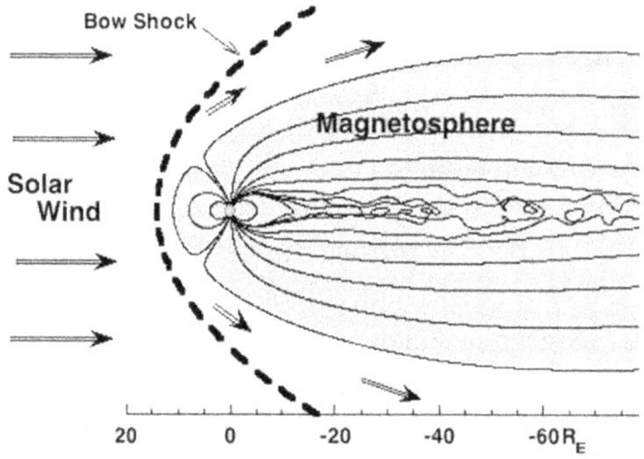

FIGURE 6.13 The solar wind and the geomagnetosphere (Borovsky and Valdivia, 2018). (Reproduced with permission from Borovsky and Valdivia (2018), Copyright 2018, Elsevier. It is licensed under the Creative Commons Attribution 4.0 International.)

Even during the Maunder Minimum, changes in the number of sunspots in the 11-year periodic cycle can be read from changes in the number of sunspots (Schröder, 1992). However, even if the activity of sunspots decreased, it is considered that there was almost no change in the amount of sunlight, so there was no significant effect on the temperature of the Earth. The relationship has not been clarified between the activity of sunspots and the climate change of the Earth.

Meanwhile, solar activity induces an explosive phenomenon called the "solar flare" that occurs abruptly in the solar active region. At this time, the solar wind plasma and the coronal MF may be blown into outer space together. This plasma ejection is termed "coronal mass ejection (CME)." This CME, unlike the solar flare which is light, causes a large mass to reach the Earth, similar to the solar wind. CMEs travel through interplanetary space with occasional serious impacts on the solar-terrestrial environment (Gutierrez et al., 2021). Therefore, CMEs are thought to have a greater effect on the GMF than solar flares (Gosling, 1991).

Most large-scale magnetic storms occur when the solar flare causes CMEs, which blow into the geomagnetosphere with a strong southward MF. Magnetic storms are more likely to occur when the solar activity with many sunspots is high because the solar flare is associated with sunspot activity. In the case of magnetic storms, a severe aurora storm often occurs, and in that case, a severe change in the GMF intensity is also observed, especially in high latitude areas.

In addition to the 11-year periodic cycle due to the solar activity, there are other periods of variation in the geomagnetic activity: 1 day cycle, due to the Earth's rotation; a 6–7-day cycle, corresponding to one sector of the solar wind; a 13–14-day cycle, corresponding to the passage of two solar wind sectors; a 27-day cycle, corresponding to the period of the Sun's rotation around its own axis (Volpe, 2003). Other rhythms are also correlated with these cycles: a 29.5-day cycle, corresponding to the synodic period of the Moon; a 1-year cycle, corresponding to the period of the Earth's revolution; some cycles of the solar activity that are not so prominent when compared to the main 11-year cycle (2-, 3-, 8-, 22-, and 35-year cycles) (Volpe, 2003).

A magnetic storm is a worldwide geomagnetic disturbance, distinct from regular diurnal variations. Geomagnetic activity is natural variations in the GMF, which is classified into quiet, unsettled, active, and magnetic storm levels (Poole, 2002; Alabdulgader et al., 2018; Table 6.1). Geomagnetic activity indices, e.g., K, K_p, A, and A_p indices, were designed to describe variation in the GMF caused by the irregular current systems (Poole, 2002; Alabdulgader et al., 2018; Table 6.1).

K index is a quasi-logarithmic local index of the 3-hourly range in magnetic activity relative to an assumed quiet-day curve for a single geomagnetic observatory site. First introduced by a German geophysicist, Julius Bartels in 1949, it consists of a single-digit 0 through 9 for each 3-hour interval of the universal time day (Bartels, 1949).

K_p index is the planetary 3-hour-range index, which is the mean standardized K index from 13 geomagnetic observatories between 44° and 60° Northern or southern geomagnetic latitudes. The scale is zero to nine expressed in thirds of a unit; e.g., 5− is $4^{2/3}$, 5 is 5, and 5+ is $5^{1/3}$. This planetary index is designed to measure solar particle emissions by their magnetic effects.

TABLE 6.1 The General Relationship between A_p and K Values

Category	A_p Index Range	K Index Range
Quiet	<8	<2
Unsettled	8–16	2–3
Active	16–30	3–5
Minor storm	30–50	5–6
Major storm	50–100	6–7
Severe storm	>100	>7

Source: Compiled from Poole (2002) and Alabdulgader et al. (2018).

a_p index is a measure of the general level of geomagnetic activity over the globe for a given day. A mean, 3-hourly "equivalent amplitude" of magnetic activity based on K index data from 11 Northern and 2 Southern Hemisphere magnetic observatories between the geomagnetic latitudes of 46° and 63° (Lockwood et al., 2019). ap values are given in units of 2 nT.

A_p index is a daily index determined from eight ap index values. That is, the A_p index is defined as the earliest occurring maximum 24-hour value obtained by computing an eight-point running average of successive 3-hour a_p index values during a magnetic storm event and is uniquely associated with the storm event. A_p values are given in units of 2 nT.

Even in the absence of magnetic storms, the GMF could be changed on various time scales longer than decades as described above. Most of the shorter-scale temporal changes of the GMF are estimated to be due to the effects of the ionosphere and magnetosphere outside the Earth. A British clockmaker, inventor, and geophysicist, George Graham in 1722 (Graham, 1724a,b) discovered that the diurnal cycle of GMF is a regular change in the daily cycle. The diurnal cycle can be clearly seen on days when the GMF perturbation by the solar wind is small, so it is termed a "geomagnetic quiet day."

The cause of diurnal changes on geomagnetic quiet days is the ionosphere at an altitude of 100 km or more from the Earth's surface. The ionosphere is warmed by radiant heat from the Sun during the daytime, cooled at nighttime, and the wind is generated inside the ionosphere. Then, the ionized gas is also carried with the wind. This interaction between the motion of the ionized gas and the GMF induces a large-scale "eddy current" that is counterclockwise in the Northern Hemisphere and clockwise in the Southern Hemisphere during the daytime when viewed from outside the Earth. This eddy current always appears on the side of the Earth-facing the Sun. The GMF emanating from the Earth' surface passes under this eddy current and thereby changes every day. This is the generation mechanism of the diurnal cycle of the GMF.

Gmitrov and Gmitrova (2004) investigated how changes in the GMF affect cardiovascular regulation under laboratory conditions. Baroreflex sensitivity estimated from BP and HR responses to intravenous injection of phenylephrine and nitroprusside, showed significant negative correlations between increasing the geomagnetic disturbance and baroreflex sensitivity, heart rate variability (HRV), and arterial BP (Gmitrov and Gmitrova, 2004). These findings support the theory that geomagnetic disturbances affect neurocardiovascular regulatory centers (Gmitrov and Gmitrova, 2004). Gmitrov (2005) further reported that the geomagnetic disturbance worsens microcirculation impairing arterial baroreflex vascular regulatory mechanism. The reduced baroreflex sensitivity may increase mortality after myocardial infarction (Garcia et al., 2014). Gmitrov (2005) recommends on days with intense geomagnetic activity, and especially during geomagnetic storms, to intensify the therapy of ischemic cerebral and heart disease to improve microcirculation in brain tissue and in the myocardium to decrease the risk of cerebral strokes and myocardial infarctions.

Human studies have shown that changes in time-varying MFs above 80 nT over a 3-hour period significantly reduce melatonin levels in the body (Weydahl et al., 2001). In addition, a greater reduction of excretion in a melatonin metabolite 6-OHMS was observed when increased geomagnetic activity was combined with an elevated 60 Hz MF or reduced ambient light exposures (Burch et al., 1999, 2008). As for a plausible mechanism of the decrease of melatonin by time-varying MFs, a possible change in the spatial structure of the photoreceptor pigment rhodopsin due to the EF (eddy current) induced by the time-varying MF has been proposed and MFs might also change either the electric activity of the pinealocytes or their ability to produce melatonin or both (Selmaoui and Touitou, 1995; Touitou et al., 2010). These reduced levels of melatonin, along with other individual factors, may be involved in the development of myocardial ischemia due to the GMF changes (Raygan et al., 2019). Rapid or sudden fluctuations in geomagnetic and solar activity, as well as magnetic storms, appear to be able to act as stressors to alter regulatory processes such as melatonin/serotonin balance (Bergiannaki et al., 1996; Rapoport et al., 1997; Burch et al., 1999), BP, respiratory, reproductive, immune, neurological, and cardiac processes (Ghione et al., 1998; Chernouss, 2001; Cherry, 2002). Geomagnetic disturbances are associated with significant increases in depression, psychiatric disorders, psychiatric hospitalizations, suicide attempts, homicides, and traffic accidents (Nikolaev et al., 1976; Kay, 1994, 2004; Halberg et al., 2005; Berk et al., 2006). Disturbed geomagnetic activity can also exacerbate existing disorders and is correlated with significant increases in cardiac arrhythmias, cardiovascular disease, the incidence of myocardial infarction-related deaths, blood flow changes, BP elevation, and epileptic seizures (Malin and Srivastava, 1979; Knox et al., 1979; Stoupel, 1993; Stoupel et al., 1995; Persinger and Psych, 1995; Ghione et al., 1998; Cornélissen et al., 2002; Caswell et al., 2016).

More recently, several researchers have found a significant correlation between HRV and the GMF strength (McCraty et al., 2017; Timofejeva et al., 2017; Alabdulgader et al., 2018). Their studies have shown that the autonomic nervous system responds to solar and geomagnetic activity (McCraty et al., 2017; Timofejeva et al., 2017; Alabdulgader et al., 2018). Moreover, it was found that the number of acute myocardial infarction events after low geomagnetic activity and high GCR days increased by a fifth, the obtained results indicate that GMF may be related to the development of myocardial infarction (Stoupel, 2008; Stoupel et al., 2012; Podolská, 2018; Jaruševičius et al., 2018; Žiubrytė et al., 2018).

Podolská (2018) investigated the impact of the Earth's ionospheric and the GMF changes on mortality from cardiovascular causes of death in the period 1994–2011 in the Czech Republic. Changes in the ionosphere caused by solar activity are described by ionospheric parameters. As for the ionospheric parameters, critical frequency of the ionospheric F2 layer (foF2) and Total Electron Content (TEC) were analyzed. The TEC parameter, measured since 2008, quantifies the content of free electrons in the ionosphere and was measured by the delay in GPS satellite signals. The results confirmed the hypothesis that there is no direct correlation between the geomagnetic solar index, K_p, and the number of deaths from acute myocardial infarction (code I21) or brain stroke (code I64) during the maxima of the solar cycle. In contrast, the ionospheric parameters of foF2 and TEC explained a greater part of the variability in the number of deaths for acute myocardial infarction or brain stroke than the model with solar parameters. The analysis showed that, because the values are geographically specific, the ionospheric parameters which may describe the variability in the number of deaths from cardiovascular diseases are better than the solar indices. Cardiovascular diseases thus respond to the changes in the solar activity and to abnormal solar events indirectly through a concentration of electrical charges in the Earth's environment.

In addition, Jaruševičius et al. (2018) investigated correlations between the strength of time-varying aspects of the local GMF in different frequencies and incidence of myocardial infarctions in the period January 1, 2016–December 31, 2016, in Lithuania. As shown in Figures 6.14–6.16, this Lithuania

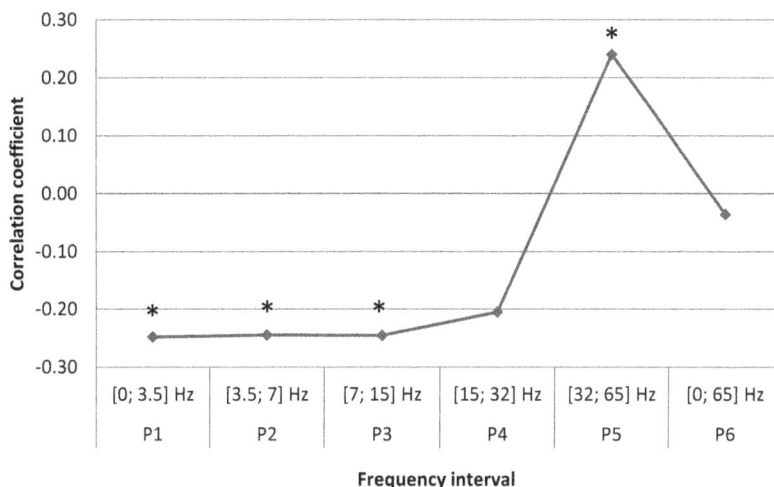

FIGURE 6.14 The correlation between number of weekly myocardial infarction cases in women and mean magnetic power in different frequency ranges spanning a time of one week (Jaruševičius et al., 2018). *$p < 0.05$. (Reproduced with permission from Jaruševičius et al. (2018), Copyright 2018, MDPI.)

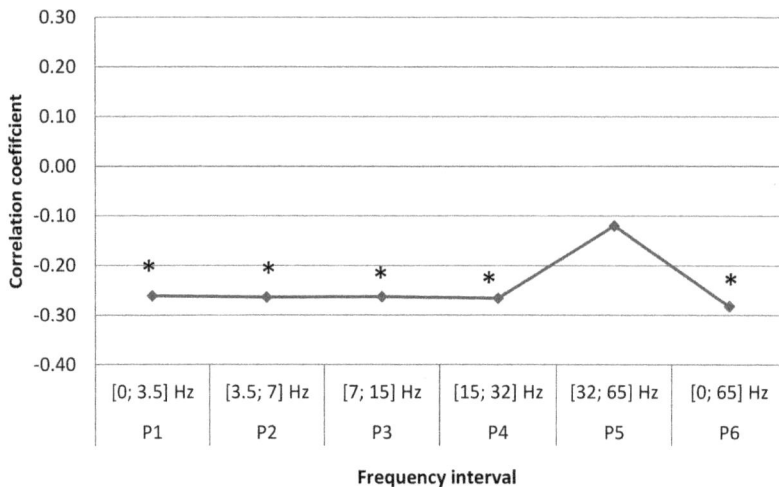

FIGURE 6.15 The correlation between weekly myocardial infarction number in men and mean magnetic power in different frequency ranges spanning a time of one week (Jaruševičius et al., 2018). $*p<0.05$. (Reproduced with permission from Jaruševičius et al. (2018), Copyright 2018, MDPI.)

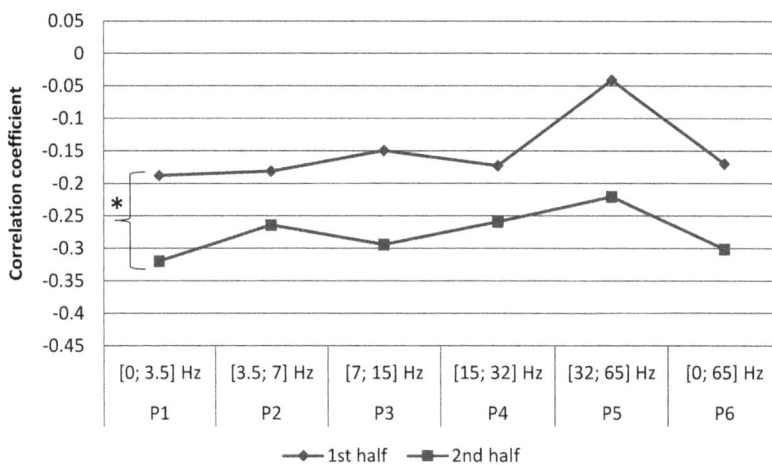

FIGURE 6.16 The correlation between the number of weekly cases of myocardial infarction in men and mean magnetic power in different frequencies for younger (<63 years) men group for the first and second half for the year 2016 (Jaruševičius et al., 2018). $*p<0.05$. (Reproduced with permission from Jaruševičius et al. (2018), Copyright 2018, MDPI.)

research team found a significant relationship between the number of acute myocardial infarctions with ST-segment elevation cases (STEMI) of electrocardiogram signal per week and the average weekly GMF strength in different frequency ranges.

As shown in Figure 6.14 (Jaruševičius et al., 2018, data from the Hospital of Lithuanian University of Health Sciences, Cardiology Clinic, 2016) in the female group, the Lithuanian research team found a single positive correlation coefficient Sγ, which indicates that higher MF intensity in this frequency range is significantly associated with increased number of STEMI cases. In other low-frequency ranges, they observed negative correlation coefficients.

As shown in Figure 6.15 (Jaruševičius et al., 2018), in the male group, the weekly correlation coefficients are similar to the female group, however, all coefficients are negative. Unlike females, they observed non-significant changes in Sγ range, which may indicate slightly different sensitivity of different sexes to the GMF changes.

As shown in Figure 6.16 (Jaruševičius et al., 2018), they found that a correlation between the number of STEMI cases and <63 y.o. males was maintained in all MF frequency ranges. Differences between both halves of the year were significant. These results revealed that different frequency ranges have different correlations with the presence of myocardial infarctions and the correlations varied in different age groups as well as in males and females, which may indicate diverse organism sensitivity to the GMF. They supposed that different people may have different sensitivity to different frequencies of MFs, but there may be potential differences due to age, gender, and health status.

In addition, the same Lithuania research team reported the following conclusions: (1) significant correlation between acute coronary syndrome and the local GMF changes was revealed; (2) the acute coronary syndrome is positively correlated with the local GMF in Sγ range in females through the year; (3) a higher MF in Sβ and Sγ ranges is associated with higher incidences of acute coronary syndrome through the year in females; (4) the higher MF in Sγ range is associated with higher incidences of acute coronary syndrome through the year in females and through the second half of the year in males (Žiubrytė et al., 2018).

In another aspect, in international stroke incidence studies, a joint research team examined the relationship between the number of first strokes and geomagnetic activity (Feigin et al., 2014). Data for stroke patients (over 11,000 total patients) were pooled from multiple studies in Europe, Australia, and New Zealand between 1981 and 2004 (Feigin et al., 2014). Geomagnetic activity data were obtained from the US National Oceanic and Atmospheric Administration (NOAA) (Feigin et al., 2014). As for main results, it has been suggested that the risk increases to an average of 1.9 times in those over 75 years old, especially in the case of a strong magnetic storm (Feigin et al., 2014). It is unclear why magnetic storms increase strokes, but previous studies have shown that magnetic storms affect the human body such as increased BP (Stoupel et al., 1995; Ghione et al., 1998), fluctuations in HR (Stoupel et al., 1994), and blood viscosity/coagulability related to the risk of blood clots (Stoupel, 2002). All of which are risk factors for stroke. The research project team proposes that people with a high risk of stroke should be cautious during severe/extremely magnetic storms, such as avoiding heavy or excessive drinking and being careful not to become dehydrated (Feigin et al., 2014).

With reference to the correlation between the geomagnetic storm and BP, Watanabe et al. (2001) reported correlations of geomagnetic activity with systolic BP (SBP) and DBP at Daini hospital in Tokyo. The authors observed an inverse relationship between Wolf Number (WN) and the variability in SBP and, to a lesser extent, DBP. The WN, or relative sunspot number, is an index of the entire visible disk of the Sun, which is determined each day without reference to previous days. The relative sunspot number is defined as:

$$R = k\left(f + 10g\right) \tag{6.1}$$

where g is the number of groups of sunspots and f is the total number of distinct spots. The scaling factor k depends on the sensitivity of the observing equipment and is usually less than unity (Hargreaves, 1992). The data used in this study were from one individual who self-measured his BP and HR at intervals of 15–30 minutes from August 1987 to July 1998, and beyond. The subject was 35 at the start of the 11-year data set and was clinically healthy. A direct association between HR and WN was found to be solar cycle stage-dependent, while an inverse relationship was consistently found between WN and HR variability (HRV) (Watanabe et al., 2001). This seems to suggest that high levels of solar activity, and therefore high levels of geomagnetic disturbance, cause HRV to decrease. Cornélissen et al. (2002) also report that magnetic storms cause HRV to decrease; Chernouss et al. (2001) confirm this result but state that the response varies significantly between different individuals.

Cornélissen et al. (2002) claim that the incidence of mortality due to myocardial infarction increases in Minnesota, USA, by 5% during years of maximal solar activity compared with years of minimum activity. This corresponds to an extra 220 deaths per year (in a population of about 5 million) during the period of high solar activity. Also reported is a positive correlation of HR and a negative correlation of HRV with solar activity, measured via the sunspot number. Interestingly, the correlation of HR with sunspot number is significant only for the ascending part of the solar cycle. Palmer et al. (2006) suggested that the increase in HR and decrease in HRV could be a result of aging of the test subject(s). However, HR normally decreases with age, and there are many other variables that affect HR. The claim would be more convincing if there was also an association during the descending part of the cycle.

Cornélissen et al. (2002) also report a 7.6% increase in the incidence of myocardial infarction after a reversal of the direction of the N-S component (B_z) of the interplanetary MF. On the following day, a 5.9% decrease is observed (Cornélissen et al., 2002). A N-S B_z "flip" is known to trigger auroras and magnetospheric storms (Arnoldy, 1971), with a typical delay of 2 or 3 hours. During such a reversal of the interplanetary MF, the total current in the magnetosphere increases by a factor of 10, while the potential difference across the geomagnetic tail and across the polar cap increases from typically 10 to 100 kV.

Cornélissen et al. (2002) conclude that there is the additional risk of myocardial infarction at solar maximum compared with the solar minimum, which implies an additional risk for high levels of geomagnetic activity. Other studies report an additional risk whenever the level of geomagnetic activity deviates from its "nominal" level. For example, Shumilov et al. (2003) report that not only high levels but also extremely low levels of geomagnetic activity have an adverse effect on fetal state in Murmansk at ~65°N magnetic latitude.

More recently, Azcárate and Mendoza (2017) conducted the correlation analysis in a group of eight adult hypertensive volunteers, four men and four women, with ages between 18 and 27 years in Mexico City during a geomagnetic storm in 2014, and the results indicated a significant correlation between the SBP and DBP and the atmospheric pressure and the horizontal GMF component, being the largest during the night (Azcárate and Mendoza, 2017). The analysis showed that the largest correlations are between the DBP and DBP and the horizontal GMF component (Azcárate and Mendoza, 2017). Thus, the superposed epoch analysis showed that the largest number of significant changes in the BP under the influence of the GMF occurred in the SBP for men (Azcárate and Mendoza, 2017).

Moreover, Gurfinkel et al. (2017) suggested that the geomagnetic storm impact leads to the decrease of capillary blood velocity in healthy volunteers from the daytime until the late evening in comparison with the quiet ambient magnetic conditions. Actually, the same effect of the decrease of capillary blood velocity had also been observed in patients suffering from myocardial infarction during the period of the geomagnetic storm (Gurfinkel et al., 1995).

More recently, Pishchalnikov et al. (2019) analyzed the correlation of a daily sequence of interbeat (RR) intervals combined with measurements of capillary blood velocity in experiments using artificial geomagnetic storms. The results revealed significant cardiovascular response specific to geomagnetic storms independently from weather conditions (Pishchalnikov et al., 2019). Significant correlations of RR intervals with atmospheric pressure at night time for all modes of MF generation and with temperature for the Storm mode were found (Pishchalnikov et al., 2019). The averaged overall subjects' capillary blood velocity for all modes appeared not insensitive to geomagnetic storms (Pishchalnikov et al., 2019). However, the reaction of a few subjects on geomagnetic storm exposure was quite strong and statistically significant. There was a hardly noticeable increase in either BP parameter during the Storm regime (Pishchalnikov et al., 2019). It was suggested that the daily low-frequency (LF) and high-frequency (HF) dynamics of RR intervals stored as a matrix in time-frequency representation is unique for each subject and does not depend on the day of the experiment and the weather conditions (Pishchalnikov et al., 2019). The averaged correlation coefficients of RR intervals with the B_x and B_y components appeared significantly higher during the Storm mode in comparison with control (Pishchalnikov et al., 2019). Thus, it is shown that artificial geomagnetic storms can cause a detectable cardiovascular response (Pishchalnikov et al., 2019).

Krylov (2017) reviewed and proposed that the understanding of mechanisms by which geomagnetic storms affect organisms will help to minimize their harmful impact on human health. More specifically, Krylov (2017) made the following proposals as future research to be solved: It is necessary to study correlations between geomagnetic activity and biological parameters at a higher temporal resolution (hour or less); one must determine relationships between the magnitude of the biological response, the local time when a geomagnetic storm was registered, and the dynamics of a storm's fluctuations with respect to the usual dynamics of diurnal geomagnetic variation; one must investigate the similarities and differences between the mechanisms of the impact of geomagnetic activity on organisms and magnetic orientation. The following research could be carried out to solve this problem: studying changes in the concentrations and functions of melatonin, CRYs, and protein-coding by the CG8198 gene, which Qin et al. (2016) called the magnetoreceptor protein (MagR) as described above, in response to simulated geomagnetic storms; and studying changes in the expression of CRY and CG8198 genes, as well as genes coded for enzymes involved in melatonin synthesis in response to simulated geomagnetic storms. Based on the idea suggested by Breus et al. (2016), the production of these molecules may be the biochemical integrator averaging and cumulating the signal of GMF fluctuations (Krylov, 2017). Experiments, where desynchronization of circadian Zeitgebers (diurnal geomagnetic variation which is shifted relative to the alternation of day and night) is used as treatment may also contribute to the confirmation or rejection of this proposed hypothesis (Krylov, 2017).

6.4.3 Schumann Resonance

Apart from the above-mentioned influence of the GMF cycles, other quasi-periodic ELF processes are caused by oscillations in the plasmasphere and magnetosphere of solar wind and by the resonant oscillations, so-called "Schumann resonance" of the ionosphere of the Earth (Vladimirsky et al., 1980; Temurjants et al., 1992). Schumann resonance (SR), which is globally propagating ELF waves, is hypothesized to be "the possible biological mechanism" that explains biological and human health effects of geomagnetic activity (Cherry, 2002). SR is a background stationary EM noise that propagates in the cavity between the Earth's surface and the lower boundary of the ionosphere at altitudes of 45–50 km, in the frequency range between 5 and 50 Hz (Schumann, 1952; Bliokh et al., 1980; Sentman, 1995). In brief, SR is a series of ELF-EMF resonances caused by lightning discharges in the atmosphere (Nickolaenko and Hayakawa, 2014). The phenomenon was named after Winfried Otto Schumann who is a German physicist at the Technical University of Munich and first predicted and discussed it in the 1950s (Schumann, 1952). In the 1950s, Winfried Otto Schumann and Herbert König first measured frequencies that were similar to a mathematical model that predicted a magnetic wave resonance between the Earth and ionosphere (Schumann and König, 1954). Herbert König, who became Schumann's follower at the University of Munich, further showed a clear connection between SR and brain rhythms. He compared human EEG recordings with the natural ELF-EMFs of the environment and found that the main frequency produced by Schumann oscillations is extremely close to the frequency of α rhythms (König and Ankermüller, 1960; König et al., 1981). Schumann (1952) showed that the resonance frequencies are given by an equation of the form:

$$f_n = 7.49\left(n(n+1)\right)^{1/2} \tag{6.2}$$

This formula predicts a fundamental mode frequency ($n=1$) of $f_1=10.6$ Hz with overtones (or harmonics) at 18.4, 26.0, 33.5, and 41.1 Hz (Barr et al., 2000). The first definite experimental confirmation of Schumann's prediction was the ELF noise spectral analysis of Balser and Wagner (1960). Thus, these resonance frequencies were predicted by Schumann (1952) and detected by Balser and Wagner (1960). As illustrated by Polk (1983), these spectra actually have maxima near 7.8, 14.2, 19.6, 25.9, and 32 Hz corresponding to the first five modes in Schumann's formula.

It is speculated that SR and ELF background fields played an important role in the evolution of biological systems and are used by them as a means of stochastic synchronization for various biorhythms (Cole and Graf, 1974). The SR frequencies are mainly controlled by the Earth's radius, which has remained constant over billions of years (Morente et al., 2003). Therefore, these frequencies can play a special role for the regulatory pathways of living organisms, the SR providing a synchronization reference signal, a Zeitgeber (time giver) (Cherry, 2002).

It has been clarified that the first SR frequency is 7.83 Hz, with a day/night variation of around ±0.5 Hz (Sentman, 1995). The higher frequencies are ~14, 20, 26, 33, 39, and 45 Hz due to frequency-related, ionospheric propagation loss (Schumann, 1952; Bliokh et al., 1980; Sentman, 1995), all of which closely overlay with α (8–12 Hz), β (12–30 Hz), and γ (30–100 Hz) brain waves (Cherry, 2002). The similarity of the EEG with the SRs was recognized early on, and the ability of the EEG rhythm to synchronize with SR activity was observed (König et al., 1981).

Altered EEG rhythms in response to changing the GMF have been observed with ELF magnetic oscillations (~3 Hz) having a sedative effect (Belov et al., 1998). It has also been demonstrated that autonomic nervous system activity not only reacts to shift in the geomagnetic and solar activity, and it can also synchronize with rhythms in the time-varying MFs related to the SR and the GMF-line resonances (McCraty et al., 2017; Timofejeva et al., 2017). The SR and the EEG have been examined in a group of participants over 6 weeks, and it was found that the EEG changes during the daily cycle were similar to variations in the SR (Pobachenko et al., 2006). The highest correlations between SR and EEG were found when the magnetic activity was increased. SR and EEG activities have also been studied in real-time and it has been shown that many of the SR frequencies can be observed in the power spectrums of most EEG activity (Saroka and Persinger, 2014; Persinger and Saroka, 2015). It has also been shown that the spectral profiles within the EEG activity displayed recurrent transient segments of real-time coherence (synchronization) with the first three resonant frequencies of the SR (7–8 Hz, 13–14 Hz, and 19–20 Hz). These findings suggest that under certain conditions variables affecting the SR parameters (such as solar wind) may affect EEG activity, such as modifications of perception and dream-related memory consolidation (Persinger and Saroka, 2015).

Hainsworth (1983) noted that the average frequency at which there is minimum power circulating in the Earth-ionosphere cavity is the same frequency as the dominant human brain-wave rhythm—10.5 Hz. Cannon and Rycroft (1982) and more recently Schlegel and Füllekrug (1999) reported the effects on SR produced by ionospheric disturbances induced by solar activity. "Solar Proton Events (SPEs)" have been found to decrease the frequency of the SR modes. Roldugin et al. (2001) found that, during the peak of four SPEs, the frequency of the first SR mode decreased by about 0.15 Hz, as measured in the Kola Peninsula of Russia. Roldugin et al. (2001) also reported an increase in the frequency and a resonance bandwidth decrease of about 0.2 Hz of the first Schumann mode as a result of a very intense solar X-ray burst. Any change in the SR signals due to ionospheric disturbances will be superposed on the diurnal (i.e., circadian) variations due to solar heating and ionization on the dayside. As visual and auditory stimulation produce biological effects, Hainsworth (1983) argued that EM signals at frequencies in the brain wave spectrum can be expected to produce biological effects too. Hainsworth (1983) also argues that the association with the human α-rhythm near 10 Hz with the frequency of minimum energy in the SR spectrum and, therefore, of minimum natural interference is unlikely to be a coincidence.

Hainsworth (1983) suggested that other factors that could affect the apparent connection between geophysical parameters and biological effects are links with geographical considerations. For instance, wind eddies carrying ionized air can produce oscillatory signals in the range of 3–6 Hz. These could be associated with thunderstorm activity or with winds such as the Fohn wind in Austria and could have localized biological effects.

It has been shown that the intensity of the SR signals is affected by the air temperature. Williams (1992) demonstrated a positive correlation between the monthly means of the tropical surface-air-temperature anomaly and the MF amplitude for the fundamental SR mode. A 2 K change in temperature was shown to produce a 20-fold change in lightning activity (Barr et al., 2000). Williams (1992) also reported that El Nino/La Nina conditions produce corresponding increases and decreases in SR signal intensity, which may have implications for future global climate change. Rycroft et al. (2000) discuss the topic of the global atmospheric electric circuit and climate change in greater depth.

The idea that the flux of GCRs can affect the parameters of SR and the ELF-EMF background can be tested using the "Forbush effect" (Beloglazov et al., 2006). During solar flares, the flux of GCRs decreases rapidly (over a day or less) due to modification of the near-Earth interplanetary MF. This so-called Forbush decrease is transient and is followed by a gradual recovery over several days (Lockwood, 1971; Cane, 2000). Based on the measurements in the Kola Peninsula of Russia, it was demonstrated that in all ten events of significant Forbush-decreases, the intensity of the ELF-atmospherics decreased (down to their complete disappearance) (Beloglazov et al., 2006). It was hypothesized that this phenomenon is caused by a decrease in the intensity of discharges of a special type (sprites and jets) as a result of a decrease in atmospheric ionization at altitudes of 10–30 km during the Forbush decrease in the flux of GCRs (Beloglazov et al., 2006).

Regarding the relationship between BP and SR, more participants showed lower BP on enhanced SR days (Mitsutake et al., 2005). That is, 32.1% of the 56 participants showed lower SBP on enhanced SR days, whereas only 3.6% showed higher SBP on those days (Mitsutake et al., 2005). In contrast, in DBP, 26.8% of the participants showed lower DBP on enhanced SR days, whereas only 3.6% showed higher DBP on those days (Mitsutake et al., 2005).

In SR frequencies, studies of rats exposed to MFs in the frequency band of 0.01–100 Hz (with magnitudes of 5, 50, and 5,000 nT) have revealed that MFs at frequencies of 0.02, 0.5–0.6, 5–6, and 8–11 Hz had the greatest impact on the circulatory system (Ptitsyna et al., 1998). There is also an intriguing report of a BP-lowering effect in humans with mild-to-moderate hypertension after exposure to 6–8 Hz MFs at 1 µT (Nishimura et al., 2011).

A recent long-term study examined the relationships between the solar and magnetic factors and the time course and lags of autonomic nervous system responses to changes in solar and geomagnetic activity (Alabdulgader et al., 2018). In this study, the inter-beat-interval (IBI), Total power, low-frequency (LF) and high-frequency (HF) powers, the LF/HF ratio, and very low-frequency (VLF) power were used as parameters of HRV measures (Alabdulgader et al., 2018; Table 6.2). Here, the power spectral density values reflect the area under the curve within the specific bandwidth of the spectrum. The interactions between autonomic neural activity, BP, respiration, and higher-level control centers in the brain produce both short- and longer-term rhythms in HRV measurements (McCraty and Shafer, 2015). The heart rhythm fluctuations are separated into three primary frequency bands: HF, LF, and VLF (Task Force of the European Society of Cardiology and the North American Society of Pacing and Electrophysiology, 1996). The "ultralow-frequency (ULF)" is generally described as geomagnetic activity <3.5 Hz. The GMF-line resonances are the most common source of ULF wave energy measured on the ground and exhibit the largest wave amplitudes compared to other oscillations that occur in the

TABLE 6.2 Summary of Magnetic and HRV Frequency Ranges Used in Measurements

Category	Hz
Schumann resonance power	3.5–36
Magnetic field ULF power	0.002–3.5
HRV	
Total power	0–0.4
Very low frequency	0.003–0.04
Low frequency	0.04–0.15
High frequency	0.15–0.4

Source: Compiled from Alabdulgader et al. (2018).

magnetosphere (Southwood, 1974). IBI is the time in milliseconds between consecutive heartbeats. Total power is a measure of all the HRV bands combined, and therefore is a measure of the overall HRV from all physiological sources, although it is highly affected by the VLF power. Space weather and environmental measures were obtained from three sources, comprising nine measures. The solar wind speed, K_p index, A_p index, number of sunspots, F10.7 index, and the geomagnetic polar cap index (PCN) were downloaded from NASA Goddard Space Flight Center's Space Physics Data Facility as part of the Omni 2 data set (Alabdulgader et al., 2018). Here, the F10.7 index is a "solar radio flux" at 10.7 cm (2,800 MHz) (Tapping, 1987). That is a measure of the solar flux unit (sfu) frequency at a wavelength of 10.7 cm, near the peak of the observed solar radio emission. F10.7 is often expressed in sfu (1 sfu $= 10^{-22}$ W/m^2 Hz). It represents a measure of diffuse, nonradiative coronal plasma heating. It is an excellent indicator of overall solar activity levels and correlates well with solar UV emissions. GCR counts were downloaded from Finland's University of Oulu's Sodankyla Geophysical Observatory's website (Alabdulgader et al., 2018). Power in the time-varying MF in two frequency bands, SR Power, 3.5–36 Hz and ULF power, 2 mHz to 3.5 Hz were obtained from a recording site located in Boulder Creek, California (Alabdulgader et al., 2018).

In agreement with previous studies (Crooker et al., 1977; Richardson et al., 1996; Mathie and Mann, 2000), solar wind speed was highly correlated with K_p and A_p indexes, and ULF power was negatively correlated with GCR counts. As expected, the solar radio flux (F10.7) was also highly correlated with the number of sunspots. The SR power was negatively and highly correlated with GCR counts. ULF power was positively correlated with solar wind speed, K_p, Ap, and PCN indexes. The correlations among the HRV variables were as expected and in agreement with other studies with the exception of HF power, where correlations to IBI, Total power, and VLF power were all higher than seen in individual 24-hour recordings. Thus, this study revealed that these HRV measures were correlated with solar and geomagnetic variables and daily autonomic nervous system activity responds to changes in geomagnetic and solar activity during the period of normal undisturbed activity (Alabdulgader et al., 2018). An increase in GCRs, solar radio flux, and SR power was all associated with increased HRV and parasympathetic activity (Alabdulgader et al., 2018).

It is clear that a major driver of changes and disturbances in the GMF environment are the Sun and solar wind (Crooker et al., 1977; McPherron, 2005). Consistent with these findings, the solar wind speed was highly correlated with K_p, A_p, and the PCN, all of which reflect MF disturbances. It was also found that ULF power, which is related to MF-line resonances, was positively correlated with solar wind speed, and indices of field disturbance were negatively correlated with GCR counts, which is consistent with the well-known inverse action of solar and geomagnetic activity and GCR counts at the Earth's surface (Richardson et al., 1996).

Regarding HRV responses, IBIs have an inverted relationship to HR where larger IBIs equated to a lower HR. HR and IBIs are an ideal indicator of changes in the relative balance between parasympathetic and sympathetic activity and how the autonomic system responds and adapts to various types of stressors or challenges (McCraty and Shafer, 2015). If an environmental variable is negatively correlated with IBIs, it indicates that HR increases with increases in that variable, which suggests a physiological stress reaction occurred. On the other hand, a positive correlation with IBIs indicates a lower HR. There were robust positive correlations between IBIs and SRs, and to a lesser degree with GCRs.

The positive correlation found between HF power and solar radio flux indicates an enhancement of parasympathetic nervous system activity during the period of increased solar radio flux. This was of particular interest because a previous study with 1,643 participants in 51 countries found that the solar radio flux index was positively correlated with reduced fatigue, improved positive affect, and mental clarity while increases in solar wind speed had the opposite effects (McCraty et al., 2012). The potential beneficial effects of the solar radio flux were also observed in several studies that looked at death rates from various causes which found a strong and inverse relationship between the F10.7 and death rates (Stoupel et al., 2006, 2011). The solar radio flux may be an important mediator of the anticipatory reactions observed by Tchijevsky (1971), which can occur several days before increases in the solar wind reach Earth and create magnetic disturbances. Of course, other sources of radiation such as X-rays, GCRs, and UV-rays from the sun during CMEs are also likely aspects of the anticipatory reaction.

The other environmental variable that was strongly associated with increased HF, LF, and VLF power and total power of the HRV measures was SR power. This was accompanied by the positive correlation to IBIs (lower HR), which was also significant during most of the analysis period. In addition, supporting a beneficial effect of enhanced SR power is a study that found reduced systolic, diastolic, and mean arterial BPs during the period of higher SR power (Mitsutake et al., 2005). Persinger and colleagues have conducted a number of studies showing that not only are the base rhythms of the brain similar, but real-time coherence between SRs and EEGs can occur in participants globally, and that the intensity of the SR is linearly related to the amount of coherence (Saroka and Persinger, 2014; Saroka et al., 2016). They also have proposed that information transfer can occur between human brains and the SRs (Persinger and Saroka, 2015).

Many species exhibit, irrespective of the size and complexity of their brain, essentially similar low-frequency electrical activity (Price et al., 2021), and it is possible that the dominant frequencies of brain waves may be an evolutionary result of the presence and influence of the SR (Direnfeld, 1983). Moreover, it is conceivable that magnetoreception could be influenced by SR and ELF background fields. As mentioned above, these fields seem to have physiological effects on humans, so any animal that can sense the GMF has more sensitive and advanced magnetoreceptors than humans is more likely to be affected by the fields.

An increase in solar wind intensity was correlated with increases in HR, which might be regarded as a biological stress response (Alabdulgader et al., 2018). These findings that are both extremely high and extremely low values of geomagnetic activity are associated with the timing of increased death rates (Stoupel et al., 2013), and also suggest that the Earth's energetic environment affects people's energy levels and that low activity or disturbances can act as triggers in sensitive and unhealthy populations, and serve to motivate and facilitate human activity (Alabdulgader et al., 2018).

6.4.4 The Effects of the Solar Cycles and the Geomagnetic Field on Infectious Diseases and Human Health

Accumulated data support the idea about cycles of solar activity as the pacemaker of numerousiological phenomena including epidemics of some infectious diseases and dynamic changes in immunological parameters of living systems (Zaporozhan and Ponomarenko, 2010). Chizhevsky (1976) performed a comparative retrospective analysis of flu epidemics with respect to solar activity cycles and came to the following conclusions: (1) outbreaks of major influenza pandemics show a clear cycle of 11.3 years on average, equal to the period of fluctuations in solar activity; (2) as a rule, a serious influenza pandemic does not occur in years of minimal solar activity; (3) most major influenza pandemics occurred at time intervals that begin 2–3 years before the maximum of solar activity and end 2–3 years later. Hope-Simpson (1978) noted coincidences between the influenza pandemic and the maximum of solar activity, which occur with an ~11-year period. These observations were continued, developed, and statistically tested later on (Hoyle and Wlckramasinghe, 1990; Ertel, 1994; Yeung, 2006; Vaquero and Gallego, 2007).

Tapping et al. (2001) showed that influenza pandemics often occur during the period of maximum solar activity. The 1946, 1957, 1968, and 1977 pandemics (shown as spikes) on a plot of the 10.7 cm Solar Flux in solar flux units (sfu) are shown by Zaporozhan and Ponomarenko (2010). The flux values prior to 1947 were estimated from sunspot data. However, there were some exceptions that occurred near the minimum activity, such as the 1977 pandemic. A summarized correlation diagram for all study periods (~300 years) indicated that there is a significant correlation between solar activity and the activity of the influenza pandemic process (Tapping et al., 2001). Zaporozhan and Ponomarenko (2010) showed the distribution of influenza pandemics (according to different authors) as a function of phase offset from solar activity maximum (modified from Tapping et al., 2001). Several other authors have also noticed a correlation between the solar cycle and major influenza and other epidemics on Earth (Hope-Simpson, 1978; Ertel, 1994; Tapping et al., 2001; Yeung, 2006; Vaquero and Gallego, 2007). However, the presence of complex interactions such as anthropogenic effects (migration, vaccination, etc.) can mask the effects of regulation of the solar cycle on influenza pandemics.

The influences of the solar cycles and the GMF have been manifested not only on periodicity in global influenza infection spread, but also on human health. In fact, there are many observations that support the impact of geomagnetic perturbations on our human health. Periodic GMF fluctuations are very similar to seasonal fluctuations in the prevalence of viral respiratory illness, suggesting a possible role for GMF fluctuations as both an influenza pandemic and a predisposition to seasonal influenza (Zaporozhan and Ponomarenko, 2010). Moreover, regular semiannual variations in the GMF influence greatly resemble seasonal variations in viral respiratory diseases sickness rate, which allows speculation about the possible role of GMF fluctuations as both flu epidemics and seasonal flu-predisposing factors (Zaporozhan and Ponomarenko, 2010). Zaporozhan and Ponomarenko (2010) speculate that the data they have presented concerning the probable influence of specific external MFs on immunological processes and virus replication via the NF-κB and other signaling pathways allows us to put forward a possible role of the influence of GMF fluctuations, and correspondingly, solar activity cycles as factors capable of influencing the occurrence of influenza (and possibly other) epidemics. It is worth mentioning that several viruses, including HIV, have binding sites for NF-κB that control the expression of viral genes, which, in turn, contribute to viral replication and/or pathogenicity (Hiscott et al., 2001). In the case of HIV-1, activation of NF-κB may, at least in part, be involved in the activation of the virus from a latent, inactive state (Hiscott et al., 2001). This indicates the theoretical possibility of MF effects on the processes of HIV activation and replication (Zaporozhan and Ponomarenko, 2010).

Taking into account a probable entrainment role of solar activity fluctuations in the regulation of genome expression and influenza epidemic cycles, Zaporozhan and Ponomarenko (2010) emphasized the regulatory/entrainment role of the solar cycles for Earth life (biosphere) microevolution. It is suggested that from 2011 until 2015 solar regulatory influences and concomitant GMF fluctuations may predispose to genetic and immunological alterations favorable to influenza epidemic spread (Zaporozhan and Ponomarenko, 2010). Recently, in this context, Morchiladze et al. (2021) proposed that understanding the role of ELF-EMFs in regulating the biosphere is important in our fight against COVID-19, and research in this direction should be intensified.

Carter et al. (2020) demonstrated a quantum biological phenomenon whereby a combination of 3 mT SMF and 7 kV/m EF ameliorated hyperglycemia and enhance insulin sensitivity in three different mouse models of type 2 diabetes: Bardet-Biedl syndrome, leptin receptor-deficient, and high-fat diet (HFD) mice. Surprisingly, all mice have a 30% or greater reduction in blood glucose and these effects appeared rapidly, within 3 days of exposure (continuously 24 hours/day for 3 days, or for 7 hours/day for a total of 3 days), without adverse effects (Carter et al., 2020). However, the glucose levels were returned to the pre-exposure levels when the daily exposure was stopped (Carter et al., 2020). Interestingly, both exposures to SMF and EF were required simultaneously, and exposure to SMF alone exacerbated diabetes through unknown mechanisms (Carter et al., 2020). In addition, human liver cells treated with both SMF and EF for 6 hours demonstrated an increase in glycogen, a surrogate marker for insulin sensitivity, indicating a similar effect with mice. This enhancement of insulin sensitivity occurred due to altered redox homeostasis, mediated primarily by the hepatic mitochondrial superoxide O_2^- (Carter et al., 2020). That is, both exposures to SMF and EF promoted activation of the antioxidant enzyme SOD2 in hepatic mitochondria and enhanced superoxide scavenging activity (Carter et al., 2020). Furthermore, it was found that non-cyclooxygenase-derived prostanoids (F2-isoprostanes), which are a biomarker of oxidative stress, decreased by about 40%, and the master transcription factor NRF2 in redox signaling was elevated by both exposures (Carter et al., 2020). Thus, both exposures to SMF and EF enhanced the redox reaction by increasing the release of glutathione into the blood via the NRF2, and reduced oxidative stress to improve insulin resistance (Carter et al., 2020).

Thus, during the evolution of living organisms, glucose metabolism has also evolved within the GMF and several EFs (lightning strikes, static electricity, etc.), and it is becoming clear that these environmental factors affect the balance between oxidation and antioxidants.

6.5 Discussion and Conclusions

The relationship between the GMF strength and UV shielding has been discussed and reviewed. Past episodes with low GMF intensity affect the UV radiation levels reaching the Earth's surface, as lowering the magnetosphere shield results in lower stratospheric O_3 levels, and thus lowers UV radiation shield (Wei et al., 2014).

There are three potential sources of energetic particles that can interact with the atmosphere of the Earth, i.e., (1) "solar particles (solar wind)," (2) "γ-ray burst," and (3) "GCRs" (Wei et al., 2014). All these three sources can reduce levels of O_3 and O_2 in the atmosphere. The reduction is exacerbated when the GMF strength is reduced or altered.

The glacial-interglacial cycle is mainly due to changes in the amount of solar radiation due to changes in the Earth's orbit and axis of rotation. The phenomenon is so-called "Milankovitch cycles," which is a theory that variations in eccentricity, axial tilt, and precession of the Earth resulted in cyclical variation in the solar radiation reaching the Earth, and that this orbital forcing strongly influenced the Earth's climatic patterns (Milanković, 1941; Knezevic, 2010). It occurs in the three basic cycles of about 21, 41, and 100 kyr, but recently even the shorter thousands to hundreds of years of climate change are supposed to be affected by the GMF (Knezevic, 2010). By integrating these Milankovitch cycles with changes in the oxygen isotope ratio of marine microfossils called foraminifera in seafloor sediments (glacial-interglacial cycles), more detailed ages of seafloor sediments are determined, and consequently, the M–B boundary is estimated to be dated to 772.9 ka as mentioned above (Haneda et al., 2020).

The events supposed to be caused by the GMF reversal may be large-scale climate change. For many years, Ueno et al. (2019) have been investigating to elucidate the mechanism by which the GMF could affect climate during the M–B reversal transition. During the last GMF reversal transition, the geomagnetic strength decreases greatly, so the GCRs increase and lower clouds increase, which is the so-called "Svensmark effect," and it is hypothesized that the GCRs-induced sunshade effect will cool the climate (Ueno et al., 2019). They have examined the geological formations that recorded GMF reversals in various places in East Asia (Ueno et al., 2019). As a result, they found that cooling occurred when the GCRs increased by more than 40% from the present, and that the summer precipitation decreased and the winter monsoon strengthened (Ueno et al., 2019). Thus, when GCRs increased during the last GMF reversal transition, the umbrella effect of low-cloud cover led to high atmospheric pressure in Siberia, causing the East Asian winter monsoon to become stronger (Ueno et al., 2019). This is the first evidence that GCRs influence changes in the Earth's climate (Ueno et al., 2019).

The above-mentioned Svensmark effect is a hypothesis that the GCRs induce low cloud formation and influence the Earth's climate (Ueno et al., 2019). This is because the GCRs ionize the atmosphere, increase cloud condensation nuclei, and increase cloud cover (Molina-Cuberos et al., 2001). The tests based on recent meteorological observation data only show minute changes in the amount of GCRs and cloud cover, making it hard to prove this theory (Ueno et al., 2019). However, during the last GMF reversal transition, when the amount of GCRs increased dramatically, there was also a large increase in cloud cover, so it should be possible to detect the impact of GCRs on climate at higher sensitivity (Ueno et al., 2019).

Magnetic storms promoted the production of NO_x in the stratosphere, resulting in O_3 reduction of more than 60% at high latitudes north in early 2004 (Randall et al., 2005). Similarly, atmospheric nitrous oxide (N_2O) concentrations increase when the GMF intensity is low and the shielding effect from magnetic storms and GCRs is reduced. The Earth's surface naturally emits N_2O from the ocean and soil, and the emission is increasing due to human practices. N_2O emissions are enhanced in interglacial and stadial (warm) climatic conditions, increasing by ~50% at the end of the last glacial period (Schilt et al., 2013, 2014). The increase in atmospheric N_2O concentration during the Bølling-Allerød warm period (15–13 ka) coincided with the apparent minimum intensity of the GMF at ~13 ka, during which period the UV radiation levels reaching the Earth's surface were elevated.

Roughly 550 Ma, the GMF rapidly flipped their orientations, swapping north and south to trigger the massive extinction that ended the Ediacaran Period. Rapid GMF reversal destroyed a large part of the O_3

layer and let in a flood of UV radiation, devastating the unusual creatures of the Ediacaran Period and triggering an evolutionary event that led to the Cambrian explosion of animal groups.

Scientists have long argued over what caused the Cambrian explosion in the first place. Potential explanations have included rising levels of atmospheric O_2 because of photosynthesis, allowing for the development of more complex animals; the rise in carnivorous species and new predatory tactics, such as the flat and segmented, armor-crushing creatures known as anomalocaridids; and the breakup of the supercontinent Rodinia, which may have created new ecological niches and isolated populations as the continents drifted apart.

A geologist Joseph Meert of the University of Florida in Gainesville and his colleagues propose a different hypothesis that these evolutionary changes might have been connected to rapid GMF reversal, which occurs about once every Myr. However, in the Ediacaran, such reversals were a lot more common (Meert et al., 2016). Certain minerals in rocks can preserve a record of the direction of the GMF when the rock is formed. While studying these magnetic records 550 million years ago (550 Ma), Ediacaran-aged sedimentary rocks in the Ural Mountains in western Russia, the team discovered evidence to suggest the reversal rate then was 20 times faster than it is today. Meert et al. (2016) estimated that the GMF underwent a period of hyperactive reversals.

Why was the GMF so hyperactive during the Ediacaran–Cambrian interval? Labrosse et al. (2001) estimated the initiation period of inner core nucleation to be 1.0 ± 0.5 Ga, and thus inner core nucleation began during the Late Ediacaran or alternatively, the inner core grew past some critical size leading to instabilities in the geodynamo. Reversal frequency may be directly related to plate tectonics, variability in core heat flux, or whole mantle convection processes (Biggin et al., 2012; Pétrélis et al., 2011; Driscoll and Olson, 2011). Here, "plate tectonics" is thought to result in a high reversal rate following intervals in Earth history when the Earth's continents are asymmetrically distributed about the equator (Biggin et al., 2012). Moreover, an increase in "core heat flux" drives the modeled geodynamo from a superchron state to reversing behavior (Biggin et al., 2012).

As shown in Figure 6.17, Gallet et al. (2019) reviewed and analyzed the period between ~550 and ~500 Ma, which might be characterized by a single or more episodes of hyperactivity in the reversing process.

Duan et al. (2018) recently proposed a frequency of ~7 reversals per Myr between ~524 and 514 Ma, but the stratigraphic sequence from which the data were obtained was fragmentary; it, therefore, seems possible (but not proven) that the reversal frequency was actually higher during this period (Gallet et al., 2019). The Lower Cambrian paleomagnetic data obtained at Khorbusuonka section in northeastern Siberia by Gallet et al. (2003), also attest to a very high GMF reversal frequency during the second stage (Tommotian) and the third stage (Atdabanian) of the Cambrian, in addition to the occurrence of a major "true polar wander (TPW)" event near the end of the Lower Cambrian, in agreement with Kirschvink et al. (1997). At this stage, it is only possible to affirm that the reversal frequency was high (>5 reversals per Myr) during the Lower Cambrian, but the hyperactivity character (>15 reversals per Myr) cannot be established (Gallet et al., 2019).

Thus, there have been suggested episodes of TPW or rapid plate motion in the Ediacaran–Cambrian (Kirschvink et al., 1997; Meert and Tamrat, 2004). TPW episodes would support the argument that links TPW to MF hyperactivity (Biggin et al., 2012). It is also an intriguing observation that an overall trend in reversal frequency shows a ramp up into a hyperactive mode (Ediacaran and Jurassic) followed by a decay in reversal frequency leading into the superchrons (Ordovician Reversed Superchron, Permian-Carboniferous Reversed Superchron and Cretaceous Normal Superchron) (Biggin et al., 2012). The schematic model of MF behavior shows a ramp-decay process leading from hyperactive intervals into superchrons (Meert et al., 2016). The schematic model of MF behavior showing a ramp-decay process leading from hyperactive intervals into superchrons is shown by Meert et al. (2016). However, whether or not this ramp/decay is a real reflection of geodynamo activity remains to be seen.

Previous research has suggested that the Earth's protective MF would be weaker across such periods of frequent reversal, compromising its ability to shield life from harmful solar radiation and GCRs

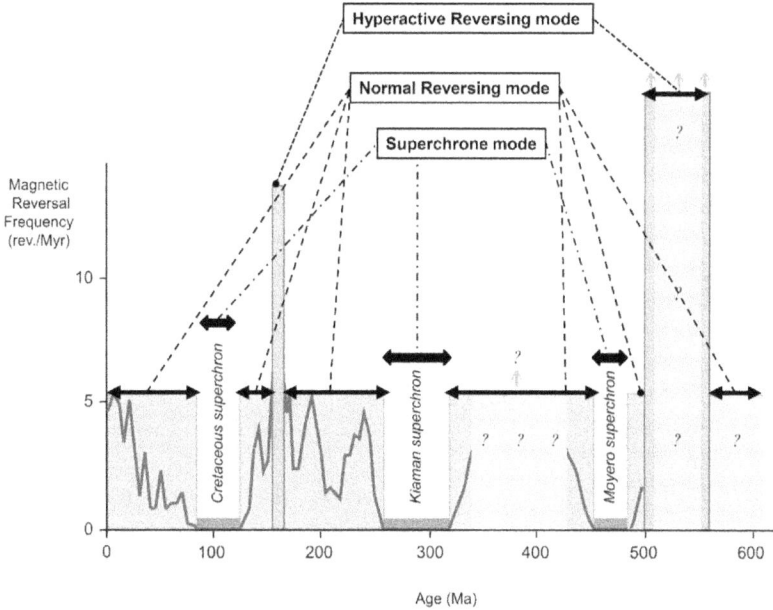

FIGURE 6.17 Synthesized sketch of the evolution in GMF reversal frequency since the end of the Precambrian. (Modified from Gallet et al. (2019).) CNS, Cretaceous Normal Superchron (~83–125 Ma); PCRS, Permian-Carboniferous (Kiaman) Reversed Superchron (~267–313 Ma); ORS, Ordovician (Moyero) Reversed Superchron (~460–490 Ma). (Reproduced with permission from Gallet et al. (2019), Copyright 2019, Elsevier.)

(Meert et al., 2016). On top of this, the duration of each individual reversal episode—thought to take an average of 7–10 kyr—would likely see the field temporarily weakened even more before growing back in the opposite direction. This weakened shielding would have allowed more energetic particles into the upper atmosphere, which would have begun to break down the O_3 layer that protects the Earth from potentially harmful UV-B radiation. A 10%–30% decrease in global O_3 levels can cause considerable disruption to living organisms exposed to increased UV-B flux (Meert et al., 2016). As an example, when 40% of O_3 coverage is lost, the biologically effective amount of UV radiation reaching the Earth's surface will double (Cockell and Blaustein, 2001; Pavlov et al., 2005). It is estimated that O_3 depletion during a reversal could reach up to 30% at higher latitudes and persist for the 7–10 kyr duration of the reversal (Winkler et al., 2008; Vogt et al., 2007). Organisms with the ability to escape UV radiation would be favored in such an environment (Meert et al., 2016). This escape from dangerous levels of UV light, therefore, might explain many of the evolutionary changes that occurred during the Late Ediacaran and Early Cambrian (Meert et al., 2016). Creatures with complex eyes to sense the light and the ability to seek shelter from the radiation, for example, by migrating into deeper waters during the daytime, would have been more successful (Meert et al., 2016). The growth of hard coatings and shells would afford additional UV protection, as would the capacity to burrow deeper into the seafloor (Meert et al., 2016).

   In turn, these morphological and functional changes may have opened up new environments to survive. The development of shells, for example, helps creatures colonize intertidal areas, protected not only from UV-rays, but also stronger radiation, and the risk of drying out. Similarly, the breakdown of the bacterial mats by early burrowing would have opened up the upper seafloor further for life. Looking forward, the researchers are now hoping to examine other Ediacaran sediments from around the globe to verify the rapid reversal signal, along with hunting for biological or chemical evidence for high doses of UV radiation in the fossil record.

Joseph Kirschvink is skeptical about the idea that increased dosing of living organisms with UV radiation is a direct result of the decreased GMF (Randall, 2016). Although this idea is long-established, its effect on the evolution of life at this time should be limited, and as the radiation would not be able to reach and damage the germline, the cells of the body used in sexual reproduction to pass genetic information to offspring (Randall, 2016). The radiation would affect the outer skin but the germ cells are usually internal and protected (Randall, 2016). As such, he argues that the idea that increased levels of UV radiation significantly affected the evolution of life in the Ediacaran is problematic (Randall, 2016).

Niitsuma and Fujii (1984) examined the foraminiferal fossil assemblages at the boundary around 730 ka during Chibanian on the Boso Peninsula. As a result, it was found that the planktonic foraminifera fossil assemblages decreased during Chibanian, and that the decrease was larger in the surface layer. This is because the geomagnetosphere became unstable due to GMF reversals, charged particles from the solar wind rushed into the upper atmosphere and destroyed the O_3 layer, and thus increasing the amount of potentially harmful UV-B reaching the Earth's surface. Therefore, it is inferred that a large number of organisms on the ground and on the ocean surface died.

However, Aubert et al. (2010) argued that only small changes that occurred in the GMF were too small to have major effects on the geomagnetosphere shielding efficiency since the early Earth. In another aspect, it has been reported that the most abundant and readily available radiation shield is water (Simonsen and Nealy, 1991), and the shielding effects of water on UV-B (280–315 nm) radiation is also important (Tedetti and Sempéré, 2006).

These hypothetical reports are very intriguing, but the synchronization that the geomagnetic escalation and the mass extinction of species occurred at the same time has not clarified the relationship between the two events, and eventually, the increase in UV radiation may not have led to the mass extinction (Suganuma, 2020). Thus, it seems likely that many researchers are skeptical about this hypothesis (Suganuma, 2020). In addition, it is speculated that the frequency of GMF reversals occurred much more frequently than that of catastrophic extinctions. There were certainly GMF reversals that occurred almost at the same period as the catastrophic extinctions, but in most other cases it had nothing to do with the catastrophic extinctions. It is supposed that there is no statistical correlation between GMF reversals and extinctions.

Regarding the direct impact of the GMF on life and humans on Earth, it is impossible to predict exactly what will happen because humans did not exist at the time of the last complete GMF reversal. Whether or not the GMF reduction is directly related to human extinction, it is predicted that humans do not have to panic about the GMF reversals and the GMF reduction. This is because there is no evidence that disasters have occurred even after the weakening of the GMF several times in the past, and if there are disasters, it may affect the electric power system or there may be an increase in patients with some diseases related to reduce the GMF intensity. It implies that there is no need to think directly about the extinction of humans.

It has been considered that the existence of the GMF is indispensable for the origin and evolution of life. The GMF plays a crucial role in shielding harmful charged particles and GCRs. It is predicted that the magnetic pole may be moved and exchanged, and the GMF reversal transition can occur unexpectedly in a short time scale. This GMF reversal transition weakens the GMF and might increase the frequency of mutations caused by the stronger solar wind and GCRs. At the individual level, mutations can cause carcinogenesis, dysfunction, etc., and can be detrimental to survival. However, at the population level, mutations can create individuals with new morphologies and functions, which may be a driving force for evolution. In the future, by improving the accumulation of paleomagnetic records and the accuracy for their analysis technology, and the accuracy for the determination of the period of the extinction and evolution of life on Earth, the effects of not only geomagnetic excursions but also GMF reversals, which are more dynamic GMF fluctuations, on the extinction and evolution of life on Earth will be clarified.

References

Agliassa, C., Narayana, R., Bertea, C.M., Rodgers, C., and Maffei, M.E. 2018a. Reduction of the geo-magnetic field delays *Arabidopsis thaliana* flowering time through downregulation of flowering-related genes. *Bioelectromagnetics* **39**(5):361–374.

Agliassa, C., Narayana, R., Christie, J.M., and Maffei, M.E. 2018b. Geomagnetic field impacts on crypto-chrome and phytochrome signaling. *J Photochem Photobiol B Biol* **185**:32–40.

Ahmad, M., and Cashmore, A.R. 1993. HY4 gene of *A. thaliana* encodes a protein with characteristics of a blue-light photoreceptor. *Nature* **366**(6451):162–166.

Ahmad, M., Galland, P., Ritz, T., Wiltschko, R., and Wiltschko, W. 2007. Magnetic intensity affects cryptochrome-dependent responses in *Arabidopsis thaliana*. *Planta* **225**:615–624.

Aitta, A. 2006. Iron melting curve with a tricritical point. *J Statis Mech* **12**:12015–12030.

Aitta, A. 2008. Light matter in the core of the Earth: its identity, quantity and temperature using tricriti-cal phenomena. arXiv:0807.0187.

Alabdulgader, A., McCraty, R., Atkinson, M., Dobyns, Y., Vainoras, A., Ragulskis, M., and Stolc, V. 2018. Long-term study of heart rate variability responses to changes in the solar and geomagnetic environment. *Sci Rep* **8**:2663.

Alboussière, T., Deguen, R., and Melzani, M. 2010. Melting-induced stratification above the Earth's inner core due to convective translation. *Nature* **466**(7307):744–747.

Alroy, J. 2010. The shifting balance of diversity among major marine animal groups. *Science* **329**:1191–1194.

Amos, J. 2016. Iron 'jet stream' detected in Earth's outer core. BBC. https://www.bbc.com/news/science-environment-38372342.

Anand, A., Nagarajan, S., Verma, A., Joshi, D., Pathak, P., and Bhardwaj, J. 2012. Pre-treatment of seeds with static magnetic field ameliorates soil water stress in seedlings of maize (*Zea mays* L.). *Indian J Biochem Biophys* **49**(1):63–70.

Anderson, D.L. 1989. Chapter 4: The lower mantle and core. In: *Theory of the Earth*, Hutton, J. (Ed.) Blackwell Scientific Publications, Boston, MA, pp. 63–78.

Anderson, J.M., Clegg, T.M., Véras, L.V.M.V.Q., and Holland, K.N. 2017. Insight into shark magnetic field perception from empirical observations. *Sci Rep* **7**(1):11042.

Arnason, B.T., Hart, L.A., and O'Connell-Rodwell, C.E. 2002. The properties of geophysical fields and their effects on elephants and other animals. *J Comp Psychol* **116**:123–132.

Aubert, J. 2018. Geomagnetic acceleration and rapid hydromagnetic wave dynamics in advanced numer-ical simulations of the geodynamo. *Geophys J Int* **214**:531–547.

Aubert, J., Tarduno, J.A., and Johnson, C.L. 2010. Observations and models of the longterm evolution of Earth's magnetic field. *Space Sci Rev* **155**:337–370.

Azcárate, T., and Mendoza, B. 2017. Influence of geomagnetic activity and atmospheric pressure in hypertensive adults. *Int J Biometeorol* **61**(9):1585–1592.

Bae, J.E., Bang, S., Min, S., Lee, S.H., Kwon, S.H., Lee, Y., Lee, Y.H., Chung, J., and Chae, K.S. 2016. Positive geotactic behaviors induced by geomagnetic field in *Drosophila*. *Mol Brain* **9**(1):55.

Baek, S., Quan, X., Kim, S., Lengner, C., Park, J.K., and Kim, J. 2014. Electromagnetic fields mediate efficient cell reprogramming into a pluripotent state. *ACS Nano* **8**(10):10125–10138.

Baek, S., Choi, H., Park, H., Cho, B., Kim, S., and Kim, J. 2019. Effects of a hypomagnetic field on DNA methylation during the differentiation of embryonic stem cells. *Sci Rep* **9**:1333.

Baez, J. 2006. Extinction. https://math.ucr.edu/home/baez/extinction/.

Balser, M., and Wagner, C.A. 1960. Observations of Earth-ionosphere cavity resonances. *Nature* **188**:638–641.

Barr, R., Jones, D.L., and Rodger, C.J. 2000. ELF and VLF radio waves. *J Atmos Sol Terr Phys* **62**(17–18):1689–1718.

Bartels, J. 1949. The standardized index K_s and the planetary index K_p. *IATME Bulletin* **12b**:97.

Bazalova, O., Kvicalova, M., Valkova, T., Slaby, P., Bartos, P., Netusil, R., Tomanova, K., Braeunig, P., Lee, H.J., Sauman, I., Damulewicz, M., Provaznik, J., Pokorny, R., Dolezel, D., and Vácha, M. 2016. Cryptochrome 2 mediates directional magnetoreception in cockroaches. *Proc Natl Acad Sci U S A* **113**(6):1660–1665.

Beason, R.C., and Semm, P. 1987. Magnetic responses of the trigeminal nerve system of the bobolink (*Dolichonyx oryzivorus*). *Neurosci Lett* **80**:229–234.

Beck, W., and Wiltschko, W. 1988. Magnetic factors control the migratory direction of Pied Flycatchers (*Ficedula hypoleuca* PALLAS). In: *Acta : XIX Congressus Internationalis Ornithologici, Ottawa 1986*, Ouellet, H. (Ed.) University of Ottawa Press, Ottawa, Canada. Pp. 1955–1962.

Bedore, C.N., and Kajiura, S.M. 2013. Bioelectric fields of marine organisms: voltage and frequency contributions to detectability by electroreceptive predators. *Physiol Biochem Zool* **86**(3):298–311.

Begall, S., Cerveny, J., Neef, J., Vojtech, O., and Burda, H. 2008. Magnetic alignment in grazing and resting cattle and deer. *Proc Natl Acad Sci U S A* **105**(36):13451–13455.

Begall, S., Malkemper, E.P., Červený, J., Němec, P., and Burda, H. 2013. Magnetic alignment in mammals and other animals. *Mamm Biol* **78**:10–20.

Bellini, S. 1963a. *Su di un particolare comportamento di batteri d'acqua dolce (On a Unique Behavior of Freshwater Bacteria)*. Institute of Microbiology, University of Pavia, Pavia, Italy, pp. 1–7.

Bellini, S. 1963b. *Ulteriori studi sui "batteri magnetosensibili" (Further Studies on Magnetosensitive Bacteria)*. Institute of Microbiology, University of Pavia, Pavia, Italy, pp. 1–15.

Bellono, N.W., Leitch, D.B., and Julius, D. 2018. Molecular tuning of electroreception in sharks and skates. *Nature* **558** (7708):122–126.

Beloglazov, M.I., Pershakov, L.A., and Beloglazova, G.P. 2006. About the change of atmospherics intensity in ELF-range during Forbush-decreases of galactic cosmic rays. In: *Proceedings of XXIX Annual Seminar "Physics of Auroral Phenomena"*. Kola Science Centre, Russian Academy of Science, Apatity, Russia, pp. 277–280.

Belov, D.R., Kanunikov, I.E., and Kiselev, B.V. 1998. Dependence of human EEG synchronization on the geomagnetic activity on the day of experiment. *Ross Fiziol Zh Im I M Sechenova* **84**:761–774.

Benediktová, K., Adámková, J., Svoboda, J., Painter, M.S., Bartoš, L., Nováková, P., Vynikalová, L., Hart, V., Phillips, J. and Burda, H. 2020. Magnetic alignment enhances homing efficiency of hunting dogs. *eLife* **9**:e55080.

Bengtson, S. 2002. Origins and early evolution of predation. In: *The Fossil Record of Predation*, Kowalewski, M., and Kelley, P.H. (Eds.) The Paleontological Society Papers, Vol 8. The Paleontological Society, Washington, DC, pp. 289–317.

Bergiannaki, J.D., Paparrigopoulos, T.J., and Stefanis, C.N. 1996. Seasonal pattern of melatonin excretion in humans: Relationship to day length variation rate and geomagnetic field fluctuations. *Experientia* **52**(3):253–258.

Bergman, N, M., Timothy, M.L., and Andrew, J.W. 2004. COPSE: A new model of biogeochemical cycling over Phanerozoic time. *Am J Sci* **304**:397–437.

Berk, M., Dodd, S., and Henry, M. 2006. Do ambient electromagnetic fields affect behaviour? A demonstration of the relationship between geomagnetic storm activity and suicide. *Bioelectromagnetics* **27**(2):151–155.

Berner, R.A. 2009. Phanerozoic atmospheric oxygen: new results using the GEOCARBSULF model. *Am J Sci* **309**:603–606.

Berner, R.A., and Kothavala, Z. 2001. GEOCARB III: A revised model of atmospheric CO_2 over phanerozoic time. *Am J Sci* **301**:182–204.

Bertea, C.M., Narayana, R., Agliassa, C., Rodgers, C.T., and Maffei, M.E. 2015. Geomagnetic field (Gmf) and plant evolution: Investigating the effects of Gmf reversal on *Arabidopsis thaliana* development and gene expression. *J Vis Exp* **105**:53286.

Bertini, A., Toti, F., Marino, M., and Ciaranfi, N. 2015. Vegetation and climate across the early-middle pleistocene transition at Montalbano Jonico, southern Italy. *Quat Int* **383**:74–88.

Biggin, A.J., Steinberger, B., Aubert, J., Suttie, N., Holme, R., Torsvik, T.H., van der Meer, D.G., and van Hinsbergen, D.J.J. 2012. Possible links between long-term geomagnetic variations and whole-mantle convection processes. *Nat Geosci* **5**:526–533.

Binhi, V.N., and Prato, F.S. 2017. Biological effects of the hypomagnetic field: An analytical review of experiments and theories. *PLoS One* **12**(6):e0179340.

Blakemore, R.P. 1975. Magnetotactic bacteria. *Science* **190**(4212):377–379.

Blakemore, R.P. 1982. Magnetotactic bacteria. *Annu Rev Microbiol* **36**:217–238.

Blakemore, R.P., Frankel, R.B., and Kalmijn, A.J. 1980. South-seeking magnetotactic bacteria in the Southern Hemisphere. *Nature* **286**(5771):384–385.

Bliokh, P.V., Nicholaenko, A.P., and Filtippov, Y.F. 1980. Geophysical investigations connected with Schumann resonances. In: *Schumann Resonances in the Earth-Ionosphere Cavity*, Jones, D.L. (Ed.) Peter Peregrinus, Stevenage, UK, pp. 139–155.

Bochert, R., and Zettler, M.L. 2004. Long-term exposure of several marine benthic animals to static magnetic fields. *Bioelectromagnetics* **25**(7):498–502.

Boles, L.C., and Lohmann, K.J. 2003. True navigation and magnetic maps in spiny lobsters. *Nature* **421**(6918):60–63.

Bolte, P., Bleibaum, F., Einwich, A., Günther, A., Liedvogel, M., Heyers, D., Depping, A., Wöhlbrand, L., Rabus, R., Janssen-Bienhold, U., and Mouritsen, H. 2016. Localisation of the putative magnetoreceptive protein Cryptochrome 1b in the retinae of migratory birds and homing pigeons. *PLoS One* **11**(3):e0147819.

Bond, D.P.G., and Grasby, S.E. 2020. Late Ordovician mass extinction caused by volcanism, warming, and anoxia, not cooling and glaciation. *Geology* **48**(8):777–781.

Bonhommet, N., and Babkine, J. 1967. Sur la presence d'aimantation inversees dans la Chaine des Puys. *C R Acad Sci Paris* **264B**:92–94.

Borovsky, J.E., and Valdivia, J.A. 2018. The Earth's magnetosphere: A systems science overview and assessment. *Surv Geophys* **39**:817–859.

Bradlaugh, A., Munro, A., Jones, A.R., and Baines, R. 2021. Exploiting the fruitfly, *Drosophila melanogaster* to identify the molecular basis of cryptochrome-dependent magnetosensitivity. *Quantum Rep* **3**(1):127–136.

Breus, T.K., Binhi, V.N., and Petrukovich, A.A. 2016. Magnetic factor of the solar terrestrial relations and its impact on the human body: Physical problems and prospects for research. *Phys Usp* **59**:502–510.

Brown, H.R., and Ilyinsky, O.B. 1978. The ampullae of Lorenzini in the magnetic field. *J Comp Physiol A* **126**(4):333–341.

Brown, M., Korte, M., Holme, R., Wardinski, I., and Gunnarson, S. 2018. Earth's magnetic field is probably not reversing. *Proc Natl Acad Sci U S A* **115**(20):5111–5116.

Brown, M.C., Holme, R., and Bargery, A. 2007. Exploring the influence of the non-dipole field on magnetic records for field reversals and excursions. *Geophys J Int* **168**:541–550.

Brunhes, B. 1905a. Sur la direction de l'aimantation permanente dans une argile de Pontfarein. *C R Acad Sci Paris* **141**:567–568.

Brunhes, B. 1905b. L'inclinaison magnétique en Europe dans le passé et l'argile métamorphique de Pontfarein, commune de Cézens, Cantal. *Rev Haute Auvergne* **7**:398–405.

Brunhes, B. 1906. Recherches sur la direction de l'aimantation des roches volcaniques. *J Phys Théor Appl 4ème série* **5**:705–724.

Buffett, B.A. 2010. Tidal dissipation and the strength of the Earth's internal magnetic field. *Nature* **468**(7326):952–954.

Bühler, W.K. 1987. *Gauss: A Biographical Study.* Springer-Verlag, Berlin, Germany.

Bullard, E.C., and Gellman, H. 1954. Homogeneous dynamos and terrestrial magnetism. *Phil Trans Roy Soc Lond A* **247**:213–278.

Burch, J.B., Reif, J.S., and Yost, M.G. 1999. Geomagnetic disturbances are associated with reduced nocturnal excretion of a melatonin metabolite in humans. *Neurosci Lett* **266**(3):209–212.

Burch, J.B., Reif, J.S., and Yost, M.G. 2008. Geomagnetic activity and human melatonin metabolite excretion. *Neurosci Lett* **438**(1):76–79.

Burda, H., Begall, S., Cerveny, J., Neef, J., and Němec, P. 2009. Extremely low-frequency electromagnetic fields disrupt magnetic alignment of ruminants. *Proc Natl Acad Sci U S A* **106**(14):5708–5713.

Burda, H., Marhold, S., Westenberger, T., Wiltschko, R., and Wiltschko, W. 1990. Evidence for magnetic compass orientation in the subterranean rodent *Cryptomys hottentotus* (Bathyergidae). *Experientia* **46**:528–530.

Burek, P., and Wänke, H. 1988. Impacts and glacio-eustasy, plate-tectonic episodes, geomagnetic reversals: A concept to facilitate detection of impact events. *Phys Earth Planet Int* **50**(2):183–194.

Cada, G.F., Bevelhimer, M.S., Riemer, K.P., and Turner, J.W. 2011. Effects on freshwater organisms of magnetic fields associated with hydrokinetic turbines. ORNL/TM 2011/244. Oak Ridge National Laboratory, Oak Ridge, TN.

Cada, G.F., Bevelhimer, M.S., Forster, A.M., Riemer, K.P., and Schweizer, P.E. 2012. Laboratory studies of the effects of static and variable magnetic fields of freshwater fish. ORNL/TM-2012/119. Oak Ridge National Laboratory, Oak Ridge, TN.

Cadiou, H., and McNaughton, P.A. 2010. Avian magnetite-based magnetoreception: A physiologist's perspective. *J R Soc Interface* **7**(Suppl 2):S193–S205.

Cakmak, T., Cakmak, Z.E., Dumlupinar, R., and Tekinay, T. 2012. Analysis of apoplastic and symplastic antioxidant system in shallot leaves: Impacts of weak static electric and magnetic field. *J Plant Physiol* **169**(11):1066–1073.

Cane, H.V. 2000. Coronal mass ejections and Forbush decreases. *Space Sci Rev* **93**:55–77.

Cannon, P.S., and Rycroft, M.J. 1982. Schumann resonance frequency variations during sudden ionospheric disturbances. *J Atmos Terr Phys* **44**:201–206.

Capraro, L., Macrì, P., Scarponi, D., and Rio, D. 2015. The lower to Middle Pleistocene Valle di Manche section (Calabria, Southern Italy): State of the art and current advances. *Quat Int* **383**:36–46.

Capraro, L., Ferretti, P., Macrì, P., Scarponi, D., Tateo, F., Fornaciari, E., Bellini, G., and Dalan, G. 2017. The Valle di Manche section (Calabria, Southern Italy): A high-resolution record of the Early-Middle Pleistocene transition (MIS 21-MIS 19) in the Central Mediterranean. *Quat Sci Rev* **165**:31–48.

Carrubba, S., Frilot, C. 2nd., Chesson, A.L., Jr., and Marino, A.A. 2007. Evidence of a nonlinear human magnetic sense. *Neuroscience* **144**(1):356–367.

Carter, C.S., Huang, S.C., Searby, C.C., Cassaidy, B., Miller, M.J., Grzesik, W.J., Piorczynski, T.B., Pak, T.K., Walsh, S.A., Acevedo, M., Zhang, Q., Mapuskar, K.A., Milne, G.L., Hinton, A.O., Guo, D., Weiss, R., Bradberry, K., Taylor, E.B., Rauckhorst, A.J., Dick, D.W., Akurathi, V., Falls-Hubert, K.C., Wagner, B.A., Carter, W.A., Wang, K., Norris, A.W., Rahmouni, K., Buettner, G.R., Hansen, J.M., Spitz, D.R., Abel, E.D., and Sheffield, V.C. 2020. Exposure to static magnetic and electric fields treats type 2 diabetes. *Cell Metab* **32**(6):1076.

Cashmore, A.R. 2003. Cryptochromes: Enabling plants and animals to determine circadian time. *Cell* **114**(5):537–543.

Caspar, K.R., Moldenhauer, K., Moritz, R.E., Němec, P., Malkemper, E.P., and Begall, S. 2020. Eyes are essential for magnetoreception in a mammal. *J R Soc Interface* **17**(170):20200513.

Caswell, J.M., Carniello, T.N., and Murugan, N.J. 2016. Annual incidence of mortality related to hypertensive disease in Canada and associations with heliophysical parameters. *Int J Biometeorol* **60**(1):9–20.

Červený, J., Begall, S., Koubek, P., Nováková, P., and Burda, H. 2011. Directional preference may enhance hunting accuracy in foraging foxes. *Biol Lett* **7**:355–357.

Červený, J., Burda, H., Ježek, M., Kušta, T., Husinec, V., Nováková, P., Hart, V., Hartova, V., Begall, S., and Malkemper, E.P. 2017. Magnetic alignment in warthogs *Phacochoerus africanus* and wild boars *Sus scrofa*. *Mamm Rev* **47**(1):1–5.

Chae, K.S., Oh, I.T., Lee, S.H., and Kim, S.C. 2019. Blue light-dependent human magnetoreception in geomagnetic food orientation. *PLoS One* **14**(2):e0211826.

Channell, J.E.T. 1999. Geomagnetic paleointensity and directional secular variation at Ocean Drilling Program (ODP) Site 984 (Bjorn Drift) since 500 ka: Comparisons with ODP site 983 (Gardar Drift). *J Geophys Res Solid Earth* **104**:22937–22951.

Channell, J.E.T. and Kleiven, H.F. 2000. Geomagnetic palaeointensities and astrochronological ages for the Matuyama-Brunhes boundary and the boundaries of the Jaramillo Subchron: Palaeomagnetic and oxygen isotope records from ODP Site 983. *Philos Trans R Soc London Ser A* **358**:1027–1047.

Channell, J.E.T., and Vigliotti, L. 2019. The role of geomagnetic field intensity in late quaternary evolution of humans and large mammals. *Rev Geophys* **57**(3):709–738.

Channell, J.E.T., Hodell, D.A., McManus, J., and Lehman, B. 1998. Orbital modulation of the Earth's magnetic field intensity. *Nature* **394**(6692):464–468.

Chapman, J.W., Reynolds, D.R., and Wilson, K. 2015. Long-range seasonal migration in insects: Mechanisms, evolutionary drivers and ecological consequences. *Ecol Lett* **18**:287–302.

Chernetsov, N., Kishkinev, D., and Mouritsen, H. 2008. A long-distance avian migrant compensates for longitudinal displacement. *Curr Biol* **18**:188–190.

Chernetsov, N., Pakhomov, A., Kobylkov, D., Kishkinev, D., Holland, R.A., and Mouritsen, H. 2017. Migratory Eurasian reed warblers can use magnetic declination to solve the longitude problem. *Curr Biol* **27**:2647–2651.

Chernetsov, N., Pakhomov, A., Davydov, A., Cellarius, F., and Mouritsen, H. 2020. No evidence for the use of magnetic declination for migratory navigation in two songbird species. *PLoS One* **15**(4):e0232136.

Chernouss, S., Vinogradov, A., and Vlassova, E. 2001. Geophysical hazard for human health in the circumpolar auroral belt: Evidence of a relationship between heart rate variation and electromagnetic disturbances. *Nat Hazards* **23**(2–3):121–135.

Cherry, R. 2002. Schumann resonances, a plausible biophysical mechanism for the human heath effects of solar/geomagnetic activity. *Nat Hazards* **26**:279–331.

Chizhevsky, A.L. 1976. *The Terrestrial Echo of Solar Storms*. Mysl, Moscow, Russia. p. 366.

Choleris, E., Del Seppia, C., Thomas, A.W., Lushi, P., Ghione, G., Moran, G.R., and Prato, F.S. 2002. Shielding, but not zeroing of the ambient magnetic field reduces stress-induced analgesia in mice. *Proc Biol Sci* **269**(1487):193–201.

Chou, Y.M., Jiang, X., Liu, Q., Hu, H.M., Wu, C.C., Liu, J., Jiang Z.X., Lee, T.Q., Wang, C.C., Song, Y.F., Chiang, C.C., Tan, L., Lone, M.A., Pan, Y., Zhu, R., He, Y., Chou, Y.C., Tan, A.H., Roberts, A.P., Zhao, X., and Shen, C.C. 2018. Multidecadally resolved polarity oscillations during a geomagnetic excursion. *Proc Natl Acad Sci U S A* **115**(36):8913–8918.

Clague, J., and the INQUA Executive Committee. 2006. Open letter by INQUA executive committee. *Quart Perspect* **16**(1):1–2.

Clites, B.L, and Pierce J.T. 2017. Identifying cellular and molecular mechanisms for magnetosensation. *Annu Rev Neurosci* **40**:231–250.

Cockell, C.S., and Blaustein, A.R., 2001. Evolution and ultraviolet radiation. In: *Ecosystems*, Cockell, C.S., and Blaustein, A.R. (Eds.) Springer, New York. p. 226.

Cole, F.E., and Graf, E.R. 1974. Precambrian ELF and abiogenesis. In: *ELF and VLF Electromagnetic Field Effects*, Persinger, M.A. (Ed.) Springer US, Boston, MA.

Cooper, A., Turney, C.S.M., Palmer, J., Hogg, A., McGlone, M., Wilmshurst, J., Lorrey, A.M., Heaton, T.J., Russell, J.M., McCracken, K., Anet, J.G., Rozanov, E., Friedel, M., Suter, I., Peter, T., Muscheler, R., Adolphi, F., Dosseto, A., Faith, J.T., Fenwick, P., Fogwill, C.J., Hughen, K., Lipson, M., Liu, J., Nowaczyk, N., Rainsley, E., Bronk Ramsey, C., Sebastianelli, P., Souilmi, Y., Stevenson, J., Thomas, Z., Tobler, R., and Zech, R. 2021. A global environmental crisis 42,000 years ago. *Science* **371**(6531):811–818.

Cornélissen, G.A., Halberg, F., Breus, T.K., Syutkina, E.V., Baevsky, R., Weydahal, A., Watanabe, Y., Otsuka, K., Siegelova, J., Fiser, B., and Bakken, E.E. 2002. Non-photic solar associations of heart rate variability and myocardial infarction. *J Atmos Sol Terr Phys* **64**(5):707–720.

Cornell, R.M., and Schwertmann, U. 2003. *The Iron Oxides: Structure, Properties, Reactions, Occurrences and Uses, 2nd Edition.* Wiley-VCH, Weinheim, Germany p. 32.

Cox, A.V., Doell, R.R., and Dalrymple, G.B. 1964. Reversals of the Earth's magnetic field. *Science* **144**(3626):1537–1543.

Cranfield, C.G., Dawe, A., Karloukovski, V., Dunin-Borkowski, R.E., de Pomerai, D., and Dobson, J. 2004. Biogenic magnetite in the nematode *Caenorhabditis elegans. Proc Biol Sci* **271**(Suppl 6):S436–S439.

Crooker, N., Feynman, J., and Gosling, J. 1977. On the high correlation between long-term averages of solar wind speed and geomagnetic activity. *J Geophys Res* **82**:1933–1937.

Datta, S., and Das, S. 2020. Study of thunderstorm induced ionospheric irregularity and plasma bubbles using NavIC data. *XXXIIIrd General Assembly and Scientific Symposium of the International Union of Radio Science (URSI GASS 2020)*, 29 August–5 September 2020, Rome, Italy.

De Michelis, P., Tozzi, R., and Consolini, G. 2017. Statistical analysis of geomagnetic field intensity differences between ASM and VFM instruments onboard Swarm constellation. *Earth, Planets Space* **69**:24.

De Santis, A., Tozzi, R., and Gaya-Pique, L.R. 2004. Information content and Kentropy of the present geomagnetic field. *Earth Planet Sci Lett* **218**(3–4):269–275.

De Souza, A., Garcia, D., Sueiro, L., Gilart, F., Porras, E., and Licea, L. 2006. Presowing magnetic treatments of tomato seeds increase the growth and yield of plants. *Bioelectromagnetics* **27**(4):247–257.

Dennis, T.E., Rayner, M.J., and Walker, M.M. 2007. Evidence that pigeons orient to geomagnetic intensity during homing. *Proc Biol Soc* **274**:1153–1158.

Denzau, S., Nießner, C., Rogers, L.J., and Wiltschko, W. 2013a. Ontogenetic development of magnetic compass orientation in domestic chickens (*Gallus gallus*). *J Exp Biol* **216**:3143–3147.

Denzau, S., Nießner, C., Wiltschko, R., and Wiltschko, W. 2013b. Different responses of two strains of chickens to different training procedures for magnetic directions. *Anim Cogn* **16**(3):395–403.

Deutschlander, M.E., Philipps, J.B., and Munro, U. 2012. Age-dependent orientation to magnetically-simulated geographic displacement in migratory Australian silvereyes (*Zosterops l. lateralis*). *Wilson J Ornithol* **124**(3):467–477.

Díaz-Jaimes, P., Bayona- Vásquez, N.J., Escatel-Luna, E., Uribe-Alcocer, M., Pecoraro, C., Adams, D.H., Frazier, B.S., Glenn, T.C., and Babbucci, M. 2020. Population genetic divergence of bonnethead sharks *Sphyrna tiburo* in the western North Atlantic: Implications for conservation. *Aquat Conserv* **31**(1):83–98.

Diebel, C.E., Proksch, R., Green, C.R., Neilson, P., and Walker, M.M. 2000. Magnetite defines a vertebrate magnetoreceptor. *Nature* **406**(6793):299–302.

Direnfeld, L.K. 1983. The genesis of the EEG and its relation to electromagnetic radiation. *J Bioelectricity* **2**:111–121.

Dodson, C.A., Hore, P.J., and Wallace, M.I. 2013. A radical sense of direction: Signalling and mechanism in cryptochrome magnetoreception. *Trends Biochem Sci* **38**:435–446.

Doglioni, C., Pignatti, J., and Coleman, M. 2016. Why did life develop on the surface of the Earth in the Cambrian? *Geosci Front* **7**(6):865–873.

Dommer, D.H., Gazzolo, P.J., Painter, M.S., and Phillips, J.B. 2008. Magnetic compass orientation by larval *Drosophila melanogaster. J Insect Physiol* **54**(4):719–726.

Driscoll, P., and Olson, P. 2011. Superchron cycles driven by variable core heat flow. *Geophys Res Lett* **38**(9):L09304.

DTU Space. 2016. Strange blackouts hits space satellites near equator. https://www.space.dtu.dk/english/news/2016/09/swarm?id=ef98dd63-fe6a-4a7c-a26a-781158d3df50.

Duan, Z., Liu, Q., Li, L., Deng, X., and Liu, J. 2018. Magnetic reversal frequency in the lower Cambrian niutitang formation inferred from Ciye 1 Hole, Hunan Province, South China. *Geophys J Int* **214**:1301–1312.

Dumberry, M., and Mound, J. 2010. Inner core mantle gravitational locking and the super-rotation of the inner core. *Geophys J Int* **181**(2):806–817.

Dunlop, D.J. 1997. *Rock Magnetism*. Cambridge University Press, Cambridge, UK.

Eder, S.H., Cadiou, H., Muhamad, A., McNaughton, P.A., Kirschvink, J.L., and Winklhofer, M. 2012. Magnetic characterization of isolated candidate vertebrate magnetoreceptor cells. *Proc Natl Acad Sci U S A* **109**(30):12022–12027.

Eide, E.A., and Torsvik, T.H. 1996. Paleozoic supercontinental assembly, mantle flushing, and genesis of the Kiaman Superchron. *Earth Planet Sci Lett* **144**(3–4):389–402.

Eitel, M., Gilder, S., Kunzmann, T., and Pohl, J. 2014. Rochechouart impact crater melt breccias record no geomagnetic field reversal. *Earth Planet Sci Lett* **387**:97–106.

Eitel, M., Gilder, S., Spray, J., Thompson, L., and Pohl, J. 2016. A paleomagnetic and rock magnetic study of the Manicouagan impact structure: Implications for crater formation and geodynamo effects. *J Geophys Res Solid Earth* **121**:1–19.

Elsasser, W.M. 1950. The Earth's interior and geomagnetism. *Rev Mod Phys* **22**:1.

Engels, S., Schneider, N.L., Lefeldt, N., Hein, C.M., Zapka, M., Michalik, A., Elbers, D., Kittel, A., Hore, P.J., and Mouritsen, H. 2014. Anthropogenic electromagnetic noise disrupts magnetic compass orientation in a migratory bird. *Nature* **509**(7500):353–356.

Erdmann, W., Kmita, H., Kosicki, J.Z., and Kaczmarek, Ł. 2021. How the geomagnetic field influences life on Earth: An integrated approach to geomagnetobiology. *Orig Life Evol Biosph.* doi: 10.1007/s11084-021-09612-5.

Ertel, S. 1994. Influenza pandemics and sunspots: Easing the controversy. *Naturwissenschaften* **81**(7):308–311.

Erwin, D.H. 2006. *Extinction: How Life on Earth Nearly Ended 250 Million Years Ago*. Princeton University Press, Princeton, NJ.

ESA. 2012. South Atlantic Anomaly. http://www.esa.int/ESA_Multimedia/Images/2012/09/South_Atlantic_Anomaly.

ESA. 2016a. Swarm reveals why satellites lose track. https://www.esa.int/ESA_Multimedia/Images/2016/10/GPS_losses.

ESA. 2016b. Magnetic oceans and electric Earth. https://www.esa.int/ESA_Multimedia/Images/2012/10/Magnetic_field_sources.

Escatel-Luna, E., Adams, D.H., Uribe-Alcocer, M., Islas-Villanueva, V., and Díaz-Jaimes, P. 2015. Population genetic structure of the Bonnethead shark, *Sphyrna tiburo*, from the western north Atlantic Ocean based on mtDNA sequences. *J Hered* **106**:355–365.

European Wind Energy Association. 2009. *Wind Energy - The Facts: A Guide to the Technology, Economics and Future of Wind Power*. Earthscan, London, UK.

Falkenberg, G., Fleissner, G., Schuchardt, K., Kuehbacher, M., Thalau, P., Mouritsen, H., Heyers, D., Wellenreuther, G., and Fleissner, G. 2010. Avian magnetoreception: Elaborate iron mineral containing dendrites in the upper beak seem to be a common feature of birds. *PLoS One* **5**(2):e9231.

Fastovsky, D.E., and Sheehan, P.M. 2005. The extinction of the dinosaurs in North America. *GSA Today* **15**(3):4–10.

Fedele, G., Edwards, M.D., Bhutani, S., Hares, J.M., Murbach, M., Green, E.W., Dissel, S., Hastings, M.H., Rosato, E., and Kyriacou, C.P. 2014a. Genetic analysis of circadian responses to low frequency electromagnetic fields in *Drosophila melanogaster*. *PLoS Genet* **10**(12):e1004804.

Fedele, G., Green, E.W., Rosato, E., and Kyriacou, C.P. 2014b. An electromagnetic field disrupts negative geotaxis in *Drosophila* via a CRY-dependent pathway. *Nat Commun* **5**:4391.

Feigin, V.L., Parmar, P.G., Barker-Collo, S., Bennett, D.A., Anderson, C.S., Thrift, A.G., Stegmayr, B., Rothwell, P.M., Giroud, M., Bejot, Y., Carvil, P., Krishnamurthi, R., Kasabov, N., and International Stroke Incidence Studies Data Pooling Project Collaborators. 2014. Geomagnetic storms can trigger stroke: Evidence from 6 large population-based studies in Europe and Australasia. *Stroke* **45**(6):1639–1645.

Ferk, A., and Leonhardt, R. 2009. The Laschamp geomagnetic field excursion recorded in Icelandic lavas. *Phys Earth Planet Inter* **177**(1–2):19–30.

Fields, A.T., Feldheim, K.A., Gelsleichter, J., Pfoertner, C., and Chapman, D.D. 2016. Population structure and cryptic speciation in bonnethead sharks Sphyrna tiburo in the south-eastern U.S.A. and Caribbean. *J Fish Biol* **89**(5):2219–2233.

Fischer, J.H., Freake, M.J., Borland, S.C., and Phillips, J.B. 2001. Evidence for the use of magnetic map information by an amphibian. *Anim Behav* **62**:1–10.

Fleissner, G., Holtkamp-Rotzler, E., Hanzlik, M., Winklhofer, M., Fleissner, G., Petersen, N., and Wiltschko, W. 2003. Ultrastructural analysis of a putative magnetoreceptor in the beak of homing pigeons. *J Comp Neurol* **458**(4):350–360.

Fleissner, G., Stahl, B., Thalau, P., Falkenberg, G., and Fleissner, G. 2007. A novel concept of Fe-mineral-based magnetoreception: Histological and physicochemical data from the upper beak of homing pigeons. *Naturwissenschaften* **94**(8):631–642.

Foley, L.E., Gegear, R.J., and Reppert, S.M. 2011. Human cryptochrome exhibits light-dependent magnetosensitivity. *Nat Commun* **2**:356.

Formicki, K., Korzelecka-Orkisz, A., and Tandki, A. 2019. Magnetoreception in fish. *J Fish Biol* **95**:73–91.

Frankel, R.B. 2009. The discovery of magnetotactic/magnetosensitive bacteria. *Chin J Oceanol Limnol* **27**(1):1–2.

Frankel, R.B., Blakemore, R.P., and Wolfe, R.S. 1979. Magnetite in freshwater magnetotactic bacteria. *Science* **203**(4387):1355–1356.

Fransson, T., Jakobsson, S., Johansson, P., Kullberg, C., Lind, J., and Vallin, A. 2001. Magnetic cues trigger extensive refuelling. *Nature* **414**(6859):35–36.

Freire, R., Munro, U.H., Rogers, L.J., Wiltschko, R., and Wiltschko, W. 2005. Chickens orient using a magnetic compass. *Curr Biol* **15**:R620–R621.

Freire, R., Munro, U., Rogers, L.J., Sagasser, S., Wiltschko, R., and Wiltschko, W. 2008. Different responses in two strains of chickens (*Gallus gallus*) in a magnetic orientation test. *Anim Cogn* **11**(3):547–552.

French, B. 1998. Traces of Catastrophe: A Handbook of Shock-Metamorphic Effects in Terrestrial Meteorite Impact Structures, LPI Contribution. No. 954. Lunar and Planetary Institute, Houston, TX, p. 120.

Frost, D.A., Lasbleis, M., Chandler, B., and Romanowicz, B. 2021. Dynamic history of the inner core constrained by seismic anisotropy. *Nat Geosci* **14**:531–535.

Furuyama, K., Nagao, K., Kasatani, K., and Mitsui, S. 1993. K-Ar ages of the Kannabe Volcano Group and the adjacent basaltic monogenetic volcanoes, east San-in district. *Earth Sci (Chikyu Kagaku)* **47**(5):377–390.

Gallet, Y., Pavlov, V., and Courtillot, V. 2003. Magnetic reversal frequency and apparent polar wander of the Siberian platform in the earliest Palaeozoic, inferred from the Khorbusuonka river section (northeastern Siberia). *Geophys J Int* **154**:829–840.

Gallet, Y., Pavlov, V., and Korovnikov, I. 2019. Extreme geomagnetic reversal frequency during the Middle Cambrian as revealed by the magnetostratigraphy of the Khorbusuonka section (northeastern Siberia). *Earth Planet Sci Lett* **528**:115823.

Garcia, R., Sosner, P., Laude, D., Hadjadj, S., Herpin, D., and Ragot, S. 2014. Spontaneous baroreflex sensitivity measured early after acute myocardial infarction is an independent predictor of cardiovascular mortality: Results from a 12-year follow-up study. *Int J Cardiol* **177**(1):120–122.

Garreaud, R.D., Molina, A., and Farias, M. 2010. Andean uplift, ocean cooling and Atacama hyperaridity: A climate modeling perspective. *Earth Planet Sci Lett* **292**(1):39–50.

Gegear, R.J., Casselman, A., Waddell, S., and Reppert, S.M. 2008. Cryptochrome mediates light-dependent magnetosensitivity in *Drosophila*. *Nature* **454**(7207):1014–1018.

Gegear, R.J., Foley, L.E., Casselman, A., and Reppert, S.M. 2010. Animal cryptochromes mediate magnetoreception by an unconventional photochemical mechanism. *Nature* **463**(7282):804–807.

Ghione, S., Mezzasalma, L., Del Seppia, C., and Papi, F. 1998. Do geomagnetic disturbances of solar origin affect arterial blood pressure? *J Hum Hypertens* **12**:749–754.

Gilbert, W. 1600. *De Magnete, Magnetisque Corporoibus, et de Magno Magnete Tellure: Physiologia noua, Plurimis & Argumentis, & Experimentis Demonstrata (in Latin).* Peter Short, London, UK.

Gilbert, W. 1893. *On the Loadstone and Magnetic Bodies, and on That Great Magnet the Earth: A New Physiology, Demonstrated with Many Arguments and Experiments.* Translated by Mottelay, P.F. John Wiley & Sons, New York.

Gilder, S.A., J. Pohl, and M. Eitel. 2018. Magnetic signatures of terrestrial meteorite impact craters: A summary. In: *Magnetic Fields in the Solar System: Planets, Moons and Solar Wind Interactions,* Lühr, H., Wicht, J., Gilder, S.A., and Holschneider, M. (Eds.) Springer International Publishing, Gewerbestrasse, Switzerland, pp. 357–376.

Gill, A.B. 2005. Offshore renewable energy: Ecological implications of generating electricity in the coastal zone. *J Appl Ecol* **42**:605–615.

Glass, B.P. 1990. Tektites and microtektites: Key facts and inferences. *Tectonophysics* **171**(1–4):393–404.

Glatzmaier, G.A., and Olson, P. 2005. Probing the geodynamo. *Sci Am* **292**(4):50–57.

Glatzmaier, G.A., and Roberts, P.H. 1995. A threedimensional self-consistent computer simulation of a geomagnetic field reversal. *Nature* **377**(6546):203–209.

Gmitrov, J. 2005. Geomagnetic disturbance worsen microcirculation impairing arterial baroreflex vascular regulatory mechanism. *Electromagn Biol Med* **24**(1):31–37.

Gmitrov, J., and Gmitrova, A. 2004. Geomagnetic field effect on cardiovascular regulation. *Bioelectromagnetics* **25**(2):92–101.

Gonzales, D.E. 1969. Note on the solar wind-induced drag on comets. *Sol Phys* **9**:205–209.

Gonzalez, C., Gallagher, A.J., and Caballero, S. 2019. Conservation genetics of the bonnethead shark *Sphyrna tiburo* in Bocas del Toro Panama: preliminary evidence of a unique stock. *PLoS One* **14**:e0220737.

Gosling, J.T. 1991. In situ observations of coronal mass ejections in interplanetary space. NASA STI/ Recon Technical Report N 92:17725.

Gould, J.L. 2008a. Animal navigation: The longitude problem. *Curr Biol* **18**(5):R214–R216.

Gould, J.L. 2008b. Animal navigation: The evolution of magnetic orientation. *Curr Biol* **18**(11):R482–R484.

Gould, J.L., Kirschvink, J.L., and Deffeyes, K.S. 1978. Bees have magnetic remanence. *Science* **201**(4360):1026–1028.

Gradstein, F., Ogg, J., and Smith, A. 2004. *Geologic Timescale 2004.* Cambridge University Press, Cambridge, UK. p. 589.

Graham, G. 1724a. An account of observations made of the variation of the horizontal needle at london, in the latter part of the year 1722, and beginning of 1723. *Phil Trans* **33**(383):96–107.

Graham, G. 1724b. Observation of the dipping needle, made at London, in the beginning of the Year 1723. *Phil Trans* **33** (389):332–339.

Gubbins, D. 2008. Earth science: Geomagnetic reversals. *Nature* **452**(7184):165–167.

Günther, A., Einwich, A., Sjulstok, E., Feederle, R., Bolte, P., Koch, K.W., Solov'yov, I.A., and Mouritsen, H. 2018. Double-cone localisation and seasonal expression pattern suggest a role in magnetoreception for European robin cryptochrome 4. *Curr Biol* **28**:1–13.

Gurfinkel, Yu.I., At'kov, O.Y., Vasin, A.L., Breus, T.K., Sasonko, M.L., and Pishchalnikov, R.Y. 2016. Effect of zero magnetic field on cardiovascular system and microcirculation. *Life Sci Space Res (Amst)* **8**:1–7.

Gurfinkel, Y.I., Vasin, A.L., Pishchalnikov, R.Y., Sarimov, R.M., Sasonko, M.L., and Matveeva, T.A. 2018. Geomagnetic storm under laboratory conditions: Randomized experiment. *Int J Biometeorol* **62**(4):501–512.

Gurhan, H., Bruzon, R., Kandala, S., Greenebaum, B., and Barnes, F. 2021. Effects induced by a weak static magnetic field of different intensities on HT-1080 fibrosarcoma cells. *Bioelectromagnetics* **42**(3):212–223.

Gutierrez, M.V., Otsuji, K., Asai, A., Terrazas, R., Ishitsuka, M., Ishitsuka, J., Nakamura, N., Yoshinaga, Y., Morita, S., Ishii, T.T., Ueno, S., Kitai, R., and Shibata, K. 2021. A three-dimensional velocity of an erupting prominence prior to a coronal mass ejection. *Publ Astron Soc Jpn* **73**(2):394–404.

Guyodo, Y., and Valet, J.P. 1999. Global changes in intensity of the Earth's magnetic field during the past 800 kyr. *Nature* **399**(6733):249–252.

Hainsworth, L.B. 1983. The effect of geophysical phenomena on human health. *Speculat Sci Technol* **6**(5):439–444.

Halberg, F., Cornélissen, G., Panksepp, J., Otsuka, K., and Johnson, D. 2005. Chronomics of autism and suicide. *Biomed Pharmacother* **59**(1):S100–S108.

Hale, C.J. 1987. Paleomagnetic data suggest link between the Archean-Proterozoic boundary and inner-core nucleation. *Nature* **329**(6136):233–237.

Hall, S. 2020. Familiar culprit may have caused mysterious mass extinction: A planet heated by giant volcanic eruptions drove the earliest known wipeout of life on Earth. *The New York Times*, 10 June 2020.

Haneda, Y., Okada, M., Suganuma, Y., and Kitamura, T. 2020. A full sequence of the Matuyama–Brunhes geomagnetic reversal in the Chiba composite section, Central Japan. *Prog Earth Planet Sci* **7**:44.

Harada, Y., Taniguchi, M., Namatame, H., and Iida, A. 2001. Magnetic materials in otoliths of bird and fish lagena and their function. *Acta Otolaryngol* **121**(5):590–595.

Hargreaves, J.K. 1992. *The Solar-Terrestrial Environment.* Cambridge University Press, Cambridge, UK. p. 420.

Hart, V., Kušta, T., Němec, P., Bláhová, V., Ježek, M., Nováková, P., Begall, S., Cervený, J., Hanzal, V., Malkemper, E.P., Stípek, K., Vole, C., and Burda, H. 2012. Magnetic alignment in carps: Evidence from the Czech Christmas fish market. *PLoS One* **7**(12):e51100.

Hart, V., Nováková, P., Malkemper, E.P., Begall, S., Hanzal, V., Ježek, M., Kušta, T., Němcová, V., Adámková, J., Benediktová, K., Červený, J., and Burda, H. 2013. Dogs are sensitive to small variations of the Earth's magnetic field. *Front Zool* **10**(1):80.

Henbest, K.B., Maeda, K., Hore, P.J., Joshi, M., Bacher, A., Bittl, R., Weber, S., Timmel, C.R., and Schleicher, E. 2008. Magnetic-field effect on the photoactivation reaction of *Escherichia coli* DNA photolyase. *Proc Natl Acad Sci U S A* **105**:14395–14399.

Henshaw, I., Fransson, T., Jakobsson, S., Lind, J., Vallin, A., and Kullberg, C. 2008. Food intake and fuel deposition in a migratory bird is affected by multiple as well as single-step changes in the magnetic field. *J Exp Biol* **211**(5):649–653.

Henshaw, I., Fransson, T., Jakobsson, S., and Kullberg, C. 2010. Geomagnetic field affects spring migratory direction in a long distance migrants. *Behav Ecol Sociobiol* **64**:1317–1323.

Hervé, G., Gilder, S., Marion, C., Osinski, G., Pohl, J., Petersen, N., and Sylvester, P. 2015. Paleomagnetic and rock magnetic study of the Mistastin Lake impact structure (Labrador, Canada): Implications for geomagnetic perturbations and shock effects. *Earth Planet Sci Lett* **417**:151–163.

Heyers, D., Manns, M., Luksch, H., Güntürkün, O., and Mouritsen, H. 2007. A visual pathway links brain structures active during magnetic compass orientation in migratory birds. *PLoS One* **2**:e937.

Heyers, D., Zapka, M., Hoffmeister, M., Wild, J.M., and Mouritsen, H. 2010. Magnetic field changes activate the trigeminal brainstem complex in a migratory bird. *Proc Natl Acad Sci U S A* **107**(20):9394–9399.

Hill, M.J., Shaw, J., and Herrero-Bervera, E. 2006. Determining palaeointensity from the Gilbert Gauss Reversal recorded in the Pu'u Heleakala lava section, Wai'anae Volcano, Oahu, Hawaii. *Earth Planet Sci Lett* **245**(1):29–38.

Hiscock, H.G., Worster, S., Kattnig, D.R., Steers, C., Jin Y., Manolopoulos, D.E., Mouritsen, H., and Hore, P.J. 2016. The quantum needle of the avian magnetic compass. *Proc Natl Acad Sci U S A* **113**(17):4634–4639.

Hiscott, J., Kwon, H., and Genin, P. 2001. Hostile takeovers: Viral appropriation of the NF-κB pathway. *J Clin Invest* **107**:143–151.

Hochstoeger, T., Al Said, T., Maestre, D., Walter, F., Vilceanu, A., Pedron, M., Cushion, T.D., Snider, W., Nimpf, S., Nordmann, G.C., Landler, L., Edelman, N., Kruppa, L., Durnberger, G., Mechtler, K., Schuechner, S., Ogris, E., Malkemper, E.P., Weber, S., Schleicher, E., and Keays, D.A. 2020. The biophysical, molecular, and anatomical landscape of pigeon Cry4: A candidate light-based quantal magnetosensor. *Sci Adv* **6**:eabb9110.

Hoffman, P.F., Kaufman, A.J., Halverson, G.P., and Schrag, D.P. 1998. A neoproterozoic snowball Earth. *Science* 281:1342–1346.

Hofmann, C., Feraud, G., and Courtillot, V. 2000.^{40}Ar/^{39}Ar dating of mineral separates and whole rocks from the western Ghats lava pile: Further constraints on duration and age of Deccan traps. *EPSL* 180(1–2):13–27.

Holser, W.T., Schoenlaub, H.P., Attrep, M., Jr., Boeckelmann, K., Klein, P., Margaritz, M., Orth, C.J., Fenninger, A., Jenny, C., Kralik, M., Mauritsch, H., Pak, E., Schramm, J.M., Stattegger, K., and Schmoeller, R. 1989. A unique geochemical record at the Permian/Triassic boundary. *Nature* 337(6202):39–44.

Hope-Simpson, R.E. 1978. Sunspots and flu: A correlation. *Nature* 275(5676):86.

Horvath, J.E. 2003. On gamma-ray bursts and their biological effects: A case for an extrinsic trigger for the Cambrian explosion? *Astrophys Abstr,* arXiv:astro-Ph/0310034v1:1-9.

Howell, B.F. 1990. *An Introduction to Seismological Research, History and Development.* Cambridge University Press, Cambridge, UK.

Hoyle, F., and Wlckramasinghe, N.C. 1990. Sunspots and influenza. *Nature* 343:304.

Hyodo, M., Katoh, S., Kitamura, A., Takasaki, K., Matsushita, H., Kitaba, I., Tanaka, I., Nara, M., Matsuzaki, M., Dettman, D.L., and Okada, M. 2016. High resolution stratigraphy across the early-middle pleistocene boundary from a core of the Kokumoto Formation at Tabuchi, Chiba Prefecture, Japan. *Quat Int* 397:16–26.

Isozaki, Y. 1997. Permo-Triassic boundary superanoxia and stratified superocean: Records from lost deep-sea. *Science* 276:235–238.

Isozaki, Y. 2007a. Plume winter scenario for biospheric catastrophe: the Permo-Triassic boundary case. In: *Superplume: Beyond Plate Tectonics*, Yuen, D., Maruyama, S., Karato, S., and Windley, B.F. (Eds.) Springer, Dordrecht, Netherlands, pp. 409–440.

Isozaki, Y. 2007b. Guadalupian–Lopingian boundary event in mid-Panthalassa: Correlation of accreted deep-sea chert and mid-oceanic atoll carbonates. In: *Proceedings of the XVth International Congress on Carboniferous and Permian Stratigraphy 2003*, Wong, T.E. (Ed.) Royal Netherlands Academy of Arts and Sciences, Special Publication, Amsterdam, Netherlands, pp. 111–124.

Isozaki, Y. 2009. Illawarra reversal: The fingerprint of a superplume that triggered Pangean breakup and the end-Guadalupian (Permian) mass extinction. *Gondwana Res* 15(3–4):421–432.

Isozaki, Y., and Ota, A. 2001. Middle/upper permian (Maokouan/Wuchiapingian) boundary in mid-oceanic paleo-atoll limestone in Kamura and Akasaka, Japan. *Proc Jpn Acad* 77(B):104–109.

Isozaki, Y., Kawahata, H., and Minoshima, K. 2007a. The Capitanian (Permian) Kamura cooling event: The beginning of the Paleozoic–Mesozoic transition. *Palaeoworld* 16:16–30.

Isozaki, Y., Kawahata, H., and Ota, A. 2007b. A unique carbon isotope record across the Guadalupian-Lopingian (Middle-Upper Permian) boundary in mid-oceanic paleo-atoll carbonates: the high-productivity "Kamura event" and its collapse in Panthalassa. *Global Planet Change* 55:21–38.

Jackson, A., and Livermore, P. 2009. On Ohmic heating in the Earth's core I: Nutation constraints. *Geophys J Int* 177:367–382.

Jackson, A,. Jonkers, A.R.T., and Walker, M.R. 2000. Four centuries of geomagnetic secular variation from historical records. *Phil Trans R Soc Lond A* 358:957–990.

Jackson A., Livermore P., and Ierley, G. 2011. On Ohmic heating in the Earth's core II: Poloidal magnetic fields obeying Taylor's constraint. *Phys Earth Planet Inter* 187(3):322–327.

Janoutova, J., Blaha, J., and Jelinek, R. 2000. The absence of geomagnetic field does not influence the development of the chick embryo. *Biologia* 54:151–156.

Jaruševičius, G., Rugelis, T., McCraty, R., Landauskas, M., Berškienė, K., and Vainoras, A. 2018. Correlation between changes in local Earth's magnetic field and cases of acute myocardial infarction. *Int J Environ Res Public Health* 15(3):399.

Jia, B., Xie, L., Zheng, Q., Yang, P.F., Zhang, W.J., Ding, C., Qian, A.R., and Shang, P. 2014. A hypo-magnetic field aggravates bone loss induced by hindlimb unloading in rat femurs. *PLoS One* 9(8):e105604.

Johnsen, S., and Lohmann, K.J. 2005. The physics and neurobiology of magnetoreception. *Nat Rev Neurosci* **6**:703–712.

Johnsen, S., and Lohmann, K.J. 2008. Magnetoreception in animals. *Phys Today* **61**:29–35.

Jones, C.A. 2015. Thermal and compositional convection in the outer core. In: *Treatise on Geophysics: Core Dynamics*, 2nd Edition, Vol. 8, Schubert, G. (Ed.) Elsevier, Amsterdam, Netherlands. p. 131–185.

Juutilainen, J., Herrala, M., Luukkonen, J., Naarala, J., and Hore, P.J. 2018. Magnetocarcinogenesis: Is there a mechanism for carcinogenic effects of weak magnetic fields? *Proc Biol Sci* **285**(1879):20180590.

Kageyama, A., and Sato, T. 1997. Generation mechanism of a dipole field by a magnetohydrodynamic dynamo. *Phys Rev E* **55**:4617.

Kageyama, A., Sato, T., Watanabe, K., Horiuchi, R., Hayashi, T., Todo, Y., Watanabe, T.H., and Takamaru, H. 1995. Computer simulation of a magnetohydrodynamic dynamo. II. *Phys Plasmas* **2**(5):1421–1431.

Kalmijn, A.J. 1981. Biophysics of geomagnetic field detection. *IEEE Trans Magn* **17**:1113–1124.

Kalmijn, A.J. 1982. Electric and magnetic field detection in elasmobranch fishes. *Science* **218** (4575):916–918.

Kalmijn, A.J., and Blakemore, R.P. 1978. The magnetic behavior of mud bacteria. In: *Animal Migration, Navigation and Homing*, Schmidt-Koenig, K. and Keeton, W.T. (Eds.) Springer Verlag, Berlin, Germany, pp. 354–355.

Kantserova, N.P., Krylov, V.V., Lysenko, L.A., Ushakova, N.V., and Nemova, N.N. 2017. Effects of hypomagnetic conditions and reversed geomagnetic field on calcium-dependent proteases of invertebrates and fish. *Izv Atmos Ocean Phys* **53**(7):719–723.

Kay, R.W. 1994. Geomagnetic storms: Association with incidence of depression as measured by hospital admission. *Br J Psychiatry* **164**:403–409.

Kay, R.W. 2004. Schizophrenia and season of birth: Relationship to geomagnetic storms. *Schiz Res* **66**:7–20.

Kazaoka, O., Suganuma, Y., Okada, M., Kameo, K., Head, M.J., Yoshida, T., Sugaya, N., Kameyama, S., Ogitsu, I., Nirei, H., Aida, N., and Kumai, H. 2015. Stratigraphy of the Kazusa Group, Chiba Peninsula, Central Japan: An expanded and highly-resolved marine sedimentary record from the Lower and Middle Pleistocene. *Quat Int* **383**:116–135.

Keller, B.A., Putman, N.F., Grubbs, R.D., Portnoy, D.S., and Murphy, T.P. 2021. Map-like use of Earth's magnetic field in sharks. *Curr Biol* **S0960-9822**(21):00476.

Kirkby, J. 2007. Cosmic Rays and Climate. *Surv Geophys* **28**(5–6):333–375.

Kirschvink, J.L. 1980. South-seeking magnetic bacteria. *J Exp Biol* **86**:345–347.

Kirschvink, J.L. 1992. Late proterozoic low-latitude global glaciation: The snowball Earth. In: *The Proterozoic Biosphere: A Multidisciplinary Study*, Schopf, J.W., and Klein, C. (Eds.) Cambridge University Press, Cambridge, UK, pp. 51–52.

Kirschvink, J.L., and Gould, J.L. 1981. Biogenic magnetite as a basis for magnetic field detection in animals. *BioSystems* **13**(3):181–201.

Kirschvink, J.L., and Lowenstam, H.A. 1979. Mineralization and magnetization of chiton teeth: Paleomagnetic, sedimentologic, and biologic implications of organic magnetite. *Earth Planet Sci Lett* **44**:193–204.

Kirschvink, J.L., and Walker, M.M. 1985. Particle-size considerations for magnetite-based magnetoreceptors. In: *Magnetite Biomineralization and Magnetoreception in Animals: A New Biomagnetism*, Kirschvink, J.L., Jones, D.S., and MacFadden, B.J. (Eds.) Plenum Press, New York, pp. 243–254.

Kirschvink, J.L., Kobayashi-Kirschvink, A., and Woodford, B.J. 1992a. Magnetite biomineralization in the human brain. *Proc Natl Acad Sci U S A* **89**(16):7683–7687.

Kirschvink, J.L., Kobayashi-Kirschvink, A., Diaz-Ricci, J.C., and Kirschvink, S.J. 1992b. Magnetite in human tissues: A mechanism for the biological effects of weak ELF magnetic fields. *Bioelectromagnetics* **13**(Suppl 1):101–113.

Kirschvink, J.L., Ripperdan, R.L., and Evans, D.A. 1997. Evidence for large-scale reorganization of early Cambrian continental masses by inertial interchange true polar wander. *Science* 277(5325):541–545.

Kishkinev, D., Chernetsov, N., Heyers, D., and Mouritsen, H. 2013. Migratory reed warblers need intact trigeminal nerves to correct for a 1,000 km eastward displacement. *PLoS One* 8(6):e65847.

Kishkinev, D., Chernetsow, N., Pakhomov, A., Heyers, D., and Mouritsen, H. 2015. Eurasian reed warblers compensate for virtual magnetic displacement. *Curr Biol* 25:R822–R824.

Knezevic, Z. 2010. Milutin Milankovic and the astronomical theory of climate changes *Europhys News* 41(3):17–20.

Knox, E.G., Armstrong, E., Lancashire, R., Wall, M., and Hayes, R. 1979. Heart attacks and geomagnetic activity. *Nature* 281(5732):564–565.

Koch, S., Gilder, S., Pohl, J., and Trepmann, C. 2012. Geomagnetic field intensity recorded after impact in the Ries meteorite crater. *Germany Geophys J Int* 189:383–390.

Köller, J., Köppel, J., and Peters, W. 2006. *Offshore Wind Energy Research on Environmental Impacts.* Springer-Verlag, Berlin/Heidelberg, Germany.

König, H., and Ankermüller, F. 1960. Über den Einfluß besonders niederfrequenter elektrischer Vorgänge in der Atmosphäre auf den Menschen. *Naturwissenschaften* 47:486–490.

König, H.L., Krüger, A.P., Lang, S., and Sönning, W. 1981. *Biologic Effects of Environmental Electromagnetism.* Springer-Verlag, New York.

Kono, M. 1971. Intensity of the Earth's magnetic field in Pliocene and Pleistocene in relation to the amplitude of mid-ocean ridge magnetic anomalies. *Earth Planet Sci Lett* 11:10–17.

Kono, M., and Nagata, T. 1967. Intensity of the geomagnetic field during a reversed polarity. *Nature* 212(5059):274–275.

Korte, C., Jasper, T., Kozur, H.W., and Veizer, J. 2005. $\delta^{18}O$ and $\delta^{13}C_{carb}$ of Permian brachiopods: A record of seawater evolution and continental glaciation. *Palaeogeogr Palaeoclimatol Palaeoecol* 224:333–351.

Korte, M., and Constable, C. 2011. Improving geomagnetic field reconstructions for 0–3 ka. *Phys Earth Planet Int* 188:247–259.

Korte, M., and Constable, C.G. 2005a. The geomagnetic dipole moment over the last 7000 years: New results from a global model. *Earth Planet Sci Lett* 236:348–358.

Korte, M., and Constable, C.G. 2005b. Continuous geomagnetic field models for the past 7 millennia: 2. CALS7K. *Geochem Geophys Geosyst* 6:Q02H16.

Krylov, V.V. 2017. Biological effects related to geomagnetic activity and possible mechanisms. *Bioelectromagnetics* 38(7):497–510.

Kullberg, C., Henshaw, I., Jakobsson, S., Johansson, P., and Fransson, T. 2007. Fuelling decisions in migratory birds: Geomagnetic cues override the seasonal effect. *Proc R Soc B* 274:2145–2151.

Kuz'mina, V.V., Ushakova, N., and Krylov, V.V. 2015. The effect of magnetic fields on the activity of proteinases and glycosidases in the intestine of the crucian carp *Carassius carassius*. *Biol Bull* 42(1):61–66.

Labrosse, S., Poirer, J., and LeMouel, J. 2001. The age of the inner core. *Earth Planet Sci Lett* 190(3–4):111–123.

Laflamme, M., Darroch, S.A.F., Tweedt, S.A., Peterson, K.J., and Erwin, D.H. 2013. The end of the Ediacaran biota: Biotic replacement or Cheshire cat? *Gondwana Res* 23:558–573.

Laj, C., Guillou, H., and Kissel, C. 2014. Dynamics of the earth magnetic field in the 10–75 kyr period comprising the Laschamp and Mono Lake excursions: New results from the French Chaine des Puys in a global perspective. *Earth Planet Sci Lett* 387:184–197.

Larson, R.L. 1991. Latest pulse of Earth: Evidence for a mid-Cretaceous superplume. *Geology* 19(6):547–550.

Lay, E.H., Shao, X.M., and Carrano, C.S. 2013. Variation in total electron content above large thunderstorms. *Geophys Res Lett* 40(10):1945–1949.

Leão, P., Teixeira, L.C., Cypriano, J., Farina, M., Abreu, F., Bazylinski, D.A., and Lins, U. 2016. North-seeking magnetotactic gammaproteobacteria in the Southern Hemisphere. *Appl Environ Microbiol* 82(18):5595–5602.

Lebedev, S.I., Baranskiy, P.I., Litvinenko, L.G., and Shiyan, L.T. 1977. Barley growth in superweak magnetic field. *Electron Treat Mater* **3**:71–73.

Lee, S.H., Oh, I.T., Lee, M.G., Ju, Y.G., Kim, S.C., and Chae, K.S. 2018. A geomagnetic declination compass for horizontal orientation in fruit flies. *Entomol Res* **48**(1):32–40.

Lefeldt, N., Heyers, D., Schneider, N.L., Engels, S., Elbers, D. and Mouritsen, H. 2014. Magnetic field-driven induction of ZENK in the trigeminal system of pigeons (*Columba livia*). *J R Soc Interface* **11**(100):20140777.

Liedvogel, M., Maeda, K., Henbest, K., Schleicher, E., Simon, T., Timmel, C.R., Hore, P.J., and Mouritsen, H. 2007. Chemical magnetoreception: Bird cryptochrome 1a is excited by blue light and forms long-lived radical-pairs. *PLoS One* **2**(10):e1106.

Lin, C., and Todo, T. 2005. The cryptochromes. *Genome Biol* **6**(5):220.1–220.9.

Lister, J.R., and Buffett, B.A. 1995. The strength and efficiency of thermal and compositional convection in the geodynamo. *Phys Earth Planet Inter* **91**:17–30.

Liu, J., Nowaczyk, N.R., Panovska, S., Korte, M., and Arz, H.W. 2020. The Norwegian-Greenland Sea, the Laschamps, and the Mono Lake excursions recorded in a Black Sea sedimentary sequence spanning from 68.9 to 14.5 ka. *J Geophys Res Solid Earth* **125**:e2019JB019225.

Livermore, P.W., Hollerbach, R., and Finlay, C.C. 2017. An accelerating high-latitude jet in Earth's core. *Nat Geosci* **10**:62–68.

Livermore, P.W., Finlay, C.C., and Bayliff, M. 2020. Recent north magnetic pole acceleration towards Siberia caused by flux lobe elongation. *Nat Geosci* **13**:387–391.

Lockwood, J.A. 1971. Forbush decreases in the cosmic radiation. *Space Sci Rev* **12**:658–715.

Lockwood, M., Chambodut, A., Finch, I.D., Barnard, L.A., Owens, M.J., and Haines, C. 2019. Time-of-day/time-of-year response functions of planetary geomagnetic indices. *J Space Weather Space Clim* **9**:A20.

Lohmann, K.J. 2010. Q&A: Animal behaviour: Magnetic-field perception. *Nature* **464**(7292):1140–1142.

Lohmann, K.J., and Lohmann, C.M.F. 1998. Migratory guidance mechanisms in marine turtles. *J Avian Biol* **29**:585–596.

Lohmann, K.J., Hester, J.T., and Lohmann, C.M.F. 1999. Long-distance navigation in sea turtles. *Ethol Ecol Evol* **11**:1–23.

Lohmann, K.J., Cain, S.D., Dodge, S.A., and Lohmann, C.M.F. 2001. Regional magnetic fields as navigational markers for sea turtles. *Science* **294**(5541):364–366.

Lohmann, K.J., Lohmann, C.M.F., Ehrhart, L.M., Bagley, D.A., and Swing, T. 2004. Geomagnetic map used in sea-turtle navigation. *Nature* **428**(6986):909–910.

Lohmann, K.J., Lohmann, C.M.F., and Putman, N.F. 2007. Magnetic maps in animals: Nature's GPS. *J Exp Biol* **210**(21):3697–3705.

Lohmann, K.J., Lohmann, C.M.F., Brothers, J.R., and Putman, N.F. 2013. *The Biology of Sea Turtles,* Vol. 3, Wyneken, J., Lohmann, K.J., and Musick, J.A. (Eds.) CRC Press, Boca Raton, FL, pp. 59–77.

Lowenstam, H.A. 1962. Magnetite in denticle capping in recent chitons (Polyplacophora). *Geol Soc Am Bull* **73**:435–438.

Luschi, P., Benhamou, S., Girard, C., Ciccioe, S., Roos, D., Sudre, J., and Benvenuti, S. 2007. Marine turtles use geomagnetic cues during open see homing. *Curr Biol* **17**(2):126–133.

Lythgoe, K.H., Rudge, J.F., Neufeld, J.A., and Deuss, A. 2015. The feasibility of thermal and compositional convection in Earth's inner core. *Geophys J Int* **201**:764–782.

Macri, P., Capraro, L., Ferretti, P., and Scarponi, D. 2018. A high-resolution record of the Matuyama–Brunhes transition from the Mediterranean region: The Valle di Manche section (Calabria, Southern Italy). *Phys Earth Planet Inter* **278**:1–15.

Maeda, K., Henbest, K.B., Cintolesi, F., Kuprov, I., Rodgers, C.T., Liddell, P.A., Gust, D., Timmel, C.R., and Hore, P.J. 2008. Chemical compass model of avian magnetoreception. *Nature* **453**(7193):387–390.

Maeda, K., Robinson, A.J., Henbest, K.B., Hogben, H.J., Biskup, T., Ahmad, M., Schleicher, E., Weber, S., Timmel, C.R., and Hore, P.J. 2012. Magnetically sensitive light-induced reactions in cryptochrome are consistent with its proposed role as a magnetoreceptor. *Proc Natl Acad Sci U S A* **109**(13):4774–4779.

Maffei, M.E. 2014. Magnetic field effects on plant growth, development, and evolution. *Front Plant Sci* **5**:445.

Maiorano, P., Bertini, A., Capolongo, D., Eramo, G., Gallicchio, S., Girone, A., Pinto, D., Toti, F., Ventruti, G., Marino, M. 2016. Climate signatures through Marine Isotope Stage 19 in the Montalbano Jonico section (southern Italy): A landesea perspective. *Palaeogeogr Palaeoclimatol Palaeoecol* **461**:341–361.

Malin, S.R., and Srivastava, B.J. 1979. Correlation between heart attacks and magnetic activity. *Nature* **277** (5698):646–648.

Mankinen, E.A., and Dalrymple, G.B. 1979. Revised geomagnetic polarity time scale for the interval 0–5 m.y. B.P. *J Geophys Res Solid Earth* **84**:615–626.

Malkemper, E.P., Eder, S.H., Begall S., Phillips, J.B., Winklhofer, M., Hart, V., and Burda, H. 2015. Magnetoreception in the wood mouse (*Apodemus sylvaticus*): Influence of weak frequency-modulated radio frequency fields. *Sci Rep* **4**:9917.

Mankinen, E.A., and Wentworth, C.M. 2003. Preliminary paleomagnetic results from the Coyote Creek Outdoor Classroom drill hole, Santa Clara Valley, California. *Open-File* Report 03-187. USGS, Reston, VA.

Marino, M., Bertini, A., Ciaranfi, N., Aiello, G., Barra, D., Gallicchio, S., Girone, A., La Perna, R., Lirer, F., Maiorano, P., Petrosino, P., and Toti, F. 2015. Paleoenvironmental and climatostratigraphic insights for Marine Isotope Stage 19 (Pleistocene) at the Montalbano Jonico succession, South Italy. *Quat Int* **383**:104–115.

Marley, R., Giachello, C.N., Scrutton, N.S., Baines, R.A., and Jones, A.R. 2014. Cryptochrome-dependent magnetic field effect on seizure response in *Drosophila* larvae. *Sci Rep* **4**:5799.

Marsh, N.D., and Svenmark, H. 2000. Low cloud properties influenced by cosmic rays. *Phys Rev Lett* **15**(23):5004–5007.

Marshall, C.R. 2006. Explaining the Cambrian "explosion" of animals. *Annu Rev Earth Planet Sci* **34**(1):355–384.

Martin, B. 2009. *Darwin's Lost World: The Hidden History of Animal Life.* Oxford University Press, London, UK.

Martini, S., Begall, S., Findeklee, T., Schmitt, M., Malkemper, E.P., and Burda, H. 2018. Dogs can be trained to find a bar magnet. *PeerJ* **6**:e6117.

Martino, C.F., Portelli, L., McCabe, K., Hernandez, M., and Barnes, F. 2010. Reduction of the Earth's magnetic field inhibits growth rates of model cancer cell lines. *Bioelectromagnetics* **31**:649–655.

Maruyama, S. 1994. Plume tectonics. *J Geol Soc Jpn* **100**:24–49.

Maruyama, S., Liou, J.G., and Seno, T. 1989. Mesozoic and Cenozoic evolution of Asia. In: *The Evolution of Pacific Ocean Margins*, Ben-Avraham, Z. (Ed.) Oxford University Press, New York, pp. 75–99.

Mason, R.A. 2015. The sixth mass extinction and chemicals in the environment: Our environmental deficit is now beyond nature's ability to regenerate. *J Biol Phys Chem* **15**(3):160–176.

Massey, H.S.W. 1980. Obituary: Sir Edward Bullard. *Phys Today* **33**(8):67–68.

Mather, J.G., and Baker, R.R. 1981. Magnetic sense of direction in woodmice for route-based navigation. *Nature* **284**:259–262.

Mathie, R., and Mann, I. 2000. A correlation between extended intervals of ULF wave power and short-time geosynchronous relativistic electron flux enhancements. *Geophys Res Lett* **27**(20):3261–3264.

Matuyama, M. 1929. On the direction of magnetisation of basalt in Japan, Tyosen and Manchuria. *Proc Imp Acad Jpn* **5**:203–205.

McCraty, R., and Shafer, F. 2015. Heart rate variability: New perspectives on physiological mechanisms, assessment of self-regulatory capacity, and health risk. *Glob Adv Health Med* **4**(1):46–61.

McCraty, R., Deyhle, A., and Childre, D. 2012. The global coherence initiative: Creating a coherent planetary standing wave. *Glob Adv Health Med* **1**(1):64–77.

McCraty, R., Atkinson, M., Stolc, V., Alabdulgader, A.A., Vainoras, A., and Ragulskis, M. 2017. Synchronization of human autonomic nervous system rhythms with geomagnetic activity in human subjects. *Int J Environ Res Public Health* **14**(7):770.

McElhinny, M.W. 1971. Geomagnetic reversals during the Phanerozoic. *Science* **172**(3979):157–159.

McPherron, R.L. 2005. Magnetic pulsations: Their sources and relation to solar wind and geomagnetic activity. *Surv Geophys* **26**:545–592.

Meert, J.G., and Tamrat, E. 2004. A mechanism for explaining rapid continental motion in the late neoproterozoic. *In: The Precambrian Earth: Tempos and Events (Developments in Precambrian Geology 12)*, Eriksson, P.G., Altermann, W., Nelson, D.R., Mueller, W.U., and Catuneanu, O. (Eds.) Elsevier, Amsterdam, Netherlands, pp. 255–266.

Meert, J.G., Levashova, N.M., Bazhenov, M.L., and Landing, E. 2016. Rapid changes of magnetic field polarity in the late Ediacaran: Linking the Cambrian evolutionary radiation and increased UV-B radiation. *Gondwana Res* **34**:149–157.

Meißner, K. 2006. Impacts of submarine cables on the marine environment: A literature review. http://www.naturathlon.eu/fileadmin/BfN/meeresundkuestenschutz/Dokumente/BfN_Literaturstudie_Effekte_marine_Kabel_2007-02_01.pdf.

Melott, A.L., Lieberman, B.S., Laird, C.M., Martin, L.D., Medvedev, M.V., Thomas, B.C., Cannizzo, J.K., Gehrels, N., and Jackman, C.H. 2004. Did a gamma-ray burst initiate the Ordovician mass extinction? *Int J Astrobiol* **3**:55–61.

Merrill, R.T., and McElhinny, M.W. 1983. *The Earth's Magnetic Field: Its History, Origin and Planetary Perspective*, 2nd printing ed. Academic Press, San Francisco, CA.

Merrill, R.T., and McFadden, P.L. 1999. Geomagnetic polarity transitions. *Rev Geophys* **37**:201–226.

Merrill, R.T., McElhinny, M.W., and McFadden, P.L. 1996. *The Magnetic Field of the Earth: Paleomagnetism, the Core, and the Deep Mantle*. Academic Press, San Diego, CA.

Milanković, M. 1941. *Kanon der Erdbestrahlung und seine Anwendung auf das Eiszeitproblem,* Vol. 133. Königlich Serbische Academie Publication, Königlich Serbische Akademie, Belgrad, Serbia.

Mitsutake, G., Otsuka, K., Hayakawa, M., Sekiguchi, M., Cornélissen, G., and Halberg, F. 2005. Does Schumann resonance affect our blood pressure? *Biomed Pharmacother* **59**(Suppl 1):S10–S14.

Moench, T.T. and Konetzka, W.A. 1978. A novel method for the isolation and study of a magnetotactic bacterium. *Arch Microbiol* **119**(2):203–212.

Molina-Cuberos, G.J., López-Moreno, J.J., Rodrigo, R., Lichtenegger, H., and Schwingenschuh, K. 2001. A model of the martian ionosphere below 70 km. 2001. *Adv Space Res* **27**(11):1801–1806.

Möller, A., Sagasser, S., Wiltschko, W., and Schierwater, B. 2004. Retinal cryptochrome in a migratory passerine bird: A possible transducer for the avian magnetic compass. *Naturwissenschaften* **91**(12):585–588.

Molteno, T.C.A., and Kennedy, W.L. 2009. Navigation by induction-based magnetoreception in elasmobranch fishes. *J Biophys* **2009**:380976.

Monnereau, M., Calvet, M., Margerin, L. and Souriau, A. 2010. Lopsided growth of Earth's inner core. *Science* **328**(5981):1014–1017.

Mora, C.V., Davison, M., Wild, J.M., and Walker, M.M. 2004. Magnetoreception and its trigeminal mediation in the homing pigeon. *Nature* **432**(7016):508–511.

Morchiladze, M.M., Silagadze, T.K., and Silagadze, Z.K. 2021. What sunspots are whispering about covid-19? *Med Hypotheses* **147**:110487.

Morente, J.A., Molina-Cuberos, G.J., Porti, J.A., Besser, B.P., Salinas, A., Schwingenschuch, K., and Lichtenegger, H. 2003. A numerical simulation of Earth's electromagnetic cavity with the Transmission Line Matrix method: Schumann resonances. *J Geophys Res* **108**(A5):1195–1205.

Moskowitz, B.M., Jackson, M., and Chandler, V. 2015. Geophysical properties of the near-surface Earth: Magnetic properties. In: *Treatise on Geophysics: Resources in the Near-Surface Earth*, 2nd Edition, Vol. 8, Schubert, G. (Ed.) Elsevier, Amsterdam, Netherlands, pp. 139–174.

Mouritsen, H. 2012. Sensory biology: Search for the compass needles. *Nature* **484**(7394):320–321.

Mouritsen, H. 2018. Long-distance navigation and magnetoreception in migratory animals. *Nature* **558**(7708):50–59.

Mouritsen, H., and Hore, P.J. 2012. The magnetic retina: Light-dependent and trigeminal magnetoreception in migratory birds. *Curr Opin Neurobiol* **22**(2):343–352.

Mouritsen, H., Janssen-Bienhold, U., Liedvogel, M., Feenders, G., Stalleicken, J., Dirks, P., and Weiler, R. 2004. Cryptochromes and neuronal-activity markers colocalize in the retina of migratory birds during magnetic orientation. *Proc Natl Acad Sci U S A* **101**(39):14294–14299.

Mouritsen, H., Feenders, G., Liedvogel, M., Wada, K., and Jarvis, E.D. 2005. Night-vision brain area in migratory songbirds. *Proc Natl Acad Sci U S A* **102**(23):8339–8344.

Mulligan, B.P., and Persinger, M.A. 2012. Experimental simulation of the effects of sudden increases in geomagnetic activity upon quantitative measures of human brain activity: Validation of correlational studies. *Neurosci Lett* **516**(1):54–56.

Murray, R.W. 1960. Electrical sensitivity of the ampullae of Lorenzini. *Nature* **187**(4741):957.

Musashi, M., Isozaki, Y., Koike, T., and Kreulen, R. 2001. Stable carbon isotope signature in mid-Panthalassa shallow-water carbonates across the Permo-Triassic boundary: Evidence for ^{13}C-depleted ocean. *Earth Planet Sci Lett* **196**:9–20.

Nagata, T., and Kokubun, S. 1962. A particular geomagnetic daily variation (Sq^p) in the polar regions during geomagnetically quiet days. *Nature* **195**(4841):555–557.

Nagata, T., Uyeda, S., Akimoto, S., and Kawai, N. 1952. Self-reversal of thermoremanent magnetism of igneous rocks (II). *J Geomag Geoelec* **4**:102–107.

Nagata, T., Akimoto, S., and Uyeda, S. 1953. Selfreversal of thermoremanent magnetism of igneous rocks (III). *J Geomag Geoelec* **5**:168–184.

Nakagawa, N. 1976. Magnetic field deficiency syndrome and magnetic treatment. *Jpn Med J (Nippon Iji Shimpo)* **2745**:24–32. http://4data.ca/ottawa/archive/health/biomagnetic.html.

Nanushyan, E.R. and Murashov, V.V. 2001. Plant meristem cell response to stress factors of the geomagnetic field (GMF) fluctuations. In: *Plant Under Environmental Stress*. Friendship University of Russia, Moscow, Russia, pp. 204–205.

Narayana, R., Fliegmann, J., Paponov, I., and Maffei, M.E. 2018. Reduction of geomagnetic field (GMF) to near null magnetic field (NNMF) affects *Arabidopsis thaliana* root mineral nutrition. *Life Sci Space Res (Amst)* **19**:43–50.

National Geographic Science News 2019. Earth's magnetic field flips much more frequently than we thought. https://www.nationalgeographic.com/science/article/earths-magnetic-field-flipped-more-times-scientists-thought.

Nedukha, O., Kordyum, E., Bogatina, N., Sobol, M., Vorobyeva, T., and Ovcharenko, Y. 2007. The influence of combined magnetic field on the fusion of plant protoplasts. *J Gravit Physiol* **14**(1):117–118.

Newton, K.C., and Kajiura, S.M. 2017. Magnetic field discrimination, learning, and memory in the yellow stingray (*Urobatis jamaicensis*). *Anim Cogn* **20**:603–614.

Newton, K.C., and Kajiura, S.M. 2020a. The yellow stingray (*Urobatis jamaicensis*) can use magnetic field polarity to orient in space and solve a maze. *Mar Biol* **167**:36.

Newton, K.C., and Kajiura, S.M. 2020b. The yellow stingray (*Urobatis jamaicensis*) can discriminate the geomagnetic cues necessary for a bi-coordinate magnetic map. *Mar Biol* **167**:151–163.

Nickolaenko, A., and Hayakawa, M. 2014. *Schumann Resonance for Tyros*. Springer, Tokyo, Japan.

Nießner, C., Denzau, S., Gross, J.C., Peichl, L., Bischof, H.J., Fleissner, G., Wiltschko, W., and Wiltschko, R. 2011. Avian ultraviolet/violet cones identified as probable magnetoreceptors. *PLoS One* **6**(5):e20091.

Nießner, C., Gross, J.C., Denzau, S., Peichl, L., Fleissner, G., Wiltschko, W., and Wiltschko, R. 2016. Seasonally changing cryptochrome 1b expression in the retinal ganglion cells of a migrating passerine bird. *PLoS One* **11**:e0150377.

Niitsuma, N., and Fujii, N. 1984. Analysis on the changes in the paleoenvironment and biomass during geomagnetic reversal by means of oxygen and carbon isotope (in Japanese with abstract in English). *Geosci Repts Shizuoka Univ* **10**:123–132.

Nikolaev, Y.S., Rudakov, Y.Y., Mansurov, S.M., and Mansurova, L.G. 1976. Interplanetary magnetic field sector structure and disturbances of the central nervous system activity. *In: Reprint N 17a, Acad Sci USSR, IZMIRAN*, Moscow, Russia. p. 29.

Nimpf, S., Nordmann, G.C., Kagerbauer, D., Malkemper, E.P., Landler, L., Papadaki-Anastasopoulou, A., Ushakova, L., Wenninger-Weinzierl, A., Novatchkova, M., Vincent, P., Lendl, T., Colombini, M., Mason, M.J., and Keays, D.A. 2019. A putative mechanism for magnetoreception by electromagnetic induction in the pigeon inner ear. *Curr Biol* **29**(23):4052–4059.

Nishimura, T., Tada, H., Guo, X., Murayama, T., Teramukai, S., Okano, H., Yamada, J., Mohri, K., and Fukushima, M. 2011. A 1-μT extremely low-frequency electromagnetic field vs. sham control for mild-to-moderate hypertension: A double-blind, randomized study. *Hypertens Res* **34**(3):372–377.

Nordmann, G.C., Hochstoeger, T., and Keays, D.A. 2017. Magnetoreception: A sense without a receptor. *PLoS Biol* **15**(10):e2003234.

Nordt, L., Atchley, S., and Dworkin, S. 2003. Terrestrial evidence for two greenhouse events in the latest Cretaceous. *GSA Today* **13**(12):4–9.

Norimoto, H., and Ikegaya, Y. 2015. Visual cortical prosthesis with a geomagnetic compass restores spatial navigation in blind rats. *Curr Biol* **25**(8):1091–1095.

Nymmik, R.A. 2011. Some problems with developing a standard for determining solar energetic particle fluxes. *Adv Space Res* **47**:622–628.

Oberbauer, A.M., Lai, E., Kinsey, N.A., and Famula, T.R. 2021. Enhancing student scientific literacy through participation in citizen science focused on companion animal behavior. *Transl Anim Sci* **5**(3):txab131.

Occhipinti, A., De Santis, A., and Maffei, M.E. 2014. Magnetoreception: An unavoidable step for plant evolution? *Trends Plant Sci* **19**(1):1–4.

Oda, H. 2005. Recurrent geomagnetic excursions: A review for the Brunhes normal polarity chron (in Japanese with abstract in English). *J Geogr* **114**(2):174–193.

Ogg, J.G., Ogg, G., and Gradstein, F.M. 2008. *The Concise Geologic Time Scale*. Cambridge University Press, Cambridge, UK.

Oh, I.T., Kwon, H.J., Kim, S.C., Kim, H.J., Lohmann, K.J., and Chae, K.S. 2020. Behavioral evidence for geomagnetic imprinting and transgenerational inheritance in fruit flies. *Proc Natl Acad Sci U S A* **117**(2):1216–1222.

Okada, M., Suganuma, Y., Haneda, Y., Kazaoka, O. 2017. Paleomagnetic direction and paleointensity variations during the Matuyama-Brunhes polarity transition from a marine succession in the Chiba composite section of the Boso Peninsula, central Japan. *Earth Planets Space* **69**:45.

Olson, P. 2016. Mantle control of the geodynamo: Consequences of top-down regulation. *Geochem Geophys Geosys* **17**:1935–1956.

Olson, P., Deguen, R., Hinnov, L.A., and Zhong, S. 2013. Controls on geomagnetic reversals and core evolution by mantle convection in the Phanerozoic. *Phys Earth Planet Inter* **214**:87–103.

Opdyke, N.D., and Channell, J.E.T. 1996. *Magnetic Stratigraphy. International Geophysics Series*, Vol. 64. Academic Press, San Diego, CA.

Osipova, E.A., Pavlova, V.V., Nepomnyashchikh, V.A., and Krylov, V.V. 2016. Influence of magnetic field on zebrafish activity and orientation in a plus maze. *Behav Proc* **122**:80–86.

Ota, A., and Isozaki, Y. 2006. Fusuline biotic turnover across the Guadalupian–Lopingian (Middle-Upper Permian) boundary in mid-oceanic carbonate buildups: Biostratigraphy of accreted limestone in Japan. *J Asian Earth Sci* **26**:353–368.

Palmer, S.J., Rycroft, M.J., and Cermack, M. 2006. Solar and geomagnetic activity, extremely low frequency magnetic and electric fields and human health at the Earth's surface. *Surv Geophys* **27**(5):557–595.

Pang, K., You, H., Chen, Y., Chu, P., Hu, M., Shen., J., Guo, W., Xie., C, and Lu, B. 2017. MagR alone is insufficient to confer cellular calcium responses to magnetic stimulation. *Front Neural Circuits* **11**:11.

Parker, A.R. 1998. Colour in Burgess Shale animals and the effect of light on evolution in the Cambrian. *Proc R Soc Lond B* **265**:967–972.

Paulin, M. 1995. Electroreception and the compass sense of sharks. *J Teor Biol* **174**:325–339.

Pavlov, V., and Gallet, Y. 2005. A third superchron during the early palaeozoic. *Episodes* **28**:1–7.

Pavlov, V., and Gallet, Y. 2010. Variations in geomagnetic reversal frequency during the Earth's middle age. *Geochem Geophys Geosyst* **11**:Q01Z10.

Pavlov, A.A., Pavlov, A.K., Mills, M.J., Ostrykov, V.M., Vasilyev, G.I., and Toon, O.B. 2005. Catastrophic ozone loss during passage of the solar system through an interstellar cloud. *Geophys Res Lett* **32**(1):L01815.

Peng, W., Hu, L., Zhang, Z., and Hu, Y. 2012. Causality in the association between P300 and alpha event-related desynchronization. *PLoS One* **7**(4):e34163.

Persinger, M.A., and Psych, C. 1995. Sudden unexpected death in epileptics following sudden, intense, increases in geomagnetic activity: Prevalence of effect and potential mechanisms. *Int J Biometeorol* **38**(4):180–187.

Persinger, M.A., and Saroka, K.S. 2015. Human quantitative electroencephalographic and Schumann resonance exhibit real-time coherence of spectral power densities: Implications for interactive information processing. *J Signal Inf Proc* **6**(2):153–164.

Pétrélis, F., Besse, J., and Valet, J.P. 2011. Plate tectonics may control geomagnetic reversal frequency. *Geophys Res Lett* **38**(19):L19303.

Phillips, J.B., Muheim, R., and Jorge, P.E. 2010. A behavioral perspective on the biophysics of the light-dependent magnetic compass: A link between directional and spatial perception? *J Exp Biol* **213**:3247–3255.

Pinzon-Rodriguez, A., and Muheim, R. 2017. Zebra finches have a light-dependent magnetic compass similar to migratory birds. *J Exp Biol* **220**:1202–1209.

Pinzon-Rodriguez, A., Bensch, S., and Muheim, R. 2018. Expression patterns of cryptochrome genes in avian retina suggest involvement of Cry4 in light-dependent magnetoreception. *J R Soc Interface* **15**(140):20180058.

Pishchalnikov, R.Y., Gurfinkel, Y.I., Sarimov, R.M., Vasin, A.L., Sasonko, M.L., Matveeva, T.A., Binhi, V.N., and Baranov, M.V. 2019. Cardiovascular response as a marker of environmental stress caused by variations in geomagnetic field and local weather. *Biomed Signal Process Control* **51**:401–410.

Pobachenko, S.V., Kolesnik, A.G., Borodin, A.S., and Kalyuzhin, V.V. 2006. The contingency of parameters of human encephalograms and Schumann resonance electromagnetic fields revealed in monitoring studies. *Complex Syst Biophys* **51**:480–483.

Podolská, K. 2018. The impact of ionospheric and geomagnetic changes on mortality from diseases of the circulatory system. *J Stroke Cerebrovasc Dis* **27**(2):404–417.

Polk, C. 1983. Natural and man-made noise in the Earth-ionosphere cavity at extremely low frequencies Schumann resonances and man-made interference. *Space Sci Rev* **35**:83–89.

Poole, I. 2002. Understanding solar indices. *QST*, 38–40.

Prato, F.S., Robertson, J.A., Desjardins, D., Hensel, J., and Thomas, A.W. 2005. Daily repeated magnetic field shielding induces analgesia in CD-1 mice. *Bioelectromagnetics* **26**(2):109–117.

Price, C., Williams, E., Elhalel, G., and Sentman, D. 2021. Natural ELF fields in the atmosphere and in living organisms. *Int J Biometeorol* **65**(1):85–92.

Prölss, G.W. 2004. *Physics of the Earth's Space Environment*. Springer, Berlin, Heidelberg, Germany.

Ptitsyna, N.G., Villoresi, G., Dorman, L.I., Iucci, N., and Tyasto, M.I. 1998. Natural and man-made low-frequency magnetic fields as a potential health hazard. *Physics—Uspekhi* **41**:687–709.

Qin, S,. Yin, H., Yang, C., Dou, Y., Liu, Z., Zhang, P., Yu, H., Huang, Y., Feng, J., Hao, J., Hao, J., Deng, L., Yan, X., Dong, X., Zhao, Z., Jiang, T., Wang, H.W., Luo, S.J., and Xie, C. 2016. A magnetic protein biocompass. *Nat Mater* **15**(2):217–226.

Rabøl, J., and Thorup, K. 2006. Migratory direction established in inexperienced bird migrants in the absence of magnetic field references in their pre-migratory period and during testing. *Ethol Ecol Evol* **18**:43–51.

Randall, I. 2016. Hyperactive magnetic field may have led to one of Earth's major extinctions. https://www.science.org/news/2016/02/hyperactive-magnetic-field-may-have-led-one-earth-s-major-extinctions.

Randall, C.E., Harvey, V.L., Manney, G.L., Orsolini, Y., Codrescu, M., Sioris, C., Brohede, S., Haley, C.S., Gordley, L.L., Zawodny, J.M., and Russell, J.M. 2005. Stratospheric effects of energetic particle precipitation in 2003–2004. *Geophys Res Lett* **32**:L05802.

Rapoport, S.I., Malinovskaia, N.K., Oraevskiĭ, V.N., Komarov, F.I., Nosovskii, A.M., and Vetterberg, L. 1997. Effects of disturbances of natural magnetic field of the Earth on melatonin production in patients with coronary heart disease. *Klin Med (Mosk)* **75**(6):24–26.

Raup, D.M, and Sepkoski, J.J., Jr. 1982. Mass extinctions in the marine fossil record. *Science* **215**(4539):1501–1503.

Rawer, K. 1993. *Wave Propagation in the Ionosphere*. Kluwer Acadademy Publishers, Dordrecht, Netherlands.

Raygan, F., Ostadmohammadi, V., Bahmani, F., Reiter, R.J., and Asemi, Z. 2019. Melatonin administration lowers biomarkers of oxidative stress and cardio-metabolic risk in type 2 diabetic patients with coronary heart disease: A randomized, double-blind, placebo-controlled trial. *Clin Nutr* **38**(1):191–196.

Renne, P.R., Deino, A.L., Hilgen, F.J., Kuiper, K.F., Mark, D.F., Mitchell, W.S., Morgan, L.E., Mundil, R., and Smit, J. 2013. Time scales of critical events around the Cretaceous-Paleogene boundary. *Science* **339**(6120):684–687.

Ribas, I., Guinan, E.F., Güdel, M., and Audard, M. 2005. Evolution of the solar activity over time and effects on planetary atmospheres. I. High-energy irradiances (1–1700 Å). *Astrophys J* **622**(1):680–694.

Ribas, I., De Mello, G.P., Ferreira, L.D., Hebrard, E., Selsis, F., Catalan, S., Garces, A., do Nascimento, J.D., Jr., and De Medeiros, J.R. 2010. Evolution of the solar activity over time and effects on planetary atmospheres. II. k1 Ceti, an analog of the Sun when life arose on Earth. *Astrophys J* **714**(1):384–395.

Richards, I.R.J., Raoult, V., Powter, D.M., and Gaston, T.F. 2018. Permanent magnets reduce bycatch of benthick sharks in an ocean trap fishery. *Fish Res* **208**:16–21.

Richardson, I., Wibberenz, G., and Cane, H. 1996. The relationship between recurring cosmic ray depressions and corotating solar wind streams at ≤1 AU: IMP 8 and Helios 1 and 2 anticoincidence guard rate observations. *J Geophys Res Space Phys* **101**:13483–13496.

Riley, P. 2012. On the probability of occurrence of extreme space weather events. *Space Weather* **10**(2):S02012.

Ritz, T., Adem, S., and Schulten, K. 2000. A model for photoreceptor-based magnetoreception in birds. *Biophys J* **78**(2):707–718.

Ritz, T., Wiltschko, R., Hore, P.J., Rodgers, C.T., Stapput, K., Thalau, P., Timmel, C.R., and Wiltschko, W. 2009. Magnetic compass of birds is based on a molecule with optimal directional sensitivity. *Biophys J* **96**(8):3451–3457.

Roberts, R.G. 2016. Living life on a magnet. *PLoS Biol* **14**(8):e2000613.

Rodgers, C.T., and Hore, P.J. 2009. Chemical magnetoreception in birds: The radical pair mechanism. *Proc Natl Acad Sci U S A* **106**(2):353–360.

Rohde, R.A. 2006. Phanerozoic carbon dioxide (Global Warming Art project). https://commons.wikimedia.org/wiki/File:Phanerozoic_Carbon_Dioxide.png.

Roldugin, V.C., Maltsev, Y.P., Petrova, G.A., and Vasiljev, A.N. 2001. Decrease in the first Schumann resonance frequency during solar proton events. *J Geophys Res Space Phys* **106**(A9):18555–18562.

Rothman, D.H. 2002. Atmospheric carbon dioxide levels for the last 500 million years. *Proc Natl Acad Sci U S A* **99**(7):4167–4171.

Royer, D.L., Berner, R.A., Montañ̄ez, I.P., Tabor, N.J., and Beerling, D.J. 2004. CO_2 as a primary driver of Phanerozoic climate. *GSA Today* **14**(3):4–10.

Runcorn, S.K. 1969. The paleomagnetic vector field. In: *The Earth's Crust and Upper Mantle*, Vol. 13, Hart, P.J. (Ed.) The American Geophysical Union, Washington, DC, pp. 447–457.

Rycroft, M.J., Israelsson, S., and Price, C. 2000. The global atmospheric electric circuit, solar activity and climate change. *J Atmos Sol Terr Phys* **6**(17–18):1563–1576.

Sagan, L. 1967. On the origin of mitosing cells. *J Theor Biol* **14**(3):255–274.

Saltzman, M.R. 2005. Phosphorous, nitrogen, and redox evolution of the Paleozoic oceans. *Geology* **33**:573–576.

Saltzman, M.R., and Young, S.A. 2005. Long-lived glaciation in the Late Ordovician? Isotopic and sequence-stratigraphic evidence from western Laurentia. *Geology* **33**(2):109–112.

Saroka, K.S., and Persinger, M.A. 2014. Quantitative evidence for direct effects between Earth-ionosphere Schumann resonances and human cerebral cortical activity. *Int Lett Chem Phys Astron* **20**:166.

Saroka, K.S., Vares, D.E., and Persinger, M.A. 2016. Similar spectral power densities within the Schumann resonance and a large population of quantitative electroencephalographic profiles: Supportive evidence for Koenig and Pobachenko. *PLoS One* **11**(1):e0146595.

Schilt, A., Baumgartner, M., Eicher, O., Chappellaz, J., Schwander, J., Fischer, H., and Stocker, T.F. 2013. The response of atmospheric nitrous oxide to climate variations during the last glacial period. *Geophys Res Let* **40**(9):1888–1893.

Schilt, A., Brook, E.J., Bauska, T.K., Baggenstos, D., Fischer, H., Joos, F., Petrenko, V.V., Schaefer, H., Schmitt, J., Severinghaus, J.P., Spahni, R., and Stocker, T.F. 2014. Isotopic constraints on marine and terrestrial N_2O emissions during the last deglaciation. *Nature* **516**(7530):234–237.

Schlegel, K., and Füllekrug, M. 1999. Schumann resonance parameter changes during high energy particle precipitation. *J Geophys Res Space Phys* **104**(A5):10111–10118.

Schröder, W. 1992. On the existence of the 11-year cycle in solar and auroral activity before and during the so-called Maunder Minimum. *J Geomagn Geoelectr* **44**(2):119–128.

Schüler, D. 2002. The biomineralization of magnetosomes in *Magnetospirillum gryphiswaldense*. *Int Microbiol* **5**(4):209–214.

Schulte, P., Alegret, L., Arenillas, I., Arz, J.A., Barton, P.J., Bown, P.R., Bralower, T.J., Christeson, G.L., Claeys, P., Cockell, C.S., Collins, G.S., Deutsch, A., Goldin, T.J., Goto, K., Grajales-Nishimura, J.M., Grieve, R.A.F., Gulick, S.P.S., Johnson, K.R., Kiessling, W., Koeberl, C., Kring, D.A., MacLeod, K.G., Matsui, T., Melosh, J., Montanari, A., Morgan, J.V., Neal, C.R., Nichols, D.J., Norris, R.D., Pierazzo, E., Ravizza, G., Rebolledo-Vieyra, M., Reimold, W.U., Robin, E., Salge, T., Speijer, R.P., Sweet, A.R., Urrutia-Fucugauchi, J., Vajda, V., Whalen, M.T., and Willumsen, P.S. 2010. The Chicxulub asteroid impact and mass extinction at the Cretaceous-Paleogene boundary. *Science* **327**(5970):1214–1218.

Schumann, W.O. 1952. Uber die strahlungslosen Eigenschwingungen einer leitenden Kugel, die von einer Luftshicht und einer Ionosphairenhulle umgeben ist. *Z Naturforsch* **7A**:149.

Schumann, W., and König, H. 1954. Uber die beobachtung von "atmospherics" bei geringsten frequenzen. *Die Naturwissenschaften* **41**:183–184.

Selkin, P.A., Gee, J.S., Tauxe, L., Meurer, W.P., and Newell, A.J. 2000. The effect of remanence anisotropy on paleointensity estimates: A case study from the archean stillwater complex. *Earth Planet Sci Lett* **183**:403–416.

Selmaoui, B., and Touitou, Y. 1995. Sinusoidal 50-Hz magnetic fields depress rat pineal NAT activity and serum melatonin. Role of duration and intensity of exposure. *Life Sci* **57**:1351–1358.

Semm, P., and Beason, R.C. 1990. Responses to small magnetic variations by the trigeminal system of the Bobolink. *Brain Res Bull* **25**:735–740.

Sentman, D.D. 1995. Schumann resonances. In: *Handbook of Atmospheric Electrodynamics*, Vol. 1, Volland, H. (Ed.) CRC Press, Boca Raton, FL, pp. 267–298.

Sepkoski, J.J., Jr. 1984. A kinetic model of phanerozoic taxonomic diversity, III. Post paleozoic families and mass extinctions. *Paleobiology* **10**:246–267.

Shaw, J., Boyd, A., House, M., Woodward, R., Mathes, F., Cowin, G., Saunders, M., and Baer, B. 2015. Magnetic particle-mediated magnetoreception. *J R Soc Interface* **12**(110):0499.

Shcherbakov, V.P., Solodovnikov, G.M., and Sycheva, N.K. 2002. Variations in the geomagnetic dipole during the past 400 million years (volcanic rocks). *Izvestiya Phys Solid Earth* **38**(2):113–119.

Sherbakov, D., Winklehofer, M., Peterson, N., Steidle, J., Hilbig, R., and Blum, M. 2005. Magnetosensation in zebrafish. *Curr Biol* **15**:161–162.

Sieh, K., Herrin, J., Jicha, B., Schonwalder-Angel, D., Moore, J.D.P., Banerjee, P., Wiwegwin, W., Sihavong, V., Singer, B., Chualaowanich, T., and Charusiri, P. 2019. Australasian impact crater buried under the Bolaven volcanic field, Southern Laos. *Proc Natl Acad Sci U S A* **117**(3):1346–1353.

Simmons, S.L., Bazylinski, D.A., and Edwards, K.J. 2006. South-seeking magnetotactic bacteria in the Northern Hemisphere. *Science* **311**(5759):371–374.

Simon, Q., Suganuma, Y., Okada, M., Haneda, Y., and ASTER Team. 2019. High-resolution [10]Be and paleomagnetic recording of the last polarity reversal in the Chiba composite section: Age and dynamics of the Matuyama–Brunhes transition. *Earth Planet Sci Lett* **519**:92–100.

Simonsen, L.C., and Nealy, J.E. 1991. Radiation protection for human missions to the Moon and Mars. *NASA Tech Paper* **3079**:1–25.

Sisneros, J.A., and Tricas, T.C. 2002. Neuroethology and life history adaptations of the elasmobranch electric sense. *J Physiol Paris* **96**(5–6):379–389.

Skauli, K.S., Reitan, J.B., and Walther, B.T. 2000. Hatching in zebrafish (*Danio rerio*) embryos exposed to a 50 Hz magnetic field. *Bioelectromagnetics* **21**(5):407–410.

Skiles, D.D. 1985. The geomagnetic field: Its nature, history, and biological relevance. In: *Magnetite Biomineralization and Magnetoreception in Organisms: A New Biomagnetism*, Kirschvink, J.L., Jones, D.S., and MacFadden, B.J. (Eds.) Plenum Press, New York, pp. 43–102.

Slaby, P., Bartos, P., Karas, J., Netusil, R., Tomanova, K., and Vácha, M. 2018. How swift is cry-mediated magnetoreception? Conditioning in an American cockroach shows sub-second response. *Front Behav Neurosci* **12**:107.

Smirnov, A.V., Tarduno, J.A., and Pisakin, B.N. 2003. Paleointensity of the early geodynamo (2.45 Ga) as recorded in Karelia: A single-crystal approach. *Geology* **31**:415–418.

Solov'yov, I.A., and Greiner, W. 2007. Theoretical analysis of an iron mineral-based magnetoreceptor model in birds. *Biophys J* **93**(5):1493–1509.

Solov'yov, I.A., and Greiner, W. 2009. Micromagnetic insight into a magnetoreceptor in birds: Existence of magnetic field amplifiers in the beak. *Phys Rev E* **80**:041919.

Soltani, F., Kashi, A., and Arghavani, M. 2006. Effect of magnetic field on *Asparagus officinalis* L. seed germination seedling growth. *Seed Sci Technol* **34**:349–353.

Southwood, D. 1974. Some features of field line resonances in the magnetosphere. *Planet Space Sci* **22**:483–491.

Specktor, B. 2021. Earth's core is growing 'lopsided' and scientists don't know why. https://www.livescience.com/earth-inner-core-lopsided-crystal-growth.html.

Sreenivasan, B. 2010. Modelling the geodynamo: Progress and challenges. *Curr Sci* **99**(12):1739–1750.

St. Fleur, N. 2017. After Earth's worst mass extinction, life rebounded rapidly, fossils suggest. *The New York Times,* 16 February 2017.

Staubwasser, M., Drăguşin, V., Onac, B.P., Assonov, S., Ersek, V., Hoffmann, D.L., and Veres, D. 2018. Impact of climate change on the transition of Neanderthals to modern humans in Europe. *Proc Natl Acad Sci U S A* **115**:9116–9121.

Stoner, A.W., and Kaimmer, S.M. 2008. Reducing elasmobranch bycatch: Laboratory investigation of rare earth metal and magnetic deterrents with spiny dogfish and Pacific halibut. *Fish Res* **92**:162–168.

Stoupel, E. 1993. Sudden cardiac deaths and ventricular extrasystoles on days of four levels of geomagnetic activity. *J Basic Physiol Pharmacol* **4**(4):357–366.

Stoupel, E. 2002. The effect of geomagnetic activity on cardiovascular parameters. *Biomed Pharmacother* **56**(suppl 2):247s–256s.

Stoupel, E. 2008. Atherothrombosis: Environmental links. *J Basic Clin Physiol Pharmacol* **19**(1):37–47.

Stoupel, E., Martfel, J.N., and Rotenberg, Z. 1994. Paroxysmal atrial fibrillation and stroke (cerebrovascular accidents) in males and females above and below age 65 on days of different geomagnetic activity levels. *J Basic Clin Physiol Pharmacol* **5**:315–329.

Stoupel, E., Wittenberg, C., Zabludowski, J., and Boner, G. 1995. Ambulatory blood pressure monitoring in patients with hypertension on days of high and low geomagnetic activity. *J Hum Hypertens* **9**:293–294.

Stoupel, E., Babayev, E.S., Mustafa, F.R., Abramson, E., Israelevich, P., and Sulkes, J. 2006. Clinical cosmobiology-sudden cardiac death and daily/monthly geomagnetic, cosmic ray and solar activity-the Baku study (2003–2005). *Sun Geosphere* **1**:13–16.

Stoupel, E., Kalediene, R., Petrauskiene, J., Starkuviene, S., Abramson, E., Israelevich, P., and Sulkes, J. 2011. Twenty years study of solar, geomagnetic, cosmic ray activity links with monthly deaths number (n-850304). *Biomed Sci Eng* **4**(6):426–434.

Stoupel, E., Tamoshiunas, A., Radishauskas, R., Bernotiene, G., Abramson, E., and Israelevich, P. 2012. Acute myocardial infarction (AMI) (n-11026) on days of zero geomagnetic activity (GMA) and the following week: Differences at months of maximal and minimal solar activity (SA) in solar cycles 23 and 24. *J Basic Clin Physiol Pharmacol* **23**(1):5–9.

Stoupel, E., Babayev, E., Abramson, E., and Sulkes, J. 2013. Days of "Zero" level geomagnetic activity accompanied by the high neutron activity and dynamics of some medical events: Antipodes to geomagnetic storms. *Health* **5**(5):855–861.

Suganuma, Y. 2020. The geomagnetic reversal and "Chibanian" (written in Japanese). Blue Backs. B-2132. Kodansha, Tokyo, Japan.

Suganuma, Y., Haneda, Y., Kameo, K., Kubota, Y., Hayashi, H., Itaki, T., Okuda, M., Head, M.J., Sugaya, M., Nakazato, H., Igarashi, A., Shikoku, K., Hongo, M., Watanabe, M., Satoguchi, Y., Takeshita, Y., Nishida, N., Izumi, K., Kawamura, K., Kawamata, M., Okuno, J., Yoshida, T., Ogitsu, I., Yabusaki, H., and Okada, M. 2018. Paleoclimatic and paleoceanographic records of Marine Isotope Stage 19 at the Chiba composite section, central Japan: A reference for the Early–Middle Pleistocene boundary. *Quat Sci Rev* **191**:406–430.

Suganuma, Y., Okada, M., Head, M.J., Kameo, K., Haneda, Y., Hayashi, H., Irizuki, T., Itaki, T., Izumi, K., Kubota, Y., Nakazato, H., Nishida, N., Okuda, M., Satoguchi, Y., Simon, Q., and Takeshita, Y. 2121. Formal ratification of the Global Boundary Stratotype Section and Point (GSSP) for the Chibanian Stage and Middle Pleistocene Subseries of the Quaternary System: the Chiba Section, Japan. *Episodes* **44**(3):317–347.

Svensmark, H., 2006. Cosmic rays and the biosphere over 4 billion years. *Astron Nachr* **327**:871–879.

Svensmark, H., and Friis-Christensen, E. 1997. Variation of cosmic rays flux and global cloud coverage: A missing link in solar-climate relationships. *J Atmos Sol Terr Phys* **59**:1225–1232.

Takebe, A., Furutani, T., Wada, T., Koinuma, M., Kubo, Y., Okano, K., and Okano, T. 2012. Zebrafish respond to the geomagnetic field by bimodal and group-dependent orientation. *Sci Rep* **2**:727.

Takeshita, Y., Matsushima, N., Teradaira, H., Uchiyama, T., and Kumai, H. 2016. A marker tephra bed close to the lower-middle pleistocene boundary: Distribution of the Ontake-Byakubi Tephra Bed in central Japan. *Quat Int* **397**:27–38.

Tapping, K.F. 1987. Recent solar radio astronomy at centimeter wavelength: The temporal variability of the 10.7-cm flux. *J Geophys Res* **92**(D1):829–838.

Tapping, K.F., Mathias, R.G., and Surkan, D.L. 2001. Influenza pandemics and solar activity. *Can J Infect Dis* **12**:61–62.

Tarduno, J.A., Cottrell, R.D., Watkeys, M.K. and Bauch, D. 2007. Geomagnetic field strength 3.2 billion years ago recorded by single silicate crystals. *Nature* **446**(7136):657–660.

Tarduno, J.A., Cottrell, R.D., Watkeys, M.K., Hofmann, A., Doubrovine, P.V., Mamajek, E., Liu, D., Sibeck, D.G., Neukirch, L.P., and Usui, Y. 2010. Geodynamo, solar wind and magnetopause 3.45 billion years ago. *Science* **327**(5970):1238–1240.

Task Force of the European Society of Cardiology and the North American Society of Pacing and Electrophysiology. 1996. Heart rate variability standards of measurement, physiological interpretation, and clinical use. *Circulation* **93**(5):1043–1065.

Tateno, S., Hirose, K., Ohishi, Y., and Tatsumi, Y. 2010. The structure of iron in Earth's inner core. *Science* **330**:359–361.

Tchijevsky, A.L. (de Smitt, V.P. translation). 1971. Physical factors of the historical process. *Cycles* **22**:11–27.

Tedetti, M., and Sempéré R. 2006. Penetration of ultraviolet radiation in the marine environment: A review. *Photochem Photobiol* **82**:389–397.

Temurjants, N.A., Vladimirsy, B.M., and Tishkin, O.G. 1992. *Extremely Low-Frequency Signals in Biological World.* Naukova Dumka, Kiev, Ukraine.

The Geological Society of Japan. 2010. The Quaternary part of the International Chronostratigraphic Chart. https://www.meti.go.jp/english/mobile/2020/20200514001en.html.

Thébault, E., Finlay, C.C., Beggan, C.D., Alken, P., Aubert, J., Barrois, O., Bertrand, F., Bondar, T., Boness, A., Brocco, L., Canet, E., Chambodut, A., Chulliat, A., Coïsson, P., Civet, F., Du, A., Fournier, A., Fratter, I., Gillet, N., Hamilton, B., Hamoudi, M., Hulot, G., Jager, T., Korte, M., Kuang, W., Lalanne, X., Langlais, B., Léger, J.M., Lesur, V., Lowes, F.J., Macmillan, S., Mandea, M., Manoj, C., Maus, S., Olsen, N., Petrov, V., Ridley, V., Rother, M., Sabaka, T.J., Saturnino, D., Schachtschneider, R., Sirol, O., Tangborn, A., Thomson, A., Tøffner-Clausen, L., Vigneron, P., Wardinski, I., and Zvereva, T. 2015. International geomagnetic reference field: The 12th generation. *Earth, Planets Space* **67**:79.

Thoss, F., Bartsch, B., Fritzsche, B., Tellschaft, D., and Thoss, M. 2000. The magnetic field sensitivity of the human visual system shows resonance and compass characteristic. *J Comp Physiol A* **186**(10):1007–1010.

Thoss, F., Bartsch, B., Tellschaft, D., and Thoss, M. 2002. The light sensitivity of the human visual system depends on the direction of view. *J Comp Physiol A* **188**(3):235–237.

Timofejeva, I., McCraty, R., Atkinson, M., Joffe, R., Vainoras, A., Alabdulgader, A., and Ragulskis, M. 2017. Identification of a group's physiological synchronization with Earth's magnetic field. *Int J Environ Res Public Health* **14**(9):998.

Touitou Y, Coste O, Dispersyn G, and Pain L. 2010. Disruption of the circadian system by environmental factors: Effects of hypoxia, magnetic fields and general anesthetics agents. *Adv Drug Deliv Rev* **62**:928–945.

Trański, A., Formicki, K., Śmietana, P., Sadowski, M., and Winnicki, A. 2005. Sheltering behaviour of spinycheek crayfish (*Orconectes limosus*) in the presence of an artificial magnetic field. *Bull Fr Pêche Piscic* **376–377**:787–793.

Treiber, C.D., Salzer, M.C., Riegler, J., Edelman, N., Sugar, C., Breuss, M., Pichler, P., Cadiou, H., Saunders, M., Lythgoe, M., Shaw, J., and Keays, D.A. 2012. Clusters of iron-rich cells in the upper beak of pigeons are macrophages not magnetosensitive neurons. *Nature* **484**(7394):367–370.

Tsunakawa, H., and Shaw, J. 1994. The Shaw method of palaeointensity determinations and its application to recent volcanic rocks. *Geophys J Int* **118**(3):781–787.

Tsunakawa, H., Shimura, K., and Yamamoto, Y. 1997. *Application of Double Heating Technique of the Shaw Method to the Brunhes Epoch Volcanic Rocks (Abstract),* 8th edition. Scientific Assembly of IAGA, Uppsala, Sweden.

Uebe, R., and Schüler, D. 2016. Magnetosome biogenesis in magnetotactic bacteria. *Nat Rev Microbiol* **14**:621–637.

Ueno, Y, Hyodo, M, Yang, T.S, and Katoh, S. 2019. Intensified East Asian winter monsoon during the last geomagnetic reversal transition. *Sci Rep* **9**:9839.

Ujihara, A., 1986. Pelagic gastropoda assemblages from the Kazusa Group of the Boso peninsula, Japan and Plio-Pleistocene climatic changes. *J Geol Soc Jpn* **92**:639e651 (in Japanese with English abstract).

USGS FAQs. 2013. How does the Earth's core generate a magnetic field? United States Geological Survey. https://en.wikipedia.org/wiki/Dynamo_theory#/media/File:Dynamo_Theory_-_Outer_core_convection_and_magnetic_field_geenration.svg.

Vácha, M. 2006. Laboratory behavioural assay of insect magnetoreception: Magnetosensitivity of *Periplaneta americana. J Exp Biol* **209**(19):3882–3886.

Vácha, M. 2017. Magnetoreception of invertebrates. In: *The Oxford Handbook of Invertebrate Neurobiology,* Byrne, J.H. (Ed.) Oxford University Press, London, UK, pp. 366–388.

Vácha, M. 2020. Invertebrate magnetoreception–In between orientation and general sensitivity. In: *The Senses: A Comprehensive Reference*, 2nd Edition. Vol. 7, Mechanosensory Lateral Line, Electroreception, Magnetoreception, Fritzsch, B. (Ed.) Elsevier, Amsterdam, Netherlands, pp. 445–458.

Vácha, M., Půzová, T., and Drstková, D. 2008. Ablation of antennae does not disrupt magnetoreceptive behavioural reaction of the American cockroach to periodically rotated geomagnetic field. *Neurosci Lett* **435**(2):103–107.

Vácha, M., Půzová, T., and Kvíćalová, M. 2009. Radio frequency magnetic fields disrupt magnetoreception in American cockroach. *J Exp Biol* **212**(21):3473–3477.

Valet, J.P., and Plenier, G. 2008. Simulations of a time-varying non dipole field during geomagnetic reversals and excursions. *Phys Earth Planetary Inter* **169**:178–193.

Valet, J.P., Fournier, A., Courtillot, V., and Herrero-Bervera, E. 2012. Dynamical similarity of geomagnetic field reversals. *Nature* **490**(7418):89–94.

Valet, J.P., Bassinot, F., Simon, Q., Savranskaia, T., Thouveny, N., Bourlés, D.L., and Villedieu, A. 2019. Constraining the age of the last geomagnetic reversal from geochemical and magnetic analyses of Atlantic, Indian, and Pacific Ocean sediments. *Earth Planet Sci Lett* **506**:323–331.

Vanderstraeten, J., Gailly, P., and Malkemper, E.P. 2018. Low-light dependence of the magnetic field effect on cryptochromes: Possible relevance to plant ecology. *Front Plant Sci* **9**:121.

Vaquero, J.M., and Gallego, M.C. 2007. Sunspot numbers can detect pandemic influenza A: The use of different sunspot numbers. *Med Hypotheses* **68**:1189–1190.

Veizer, J., Ala, D., Azmy, K., Bruckschen, P., Buhl, D., Bruhm, F., Carden, G.A.F., Diener, A., Ebneth, S., Godderis, Y., Jasper, T., Korte, C., Pawellek, F., Podlaha, O.G., and Strauss, H. 1999. $^{87}Sr/^{86}Sr$, $\delta^{13}C_{carb}$ and $\delta^{18}O_{carb}$ evolution of Phanerozoic seawater. *Chem Geol* **161**:59–88.

Vidal-Gadea, A., Ward, K., Beron, C., Ghorashian, N., Gokce, S., Russell, J., Truong, N., Parikh, A., Gadea, O., Ben-Yakar, A., and Pierce-Shimomura, J. 2015. Magnetosensitive neurons mediate geomagnetic orientation in *Caenorhabditis elegans. eLife* **4**:e07493.

Vladimirsky, B.M. 1980. Biological rhythms and the solar activity. In: *Problems of Cosmic Biology*, Vol. 41, Chernigovsky, V.N. (Ed.) Nauka, Moscow, Russia, pp. 289–315.

Vogt, J., Zieger, B., Glassmer, K.H., Stadelmann, A., Kallenrode, M.B., Sinnhuber, M., and Winkler, H. 2007. Energetic particles in the paleomagnetosphere: Reduced dipole configurations and quadrupolar contributions. *J Geophys Res Space Phys* **112**(A6):A06216.

Volpe, P. 2003. Interactions of zero-frequency and oscillating magnetic fields with biostructures and biosystems. *Photochem Photobiol Sci* **2**(6):637–648.

Voosen, P. 2021. Kauri trees mark magnetic flip 42,000 years ago. *Science* **371**(6531):766.

Voss, J., Keary, N., and Bischof, H.J. 2007. The use of the geomagnetic field for short-distance orientation in zebra finches. *Behaviour* **18**:1053–1057.

Wajnberg, E., Acosta-Avalos, D., Alves, O.C., de Oliveira, J.F., Srygley, R.B., and Esquivel, D.M. 2010. Magnetoreception in eusocial insects: an update. *R Soc Interface* **7**(Suppl 2):S207–S225.

Walker, M.M., Diebel, C.E., Haugh, C.V., Pankhurst, P.M., Montgomery, J.C., and Green, C.R. 1997. Structure and function of the vertebrate magnetic sense. *Nature* **390**(6658):371–376.

Wallraff, H.G. 1999. The magnetic map of homing pigeons: An evergreen phantom. *J Theor Biol* **197**:265–269.

Wan, G.J., Jiang, S.L., Zhao, Z.C., Xu, J.J., Tao, X.R., Sword, G.A., Gao, Y.B., Pan, W.D., and Chen, F.J. 2014. Bio-effects of near-zero magnetic fields on the growth, development and reproduction of small brown planthopper, Laodelphax striatellus and brown planthopper, *Nilaparvata lugens. J Insect Physiol* **68**:7–15.

Wan, G., Liu, R., Li, C., He, J., Pan, W., Sword, G.A., Hu, G., and Chen, F. 2020a. Change in geomagnetic field intensity alters migration-associated traits in a migratory insect. *Biol Lett* **16**(4):20190940.

Wan, G.J., Jiang, S.L., Zhang, M., Zhao, J.Y., Zhang, Y.C., Pan, W.D., Sword, G.A., and Chen, F.J. 2020b. Geomagnetic field absence reduces adult body weight of a migratory insect by disrupting feeding behavior and appetite regulation. *Insect Sci* **28**(1):251–260.

Wan, G., Hayden, A.N., Iiams, S.E., and Merlin, C. 2021. Cryptochrome 1 mediates light-dependent inclination magnetosensing in monarch butterflies. *Nat Commun* **12**:771.

Wang, J., and Pantopoulos, K. 2011. Regulation of cellular iron metabolism. *Biochem J* **434**:365–381.

Wang, W., Cao, C.Q., and Wang, Y. 2004. The carbon isotope excursion on GSSP candidate section of Lopingian–Guadalupian boundary. *Earth Planet Sci Lett* **220**:57–67.

Wang, C.X., Hilburn, I.A., Wu, D.A., Mizuhara, Y., Couste, C.P., Abrahams, J.N.H., Bernstein, S.E., Matani, A., Shimojo, S., and Kirschvink, J.L. 2019. Transduction of the geomagnetic field as evidenced from alpha-band activity in the human brain. *eNeuro* **6**(2):ENEURO.0483-18.

Watanabe, Y., Cornélissen, G., Halberg, F., Otsuka, K., and Ohkawa, S.I. 2001. Associations by signatures and coherences between the human circulation and helio- and geomagnetic activity. *Biomed Pharmacother* **55**(Suppl 1):76–83.

Wei, Y., Pu, Z., Zong, Q., Wan, W., Ren, Z., Fraenz, M., Dubinin, E., Tian, F., Shi, Q., Fu, S., and Hong, M. 2014. Oxygen escape from the Earth during geomagnetic reversals: Implications to mass extinction. *Earth Planet Sci Lett* **394**:94–98.

Weydahl, A., Sothern, R.B., Cornélissen, G., and Wetterberg, L. 2001. Geomagnetic activity influences the melatonin secretion at latitude 70 degrees. *Biomed Pharmacother* **55**(Suppl 1):57s–62s.

William, L. 2007. *Fundamentals of Geophysics*. Cambridge University Press, Cambridge, UK, p. 281.

Williams, E.R. 1992. The Schumann resonance: A global tropical thermometer. *Science* **256**(5060):1184–1187.

Wiltschko, R., and Wiltschko, W. 1978. Evidence for the use of magnetic outward-journey information in homing pigeons. *Naturwissenschaften* **65**:112–113.

Wiltschko, R., and Wiltschko, W. 2012. Magnetoreception. *Adv Exp Med Biol* **739**:126–141.

Wiltschko, R., and Wiltschko, W. 2021. The discovery of the use of magnetic navigational information. *J Comp Physiol A*. doi: 10.1007/s00359-021-01507-0.

Wiltschko, R., Ritz, T., Stapput, K., Thalau, P., and Wiltschko, W. 2005. Two different types of light-dependent responses to magnetic fields in birds. *Curr Biol* **15**(16):1518–1523.

Wiltschko, W., and Gwinner, E. 1974. Evidence for an innate magnetic compass in garden warblers. *Naturwissenschaften* **61**(9):406.

Wiltschko, W., and Wiltschko, R. 1972. Magnetic compass of European robins. *Science* **176**(4030):62–64.

Wiltschko, W., and Wiltschko, R. 2005. Magnetic orientation and magnetoreception in birds and other animals. *J Comp Physiol A* **191**(8):675–693.

Wiltschko, W., Wiltschko, R., Munro, U., and Ford, H. 1998. Magnetic versus celestial cues: Cue-conflict experiments with migrating silvereyes at dusk. *J Comp Physiol A* **182**:521–529.

Wiltschko, W., Munro, U., Ford, H., and Wiltschko, R. 2006. Bird navigation: What type of information does the magnetite-based receptor provide? *Proc Biol Sci* **273**(1603):2815–2820.

Wiltschko, W., Freire, R., Munro, U., Ritz, T., Rogers, L., Thalau, P., and Wiltschko, R. 2007. The magnetic compass of domestic chickens, *Gallus gallus*. *J Exp Biol* **210**(13):2300–2310.

Wiltschko, R., Nießner, C., and Wiltschko, W. 2021. The magnetic compass of birds: The role of cryptochrome. *Front Physiol* **12**:667000.

Winkler, H., Sinnhuber, M., Notholt, J., Kallenrode, M.B., Steinhilber, F., Vogt, J., Zieger, B., Glassmeier, K.H., and Stadelmann, A. 2008. Modeling impacts of geomagnetic field variations on middle atmospheric ozone responses to solar proton events on long timescales. *J Geophys Res Atmos* **113**(D2):D02302.

Winklhofer, M., and Kirschvink, J.L. 2010. A quantitative assessment of torque-transducer models for magnetoreception. *J R Soc Interface* **7**(Suppl 2):S273–S289.

Woodward, A. 2021. Earth's core is growing lopsidedly, new study suggests - and has been for at least half a billion years. https://www.businessinsider.com/earth-core-lopsided-growing-faster-on-one-side-2021-6.

Woodward, J.R., Foster, T.J., Jones, A.R., Salaoru, A.T., and Scrutton, N.S. 2009. Timeresolved studies of radical pairs. *Biochem Soc Trans* **37**:358–362.

World Data Center for Geomagnetism, Kyoto. http://wdc.kugi.kyoto-u.ac.jp/element/eleexp.html.

Wu, L.Q., and Dickman, J.D. 2011. Magnetoreception in an avian brain in part mediated by inner ear lagena. *Curr Biol* **21**(5):418–423.

Wu, L.Q., and Dickman, J.D. 2012. Neural correlates of a magnetic sense. *Science* **336**(6084):1054–1057.

Wu, H., Scholten, A., Einwich, A., Mouritsen, H., and Koch, K.W. 2020. Protein-protein interaction of the putative magnetoreceptor cryptochrome 4 expressed in the avian retina. *Sci Rep* **10**:7364.

Xie, C., Li, Z., Sui, H., Pan, W., and Chen, F., 2011. Detection of magnetic materials in adults of the brown planthopper, *Nilaparvata lugens* (Hemiptera: Delphacidae). *Acta Entomol Sinica* **54**:1189–1193.

Xiong, C., Stolle, C., and Lühr, H. 2016. The Swarm satellite loss of GPS signal and its relation to ionospheric plasma irregularities. *Space Weather* **14**:563–577.

Xu, C.X., Yin, X., Lv, Y., Wu, C.Z., Zhang, Y.X., and Song, T. 2012. A near-null magnetic field affects cryptochrome-related hypocotyl growth and flowering in *Arabidopsis. Adv Space Res* **49**(5):834–840.

Xu, C.X., Wei, S.F., Lu, Y., Zhang, Y.X., Chen, C.F., and Song, T. 2013. Removal of the local geomagnetic field affects reproductive growth in *Arabidopsis. Bioelectromagnetics* **34**(6):437–442.

Xu, C., Lv, Y., Chen, C., Zhang, Y., and Wei, S. 2014. Blue light-dependent phosphorylations of cryptochromes are affected by magnetic fields in *Arabidopsis. Adv Space Res* **53**(7):1118–1124.

Xue, X., Ali, Y.F., Luo, W., Liu, C., Zhou, G., and Liu, N.A. 2021. Biological effects of space hypomagnetic environment on circadian rhythm. *Front Physiol* **12**:643943.

Yadav, R.K., Gastine, T., Christensen, U.R., Wolk, S.J. and Poppenhaeger, K. 2016. Approaching a realistic force balance in geodynamo simulations. *Proc Natl Acad Sci U S A* **113**(43):12065–12070.

Yamashita, M., Tomita-Yokotani, K., Hashimoto, H., Takai, M., Tsushima, M., and Nakamura, T. 2004. Experimental concept for examination of biological effects of magnetic field concealed by gravity. *Adv Space Res* **34**(7):1575–1578.

Yeung, J.W. 2006. A hypothesis: Sunspot cycles may detect pandemic influenza A in 1700–2000 A.D. *Med Hypotheses* **67**:1016–1022.

Yong, W., Secco, R.A., Littleton, J.A., and Silber, R.E. 2019. The iron invariance: Implications for thermal convection in Earth's core. *Geophys Res Lett* **46**(20):11065–11070.

Yosef, R., Raz, M., Ben-Baruch, N., Shmueli, L., Kosicki, J.Z., Fratczak, M., and Tryjanowski, P. 2020. Directional preferences of dogs' changes in the presence of a bar magnet:educational experiments in Israel. *J Vet Behav* **35**:34–37.

Yoshii, T., Ahmad, M., and Helfrich-Forster, C. 2009. Cryptochrome mediates light-dependent magnetosensitivity of *Drosophila*'s circadian clock. *PLoS Biol* **7**(4):e1000086.

Yuan, Q., Metterville, D., Briscoe, A.D., and Reppert, S.M. 2007. Insect cryptochromes: Gene duplication and loss define diverse ways to construct insect circadian clocks. *Mol Biol Evol* **24**:948–955.

Zamoshchina, T.A., Krivova, N.A., Khodanovich, M., Trukhanov, K.A., Tukhvatulin, R.T., Zaeva, O.B., Zelenskaia, A.E., and Gul, E.V. 2012. Influence of simulated hypomagnetic environment in a far space flight on the rhythmic structure of rat's behavior. *Aviakosm Ekolog Med* **46**:17–23.

Zapka, M., Heyers, D., Hein, C.M., Engels, S., Schneider, N.L., Hans, J., Weiler, S., Dreyer, D., Kishkinev, D., Wild, JM, and Mouritsen, H. 2009. Visual but not trigeminal mediation of magnetic compass information in a migratory bird. *Nature* **461**(7268):1274–1277.

Zaporozhan, V., and Ponomarenko, A. 2010. Mechanisms of geomagnetic field influence on gene expression using influenza as a model system: Basics of physical epidemiology. *Int J Environ Res Public Health* **7**(3):938–965.

Zell, H. 2020. *Earth's Atmospheric Layers*. NASA, The Rosen Publishing Group, Inc, New York.

Zhang, B., Lu, H., Xi, W., Zhou, X., Xu, S., Zhang, K., Jiang, J., Li, Y., and Guo, A. 2004. Exposure to hypomagnetic field space for multiple generations causes amnesia in *Drosophila melanogaster. Neurosci Lett* **371**(2–3):190–195.

Zhang, Y., Sekine, T., He, H., Yu, Y., Liu, F., and Zhang, M. 2016. Experimental constraints on light elements in the Earth's outer core. *Sci Rep* **6**:22473.

Zhang, Z., Xue, Y., Yang, J., Shang, P., and Yuan, X. 2021. Biological effects of hypomagnetic field: Ground-based data for space exploration. *Bioelectromagnetics* **42**(6):516–531.

Žiubrytė, G., Šiaučiūnaitė, V., Jaruševičius, G., and McCraty, R. 2018. Local earth magnetic field and ischemic heart disease: Peculiarities of interconnection. *Cardiovasc Disord Med* **3**(4):1–3.

7

Safety Guidelines on Human Exposures to Electromagnetic Fields

Kenichi Yamazaki

Tsukasa Shigemitsu

7.1 Introduction

In general, when considering the effects of electromagnetic fields on health, direct and indirect effects are considered. Direct effects include stimulation of the nervous system and muscles by the induced electric current by the electromagnetic fields, causing perception and muscle contraction, and thermal effects where the energy of electromagnetic field is absorbed in the living tissues, causing an increase in temperature in the tissue. The indirect effect is mainly the contact current that flows when person contacts with an object with different potential, due to induction from the electromagnetic field. The contact current may cause electric shocks and burns in person. It may also affect medical devices implanted in the body. There have been reports suggesting the possibility of increased risk at the level of electromagnetic fields that do not cause above direct and indirect effects.

There has been ongoing social concern and anxiety about the health risks of electromagnetic fields for about 50 years. In particular, the International Electromagnetic Fields Project (International EMF Project), initiated by the World Health Organization (WHO) in 1996, has promoted the assessment of health risks from electromagnetic fields. The project divides electromagnetic fields into direct current, low-frequency electric and magnetic fields, and high-frequency electromagnetic fields according to frequency range, and tries to summarize the environmental health criteria for each. Although the WHO

DOI: 10.1201/9781003181354-7

is aimed at health risk assessment, it also focuses on whether electromagnetic fields have effects on living organisms, animal, and plants. The earth itself is a large magnet and generates geomagnetic field. Magnetic bacteria and organisms such as honeybees, pigeons, migratory birds, salmon, etc. are known to utilize geomagnetism. On the other hand, superconducting magnets are used in magnetic resonance imaging (MRI) systems in the medical field. In this way, the opportunities for people to be exposed to DC magnetic fields have been steadily increasing.

Based on the findings from the health risk assessment of the WHO project, exposure guidelines due to scientifically established interactions between electromagnetic fields and living organisms have been considered.

There have been safety guidelines to protect excessive human exposure to electromagnetic fields (EMF) developed based on reproducible and scientifically established findings of biological effects. Widely recognized guideline setting bodies worldwide are International Commission on Non-Ionizing Radiation Protection (ICNIRP) and Institute of Electrical and Electronics Engineers (IEEE/ICES). These guidelines have been used as a scientific basis for evaluating the human safety of electromagnetic fields encountered in real living or industrial environments. Outlines of the guidelines developed by these organizations are described in the following sections.

ICNIRP and IEEE/ICES have been introducing exposure limits, respectively, as guideline/standard based on scientific findings. Guidelines for limiting exposure to electromagnetic fields issued by ICNIRP are often used as the basis for national regulations. ICNIRP issued a guideline covering time-varying electromagnetic fields up to 300 GHz in 1998 (ICNIRP, 1998). Since then, guidelines have been revised, with guidelines for 1 Hz to 100 kHz issued in 2010 (ICNIRP, 2010), and guidelines for 100 kHz to 300 GHz in 2020 (ICNIRP, 2020). For DC magnetic fields, guidelines were issued in 2009 (ICNIRP, 2009). IEEE/ICES have published IEEE standards (IEEE C95.1-2019) covering frequencies up to 300 GHz in 2019 (IEEE, 2019).

7.2 Electromagnetic Fields

According to ICNIRP statement, non-ionizing radiation refers to electromagnetic radiation and fields with photon energy lower than 10 eV, which corresponds to frequencies lower than 3 PHz (3×10^{15} Hz). The groups of static and magnetic fields (0 Hz), low frequency (LF) electromagnetic fields (1 Hz to 100 kHz), and radiofrequency (RF) electromagnetic fields (100 kHz to 300 GHz) are categorized into non-ionizing radiation. Ultraviolet (UV) radiation (wavelengths 100–400 nm), visible light (wavelength 400–780 nm), and infrared radiation (wavelengths 780–1,000 nm) are also grouped into non-ionizing radiation (ICNIRP, 2020a). Ionizing radiation causes damage to living organisms through ionization and excitation and its effect on genes is major problem. On the other hand, light such as ultraviolet, visible light, and infrared radiations, which are classified as non-ionizing radiation, has weak penetration to living organisms and is not harmful to them. For example, the near-infrared light penetrates relatively well into the body. So it is used for example detection of biological information, such as pulse oximetry, for safety reason.

Exposure to non-ionizing radiation is ubiquitous in our everyday life. There are natural sources of non-ionizing radiation generating from the earth's magnetic field (geomagnetic field), lightning storms, and the sun. Man-made non-ionizing radiation exposures in the environment include LF electric and magnetic fields from power supply and distribution infrastructure, RF fields from telecommunications and other radio transmissions.

7.2.1 Static Fields

Static electric and magnetic fields at 0 Hz are naturally found in the earth's atmosphere including the geomagnetic field. Man-made static fields occur commonly in transportation, industry, and medicine. There has been a lot of research on whether exposure to static and LF fields cause any health effects. Most of the research indicates that exposure to static and LF field normally encountered in the environment does not pose a risk to human health. At high levels, there are established acute effects of static fields including vertigo and nausea (WHO, 2006). There is no established evidence that static fields cause long-term health effects and the International Agency for Research on Cancer (IARC) has classified static fields as not classifiable as to their carcinogenicity to humans (Group 3) (IARC, 2002).

Here, a brief overview of the health issues of non-ionizing radiation will be presented. First, several reviews have been published on the biological and health effects of exposure to static electric and magnetic fields (IARC, 2002; ICNIRP, 2009b; WHO, 2006). Static electric and magnetic fields originate from both natural and man-made sources. Static electric fields are derived from the earth's atmosphere as a part of global electric circuit. The naturally originated static electric field on earth is highest near the surface from about 100–150 V/m during fair weather to several thousand V/m under thunderclouds. Static electric field depends on the temperature, relative humidity, altitude, and other weather conditions. The natural static magnetic field originates from electric current flow in the liquid outer core of the earth. This field called the geomagnetic field. The geomagnetic field is described by three components: total magnetic intensity, declination, and inclination. The total field intensity in Japan is around 50 μT. The geomagnetic field fluctuates according to diurnal, lunar, and seasonal variation. On the other hand, man-made sources of static electric and magnetic fields are found everywhere from our day life to industrial facilities and medical equipment through power transmission systems (WHO, 1987, 2006). Static electric field does not penetrate the human body and induces a surface charge. This charge may be perceived through its interaction with body hair and by other phenomena such as discharge (microshock), at sufficiently high field. The perception in humans is dependent on various factors and can range from 10 to 45 kV/m. Static magnetic field is unperturbed by the human body. There are three well-established mechanisms by which static magnetic field interact with biological systems: magnetic induction, magneto-mechanical effects, and electron spin effects (ICNIRP, 2009). Based on the evaluation of biological effect research, the WHO carried out the human health assessment of static electric and magnetic field (WHO, 2006).

As the scientific base to develop the rationale for the guidelines, a document published by WHO (WHO, 2006) on static fields within their Environmental Health Criteria Program, which contains a review of biological effects reported from exposure to static fields, is referred along with other publications (ICNIRP, 2003; McKinlay et al., 2004; Noble et al., 2005).

As the scientific evidence, three established interaction mechanisms with living matter are considered: magnetic induction, magneto-mechanical, and electronic interactions. For the magnetic induction, the following types of interaction were evaluated: electrodynamic interactions with moving electrolytes, and induced electric fields in living tissues. The electrodynamic interactions with moving charged particles can lead to an induced electric field. The change in electrocardiograms is a well-known example of this electrodynamic interaction. In the presence of a static magnetic field, the electrical potential is induced. In the blood, this is the result of Lorenz force exerted on moving charged particles (electrolytes). Kinouchi carried out the detailed theoretical treatment if the effects of magnetic fields on blood flow by using the Navier-Stokes equation (Kinouchi et al., 1996). In the case of magnetic fields perpendicular to the blood flow, they found a reduction in the flow rate of blood.

For the magneto-mechanical interaction, two types of mechanical effects that a static magnetic field can exert on biological objects are evaluated. The first type is magneto-orientation. This concerns the orientation of paramagnetic molecules in the high static magnetic field. This effect is involved in magneto-reception in certain species of animals. The second type of interaction is the magneto-mechanical translation. This occurs in the presence of a field gradient for paramagnetic or diamagnetic materials (Ueno and Iwakasa, 1994a,b).

For the electronic interaction, electron-spin interactions are evaluated. This interaction can affect the rate of recombination of pairs of free radicals in chemical reaction intermediates. It seems that this mechanisms plays a part in the navigation system of certain birds. It was given the excellent review on the role of free radicals in biology (Hayashi, 2004; Okano, 2008).

7.2.2 Low-Frequency Fields

Many studies have investigated the biological effects of low frequency electromagnetic fields up to 100 kHz (IARC, 2002; ICNIRP, 2003; NIEHS, 1998; WHO, 1987, 2007).

The coupling between the low-frequency magnetic field and the body is summarized below from the document of ICNIRP (ICNIRP, 2010). For magnetic fields, the permeability of tissue is the same as that of air so the field in tissues is the same as the external field. In this frequency ranges, the main interaction of magnetic field with the body is the Faraday induction of electric fields and associated currents in the tissues. Key features of dosimetry for exposure of humans to low frequency magnetic fields include: (1) for a given magnetic field strength and orientation, higher electric fields are induced in the bodies of larger people because the possible conduction loops are larger. (2) Induced electric field and current depend on the orientation of the external magnetic field to the body. Generally, induced fields in the body are greatest when they field is aligned from the front to the back of the body, but for some organs the highest values are for different field alignments. (3) Weakest electric fields are induced by a magnetic field oriented along the principal body axis. (4) Distribution of the induced electric field is affected by the conductivity of the various organs and tissues (Ueno and Sekino, 2016).

The only established health effects from LF also occur at high levels and include electric shock and magnetophosphenes in the eye. Some epidemiological studies have reported a possible association between prolonged exposure to LF magnetic fields and increased rates of childhood leukemia. This association is not supported by laboratory or animal studies, and no credible theoretical mechanism has been proposed. Based largely on this evidence, IARC has classified LF magnetic fields as possibly carcinogenic to humans (Group 2B) (IARC, 2002). Overall, the evidence related to childhood leukemia is not strong and research in this area is continuing.

From the definition of ICNRIP, LF is used to describe the frequency range from 1 Hz to 100 kHz (ICNIRP, 2010a,b). In this frequency range, the interaction of electric and magnetic fields with the human body is the induction of electric fields and currents in the tissues. The established effect is the induction of magnetophosphenes, a perception of faint flickering light in the periphery of the visual field. They are thought to result from the interaction of the induced electric field with electrically excitable cells in the retina. The threshold for induction of magnetophosphenes has been estimated to low between about 50 and 100 mV/m at 20 Hz. Epidemiological studies have suggested that long-term exposure to 50/60 Hz magnetic fields might be associated with an increased risk of childhood leukemia. Two pooled analysis indicate that an excess risk may exist for average exposures exceeding 0.3–0.4 μT. However, some degree of confounding and chance could possibly explain these results. No biophysical mechanism has been identified and results from animal and cellular studies do not support the mention that exposure to 50/60 Hz magnetic fields is a cause of childhood leukemia. There is no substantial evidence for an association between low frequency magnetic field exposure and Parkinson's disease, multiple sclerosis, and cardiovascular diseases. The evidence for an association between low-frequency magnetic field exposure and Alzheimer's disease and amyotrophic sclerosis is inconclusive. Overall research has not shown that long-term low-level LF magnetic field exposure has detrimental effects on health. ICNIRP's view is that the currently existing scientific evidence that prolonged exposure to LF magnetic fields is causally related with an increased risk of childhood leukemia is too weak to form the basis for exposure guidelines. The perception of surface electric charge, the direct stimulation of nerve and muscle tissue, and the induction of retinal phosphenes are the only well-established adverse effects and serve as the basis for guidance (ICNIRP, 2010b).

7.2.3 Radiofrequency Fields

Sheppard quantitatively evaluated potential mechanisms of interaction between RF electromagnetic fields and biological systems (Sheppard et al., 2008). First, the established mechanisms are dielectric relaxation and ohmic loss, which lead to elevate the temperature in tissue though heating. On the other hand, the proposed is the radical pair mechanism.

Exposure to RF electromagnetic fields can induce heating in biological tissues. Biological effects caused by RF electromagnetic fields ranging from 100 kHz to 300 GHz can be divided into two categories: thermal effects and nonthermal effects. Thermal effects are due to tissue heating and nonthermal effects are due to unknown mechanisms. Heating is classically given by a quantity of specific absorption rate (SAR) with units of watts per kilogram (W/kg). The SAR is derived from the square of electric field strength in tissue. The SAR cannot be measured directly in humans and is usually estimated from computer-based simulation models of the human body.

In ICNIRP's definition, RF electromagnetic field is used to describe the frequency range from 100 kHz to 300 GHz. Due to the gradual increase of the use of RF-operated devices such as radio-television broadcasting, mobile phone, Wi-Fi, Bluetooth, radar, smart meters, medical equipment, etc., exposure levels of RF fields have increased gradually around us. Understanding of health, biological, and environmental effects of electromagnetic field in the frequency range up to 300 GHz is now advancing rapidly.

In 2020, ICNIRP published a guideline in the frequency range from 100 kHz to 300 GHz (ICNIRP, 2020). This guideline protects against adverse health effects relating to exposure RF field including from 5G technologies.

RF is emitted from both natural and artificial sources. The artificial sources of RF that the public are most familiar with include telecommunication sources such as radio and television broadcasting Wi-Fi and mobile telephony; however, there are also medical and industrial sources of RF. The health effects from RF have been highly researched and the only established effect is heating of tissue, which can cause tissue damage. Although studies have reported biological effects at low levels, there has been no indication that such effects might constitute a human health hazard. In 2012, IARC classified RF as possibly carcinogenic to humans (Group 2B) (IARC, 2013). This classification was primarily based on evidence of an association between wireless phone use and certain brain tumors. Research performed since has found no overall increase in the incidence of brain cancers since the introduction of mobile phones (e.g. Karipidis et al., 2018).

7.3 Guidelines Setting Bodies

The major bodies setting guidelines/standards are the ICNIRP and the IEEE. The specialist committee grouped within IEEE, the International Committee on Electromagnetic Safety (ICES), develops the RF standard. ICNIRP is organized in Germany and IEEE is an organization based in the USA. ICNIRP and IEEE are well-recognized organization bodies that have already published guidelines (or standards for IEEE) to protect people from exposure to electromagnetic fields. ICNIRP has developed guidelines for exposure to static electric and extremely low-frequency magnetic fields and to protect workers and the public against established health effects (ICNIRP, 2009, 2010).

7.3.1 ICNIRP

ICNIRP is an independent scientific commission which aims to protect people and the environment against adverse effects of non-ionizing radiation and develops science-based advice on limiting exposure to non-ionizing radiation. Historically, ICNIRP was chartered in 1992 as an independent commission to continue the work of the International Non-Ionizing Radiation Committee of the International Radiation Protection Association (IRPA), and is formally recognized as an official collaborating non-state actor by the WHO and the International Labour Organization. ICNIRP's protection advice is formulated in its guidelines, and currently there are four separate ICNIRP guidelines for static magnetic fields (ICNIRP, 2009), for magnetic fields from 0 to 1 Hz (ICNIRP, 2014), for low-frequency electric and magnetic fields from 1 Hz to 100 kHz (ICNIRP, 2010), and for high-frequency electromagnetic fields from 100 kHz to 300 GHz (ICNIRP, 2020). These guidelines cover all frequency range of non-ionizing radiation.

7.3.2 IEEE/ICES

IEEE is an association designed to serve professionals involved in all aspects of the electrical, electronic, and computing fields and related areas of science and technology that underlie modern civilization. In the IEEE, IEEE Standards Association is organized to develop standards which aim to enable the creation and expansion of international markets and help protect health and public safety. Under the rules and oversight of the IEEE SA Standards Board, the ICES, which is responsible for the development of standards for the safe use of electromagnetic energy in the range of 0 Hz to 300 GHz is operating. ICES TC95 is responsible for setting IEEE standards related to environmental limits. Currently, IEEE C95.1-2019 is in active which covers frequencies from 0 Hz to 300 GHz (IEEE, 2019). The standard developed by IEEE does not provide a static magnetic (or electric) field.

7.4 Two-Tier Target Population of Exposure

In the safety guidelines for low-frequency and high-frequency electromagnetic fields, separate limit values are given for two-tier target population of the population, occupational, and general public exposures.

The term general public refers to individuals of all ages and of varying health status, which might increase the variability of individual susceptibility (ICNIRP, 2010). Therefore, it is considered appropriate that the exposure limits for the general public be more stringent than the exposure limits for workers.

In many cases, members of the public are unaware of their exposure to EMF. These considerations underlie the adoption of more stringent exposure restrictions for the public than for workers while they are occupationally exposed.

The ICNIRP guidelines distinguish between occupational exposure and general public exposure, while the IEEE C95.1-2019 uses the terms "restricted environment" (upper tier) and "unrestricted environment" (lower tier) for occupational exposure and general public exposure. This classification is equivalent to the controlled and uncontrolled environments previously used in IEEE C95.1-2005.

7.4.1 Occupational Exposure

Workers may be exposed to relatively high exposures in work environment. In such cases, adults are a healthy population and appropriate precautions are taken. Occupational exposure in these guidelines refers to adults exposed to electromagnetic fields at their workplaces, generally under known conditions, and as a result of performing their regular or assigned job activities (ICNIRP, 2010). On the other hand, the general public includes a wide variety of people including young children who are expected to be exposed and unaware of the risks and no precautions are taken against exposure.

IEEE C95.1-2019 uses restricted environment for occupational exposure and unrestricted environment for general public (IEEE, 2019). Restricted environment for occupational exposure is an environment in which exposure can result in exceeding the unrestricted environment (lower tier) dosimetric reference limit (DRL). Here, (1) implementation of an effective safety program (see IEEE Std C95.7 for the RF range) is to help ensure that persons are not exposed above the DRL or exposure reference level (ERL) for the restricted environment, (2) in some documents, exposure in restricted environments is referred to as "upper tier" or "controlled environment" or "occupational exposure," and (3) members of the general public are not permitted in restricted environments unless they become subject to the applicable safety program, at which time they are no longer considered members of the "general public."

7.4.2 General Public Exposure

By definition of IEEE of general public is that all members of the human population who have no knowledge or control of their exposure and are, consequently, not permitted in a restricted environment

(IEEE, 2019). The unrestricted tier exposure (unrestricted environment) limit applies to the general public. The general public includes, but is not limited to, children, pregnant women, individuals with impaired thermoregulatory systems, and persons using medications that can result in poor thermoregulatory system performance.

Unrestricted environment for general public is an environment in which exposure does not result in exceeding the DRL that marks the safety program initiation level, and which serves as an exposure limit for the general public. Here, (1) the exposures can occur in living quarters or workplaces where there are no expectations that the DRL or ERL for unrestricted environments would be exceeded and where the induced currents or contact currents do not exceed the limits for unrestricted environments. (2) In some documents, the unrestricted environment is referred to as a "lower tier" or an "uncontrolled environment" or a "general public exposure."

7.5 Basic Restrictions and Reference Levels

In the safety guidelines for low-frequency and high-frequency electromagnetic fields, two types of parameters are given to limit exposure. In the INCIRP guidelines, they are named basic restrictions and reference levels (ICNIRP, 2010; ICNIRP, 2020).

Basic restrictions are defined based on the physical quantities directly related to the established health effects. Depending on frequency, the physical quantities are used to specify the basic restrictions. For low frequencies up to 10 MHz, the physical quantity is the internal electric field strength (or in situ electric field in the IEEE standard) induced by electric and magnetic fields, which affects nerve cells and other electrically sensitive cells. On the other hand, for high frequencies from 100 kHz to 6 GHz, the physical quantity is the specific energy absorption rate (SAR), which is the power absorbed per unit mass and largely responsible for the heating effects of biological tissues. In addition, for frequencies above 6 GHz where electromagnetic fields are absorbed more superficially, the physical quantity is the density of absorbed power over the area and referred to as the absorbed power density.

Reference levels are provided as the strength of the environmental electric or magnetic field for practical exposure assessment purposes since the internal electric quantities are practically difficult to assess. The reference levels are derived from the corresponding basic restrictions assuming a human equivalent model and using measurement or computational techniques.

In the practical exposure situation examined, the measured or calculated values of electric or magnetic fields are first compared with the relevant reference level. Compliance with the reference level ensures compliance with the relevant basic restrictions. Even the value exceeds the reference level, it does not necessarily mean that the basic restriction will be exceeded. However, it is necessary to test compliance with the relevant basic restrictions.

It is noted that, in the IEEE standard (IEEE, 2019), the terms "dosimetric reference limit (DRL)" and "exposure reference level (ERL)" are used for the counterparts of the basic restriction and the reference level as in ICNIRP guidelines, respectively.

7.6 Guidelines for Static Magnetic Fields

According to the WHO Environmental Health Criteria for direct current electric and magnetic fields, in order to limit the exposure of workers and the general public, (1) international exposure limits based on science should be applied (general public: 40 mT for continuous exposure, occupational exposure: 200 mT for weighted average of all working hours a day, extremities (upper limit): 5 T, non-extremities (upper limit): 2 T); (2) maintain a distance from magnetic fields, enclose it, and provide protective measures for industrial and scientific use of magnetic fields; (3) consider a licensing system for MRI equipment exceeding 2 T; (4) provide research grants to compensate for the lack of knowledge on safety; and (5) collect health information on worker and patient exposure and make it available in a database.

TABLE 7.1   Limits of Exposure to Static Magnetic Fields in the ICNIRP Guidelines for Static Magnetic Fields[a]

Exposure Characteristics		Magnetic Flux Density
Occupational[b]	Exposure of head and of trunk	2 T
	Exposure of limbs[c]	8 T
General public[d]	Magnetic flux density	400 mT

Source: Reproduced from ICNIRP, *Health Physics* 96: 504–514, 2009.

[a] ICNIRP recommends that these limits should be viewed operationally as spatial peak exposure limits.

[b] For specific work applications, exposure up to 8 T can be justified, if the environment is controlled and appropriate work practices are implemented to control movement-induced effects.

[c] Not enough information is available on which to base exposure limits beyond 8 T.

[d] Because of potential indirect adverse effects, ICNIRP recognizes that practical policies need to be implemented to prevent inadvertent harmful exposure of persons with implanted electronic medical devices and implants containing ferromagnetic material, and dangers from flying objects, which can lead to much lower restriction levels such as 0.5 mT.

The ICNIRP provides guidelines for static magnetic fields (ICNIRP, 2004). The guidelines were originally established in 1994 and revised in 2004. The exposure limit values are tabulated in Table 7.1. Separate limits are given for occupational exposure and general public exposure, and the limits for occupational exposure are applied to those individuals who are exposed to static magnetic fields as a result of performing their regular or assigned job activities, while the limits for general public exposure are applied to the entire population. The limit for occupational exposure of the head and trunk should not exceed a spatial peak magnetic flux density of 2 T. Exposure up to 8 T can be permitted if the environment is controlled and appropriate work practices are implemented to control movement-induced effects.

Exposure limits for occupational exposures: It is recommended that occupational exposure of the head and trunk should not exceed a spatial peak magnetic flux density of 2 T except for the following circumstance: for working application for which exposure above 2 T are deemed necessary following, exposure up to 8 T can be permitted if the environment is controlled and appropriate work practices are implemented to control movement-induced effects. Sensory effects due to the movement in the field can be avoided by complying with basic restrictions set in the extremely low-frequency guidelines. When restricted to the limbs, maximum exposures of up to 8 T are acceptable (ICNIRP, 2010).

Exposure limits for general public exposures: Based on scientific knowledge on the direct of static fields on humans, acute exposure of the general public should not exceed 400 mT (any part of the body). However, because of potential indirect adverse effects, ICNIRP recognizes that practical policies need to be implemented to prevent inadvertent harmful exposure of people with implanted electronic medical devices and implants containing ferromagnetic materials, and injuries due to flying ferromagnetic objects, and these considerations can lead to much lower restriction levels, such as 0.5 mT (IEC, 2002). The exposure limits to be set with regard to these non-biological effects are not, however, the duty of ICNIRP (CINIRP, 2010).

ICNIRP recommends that the use of these guidelines should be accompanied by appropriate protective measures. These measures need to be considered separately for public places and for workplaces. For public places, exposures to static magnetic fields are likely to be very low and infrequent. On the other hand, strong static fields may be regularly encountered for workplaces (ICNIRP, 2009). There are three main areas of concern: for members of the public, there is a need to protect people with implanted medical devices against possible interference and against forces on implants containing ferromagnetic material. In addition, in some specific situations, there is a risk from flying ferromagnetic objects such as tools. Third, in work situations involving exposure to very high fields, there is a need for a set of site-specific work procedures intended to minimize the impact of transient symptoms such as vertigo and nausea (ICNIRP, 2009).

ICNIRP provides additional guidelines for the protection of workers moving in static magnetic fields or being exposed to magnetic fields with frequencies below 1 Hz (ICNIRP, 2014). This was aimed at the protection of workers engaged in activities related to MRI to avoid sensory effects which may be annoying and impair working ability. The guideline was set to prevent magnetophosphene, peripheral nerve stimulation, and vertigo as a consequence of movement in strong static magnetic fields. The basic restrictions have been defined for "the change in the external magnetic flux density (DB)" and for the induced internal electric field. In addition, reference levels expressed as the peak (amplitude) dB/dt have been derived. A distinction is made between controlled and uncontrolled exposures. Basic restrictions for controlled exposure are intended to be used in work environments where access is restricted to workers who have been trained to understand the biological effects that may result from exposure, and where the workers are able to control their movements to prevent annoying and disturbing sensory effects.

7.7 Guidelines for Low-Frequency Electromagnetic Fields

For the low frequency, ICNIRP and IEEE/ICES produce guidelines/standard in 2010 and 2019 (ICNIRP, 2010; IEEE, 2019). The low frequency covers the region from 1 Hz to 100 kHz. The most interesting frequency is 50/60 Hz because this frequency is used as the frequency of electric power supply.

7.7.1 ICNIRP Guidelines

As the scientific base to develop the rationale for the guidelines, review, documents published by IARC (IARC, 2002), ICNIRP (ICNIRP, 2003), and the WHO (WHO, 2007) on the biological effects of exposure to low-frequency electromagnetic fields are referred. In the guidelines, the basic restrictions are set based on the identified risks related to transient nervous system responses to peripheral and central nerve stimulation, the induction of retinal phosphenes, and possible effects on some aspects of brain function (ICNIRP, 2010).

The ICNIRP guidelines for low-frequency electric and magnetic fields were originally established in 1998 and revised in 2010 (ICNIRP, 2010). The basic restrictions expressed as internal electric fields for frequencies from 1 Hz to 10 MHz are tabulated in Table 7.2. Separate limits are given for "CNS (central nervous system) tissues of the head" and "all tissues of head and body." It is noted that in the frequency range above 100 kHz, RF-specific basic restrictions need to be considered additionally.

TABLE 7.2 Basic Restrictions in ICNIRP Guidelines for Low Frequencies

Exposure Characteristics		Frequency Range	Internal Electric Field (V/m)
Occupational exposure	CNS tissue of the head	1–10 Hz	$0.5/f$
		10–25 Hz	0.05
		25–400 Hz	$2 \times 10^{-3} f$
		400 Hz to 3 kHz	0.8
		3 kHz to 10 MHz	$2.7 \times 10^{4} f$
	All tissues of head and body	1 Hz to 3 kHz	0.8
		3 kHz to 10 MHz	$2.7 \times 10^{-4} f$
General public exposure	CNS tissue of the head	1–10 Hz	$0.1/f$
		10–25 Hz	0.01
		25–1,000 Hz	$0.4 \times 10^{-3} f$
		1,000 Hz to 3 kHz	0.4
		3 kHz to 10 MHz	$1.35 \times 10^{-4} f$
	All tissues of head and body	1 Hz to 3 kHz	0.4
		3 kHz to 10 MHz	$1.35 \times 10^{-4} f$

Source: Reproduced from ICNIRP, *Health Physics* 99: 818–836, 2010.

Notes:

1. f is the frequency in Hz.

2. All values are rms.

3. In the frequency range above 100 kHz, RF-specific basic restrictions need to be considered additionally.

TABLE 7.3 Reference Levels in ICNIRP Guidelines for Low Frequencies (Unperturbed rms Values)

Exposure Characteristics	Frequency Range	E-Field Strength (kV/m)	Magnetic Flux Density (T)
Occupational exposure	1–8 Hz	20	$0.2/f^2$
	8–25 Hz	20	$2.5 \times 10^{-2}/f$
	25–300 Hz	$5 \times 10^2/f$	1×10^{-3}
	300 Hz to 3 kHz	$5 \times 10^2/f$	$0.3/f$
	3 kHz to 10 MHz	1.7×10^{-1}	1×10^{-4}
General public exposure	1–8 Hz	5	$0.04/f^2$
	8–25 Hz	5	$0.5 \times 10^{-2}/f$
	25–50 Hz	5	0.2×10^{-3}
	50–400 Hz	$2.5 \times 10^2/f$	0.2×10^{-3}
	400 Hz to 3 kHz	$2.5 \times 10^2/f$	$0.08/f$
	3 kHz to 10 MHz	0.83×10^{-1}	0.27×10^{-4}

Source: Reproduced from ICNIRP, *Health Physics* 99: 818–836, 2010.
Notes:
1. f in Hz.
2. In the frequency range above 100 kHz, RF-specific reference levels need to be considered additionally.

TABLE 7.4 Reference Levels in ICNIRP Guidelines for Low Frequencies for Time-Varying Contact Currents from Conductive Object

Exposure Characteristic	Frequency Range	Maximum Contact Current (mA)
Occupational exposure	–2.5 kHz	1.0
	2.5 kHz to 100 kHz	$0.4f$
	100 kHz to 10 MHz	40
General public exposure	–2.5 kHz	0.5
	2.5–100 kHz	$0.2f$
	100 kHz to 10 MHz	20

Source: Reproduced from ICNIRP, *Health Physics* 99: 818–836, 2010.
Note: f is the frequency in kHz.

The reference levels in Table 7.3 are derived from the basic restrictions referring to published data of computational dosimetry results (Dimbylow, 2005, 2006). In the computation, millimeter resolution anatomical human models were used, and the conditions of maximum coupling of the field to the exposed individual, i.e., uniform field exposures, were assumed to provide maximum protection. An additional reduction factor was considered in the derivation of the reference levels to consider the dosimetric uncertainty.

As can be seen in Table 7.5, the reference levels in the ICNIRP guidelines (2010) are expressed in terms of electric field strength (kV/m) (1 Hz to 10 MHz), magnetic field strength (T) (~10 kHz), and contact current (mA) (~10 MHz).

Apart from the direct effects of EMFs, "indirect effects" are also considered in the guidelines. The indirect effects involve a human touching an object where the electric potential of the object is different from that of the human body. In the guidelines, reference levels limiting the indirect effects of EMF have also been given for frequencies up to 10 MHz expressed as contact current to avoid shock and burn hazards in Table 7.4. It should be noted that the reference levels are not intended to prevent perception but to avoid painful shocks.

7.7.2 IEEE Safety Standard

The IEEE safety standard for low-frequency electric and magnetic fields was originally established in 2002 as IEEE standard C95.6 and revised along with that of high frequency as C95.1 in 2019 (IEEE, 2019).

TABLE 7.5 Basic Restrictions in ICNIRP Guidelines for Electromagnetic Field Exposure from 100 kHz to 300 GHz, for Averaging Intervals ≥6 Minutes

Exposure Scenario	Frequency Range	Whole-Body Average SAR (W/kg)	Local Head/ TorsoSAR (W/kg)	Local Limb SAR (W/kg)	Local S_{ab} (W/m²)
Occupational	100 kHz to 6 GHz	0.4	10	20	NA
	>6 to 300 GHz	0.4	NA	NA	100
General public	100 kHz to 6 GHz	0.08	2	4	NA
	>6 to 300 GHz	0.08	NA	NA	20

Source: Reproduced from ICNIRP, *Health Physics* 118:483–524, 2020.
Notes:
1. "NA" signifies "not applicable" and does not need to be taken into account when determining compliance.
2. Whole-body average SAR is to be averaged over 30 minutes.
3. Local SAR and Sab exposures are to be averaged over 6 minutes.
4. Local SAR is to be averaged over a 10-g cubic mass.
5. Local Sab is to be averaged over a square 4 cm² surface area of the body. Above 30 GHz, an additional constraint is imposed, such that exposure averaged over a square 1 cm2 surface area of the body is restricted to two times that of the 4 cm² restriction.

DRL is defined in terms of internal electric field (in situ electric field). The exposure limit is based on dosimetric thresholds for established adverse health effects expressed as in situ electric field strength (0 Hz to 5 MHz). Separate limits are given for "brain," "heart," "limb," and "other tissues." In the frequencies above 100 kHz, RF-specific basic restrictions need to be considered additionally, SAR (100 kHz to 6 GHz), or epithelial power density (6 GHz to 300 GHz) and which provides an adequate margin of safety (IEEE, 2019).

ERL provides an adequate margin of safety against established adverse health effects. It is expressed as the metric appropriate to the frequency and temporal characteristics of the exposure under consideration. ERL is the maximum exposure level relative to ambient electric and/or magnetic field strength or power density, induced and/or contact current, or contact voltage which means that the ERL is measured, estimated, or derived from the DRL (in situ electric field, SAR, or epithelial power density).

The DRL is a limit expressed in terms of field strength in the body, SAR, or power density in the epithelium, and can be considered to be roughly equivalent to the basic limit of the ICNIRP guideline. ERL is the level indicated by measurable electric fields, magnetic field, power density, induced in the limb, contact current, and contact voltage. ERL can be considered equivalent to the reference level of the ICNIRP guidelines.

The exposure limits in the low-frequency range are determined based on the threshold of electrostimulation, which is manifested by the excitation of nerves and muscle, as well as by the alteration of neural synapse activity. The major part of the scientific base is from Reilly's comprehensive work in electrostimulation (Reilly, 1998).

The ERL for magnetic field exposure are provided separately for exposure of "head and torso" and for "the limbs." The ERLs are derived analytically from the DRL by assuming simple induction models (several sizes of homogeneous ellipsoids representing each part of the human body) that have an analytical solution for the uniform magnetic field exposure (IEEE, 2019). Safety factors that are applied to a median adverse reaction threshold are considered in the derivation of the ERLs.

The ERL for electric field exposure is provided separately for exposure of "head and torso" and for "the limbs." Unlike the magnetic field exposure, the ERLs for electric field exposure are limited by indirect electrostimulation (contact current and spark discharge). This is because the environmental electric field level for the indirect electrostimulation is significantly lower than that inducing DRL level in situ electric fields.

The ERL for contact current is also provided in the IEEE safety standard for frequencies 0 Hz to 5 MHz based on the electrostimulation effects. These contact current limits apply to a freestanding individual who is insulated from the ground while touching a grounded conductor. These limits might not protect against aversive sensations from spark discharge just prior to direct contact or upon release from the grounded conductor.

7.8 Guidelines for Radiofrequency Electromagnetic Fields

In ICNIRP's definition, RF electromagnetic field is used to describe the frequency range from 100 kHz to 300 GHz. Due to the gradual increase of the use of RF-operated devices such as mobile phone, Wi-Fi, Bluetooth, radar, smart meters, medical equipment, etc., exposure levels of RF fields have increased gradually around us. Understanding of health, biological, and environmental effects of electromagnetic field in the frequency range up to 300 GHz is now advancing rapidly. In 2020, ICNIRP published a guideline in the frequency range from 100 kHz to 300 GHz (ICNIRP, 2020). This guideline protects against adverse health effects relating to exposure RF field including from 5G technologies.

7.8.1 ICNIRP Guidelines

The ICNIRP guidelines for low RF electromagnetic fields were originally established in 1998 and revised in 2020 (ICNIRP, 2020).

7.8.1.1 Basic Restriction

The basic restrictions expressed as SAR for frequencies from 100 kHz to 300 GHz are provided in Table 7.5 to prevent whole-body heat stress and excessive localized tissue heating. In the frequency range between 100 kHz and 10 MHz, the basic restrictions on the induced electric field related to nerve stimulation in Table 7.2 are additionally applied. The following provides an overview of basic restrictions in the ICNIRP guideline (ICNIRP, 2020).

Whole-body average SAR (100 kHz-300 GHz): For frequencies from 100 kHz to 300 GHz, the whole-body average SAR limit averaged over the entire body mass and a 30-minute interval was set for an increase in body core temperature of 1°C.

Local SAR (100 kHz to 6 GHz): For local exposure of the Head and Torso for 100 kHz to 6 GHz, SAR limit averaged over a 10-g cubic mass and 6-minute interval was set for the local temperature rise of 5°C or 2°C depending on the tissue type. For the local exposure for the Limb (100 kHz to 6 GHz), SAR limit averaged over a 10-g cubic mass and 6-minute interval, for the Limbs of a 5°C rise in local temperature.

Local SAR (400 MHz to 6 GHz): In addition, for frequencies from 400 MHz to 6 GHz, ICNIRP sets an additional limit expressed as SAR for exposure intervals of <6 minutes, as a function of time, to ensure that the cumulative energy permitted by the 6-minute average local SAR basic restriction is not absorbed by tissues too rapidly in Table 7.6.

TABLE 7.6 Basic Restrictions in ICNIRP Guidelines for Electromagnetic Field Exposure from 100 kHz to 300 GHz, for Integrating Intervals >0 to <6 Minutes

Exposure Scenario	Frequency Range	Local Head/Torso (kJ/kg)	Local Limb SA (kJ/ kg)	Local U_{ab} (kJ/m²)
Occupational	100 kHz to 400 MHz	NA	NA	NA
	>400 MHz to 6 GHz	$3.6[0.05+0.95(t/360)^{0.5}]$	$7.2[0.025+0.975(t/360)^{0.5}]$	NA
	>6 to 300 GHz	NA	NA	$36[0.05+0.95(t/360)^{0.5}]$
General public	100 kHz to 400 MHz	NA	NA	NA
	>400 MHz to 6 GHz	$0.72[0.025+0.95(t/360)^{0.5}]$	$1.44[0.025+0.975(t/360)^{0.5}]$	NA
	>6 to 300 GHz	NA	NA	$7.2[0.05+0.95(t/360)^{0.5}]$

Source: Reproduced from ICNIRP, *Health Physics* 118:483–524, 2020.

Notes:

1. "NA" signifies "not applicable" and does not need to be taken into account when determining compliance.
2. t is time in seconds, and restrictions must be satisfied for all values of t between >0 and <360 s, regardless of the temporal characteristics of the exposure itself.
3. Local SA is to be averaged over a 10-g cubic mass.
4. Local U_{ab} is to be averaged over a square 4-cm² surface area of the body. Above 30 GHz, an additional constraint is imposed, such that exposure averaged over a square 1-cm² surface area of the body is restricted to $72[0.025+0.975(t/360)^{0.5}]$ for occupational and $14.4[0.025+0.975 (t/360)^{0.5}]$ for general public exposure.
5. Exposure from any pulse, group of pulses, or subgroup.

Local absorbed power density (>6 GHz to 300 GHz): For local exposure for both the Head and Torso and Limb regions (6 GHz to 300 GHz), an absorbed power density (S_{ab}) limit averaged over 6 minutes and a square 4 cm² surface area of the body was set for the local temperature rise of 5°C or 2°C depending on tissue type in Table 7.5. Furthermore, to account for focal beam exposure from 30 to 300 GHz, the absorbed power density averaged over a square 1 cm² surface area of the body must not exceed two times that of the 4 cm² basic restrictions for workers or the general public.

Local absorbed energy density (>6 GHz to 300 GHz): In addition, for frequencies from 6 to 300 GHz, ICNIRP sets an additional limit expressed as local absorbed energy density (U_{ab}) for exposure intervals of <6 minutes, as a function of time, to ensure that the cumulative energy permitted by the 6-minute average absorbed power density basic restriction is not absorbed by tissues too rapidly in Table 7.6.

In ICNIRP 2020, the basic restrictions of SAR remain unchanged; the basic restriction, which was expressed in terms of spatial power density at 10–300 GHz, is now expressed in terms of Sab, the locally absorbed power density absorbed by tissues near the body surface. Basic restriction for local exposure is now expressed in terms of local SAR for the frequency range of 100 kHz to 6 GHz and local Sab for the frequency range of 6–300 GHz. On the other hand, the basic restrictions for induced electric field into the body from 100 kHz to 10 MHz are the same as in ICNIRP 2010.

7.8.1.2 Reference Level

In ICNIRP 2020, reference levels are presented in terms of electric field strength (V/m), magnetic field strength (A/m), power density (W/m²), and induced current (mA) to the extremities, in the frequency range from 100 kHz to 300 GHz. Reference levels shown in Tables 7.7–7.9 have been derived from the basic restrictions by a combination of computational and measurement studies for worst-case exposure scenarios.

TABLE 7.7   Reference Levels in ICNIRP Guidelines for Exposure, Averaged over 30 Minutes and the Whole Body, to Electromagnetic Fields from 100 kHz to 300 GHz (Unperturbed rms Values)

Exposure Scenario	Frequency Range	Incident E-Field Strength; E_{inc} (V/m)	Incident H-Field Strength; H_{inc} (A/m)	Incident Power Density; S_{inc} (W/m²)
Occupational	0.1–30 MHz	$660/f_M^{0.7}$	$4.9/f_M$	NA
	>30–400 MHz	61	0.16	10
	>400–2000 MHz	$3f_M^{0.5}$	$0.008f_M^{0.5}$	$f_M/40$
	>2–300 GHz	NA	NA	50
General public	0.1–30 MHz	$300/f_M^{0.7}$	$2.2/f_M$	NA
	>30–400 MHz	27.7	0.073	2
	>400–2,000 MHz	$1.375f_M^{0.5}$	$0.0037f_M^{0.5}$	$f_M/200$
	>2–300 GHz	NA	NA	10

Source: Reproduced from ICNIRP, *Health Physics* 118:483–524, 2020.

Notes:

1. "NA" signifies "not applicable" and does not need to be taken into account when determining compliance.
2. f_M is frequency in MHz.
3. S_{inc}, E_{inc}, and H_{inc} are to be averaged over 30 minutes, over the whole-body space. Temporal and spatial averaging of each of E_{inc} and H_{inc} must be conducted by averaging over the relevant square values.
4. For frequencies of 100 kHz to 30 MHz, regardless of the far-field/near-field zone distinctions, compliance is demonstrated if neither E_{inc} or H_{inc} exceeds the above reference level values.
5. For frequencies of >30 MHz to 2 GHz: (a) within the far-field zone: compliance is demonstrated if either S_{inc}, E_{inc} or H_{inc}, does not exceed the above reference level values (only one is required); S_{eq} may be substituted for S_{inc}; (b) within the radiative near-field zone, compliance is demonstrated if either S_{inc}, or both E_{inc} and H_{inc}, does not exceed the above reference level values; and (c) within the reactive near-field zone: compliance is demonstrated if both E_{inc} and H_{inc} do not exceed the above reference level values; S_{inc} cannot be used to demonstrate compliance, and so basic restrictions must be assessed.
6. For frequencies of >2 GHz to 300 GHz: (a) within the far-field zone: compliance is demonstrated if S_{inc} does not exceed the reference level values; S_{eq} may be substituted for S_{inc}; (b) within the radiative near-field zone, compliance is demonstrated if S_{inc} does not exceed the above reference level values; and (c) within the reactive near-field zone, reference levels cannot be used to determine compliance, and so basic restrictions must be assessed.

TABLE 7.8 Reference Levels in ICNIRP Guidelines for Local Exposure, Averaged over 6 Minutes, to Electromagnetic Fields from 100 kHz to 300 GHz (Unperturbed rms Values)

Exposure Scenario	Frequency Range	Incident E-Field Strength; E_{inc} (V/m)	Incident H-Field Strength; H_{inc} (A/m)	Incident Power Density; S_{inc} (W/m²)
Occupational	0.1–30 MHz	$1{,}504/f_M^{0.7}$	$10.8/f_M$	NA
	>30–400 MHz	139	0.36	50
	>400–2,000 MHz	$10.58f_M^{0.43}$	$0.0274f_M^{0.43}$	$0.29f_M^{0.86}$
	>2–6 GHz	NA	NA	200
	>6 to <300 GHz	NA	NA	$275f_G^{0.177}$
	300 GHz	NA	NA	100
General public	0.1–30 MHz	$671/f_M^{0.7}$	$4.9/f_M$	NA
	>30–400 MHz	62	0.163	10
	>400–2000 MHz	$4.72f_M^{0.43}$	$0.0123f_M^{0.43}$	$0.058f_M^{0.86}$
	>2–6 GHz	NA	NA	40
	>6 to <300 GHz	NA	NA	$55f_G^{0.177}$
	300 GHz	NA	NA	20

Source: Reproduced from ICNIRP, *Health Physics* 118:483–524, 2020.

Notes:

1. "NA" signifies "not applicable" and does not need to be taken into account when determining compliance.
2. f_M is frequency in MHz; f_G is frequency in GHz.
3. S_{inc}, E_{inc}, and H_{inc} are to be averaged over 6 minutes, and where spatial averaging is specified in Notes 6–7, over the relevant projected body space. Temporal and spatial averaging of each of E_{inc} and H_{inc} must be conducted by averaging over the relevant square values.
4. For frequencies of 100 kHz to 30 MHz, regardless of the far-field/near-field zone distinctions, compliance is demonstrated if neither peak spatial E_{inc} or peak spatial H_{inc}, over the projected whole-body space, exceeds the above reference level values.
5. For frequencies of >30 MHz to 6 GHz: (a) within the far-field zone, compliance is demonstrated if one of peak spatial S_{inc}, E_{inc} or H_{inc}, over the projected whole-body space, does not exceed the above reference level values (only one is required); S_{eq} may be substituted for S_{inc}; (b) within the radiative near-field zone, compliance is demonstrated if either peak spatial S_{inc}, or both peak spatial E_{inc} and H_{inc}, over the projected whole-body space, does not exceed the above reference level values; and (c) within the reactive near-field zone: compliance is demonstrated if both E_{inc} and H_{inc} do not exceed the above reference level values; S_{inc} cannot be used to demonstrate compliance; for frequencies >2 GHz, reference levels cannot be used to determine compliance, and so basic restrictions must be assessed.
6. For frequencies of >6 to 300 GHz: (a) within the far-field zone, compliance is demonstrated if S_{inc}, averaged over a square 4 cm² projected body sursface space, does not exceed the above reference level values; S_{eq} may be substituted for S_{inc}; (b) within the radiative near-field zone, compliance is demonstrated if S_{inc}, averaged over a square 4 cm² projected body surface space, does not exceed the above reference level values; and (c) within the reactive near-field zone reference levels cannot be used to determine compliance, and so basic restrictions must be assessed.
7. For frequencies of >30 to 300 GHz, exposure averaged over a square 1 cm² projected body surface space must not exceed twice that of the square 4 cm² restrictions.

Reference levels in Table 7.7 are averaged over a 30-minute interval and correspond to the whole-body average basic restrictions. Table 7.8 (averaged over a 6-minute interval) and Table 7.9 (integrated over intervals between 0 and 6 minutes) each relate to basic restrictions that are averaged over smaller body regions.

In ICNIRP 2020 guideline, for frequencies from 100 kHz to 110 MHz, additional limb current (induced current) reference levels have been set to account for effects of grounding near human body resonance frequencies that might otherwise lead to reference levels underestimating exposures within tissue at certain frequencies (averaged over 6 minutes; Table 7.10).

Regarding the contact currents for frequencies from 100 kHz to 110 MHz, unlike the low-frequency case, only guidance is provided for the contact current. High levels of RF contact current can result in nerve stimulation or pain (and potentially tissue damage or burn), depending on the frequency, and this can be a particular concern near large RF transmitters, such as those near high power broadcasting antennas.

TABLE 7.9 Reference Levels in ICNIRP Guidelines for Local Exposure, Integrated over Intervals of between >0 and <6 Minutes, to Electromagnetic Fields from 100 kHz to 300 GHz (Unperturbed rms Values)

Exposure Scenario	Frequency Range	Incident Energy Density; U_{inc} (kJ/m²)
Occupational	100 kHz–400 MHz	NA
	>400–2,000 MHz	$0.29f_M^{0.86} \times 0.36[0.05 + 0.95(t/360)^{0.5}]$
	>2–6 GHz	$200 \times 0.36[0.05 + 0.95(t/360)^{0.5}]$
	>6 to <300 GHz	$275f_G^{0.177} \times 0.36[0.05 + 0.95(t/360)^{0.5}]$
	300 GHz	$100 \times 0.36[0.05 + 0.95(t/360)^{0.5}]$
General public	100 kHz–400 MHz	NA
	>400 to 2,000 MHz	$0.058f_M^{0.86} \times 0.36[0.05 + 0.95(t/360)^{0.5}]$
	>2 to 6 GHz	$40 \times 0.36[0.05 + 0.95(t/360)^{0.5}]$
	>6 to <300 GHz	$55f_G^{0.177} \times 0.36[0.05 + 0.95(t/360)^{0.5}]$
	300 GHz	$20 \times 0.36[0.05 + 0.95(t/360)^{0.5}]$

Source: Reproduced from ICNIRP, *Health Physics* 118:483–524, 2020.

Notes:

1. "NA" signifies "not applicable" and does not need to be taken into account when determining compliance.
2. f_M is frequency in MHz; f_G is frequency in GHz; t is time interval in seconds, such that exposure from any pulse, group of pulses, or subgroup of pulses in a train, as well as from the summation of exposures (including non-pulsed EMFs), delivered in t seconds, must not exceed these reference level values.
3. U_{inc} is to be calculated over time t, and where spatial averaging is specified in Notes 5–7, over the relevant projected body space.
4. For frequencies of 100 kHz to 400 MHz, >0 to <6 minutes restrictions are not required and so reference levels have not been set.
5. For frequencies of >400 MHz to 6 GHz: (a) within the far-field zone: compliance is demonstrated if peak spatial U_{inc}, over the projected whole-body space, does not exceed the above reference level values; U_{eq} may be substituted for U_{inc}; (b) within the radiative near-field zone, compliance is demonstrated if peak spatial U_{inc}, over the projected whole-body space, does not exceed the above reference level values; and (c) within the reactive near-field zone, reference levels cannot be used to determine compliance, and so basic restrictions must be assessed.
6. For frequencies of >6 GHz to 300 GHz: (a) within the far-field or radiative near-field zone, compliance is demonstrated if U_{inc}, averaged over a square 4 cm² projected body surface space, does not exceed the above reference level values; (b) within the reactive near-field zone, reference levels cannot be used to determine compliance, and so basic restrictions must be assessed.
7. For frequencies of >30 to 300 GHz: exposure averaged over a square 1 cm² projected body surface space must not exceed $275/f_G^{0.177} \times 0.72[0.025 + 0.975(t/360)^{0.5}]$ kJ/m² for occupational and $55/f_G^{0.177} \times 0.72[0.025 + 0.975(t/360)^{0.5}]$ kJ/m² for general public exposure.

7.8.2 IEEE Safety Standard

The IEEE safety standard for RF electromagnetic fields (IEEE C95.1) was originally published in 1966 and recently revised in 2019 (IEEE, 2019).

Exposure limits are divided into DRL and ERL. DRL is limits expressed in terms of in situ electric field strength, SAR, epithelial power density, in the body and are considered to be basic restrictions of ICNIRP. The ERL is a limit indicated by the electric field strength, magnetic field strength, power density, current flowing in the limb, and contact current, which can be measured, and is considered to be equivalent to the reference level of the ICNIRP guidelines.

For RF, the DRL is defined in terms of specific absorption rate (SAR) and epithelial power density for frequencies from 100 kHz to 6 GHz to protect against adverse heating (thermal effect). Like to the ICNIRP RF guidelines, separate DRLs are defined for whole-body exposure (30 minutes averaging) and local exposure (6 minutes averaging time, 10-g averaging mass) for "head and torso" and "limbs and pinnae" for frequencies from 100 kHz to 6 GHz. For frequencies from 6 to 300 GHz, epithelial power density DRL is defined at the body surface.

For RF, the ERL has been derived from the DRLs to provide a readily assessed quantity via measurements or computation. ERLs are provided for both whole-body exposure (30-minute averaging time)

TABLE 7.10   Reference Levels in ICNIRP Guidelines for Current Induced in Any Limb, Averaged over 6 Minutes, at Frequencies from 100 kHz to 110 MHz

Exposure Scenario	Frequency Range	Electric Current; I (mA)
Occupational	100 kHz to 110 MHz	100
General public	100 kHz to 110 MHz	45

Source: Reproduced from ICNIRP, *Health Physics* 118:483–524, 2020.
Notes:
1. Current intensity values must be determined by averaging over the relevant square values.
2. Limb current intensity must be evaluated separately for each limb.
3. Limb current reference levels are not provided for any other frequency range.
4. Limb current reference levels are only required for cases where the human body is not electrically isolated from a ground plane.

and local exposure (6-minute averaging time) in terms of electric field strength, magnetic field strength, and plane-wave-equivalent power density values, for frequencies from 100 kHz to 6 GHz. For frequencies from 6 to 300 GHz, incident power density ERLs are defined (6-minute averaging time, 4 cm² averaging on body surface).

7.9 Conclusion

The guidelines are revised from time to time as new findings are obtained on the biological effects of electromagnetic fields and evaluation methods such as dosimetry. Now that the revision of the guidelines for RF electromagnetic fields has been completed, ICNIRP is moving forward with a new effort to revise the low-frequency guidelines by examining research issues based on current findings (ICNIRP, 2020b).

The IEEE/ICES has completed the revision of RF electromagnetic fields in 2019, and will now work on the revision of electromagnetic fields in the low-frequency range. The research issues of electromagnetic fields and contact currents below 100 kHz have already been organized (Reilly and Hirata, 2016). There are three issues: induced current models, electrical stimulation models, and human exposure limit. All of them are intended to be solved by dosimetry.

References

Dimbylow P.J. (2005) Development of the female voxel phantom, NAOMI and its application to calculations of induced current density and electric fields from applied low frequency magnetic and electric fields. *Phys Med Biol* 50:1047–1070.

Dimbylow P.J. (2006) Development of pregnant female, hybrid voxel mathematical models and their application to the dosimetry of applied magnetic and electric fields at 50 Hz. *Phys Med Biol* 51:2383–2394.

Hayashi H. (2004) *Introduction to Dynamic Spin Chemistry: Magnetic Field Effects on Chemical and Biochemical Reactions.* World Scientific Publishing Co Ltd, New Jersey, Londo, Singapore, Shanghai, Hong Kong, Taipei, Bangalore.

IARC (2002) *Non-Ionizing Radiation, Part 1: Static and Extremely Low Frequency (ELF) Electric and Magnetic Fields.* IARC Monographs on the Evaluation of Carcinogenic Risks to Humans, Volume 80. International Agency for Research Cancer, Lyon, France.

ICNIRP (2003) Exposure to static and low frequency electromagnetic fields, biological effects and health consequences (0–100 kHz). Matthes R., McKinlay A.F., Bernhardt J.H., Vecchia P., Veyret B., eds. ICNIRP; Publication 13/2003, Oberschleissheim, Germany.

ICNIRP (2009) Guidelines on limits of exposure to static magnetic fields. *Health Phys* 96:504–514.

ICNIRP (2010) Guidelines for limiting exposure to time-varying electric and magnetic fields (1 Hz to 100 kHz). *Health Phys* 99:818–836.

ICNIRP (2014) Guidelines for limiting exposure to electric fields induced by movement of the human body in a static magnetic field and by time-varying magnetic fields below 1 Hz. *Health Phys* 106:418–425.

ICNIRP (2020a) ICNIRP guidelines for limiting exposure to electromagnetic fields (100 kHz to 300 GHz). *Health Phys* 118:483–524.

ICNIRP (2020b) Gaps in knowledge relevant to the "Guidelines for limiting exposure to time-varying electric and magnetic fields (1 Hz-100 kHz). *Health Phys* 118:533–542.

IEEE (2019) IEEE standard for safety levels with respect to human exposure to electric, magnetic, and electromagnetic fields, 0 Hz to 300 GHz. IEEE Std C95.1, New York.

Karipidis K, Elwood M, Benke G, Sanagou M, Tjong L and Croft RJ (2018): Moble phpne ue and incidence of brain tumour histological types, grading or anatomical location: a population-based ecological study. BMJ Open 8:e0244489.

Kinouchi Y., Yamaguchi H. and Tenforde T.S. (1996) Threshold analysis of magnetic field interactions with aortic blood flow. *Bioelectromagnetics* 17:21–32.

Kripidis K, Elwood M, Benke G, Sanagou M, Tjong L and Croft RJ (2018): Mobile phone use and incidence of brain tumour histological types, grading on anatomical location: a population-based ecological study. *BMJ 8:e024489.*

McKinlay A.F., Allen S.G., Cox R., Dimbylow P.J., Mann S.M., Muirhead C.R., Saunders R.D., Sienkiewicz Z.J., Stather J.W., and Wainwright P.R. (2004) Review of the scientific evidence for limiting exposure to electromagnetic fields (0–300 GHz). Chilton: National Radiological Protection Board; Docs NRPB, 15(3).

NIEHS (1998) Assessment of health effects from exposure to power-line frequency electric and magnetic fields. *NIH Publ. 98–3981.*

Noble D., McKinlay A., and Repacholi M. (2005) Effects of static magnetic fields relevant to human health. *Prog Biophys Mol Biol* 87:171–372.

Okano H. (2008) Effects of static magnetic fields in biology: Role of free radicals. *Front Biosci* 13:6106–61225.

Reilly J.P. (1998) *Applied Bioelectricity: From Electrical Stimulation to Electropathology.* Springer, New York.

Reilly J.P. and Hirata A. (2016) Low-frequency electrical dosimetry: Research agenda of the IEEE International Committee on Electromagnetic Safety. *Phys Med Biol* 61:R138–R149.

Sheppard A.R., Swicord M.L., and Balzano Q. (2008) Quantitative evaluations of mechanisms of radiofrequency interaction with biological molecules and processes. *Health Phys* 95:365–396.

Ueno S. and Iwasaka M. (1994a) Parting of water by magnetic fields. *IEEE Trans on Magn* 30:4698–4700.

Ueno S. and Iwasaka M. (1994b) Properties of diamagnetic fluid in high gradient magnetic fields. *J Appl Phys* 75:7177–7179.

Ueno S. and Sekino M. (eds) (2016) *Biomagnetic: Principles and Applications of Biomagnetic Stimulation and Imaging.* CRC Press, Taylor & Francis Group, Boca Raton, London, New York.

WHO (1987) Magnetic fields. Environmental Health Criteria 69.

WHO (2006) Static fields. Environmental Health Criteria No 232, Geneva.

WHO (2007) Extremely low frequency fields. Environmental Health Criteria No 238, Geneva.

Index

Note: **Bold** page numbers refer to tables and *italic* page numbers refer to figures.

For Product Safety Concerns and Information please contact our EU
representative GPSR@taylorandfrancis.com
Taylor & Francis Verlag GmbH, Kaufingerstraße 24, 80331 München, Germany

9 781032 019970